BOOKS | À LA CARTE EDITION

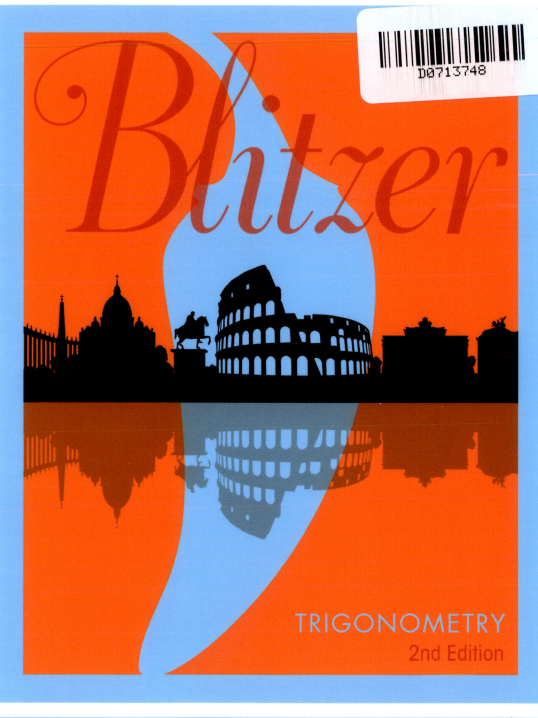

Blitzer

TRIGONOMETRY

2nd Edition

YOUR TEXTBOOK—IN A BINDER-READY EDITION!

This unbound, three-hole punched version of your textbook lets you take only what you need to class and incorporate your own notes—all at an affordable price!

ISBN-13: 978-0-13-446765-8
ISBN-10:   0-13-446765-5

EAN

9 780134 467658

90000

Pearson

A Brief Guide to **Getting the Most** from this Book

1   Read the Book

Feature	Description	Benefit
Section-Opening Scenarios	Every section opens with a scenario presenting a unique application of trigonometry in your life outside the classroom.	Realizing that trigonometry is everywhere will help motivate your learning. **(See page 71.)**
Detailed Worked-Out Examples	Examples are clearly written and provide step-by-step solutions. No steps are omitted, and each step is thoroughly explained to the right of the mathematics.	The blue annotations will help you understand the solutions by providing the reason why every trigonometric step is true. **(See page 198.)**
Applications Using Real-World Data	Interesting applications from nearly every discipline, supported by up-to-date real-world data, are included in every section.	Ever wondered how you'll use trigonometry? This feature will show you how trigonometry can solve real problems. **(See page 89.)**
Great Question!	Answers to students' questions offer suggestions for problem solving, point out common errors to avoid, and provide informal hints and suggestions.	By seeing common mistakes, you'll be able to avoid them. This feature should help you not to feel anxious or threatened when asking questions in class. **(See page 112.)**
Brief Reviews	Brief Reviews cover algebraic and geometric skills you already learned but may have forgotten.	Having these refresher boxes easily accessible will help ease anxiety about skills you may have forgotten. **(See page 191.)**
Achieving Success	NEW to this edition. Achieving Success boxes offer strategies for persistence and success in college mathematics courses.	Follow these suggestions to help achieve your full academic potential in college mathematics. **(See page 34.)**
Explanatory Voice Balloons	Voice balloons help to demystify trigonometry. They translate mathematical language into plain English, clarify problem-solving procedures, and present alternative ways of understanding.	Does math ever look foreign to you? This feature often translates math into everyday English. **(See page 81.)**
Learning Objectives	Every section begins with a list of objectives. Each objective is restated in the margin where the objective is covered.	The objectives focus your reading by emphasizing what is most important and where to find it. **(See page 192.)**
Technology	The screens displayed in the technology boxes show how graphing utilities verify and visualize trigonometric results.	Even if you are not using a graphing utility in the course, this feature will help you understand different approaches to problem solving. **(See page 145.)**

2   Work the Problems

Feature	Description	Benefit
Check Point Examples	Each example is followed by a matched problem, called a Check Point, that offers you the opportunity to work a similar exercise. The answers to the Check Points are provided in the answer section.	You learn best by doing. You'll solidify your understanding of worked examples if you try a similar problem right away to be sure you understand what you've just read. **(See page 219.)**
Concept and Vocabulary Checks	These short-answer questions, mainly fill-in-the-blank and true/false items, assess your understanding of the definitions and concepts presented in each section.	It is difficult to learn trigonometry without knowing its special language. These exercises test your understanding of the vocabulary and concepts. **(See page 35.)**
Extensive and Varied Exercise Sets	An abundant collection of exercises is included in an Exercise Set at the end of each section. Exercises are organized within several categories. Your instructor will usually provide guidance on which exercises to work. The exercises in the first category, Practice Exercises, follow the same order as the section's worked examples.	The parallel order of the Practice Exercises lets you refer to the worked examples and use them as models for solving these problems. **(See page 220.)**
Practice Plus Problems	This category of exercises contains more challenging problems that often require you to combine several skills or concepts.	It is important to dig in and develop your problem-solving skills. Practice Plus Exercises provide you with ample opportunity to do so. **(See page 179.)**
Retaining the Concepts	NEW to this edition. Beginning with Chapter 2, each Exercise Set contains review exercises under the header "Retaining the Concepts."	These exercises improve your understanding of the topics and help maintain mastery of the material. **(See page 95.)**
Preview Problems	Each Exercise Set concludes with three problems to help you prepare for the next section.	These exercises let you review previously covered material that you'll need to be successful for the forthcoming section. Some of these problems will get you thinking about concepts you'll soon encounter. **(See page 169.)**

3   Review for Quizzes and Tests

Feature	Description	Benefit
Mid-Chapter Check Points	At approximately the midway point in the chapter, an integrated set of review exercises allows you to review the skills and concepts you learned separately over several sections.	By combining exercises from the first half of the chapter, the Mid-Chapter Check Points give a comprehensive review before you move on to the material in the remainder of the chapter. **(See page 182.)**
Chapter Review Grids	Each chapter contains a review chart that summarizes the definitions and concepts in every section of the chapter. Examples that illustrate these key concepts are also referenced in the chart.	Review this chart and you'll know the most important material in the chapter! **(See page 259.)**
Chapter Review Exercises	A comprehensive collection of review exercises for each of the chapter's sections follows the grid.	Practice makes perfect. These exercises contain the most significant problems for each of the chapter's sections. **(See page 261.)**
Chapter Tests	Each chapter contains a practice test with problems that cover the important concepts in the chapter. Take the practice test, check your answers, and then watch the Chapter Test Prep Videos to see worked-out solutions for any exercises you miss.	You can use the chapter test to determine whether you have mastered the material covered in the chapter. **(See page 68.)**
Chapter Test Prep Videos	These videos contain worked-out solutions to every exercise in each chapter test and can be found in MyMathLab and on YouTube.	The videos let you review any exercises you miss on the chapter test.
Objective Videos	NEW to this edition. These fresh, interactive videos walk you through the concepts from every objective of the text.	The videos provide you with active learning at your own pace.
Cumulative Review Exercises	Beginning with Chapter 3, each chapter concludes with a comprehensive collection of mixed cumulative review exercises. These exercises combine problems from previous chapters and the present chapter, providing an ongoing cumulative review.	Ever forget what you've learned? These exercises ensure that you are not forgetting anything as you move forward. **(See page 325.)**

TRIGONOMETRY

TRIGONOMETRY

2nd EDITION

Robert Blitzer

Miami Dade College

Director, Portfolio Management: Anne Kelly
Courseware Portfolio Manager: Dawn Murrin
Portfolio Management Administrator: Joseph Colella
Content Producer: Kathleen A. Manley
Managing Producer: Karen Wernholm
Producer: Erica Lange
Manager, Courseware QA: Mary Durnwald
Manager, Content Development: Kristina Evans
Product Marketing Manager: Claire Kozar
Marketing Assistant: Jennifer Myers

Executive Marketing Manager: Peggy Lucas
Marketing Assistant: Adiranna Valencia
Senior Author Support/Technology Specialist: Joe Vetere
Production Coordination: Francesca Monaco/codeMantra
Text Design and Composition: codeMantra
Illustrations: Scientific Illustrators
Photo Research and Permission Clearance: Cenveo Publisher Services
Cover Design: Studio Montage
Cover Image: Ojal/Shutterstock

Acknowledgments of third-party content appear on page C1, which constitutes an extension of this copyright page.

Library of Congress Cataloging-in-Publication Data

Names: Blitzer, Robert.
Title: Trigonometry / Robert Blitzer, Miami Dade College.
Description: Second edition. | Hoboken, New Jersey : Pearson Prentice Hall, [2018] |
  Includes answers to selected exercises. | Includes subject index.
Identifiers: LCCN 2016042564 | ISBN 9780134469966
Subjects: LCSH: Trigonometry — Textbooks.
Classification: LCC QA531 .B635 2018 | DDC 516.24 — dc23
LC record available at https://lccn.loc.gov/2016042564

ISBN 13: 978-0-13-446996-6
ISBN 10: 0-13-446996-8

CONTENTS

PREFACE

I've written *Trigonometry*, **Second Edition**, to help diverse students, with different backgrounds and future goals, to succeed. The book has three fundamental goals:

1. To help students acquire a solid foundation in trigonometry, preparing them for other courses such as calculus and business calculus, as well as for students taking trigonometry as their final mathematics course.

2. To show students how trigonometry can model and solve authentic real-world problems.

3. To enable students to develop problem-solving skills, while fostering critical thinking, within an interesting setting.

One major obstacle in the way of achieving these goals is the fact that very few students actually read their textbook. This has been a regular source of frustration for me and for my colleagues in the classroom. Anecdotal evidence gathered over years highlights two basic reasons that students do not take advantage of their textbook:

- "I can't follow the explanations."
- "I'll never use this information."

I've written every page of the Second Edition with the intent of eliminating these two objections. The ideas and tools I've used to do so are outlined for the student in "A Brief Guide to Getting the Most from This Book," which appears at the front of the book.

What's New in the Second Edition?

- **Retaining the Concepts.** Beginning with Chapter 2, Section 2.1, each Exercise Set contains three review exercises under the header "Retaining the Concepts." These exercises are intended for students to review previously covered objectives in order to improve their understanding of the topics and to help maintain their mastery of the material. If students are not certain how to solve a review exercise, they can turn to the section and worked example given in parentheses at the end of each exercise. The Second Edition contains 66 new exercises in the "Retaining the Concepts" category.

- **Achieving Success.** The Achieving Success boxes offer strategies for persistence and success in college mathematics courses.

- **Updated Learning Guide.** Organized by the textbook's learning objectives, this updated Learning Guide helps students make the most of their textbook for test preparation. Projects are now included to give students an opportunity to discover and reinforce the concepts in an active learning environment and are ideal for group work in class.

- **Updated Graphing Calculator Screens.** All screens have been updated using the TI-84 Plus C.

- **Section 1.1 (Angles and Radian Measure)** has new examples involving radians expressed in decimal form, including converting 2.3 radians to degrees (Example 3(d)) and finding a coterminal angle for a -10.3 angle (Example 7(d)). Additional Great Question! features provide hints for locating terminal sides of angles in standard position.

- **Section 1.2 (Right Triangle Trigonometry)** has a new Discovery feature on the use of parentheses when evaluating trigonometric functions with a graphing calculator, supported by new calculator screens throughout the section. A Great Question! has been added urging students not to become too calculator dependent.

What Familiar Features Have Been Retained in the Second Edition?

- **Learning Objectives.** Learning objectives, framed in the context of a student question (What am I supposed to learn?), are clearly stated at the beginning of each section. These objectives help students recognize and focus on the section's most important ideas. The objectives are restated in the margin at their point of use.

- **Chapter-Opening and Section-Opening Scenarios.** Every chapter and every section open with a scenario presenting a unique application of trigonometry. These scenarios are revisited in the course of the chapter or section in an example, discussion, or exercise.

- **Innovative Applications.** A wide variety of interesting applications are included throughout the book. Students will be able to relate to these applications that range from their emotional cycles, their breathing cycles, the music that surrounds them, the visualization of chaos, and the role that trigonometric functions play in the genetic information necessary for the maintenance and continuation of life.

- **Detailed Worked-Out Examples.** Each worked example is titled, making clear the purpose of the example. Examples are clearly written and provide students with detailed step-by-step solutions. No steps are omitted and key steps are thoroughly explained to the right of the mathematics.

- **Explanatory Voice Balloons.** Voice balloons are used in a variety of ways to demystify mathematics. They translate trigonometric ideas into everyday English, help clarify problem-solving procedures, present alternative ways of understanding concepts, and connect problem solving to concepts students have already learned.

- **Check Point Examples.** Each example is followed by a similar matched problem, called a Check Point, offering students the opportunity to test their understanding of the example by working a similar exercise. The answers to the Check Points are provided in the answer section.

- **Brief Reviews.** The book's Brief Review boxes summarize algebraic skills that students should have learned previously, but which many students still need to review. This feature appears whenever a particular skill is first needed and eliminates the need for you to reteach algebraic topics.

- **Great Question!** This feature presents a variety of study tips in the context of students' questions. Answers to questions offer suggestions for problem solving, point out common errors to avoid, and provide informal hints and suggestions. As a secondary benefit, this feature should help students not to feel anxious or threatened when asking questions in class.

- **Blitzer Bonuses.** These enrichment essays provide historical, interdisciplinary, and otherwise interesting connections to the mathematics under study, showing students that trigonometry is an interesting and dynamic discipline.

- **Concept and Vocabulary Checks.** This feature offers short-answer exercises, mainly fill-in-the-blank and true/false items, that assess students' understanding of the definitions and concepts presented in each section. The Concept and Vocabulary Checks appear as separate features preceding the Exercise Sets.

- **Extensive and Varied Exercise Sets.** An abundant collection of exercises is included in an Exercise Set at the end of each section. Exercises are organized within nine category types: Practice Exercises, Practice Plus Exercises, Application Exercises, Explaining the Concepts, Technology Exercises, Critical Thinking Exercises, Group Exercises, Retaining the Concepts, and Preview Exercises. This format makes it easy to create well-rounded homework assignments. The order of the Practice Exercises is exactly the same as the order of the section's worked examples. This parallel order enables students to refer to the titled examples and their detailed explanations to achieve success working the Practice Exercises.

- **Practice Plus Problems.** This category of exercises contains more challenging practice problems that often require students to combine several skills or concepts. With an average of ten Practice Plus problems per Exercise Set, instructors are provided with the option of creating assignments that take Practice Exercises to a more challenging level.

- **Mid-Chapter Check Points.** At approximately the midway point in each chapter, an integrated set of Review Exercises allows students to review and assimilate the skills and concepts they learned separately over several sections.

- **Integration of Technology Using Graphic and Numerical Approaches to Problems.** Side-by-side features in the Technology boxes connect trigonometric solutions to graphic and numerical approaches to problems. Although the use of graphing utilities is optional, students can use the explanatory voice balloons to understand different approaches to problems even if they are not using a graphing utility in the course.

- **Chapter Summaries.** Each chapter contains a review chart that summarizes the definitions and concepts in every section of the chapter. Examples that illustrate these key concepts are also referenced in the chart.

- **End-of-Chapter Materials.** A comprehensive collection of Review Exercises for each of the chapter's sections follows the Summary. This is followed by a Chapter Test that enables students to test their understanding of the material covered in the chapter. Beginning with Chapter 3, each chapter concludes with a comprehensive collection of mixed Cumulative Review Exercises.

- **Discovery.** Discovery boxes, found throughout the text, encourage students to further explore trigonometric concepts. These explorations are optional and their omission does not interfere with the continuity of the topic under consideration.

I hope that my passion for teaching, as well as my respect for the diversity of students I have taught and learned from over the years, is apparent throughout this new edition. By connecting trigonometry to the whole spectrum of learning, it is my intent to show students that their world is profoundly mathematical, and indeed, π is in the sky.

Robert Blitzer

Acknowledgments

An enormous benefit of authoring a successful series is the broad-based feedback I receive from the students, dedicated users, and reviewers. Every change to this edition is the result of their thoughtful comments and suggestions. I would like to express my appreciation to all the reviewers, whose collective insights form the backbone of this revision. In particular, I would like to thank the following people for reviewing *College Algebra*, *Algebra and Trigonometry*, *Precalculus*, and *Trigonometry*.

Karol Albus, *South Plains College*

Kayoko Yates Barnhill, *Clark College*

Timothy Beaver, *Isothermal Community College*

Jaromir Becan, *University of Texas-San Antonio*

Imad Benjelloun, *Delaware Valley College*

Lloyd Best, *Pacific Union College*

David Bramlett, *Jackson State University*

Natasha Brewley-Corbin, *Georgia Gwinnett College*

Denise Brown, *Collin College-Spring Creek Campus*

David Britz, *Raritan Valley Community College*

Mariana Bujac-Leisz, *Cameron University*

Bill Burgin, *Gaston College*

Jennifer Cabaniss, *Central Texas College*

Jimmy Chang, *St. Petersburg College*

Teresa Chasing Hawk, *University of South Dakota*

Diana Colt, *University of Minnesota-Duluth*

Shannon Cornell, *Amarillo College*

Wendy Davidson, *Georgia Perimeter College-Newton*

Donna Densmore, *Bossier Parish Community College*

Disa Enegren, *Rose State College*

Keith A. Erickson, *Georgia Gwinnett College*

Nancy Fisher, *University of Alabama*

Donna Gerken, *Miami Dade College*

Cynthia Glickman, *Community College of Southern Nevada*

Sudhir Kumar Goel, *Valdosta State University*

Donald Gordon, *Manatee Community College*

David L. Gross, *University of Connecticut*

Jason W. Groves, *South Plains College*

Joel K. Haack, *University of Northern Iowa*

Jeremy Haefner, *University of Colorado*

Joyce Hague, *University of Wisconsin at River Falls*

Mike Hall, *University of Mississippi*

Mahshid Hassani, *Hillsborough Community College*

Tom Hayes, *Montana State University*

Christopher N. Hay-Jahans, *University of South Dakota*

Angela Heiden, *St. Clair Community College*

Donna Helgeson, *Johnson County Community College*

Celeste Hernandez, *Richland College*

Gregory J. Herring, *Cameron University*

Alysmarie Hodges, *Eastfield College*

Amanda Hood, *Copiah-Lincoln Community College*

Jo Beth Horney, *South Plains College*

Heidi Howard, *Florida State College at Jacksonville-South Campus*

Winfield A. Ihlow, *SUNY College at Oswego*

Nancy Raye Johnson, *Manatee Community College*

Daniel Kleinfelter, *College of the Desert*

Sarah Kovacs, *Yuba College*

Dennine Larue, *Fairmont State University*

Mary Leesburg, *Manatee Community College*

Christine Heinecke Lehman, *Purdue University North Central*

Alexander Levichev, *Boston University*

Zongzhu Lin, *Kansas State University*

Benjamin Marlin, *Northwestern Oklahoma State University*

Marilyn Massey, *Collin County Community College*

Yvelyne McCarthy-Germaine, *University of New Orleans*

David McMann, *Eastfield College*

Owen Mertens, *Missouri State University-Springfield*

James Miller, *West Virginia University*

Martha Nega, *Georgia Perimeter College-Decatur*

Priti Patel, *Tarrant County College*

Shahla Peterman, *University of Missouri-St. Louis*

Debra A. Pharo, *Northwestern Michigan College*

Gloria Phoenix, *North Carolina Agricultural and Technical State University*

Katherine Pinzon, *Georgia Gwinnett College*

David Platt, *Front Range Community College*

Juha Pohjanpelto, *Oregon State University*

Brooke Quinlan, *Hillsborough Community College*

Janice Rech, *University of Nebraska at Omaha*

Gary E. Risenhoover, *Tarrant County College*

Joseph W. Rody, *Arizona State University*

Behnaz Rouhani, *Georgia Perimeter College-Dunwoody*

Judith Salmon, *Fitchburg State University*

Michael Schramm, *Indian River State College*

Cynthia Schultz, Illinois Valley Community College

Pat Shelton, North Carolina Agricultural and Technical State University

Jed Soifer, Atlantic Cape Community College

Caroline Spillman, Georgia Perimeter College-Clarkston

Jonathan Stadler, Capital University

Franotis R. Stallworth, Gwinnett Technical College

John David Stark, Central Alabama Community College

Charles Sterner, College of Coastal Georgia

Chris Stump, Bethel College

Scott Sykes, University of West Georgia

Richard Townsend, North Carolina Central University

Pamela Trim, Southwest Tennessee Community College

Chris Turner, Arkansas State University

Richard E. Van Lommel, California State University-Sacramento

Dan Van Peursem, University of South Dakota

Philip Van Veldhuizen, University of Nevada at Reno

Philip Veer, Johnson County Community College

Jeffrey Weaver, Baton Rouge Community College

Amanda Wheeler, Amarillo College

David White, The Victoria College

Tracy Wienckowski, University of Buffalo

Additional acknowledgments are extended to Dan Miller and Kelly Barber for preparing the solutions manuals; Brad Davis for preparing the answer section, serving as accuracy checker, and writing the new learning guide; the codeMantra formatting team for the book's brilliant paging; Brian Morris and Kevin Morris at Scientific Illustrators for superbly illustrating the book; Francesca Monaco, project manager; and Kathleen Manley, production editor, whose collective talents kept every aspect of this complex project moving through its many stages.

I would like to thank my editor at Pearson, Dawn Murrin, who, with the assistance of Joseph Colella, guided and coordinated the book from manuscript through production. Finally, thanks to Peggy Lucas and Claire Kozar for their innovative marketing efforts and to the entire Pearson sales force for their confidence and enthusiasm about the book.

Robert Blitzer

Get the Most Out of
MyMathLab®

MyMathLab is the leading online homework, tutorial, and assessment program for teaching and learning mathematics, built around Pearson's best-selling content. MyMathLab helps students and instructors improve results; it provides engaging experiences and personalized learning for each student so learning can happen in any environment. Plus, it offers flexible and time-saving course management features to allow instructors to easily manage their classes while remaining in complete control, regardless of course format.

Personalized Learning

Not every student learns the same way or at the same rate. Thanks to our advances in adaptive learning technology capabilities, you no longer have to teach as if they do. *MyMathLab* courses offer a variety tools to personalize homework assignments to fit a variety of course needs and formats.

- *Skill Builder* exercises offer just-in-time adaptive practice within homework assignments to help students build the skills needed to successfully complete their work. When students are struggling to answer correctly, Skill Builder's adaptive engine delivers additional exercises chosen to help each student improve their skills until they are able to complete the assignment. Instructors can choose to make Skill Builder exercises available within homework assignments enabling the adaptive engine to track each student's performance and deliver questions to each individual that adapt to his or her level of understanding.

- MyMathLab can *personalize homework* assignments for students based on their performance on a test or quiz. This way, students can focus on just the topics they have not yet mastered.

Used by more than 37 million students worldwide, MyMathLab delivers consistent, measurable gains in student learning outcomes, retention, and subsequent course success.

www.mymathlab.com

MyMathLab Online Course for *Trigonometry* by Robert Blitzer

(access code required)

NEW! Video Program

A fresh, and all new, video program walks through the concepts from every objective of the text. Many videos provide an active learning environment where students try out their newly learned skill.

Your Turn!
Choose the option that best answers the question.

Perform the indicated operation, writing the result in standard form:
$$(-4 - 8i) - (-7 + 2i)$$

    a. $-3 - 10i$
    b. $-11 - 6i$
    c. $-11 + 6i$

NEW! Guided Visualizations

These HTML-based, interactive figures help students visualize the concepts through directed explorations and purposeful manipulation. They encourage active learning, critical thinking, and conceptual learning. They are compatible with iPad and tablet devices.

The Guided Visualizations are located in the Multimedia Library and can be assigned as homework with correlating assessment exercises. Additional Exploratory Exercises are available to help students think more conceptually about the figures and provide an excellent framework for group projects or lecture discussion.

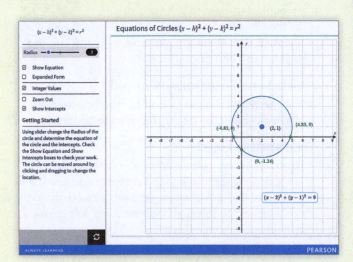

NEW! Workspace Assignments

Students can now show their work like never before! Workspace Assignments allow students to work through an exercise step-by-step, and show their mathematical reasoning as they progress. Students receive immediate feedback after they complete each step, and helpful hints and videos offer guidance when they need it. When accessed via a mobile device, Workspace exercises use handwriting recognition software that allows students to naturally write out their answers. Each student's work is automatically graded and captured in the MyMathLab gradebook so instructors can easily pinpoint exactly where they need to focus their instruction.

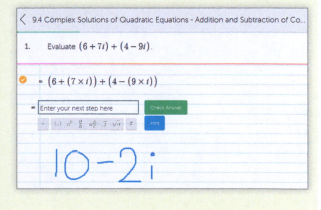

www.mymathlab.com

Resources for Success

Instructor Resources

Additional resources can be downloaded from **www.mymathlab.com** or **www.pearsonhighered.com** or hardcopy resources can be ordered from your sales representative.

Annotated Instructor's Edition

Shorter answers are on the page beside the exercises. Longer answers are in the back of the text.

Instructor's Solutions Manual

Fully worked solutions to all textbook exercises.

PowerPoint® Lecture Slides

Fully editable lecture slides that correlate to the textbook.

Mini Lecture Notes

Additional examples and helpful teaching tips for each section.

TestGen®

Enables instructors to build, edit, print, and administer tests using a computerized bank of algorithmic questions developed to cover all the objectives of the text.

Student Resources

Additional resources to help student success are available to be packaged with the Blitzer textbook and MyMathLab access code.

Objective Level Videos

An all new video program covers every objective of the text and is assignable in MyMathLab. Many videos provide an active learning environment where students try out their newly learned skill.

Chapter Test Prep Videos

Students can watch instructors work through step-by-step solutions to all the Chapter Test exercises from the textbook. These are available in MyMathLab and on YouTube.

Student Solutions Manual

Fully worked solutions to odd-numbered exercises and available to be packaged with the textbook.

Learning Guide

The note-taking guide begins each chapter with an engaging application, and provides additional examples and exercises for students to work through for a greater conceptual understanding and mastery of topics.

New to this edition: classroom projects are included for each chapter providing the opportunity for collaborative work. The Learning Guide is available in PDF and customizable Word file formats in MyMathLab. It can also be packaged with the textbook and MyMathLab access code.

MathTalk Videos

Engaging videos connect mathematics to real-life events and interesting applications. These fun, instructional videos show students that math is relevant to their daily lives and are assignable in MyMathLab. Assignable exercises are available in MyMathLab for these videos to help students apply valuable information presented in the videos.

www.mymathlab.com

TO THE STUDENT

The bar graph shows some of the qualities that students say make a great teacher. It was my goal to incorporate each of these qualities throughout the pages of this book.

Explains Things Clearly

I understand that your primary purpose in reading *Trigonometry* is to acquire a solid understanding of the required topics in your trigonometry course. In order to achieve this goal, I've carefully explained each topic. Important definitions and procedures are set off in boxes, and worked-out examples that present solutions in a step-by-step manner appear in every section. Each example is followed by a similar matched problem, called a Check Point, for you to try so that you can actively participate in the learning process as you read the book. (Answers to all Check Points appear in the back of the book.)

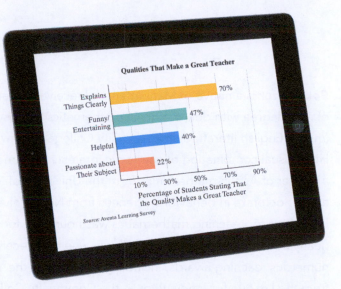

Funny & Entertaining

Who says that a trigonometry textbook can't be entertaining? From our unusual cover to the photos in the chapter and section openers, prepare to expect the unexpected. I hope some of the book's enrichment essays, called Blitzer Bonuses, will put a smile on your face from time to time.

Helpful

I designed the book's features to help you acquire knowledge of trigonometry, as well as to show you how trigonometry can solve authentic problems that apply to your life. These helpful features include:

- **Explanatory Voice Balloons:** Voice balloons are used in a variety of ways to make math less intimidating. They translate trigonometric language into everyday English, help clarify problem-solving procedures, present alternative ways of understanding concepts, and connect new concepts to concepts you have already learned.

- **Great Question!:** The book's Great Question! boxes are based on questions students ask in class. The answers to these questions give suggestions for problem solving, point out common errors to avoid, and provide informal hints and suggestions.

- **Achieving Success:** The book's Achieving Success boxes give you helpful strategies for success in learning trigonometry, as well as suggestions that can be applied for achieving your full academic potential in future college coursework.

- **Chapter Summaries:** Each chapter contains a review chart that summarizes the definitions and concepts in every section of the chapter. Examples from the chapter that illustrate these key concepts are also referenced in the chart. Review these summaries and you'll know the most important material in the chapter!

Passionate about the Subject

I passionately believe that no other discipline comes close to math in offering a more extensive set of tools for application and development of your mind. I wrote the book in Point Reyes National Seashore, 40 miles north of San Francisco. The park consists of 75,000 acres with miles of pristine surf-washed beaches, forested ridges, and bays bordered by white cliffs. It was my hope to convey the beauty and excitement of mathematics using nature's unspoiled beauty as a source of inspiration and creativity. Enjoy the pages that follow as you empower yourself with the trigonometry needed to succeed in college, your career, and your life.

Regards,

Bob
Robert Blitzer

ABOUT THE AUTHOR

Bob Blitzer is a native of Manhattan and received a Bachelor of Arts degree with dual majors in mathematics and psychology (minor: English literature) from the City College of New York. His unusual combination of academic interests led him toward a Master of Arts in mathematics from the University of Miami and a doctorate in behavioral sciences from Nova University. Bob's love for teaching mathematics was nourished for nearly 30 years at Miami Dade College, where he received numerous teaching awards, including Innovator of the Year from the League for Innovations in the Community College and an endowed chair based on excellence in the classroom. In addition to *Trigonometry*, Bob has written textbooks covering developmental mathematics, introductory algebra, intermediate algebra, college algebra, algebra and trigonometry, precalculus, and liberal arts mathematics, all published by Pearson. When not secluded in his Northern California writer's cabin, Bob can be found hiking the beaches and trails of Point Reyes National Seashore and tending to the chores required by his beloved entourage of horses, chickens, and irritable roosters.

APPLICATIONS INDEX

TRIGONOMETRY

Angles and the Trigonometric Functions

Have you had days when your physical, intellectual, and emotional potentials were all at their peak? Then there are those other days when we feel we should not even bother getting out of bed. Do our potentials run in oscillating cycles like the tides? Can they be described mathematically? In this chapter, you will get an angle on how trigonometry describes phenomena that occur in cycles.

HERE'S WHERE YOU'LL FIND THIS APPLICATION:

- The reason why trigonometry is ideally suited to describe patterns that occur in cycles is developed in Section 1.4, pages 60–61.
- Exercises 43–46 in Exercise Set 1.4 contain trigonometric functions showing repeating phenomena, including the ebb and flow of tides, as well as the emotional cycles we experience in life.

Angles and Radian Measure

What am I supposed to learn?

After studying this section, you should be able to:

1. Recognize and use the vocabulary of angles.
2. Use degree measure.
3. Use radian measure.
4. Convert between degrees and radians.
5. Draw angles in standard position.
6. Find coterminal angles.
7. Find the length of a circular arc.
8. Find the area of a sector.
9. Use linear and angular speed to describe motion on a circular path.

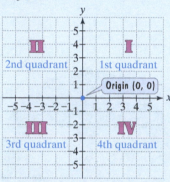

The San Francisco Museum of Modern Art was constructed in 1995 to illustrate how art and architecture can enrich one another. The exterior involves geometric shapes, symmetry, and unusual facades. Although there are no windows, natural light streams in through a truncated cylindrical skylight that crowns the building. The architect worked with a scale model of the museum at the site and observed how light hit it during different times of the day. These observations were used to cut the cylindrical skylight at an angle that maximizes sunlight entering the interior.

Angles play a critical role in creating modern architecture. They are also fundamental in trigonometry. In this section, we begin our study of trigonometry by looking at angles and methods for measuring them.

A Brief Review • The Rectangular Coordinate System

The rectangular coordinate system consists of a horizontal number line, the *x*-axis, and a vertical number line, the *y*-axis, intersecting at their zero points, the origin. Each point in the system corresponds to an ordered pair of real numbers (x, y). The first number in the pair is the *x*-coordinate; the second number is the *y*-coordinate. The axes divide the plane into four quarters called quadrants. The points located on the axes are not in any quadrant.

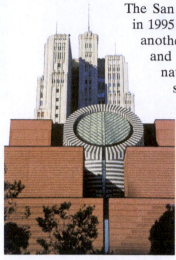

1. Recognize and use the vocabulary of angles.

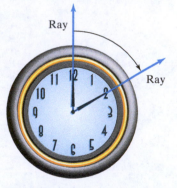

FIGURE 1.1 Clock with hands forming an angle

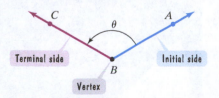

FIGURE 1.2 An angle; two rays with a common endpoint

Angles

The hour hand of a clock suggests a **ray**, a part of a line that has only one endpoint and extends forever in the opposite direction. An **angle** is formed by two rays that have a common endpoint. One ray is called the **initial side** and the other the **terminal side**.

A rotating ray is often a useful way to think about angles. The ray in **Figure 1.1** rotates from 12 to 2. The ray pointing to 12 is the **initial side** and the ray pointing to 2 is the **terminal side**. The common endpoint of an angle's initial side and terminal side is the **vertex** of the angle.

Figure 1.2 shows an angle. The arrow near the vertex shows the direction and the amount of rotation from the initial side to the terminal side. Several methods can be used to name an angle. Lowercase Greek letters, such as α (alpha), β (beta), γ (gamma), and θ (theta), are often used.

Many angles in trigonometry are located in *standard position*.

Standard Position of Angles

An angle is in **standard position** if

- its vertex is at the origin of a rectangular coordinate system and
- its initial side lies along the positive *x*-axis.

The angles in **Figure 1.3** are both in standard position.

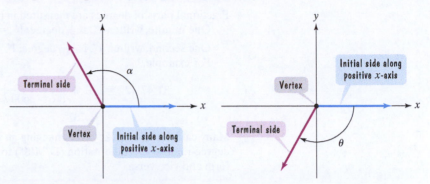

FIGURE 1.3 Two angles in standard position

(a) α in standard position; α positive

(b) θ in standard position; θ negative

When we see an initial side and a terminal side in place, there are two kinds of rotations that could have generated the angle. The arrow in **Figure 1.3(a)** indicates that the rotation from the initial side to the terminal side is in the counterclockwise direction. **Positive angles** are generated by counterclockwise rotation. Thus, angle α is positive. By contrast, the arrow in **Figure 1.3(b)** shows that the rotation from the initial side to the terminal side is in the clockwise direction. **Negative angles** are generated by clockwise rotation. Thus, angle θ is negative.

When an angle is in standard position, its terminal side can lie in a quadrant. We say that the angle **lies in that quadrant**. For example, in **Figure 1.3(a)**, the terminal side of angle α lies in quadrant II. Thus, angle α lies in quadrant II. By contrast, in **Figure 1.3(b)**, the terminal side of angle θ lies in quadrant III. Thus, angle θ lies in quadrant III.

Must all angles in standard position lie in a quadrant? The answer is no. The terminal side can lie on the x-axis or the y-axis. For example, angle β in **Figure 1.4** has a terminal side that lies on the negative y-axis. An angle is called a **quadrantal angle** if its terminal side lies on the x-axis or on the y-axis. Angle β in **Figure 1.4** is an example of a quadrantal angle.

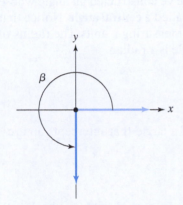

FIGURE 1.4 β is a quadrantal angle.

② Use degree measure.

A complete 360° rotation

Measuring Angles Using Degrees

Angles are measured by determining the amount of rotation from the initial side to the terminal side. One way to measure angles is in **degrees**, symbolized by a small, raised circle °. Think of the hour hand of a clock. From 12 noon to 12 midnight, the hour hand moves around in a complete circle. By definition, the ray has rotated through 360 degrees, or 360°. Using 360° as the amount of rotation of a ray back onto itself, a degree, 1°, is $\frac{1}{360}$ of a complete rotation.

Figure 1.5 shows that certain angles have special names. An **acute angle** measures less than 90° [see **Figure 1.5(a)**]. A **right angle**, one quarter of a complete rotation, measures 90° [**Figure 1.5(b)**]. Examine the right angle—do you see a small square at the vertex? This symbol is used to indicate a right angle. An **obtuse angle** measures more than 90° but less than 180° [**Figure 1.5(c)**]. Finally, a **straight angle**, one-half a complete rotation, measures 180° [**Figure 1.5(d)**].

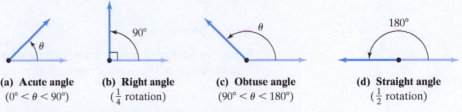

(a) Acute angle
$(0° < \theta < 90°)$

(b) Right angle
$(\frac{1}{4}$ rotation$)$

(c) Obtuse angle
$(90° < \theta < 180°)$

(d) Straight angle
$(\frac{1}{2}$ rotation$)$

FIGURE 1.5 Classifying angles by their degree measurement

We will be using notation such as $\theta = 60°$ to refer to an angle θ whose measure is 60°. We also refer to *an angle of* 60° or a 60° *angle*, rather than using the more precise (but cumbersome) phrase *an angle whose measure is* 60°.

Fractional parts of degrees are measured in minutes and seconds.

One minute, written $1'$, is $\frac{1}{60}$ degree: $1' = \frac{1}{60}°$.

One second, written $1''$, is $\frac{1}{3600}$ degree: $1'' = \frac{1}{3600}°$.

For example,

$$31°47'12'' = \left(31 + \frac{47}{60} + \frac{12}{3600}\right)°$$

$$\approx 31.787°.$$

Many calculators have keys for changing an angle from degree-minute-second notation (D°M′S″) to a decimal form and vice versa.

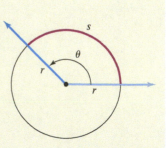

```
31°47'12"
                    31.78666667
```

③ Use radian measure.

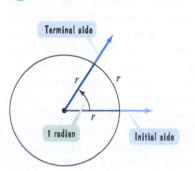

FIGURE 1.6 For a 1-radian angle, the intercepted arc and the radius are equal.

Measuring Angles Using Radians

Another way to measure angles is in *radians*. Let's first define an angle measuring **1 radian**. We use a circle of radius r. In **Figure 1.6**, we've constructed an angle whose vertex is at the center of the circle. Such an angle is called a **central angle**. Notice that this central angle intercepts an arc along the circle measuring r units. The radius of the circle is also r units. The measure of such an angle is 1 radian.

Definition of a Radian

One radian is the measure of the central angle of a circle that intercepts an arc equal in length to the radius of the circle.

The **radian measure** of any central angle is the length of the intercepted arc divided by the circle's radius. In **Figure 1.7(a)**, the length of the arc intercepted by angle β is double the radius, r. We find the measure of angle β in radians by dividing the length of the intercepted arc by the radius.

$$\beta = \frac{\text{length of the intercepted arc}}{\text{radius}} = \frac{2r}{r} = 2$$

Thus, angle β measures 2 radians.

In **Figure 1.7(b)**, the length of the intercepted arc is triple the radius, r. Let us find the measure of angle γ:

$$\gamma = \frac{\text{length of the intercepted arc}}{\text{radius}} = \frac{3r}{r} = 3.$$

Thus, angle γ measures 3 radians.

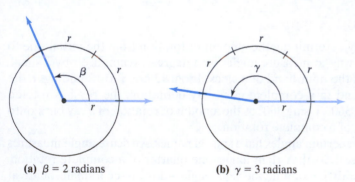

(a) $\beta = 2$ radians   **(b)** $\gamma = 3$ radians

FIGURE 1.7 Two central angles measured in radians

Radian Measure

Consider an arc of length s on a circle of radius r. The measure of the central angle, θ, that intercepts the arc is

$$\theta = \frac{s}{r} \text{ radians.}$$

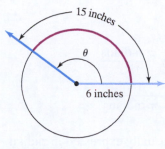

FIGURE 1.8

EXAMPLE 1   Computing Radian Measure

A central angle, θ, in a circle of radius 6 inches intercepts an arc of length 15 inches. What is the radian measure of θ?

SOLUTION

Angle θ is shown in **Figure 1.8**. The radian measure of a central angle is the length of the intercepted arc, s, divided by the circle's radius, r. The length of the intercepted arc is 15 inches: $s = 15$ inches. The circle's radius is 6 inches: $r = 6$ inches. Now we use the formula for radian measure to find the radian measure of θ.

$$\theta = \frac{s}{r} = \frac{15 \text{ inches}}{6 \text{ inches}} = 2.5$$

Thus, the radian measure of θ is 2.5.   •••

In Example 1, notice that the units (inches) cancel when we use the formula for radian measure. We are left with a number with no units. Thus, if an angle θ has a measure of 2.5 radians, we can write $\theta = 2.5$ radians or $\theta = 2.5$. We will often include the word *radians* simply for emphasis. There should be no confusion as to whether radian or degree measure is being used. Why is this so? If θ has a degree measure of, say, 2.5°, we must include the degree symbol and write $\theta = 2.5°$, and *not* $\theta = 2.5$.

GREAT QUESTION!

When determining radian measure, do the units for the length of the intercepted arc and the radius have to be the same?

Yes. Before applying the formula for radian measure, be sure that the same unit of length is used for the intercepted arc, s, and the radius, r.

GREAT QUESTION!

Why is it so important to work each of the book's Check Points?

You learn best by doing. Do not simply look at the worked examples and conclude that you know how to solve them. To be sure you understand the worked examples, try each Check Point. Check your answer in the answer section before continuing your reading. Expect to read this book with pencil and paper handy to work the Check Points.

☑ **Check Point 1** A central angle, θ, in a circle of radius 12 feet intercepts an arc of length 42 feet. What is the radian measure of θ?

④ Convert between degrees and radians.

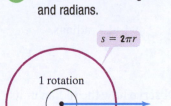

FIGURE 1.9 A complete rotation

Relationship between Degrees and Radians

How can we obtain a relationship between degrees and radians? We compare the number of degrees and the number of radians in one complete rotation, shown in **Figure 1.9**. We know that 360° is the amount of rotation of a ray back onto itself. The length of the intercepted arc is equal to the circumference of the circle. Thus, the radian measure of this central angle is the circumference of the circle divided by the circle's radius, r. The circumference of a circle of radius r is $2\pi r$. We use the formula for radian measure to find the radian measure of the 360° angle.

$$\theta = \frac{s}{r} = \frac{\text{the circle's circumference}}{r} = \frac{2\pi r}{r} = 2\pi$$

Because one complete rotation measures 360° and 2π radians,

$$360° = 2\pi \text{ radians}.$$

Dividing both sides by 2, we have

$$180° = \pi \text{ radians}.$$

Dividing this last equation by 180° or π gives the conversion rules in the box on the next page.

Conversion between Degrees and Radians

Using the basic relationship π radians $= 180°$,

1. To convert degrees to radians, multiply degrees by $\dfrac{\pi \text{ radians}}{180°}$.

2. To convert radians to degrees, multiply radians by $\dfrac{180°}{\pi \text{ radians}}$.

Angles that are fractions of a complete rotation are usually expressed in radian measure as fractional multiples of π, rather than as decimal approximations. For example, we write $\theta = \dfrac{\pi}{2}$ rather than using the decimal approximation $\theta \approx 1.57$.

EXAMPLE 2 Converting from Degrees to Radians

Convert each angle in degrees to radians:

    a. $30°$ **b.** $90°$ **c.** $-135°$.

SOLUTION

To convert degrees to radians, multiply by $\dfrac{\pi \text{ radians}}{180°}$. Observe how the degree units cancel.

    a. $30° = 30° \cdot \dfrac{\pi \text{ radians}}{180°} = \dfrac{30\pi}{180} \text{ radians} = \dfrac{\pi}{6} \text{ radians}$

    b. $90° = 90° \cdot \dfrac{\pi \text{ radians}}{180°} = \dfrac{90\pi}{180} \text{ radians} = \dfrac{\pi}{2} \text{ radians}$

    c. $-135° = -135° \cdot \dfrac{\pi \text{ radians}}{180°} = -\dfrac{135\pi}{180} \text{ radians} = -\dfrac{3\pi}{4} \text{ radians}$

> Divide the numerator and denominator by 45.

                                                             • • •

✓ **Check Point 2** Convert each angle in degrees to radians:

    a. $60°$ **b.** $270°$ **c.** $-300°$.

EXAMPLE 3 Converting from Radians to Degrees

Convert each angle in radians to degrees:

    a. $\dfrac{\pi}{3}$ radians   **b.** $-\dfrac{5\pi}{3}$ radians   **c.** 1 radian   **d.** 2.3 radians.

SOLUTION

To convert radians to degrees, multiply by $\dfrac{180°}{\pi \text{ radians}}$. Observe how the radian units cancel.

    a. $\dfrac{\pi}{3} \text{ radians} = \dfrac{\pi \text{ radians}}{3} \cdot \dfrac{180°}{\pi \text{ radians}} = \dfrac{180°}{3} = 60°$

    b. $-\dfrac{5\pi}{3} \text{ radians} = -\dfrac{5\pi \text{ radians}}{3} \cdot \dfrac{180°}{\pi \text{ radians}} = -\dfrac{5 \cdot 180°}{3} = -300°$

    c. $1 \text{ radian} = 1 \text{ radian} \cdot \dfrac{180°}{\pi \text{ radians}} = \dfrac{180°}{\pi} \approx 57.3°$

    d. $2.3 \text{ radians} = 2.3 \text{ radians} \cdot \dfrac{180°}{\pi \text{ radians}} = \dfrac{2.3 \cdot 180°}{\pi} \approx 131.8°$ **• • •**

✓ **Check Point 3** Convert each angle in radians to degrees:

    a. $\dfrac{\pi}{4}$ radians   **b.** $-\dfrac{4\pi}{3}$ radians   **c.** 6 radians   **d.** -4.7 radians.

Draw angles in standard position.

Drawing Angles in Standard Position

Although we can convert angles in radians to degrees, it is helpful to "think in radians" without having to make this conversion. To become comfortable with radian measure, consider angles in standard position: Each vertex is at the origin and each initial side lies along the positive *x*-axis. Think of the terminal side of the angle revolving around the origin. Thinking in radians means determining what part of a complete revolution or how many full revolutions will produce an angle whose radian measure is known. And here's the thing: We want to do this without having to convert from radians to degrees.

Figure 1.10 is a starting point for learning to think in radians. The figure illustrates that when the terminal side makes one full revolution, it forms an angle whose radian measure is 2π. The figure shows the quadrantal angles formed by $\frac{3}{4}$ of a revolution, $\frac{1}{2}$ of a revolution, and $\frac{1}{4}$ of a revolution.

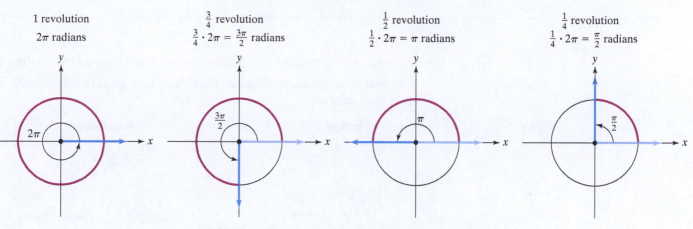

FIGURE 1.10 Angles formed by revolutions of terminal sides

EXAMPLE 4   Drawing Angles in Standard Position

Draw and label each angle in standard position:

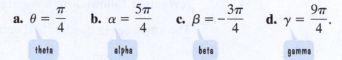

  a. $\theta = \dfrac{\pi}{4}$ **b.** $\alpha = \dfrac{5\pi}{4}$ **c.** $\beta = -\dfrac{3\pi}{4}$ **d.** $\gamma = \dfrac{9\pi}{4}$.

SOLUTION

Because we are drawing angles in standard position, each vertex is at the origin and each initial side lies along the positive *x*-axis.

a. An angle of $\dfrac{\pi}{4}$ radians is a positive angle. It is obtained by rotating the terminal side counterclockwise. Because 2π is a full-circle revolution, we can express $\dfrac{\pi}{4}$ as a fractional part of 2π to determine the necessary rotation:

$$\frac{\pi}{4} = \frac{1}{8} \cdot 2\pi.$$

$\dfrac{\pi}{4}$ is $\dfrac{1}{8}$ of a complete revolution of 2π radians.

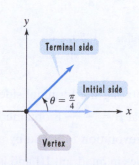

FIGURE 1.11

We see that $\theta = \dfrac{\pi}{4}$ is obtained by rotating the terminal side counterclockwise for $\dfrac{1}{8}$ of a revolution. The angle lies in quadrant I and is shown in **Figure 1.11**.

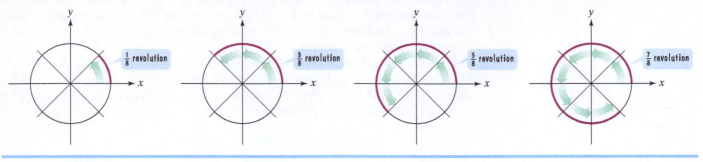

b. An angle of $\frac{5\pi}{4}$ radians is a positive angle. It is obtained by rotating the terminal side counterclockwise. Here are two ways to determine the necessary rotation:

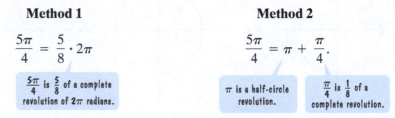

Method 1

$$\frac{5\pi}{4} = \frac{5}{8} \cdot 2\pi$$

$\frac{5\pi}{4}$ is $\frac{5}{8}$ of a complete revolution of 2π radians.

Method 2

$$\frac{5\pi}{4} = \pi + \frac{\pi}{4}.$$

π is a half-circle revolution. $\frac{\pi}{4}$ is $\frac{1}{8}$ of a complete revolution.

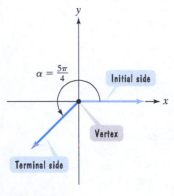

FIGURE 1.12

Method 1 shows that $\alpha = \frac{5\pi}{4}$ is obtained by rotating the terminal side counterclockwise for $\frac{5}{8}$ of a revolution. Method 2 shows that $\alpha = \frac{5\pi}{4}$ is obtained by rotating the terminal side counterclockwise for half of a revolution followed by a counterclockwise rotation of $\frac{1}{8}$ of a revolution. The angle lies in quadrant III and is shown in **Figure 1.12**.

c. An angle of $-\frac{3\pi}{4}$ is a negative angle. It is obtained by rotating the terminal side clockwise. We use $\left|-\frac{3\pi}{4}\right|$, or $\frac{3\pi}{4}$, to determine the necessary rotation.

Method 1

$$\frac{3\pi}{4} = \frac{3}{8} \cdot 2\pi$$

$\frac{3\pi}{4}$ is $\frac{3}{8}$ of a complete revolution of 2π radians.

Method 2

$$\frac{3\pi}{4} = \frac{2\pi}{4} + \frac{\pi}{4} = \frac{\pi}{2} + \frac{\pi}{4}$$

$\frac{\pi}{2}$ is a quarter-circle revolution. $\frac{\pi}{4}$ is $\frac{1}{8}$ of a complete revolution.

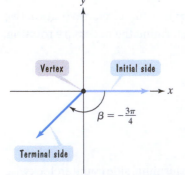

FIGURE 1.13

Method 1 shows that $\beta = -\frac{3\pi}{4}$ is obtained by rotating the terminal side clockwise for $\frac{3}{8}$ of a revolution. Method 2 shows that $\beta = -\frac{3\pi}{4}$ is obtained by rotating the terminal side clockwise for $\frac{1}{4}$ of a revolution followed by a clockwise rotation of $\frac{1}{8}$ of a revolution. The angle lies in quadrant III and is shown in **Figure 1.13**.

d. An angle of $\frac{9\pi}{4}$ radians is a positive angle. It is obtained by rotating the terminal side counterclockwise. Here are two methods to determine the necessary rotation:

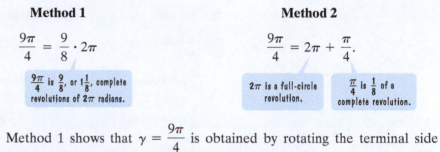

Method 1	**Method 2**

$$\frac{9\pi}{4} = \frac{9}{8} \cdot 2\pi$$

$\frac{9\pi}{4}$ is $\frac{9}{8}$, or $1\frac{1}{8}$, complete revolutions of 2π radians.

$$\frac{9\pi}{4} = 2\pi + \frac{\pi}{4}.$$

2π is a full-circle revolution.

$\frac{\pi}{4}$ is $\frac{1}{8}$ of a complete revolution.

Method 1 shows that $\gamma = \frac{9\pi}{4}$ is obtained by rotating the terminal side counterclockwise for $1\frac{1}{8}$ revolutions. Method 2 shows that $\gamma = \frac{9\pi}{4}$ is obtained by rotating the terminal side counterclockwise for a full-circle revolution followed by a counterclockwise rotation of $\frac{1}{8}$ of a revolution. The angle lies in quadrant I and is shown in **Figure 1.14**. •••

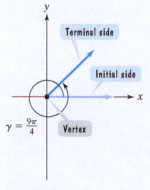

$\gamma = \frac{9\pi}{4}$

FIGURE 1.14

⊘ **Check Point 4** Draw and label each angle in standard position:

a. $\theta = -\frac{\pi}{4}$ **b.** $\alpha = \frac{3\pi}{4}$ **c.** $\beta = -\frac{7\pi}{4}$ **d.** $\gamma = \frac{13\pi}{4}$.

GREAT QUESTION!

Any connection between the radian measure denominators and the location of the terminal sides?

Yes. Denominators of 6 put the terminal sides closer to the x-axis. Denominators of 3 put the terminal sides closer to the y-axis. Denominators of 4 put the terminal sides midway between the axes.

Figure 1.15 illustrates the degree and radian measures of angles that you will commonly see in trigonometry. Each angle is in standard position, so that the initial side lies along the positive x-axis. We will be using both degree and radian measures for these angles.

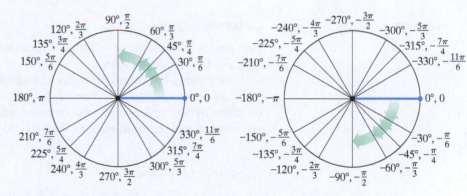

FIGURE 1.15 Degree and radian measures of selected positive and negative angles

GREAT QUESTION!

Any hints for drawing the angles in Figure 1.15?

Yes. It is helpful to first divide the rectangular coordinate system into

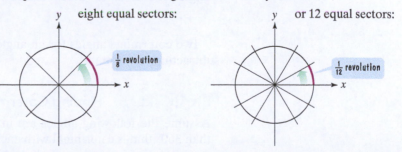

eight equal sectors:

$\frac{1}{8}$ revolution

or 12 equal sectors:

$\frac{1}{12}$ revolution

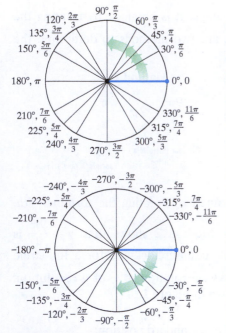

FIGURE 1.15 (repeated) Degree and radian measures of selected positive and negative angles

Table 1.1 describes some of the positive angles in **Figure 1.15** in terms of revolutions of the angle's terminal side around the origin.

Table 1.1

Terminal Side	Radian Measure of Angle	Degree Measure of Angle
$\frac{1}{12}$ revolution	$\frac{1}{12} \cdot 2\pi = \frac{\pi}{6}$	$\frac{1}{12} \cdot 360° = 30°$
$\frac{1}{8}$ revolution	$\frac{1}{8} \cdot 2\pi = \frac{\pi}{4}$	$\frac{1}{8} \cdot 360° = 45°$
$\frac{1}{6}$ revolution	$\frac{1}{6} \cdot 2\pi = \frac{\pi}{3}$	$\frac{1}{6} \cdot 360° = 60°$
$\frac{1}{4}$ revolution	$\frac{1}{4} \cdot 2\pi = \frac{\pi}{2}$	$\frac{1}{4} \cdot 360° = 90°$
$\frac{1}{3}$ revolution	$\frac{1}{3} \cdot 2\pi = \frac{2\pi}{3}$	$\frac{1}{3} \cdot 360° = 120°$
$\frac{1}{2}$ revolution	$\frac{1}{2} \cdot 2\pi = \pi$	$\frac{1}{2} \cdot 360° = 180°$
$\frac{2}{3}$ revolution	$\frac{2}{3} \cdot 2\pi = \frac{4\pi}{3}$	$\frac{2}{3} \cdot 360° = 240°$
$\frac{3}{4}$ revolution	$\frac{3}{4} \cdot 2\pi = \frac{3\pi}{2}$	$\frac{3}{4} \cdot 360° = 270°$
$\frac{7}{8}$ revolution	$\frac{7}{8} \cdot 2\pi = \frac{7\pi}{4}$	$\frac{7}{8} \cdot 360° = 315°$
1 revolution	$1 \cdot 2\pi = 2\pi$	$1 \cdot 360° = 360°$

6 Find coterminal angles.

Coterminal Angles

Two angles with the same initial and terminal sides but possibly different rotations are called **coterminal angles**.

Every angle has infinitely many coterminal angles. Why? Think of an angle in standard position. If the rotation of the angle is extended by one or more complete rotations of 360° or 2π, clockwise or counterclockwise, the result is an angle with the same initial and terminal sides as the original angle.

> **Coterminal Angles**
>
> Increasing or decreasing the degree measure of an angle in standard position by an integer multiple of 360° results in a coterminal angle. Thus, an angle of $\theta°$ is coterminal with angles of $\theta° \pm 360°k$, where k is an integer.
>
> Increasing or decreasing the radian measure of an angle by an integer multiple of 2π results in a coterminal angle. Thus, an angle of θ radians is coterminal with angles of $\theta \pm 2\pi k$, where k is an integer.

Two coterminal angles for an angle of $\theta°$ can be found by adding 360° to $\theta°$ and subtracting 360° from $\theta°$.

EXAMPLE 5   Finding Coterminal Angles

Assume the following angles are in standard position. Find a positive angle less than 360° that is coterminal with each of the following:

    a. a 420° angle   **b.** a −120° angle.

SOLUTION

We obtain the coterminal angle by adding or subtracting 360°. The requirement to obtain a positive angle less than 360° determines whether we should add or subtract.

a. For a 420° angle, subtract 360° to find a positive coterminal angle.

$$420° - 360° = 60°$$

A 60° angle is coterminal with a 420° angle. **Figure 1.16(a)** illustrates that these angles have the same initial and terminal sides.

b. For a −120° angle, add 360° to find a positive coterminal angle.

$$-120° + 360° = 240°$$

A 240° angle is coterminal with a −120° angle. **Figure 1.16(b)** illustrates that these angles have the same initial and terminal sides.

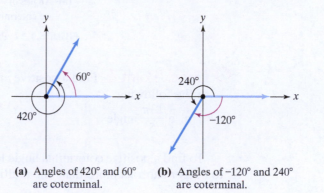

(a) Angles of 420° and 60° are coterminal.

(b) Angles of −120° and 240° are coterminal.

FIGURE 1.16 Pairs of coterminal angles

• • •

✓ **Check Point 5** Find a positive angle less than 360° that is coterminal with each of the following:

a. a 400° angle **b.** a −135° angle.

Two coterminal angles for an angle of θ radians can be found by adding 2π to θ and subtracting 2π from θ.

EXAMPLE 6   Finding Coterminal Angles

Assume the following angles are in standard position. Find a positive angle less than 2π that is coterminal with each of the following:

a. a $\dfrac{17\pi}{6}$ angle **b.** a $-\dfrac{\pi}{12}$ angle.

SOLUTION

We obtain the coterminal angle by adding or subtracting 2π. The requirement to obtain a positive angle less than 2π determines whether we should add or subtract.

a. For a $\dfrac{17\pi}{6}$ angle, note that $\dfrac{17}{6} = 2\dfrac{5}{6}$, so subtract 2π to find a positive coterminal angle.

$$\frac{17\pi}{6} - 2\pi = \frac{17\pi}{6} - \frac{12\pi}{6} = \frac{5\pi}{6}$$

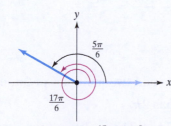

(a) Angles of $\frac{17\pi}{6}$ and $\frac{5\pi}{6}$ are coterminal.

FIGURE 1.17 Pairs of coterminal angles

A $\dfrac{5\pi}{6}$ angle is coterminal with a $\dfrac{17\pi}{6}$ angle. **Figure 1.17(a)** illustrates that these angles have the same initial and terminal sides.

b. For a $-\dfrac{\pi}{12}$ angle, add 2π to find a positive coterminal angle.

$$-\frac{\pi}{12} + 2\pi = -\frac{\pi}{12} + \frac{24\pi}{12} = \frac{23\pi}{12}$$

A $\dfrac{23\pi}{12}$ angle is coterminal with a $-\dfrac{\pi}{12}$ angle. **Figure 1.17(b)** illustrates that these angles have the same initial and terminal sides.

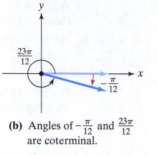

(b) Angles of $-\dfrac{\pi}{12}$ and $\dfrac{23\pi}{12}$ are coterminal.

FIGURE 1.17 Pairs of coterminal angles   • • •

Check Point **6** Find a positive angle less than 2π that is coterminal with each of the following:

a. a $\dfrac{13\pi}{5}$ angle   **b.** a $-\dfrac{\pi}{15}$ angle.

To find a positive coterminal angle less than $360°$ or 2π, it is sometimes necessary to add or subtract more than one multiple of $360°$ or 2π.

EXAMPLE 7   Finding Coterminal Angles

Find a positive angle less than $360°$ or 2π that is coterminal with each of the following:

a. a $750°$ angle   **b.** a $\dfrac{22\pi}{3}$ angle   **c.** a $-\dfrac{17\pi}{6}$ angle   **d.** a -10.3 angle.

SOLUTION

a. For a $750°$ angle, subtract two multiples of $360°$, or $720°$, to find a positive coterminal angle less than $360°$.

$$750° - 360° \cdot 2 = 750° - 720° = 30°$$

A $30°$ angle is coterminal with a $750°$ angle.

DISCOVERY

Make a sketch for each part of Example 7 illustrating that the coterminal angle we found and the given angle have the same initial and terminal sides.

b. For a $\dfrac{22\pi}{3}$ angle, note that $\dfrac{22}{3} = 7\dfrac{1}{3}$, so subtract three multiples of 2π, or 6π, to find a positive coterminal angle less than 2π.

$$\frac{22\pi}{3} - 2\pi \cdot 3 = \frac{22\pi}{3} - 6\pi = \frac{22\pi}{3} - \frac{18\pi}{3} = \frac{4\pi}{3}$$

A $\dfrac{4\pi}{3}$ angle is coterminal with a $\dfrac{22\pi}{3}$ angle.

c. For a $-\dfrac{17\pi}{6}$ angle, note that $-\dfrac{17}{6} = -2\dfrac{5}{6}$, so add two multiples of 2π, or 4π, to find a positive coterminal angle less than 2π.

$$-\frac{17\pi}{6} + 2\pi \cdot 2 = -\frac{17\pi}{6} + 4\pi = -\frac{17\pi}{6} + \frac{24\pi}{6} = \frac{7\pi}{6}$$

A $\dfrac{7\pi}{6}$ angle is coterminal with a $-\dfrac{17\pi}{6}$ angle.

d. For a -10.3 angle, it is helpful to remember that 1 radian $\approx 57°$. Therefore,

$$-10.3 \approx -10.3(57°) = -587.1°.$$

For a $-587.1°$ angle, we need to add two multiplies of $360°$ to find a positive coterminal angle less than $360°$. Equivalently, for a -10.3 angle, we need to add two multiples of 2π, or 4π, to find a positive coterminal angle less than 2π.

$$-10.3 + 2\pi \cdot 2 = -10.3 + 4\pi \approx 2.3$$

A 2.3 angle is approximately coterminal with a -10.3 angle. • • •

⊘ Check Point **7** Find a positive angle less than $360°$ or 2π that is coterminal with each of the following:

a. an $855°$ angle

b. a $\dfrac{17\pi}{3}$ angle

c. a $-\dfrac{25\pi}{6}$ angle

d. a 17.4 angle.

7 Find the length of a circular arc.

The Length of a Circular Arc

We can use the radian measure formula, $\theta = \dfrac{s}{r}$, to find the length of the arc of a circle. How do we do this? Remember that s represents the length of the arc intercepted by the central angle θ. Thus, by solving the formula for s, we have an equation for arc length.

GREAT QUESTION!

Can I apply the formula $s = r\theta$ if θ is expressed in degrees?

No. The formula can only be used when θ is expressed in radians. If θ is given in degrees, you'll need to convert from degrees to radians before using $s = r\theta$ to determine s, the length of the circular arc.

The Length of a Circular Arc

Let r be the radius of a circle and θ the nonnegative radian measure of a central angle of the circle. The length of the arc intercepted by the central angle is

$$s = r\theta.$$

s = arc length

EXAMPLE 8 Finding the Length of a Circular Arc

A circle has a radius of 10 inches. Find the length of the arc intercepted by a central angle of $120°$.

SOLUTION

The formula $s = r\theta$ can be used only when θ is expressed in radians. Thus, we begin by converting $120°$ to radians. Multiply by $\dfrac{\pi \text{ radians}}{180°}$.

$$120° = 120° \cdot \frac{\pi \text{ radians}}{180°} = \frac{120\pi}{180} \text{ radians} = \frac{2\pi}{3} \text{ radians}$$

Now we can use the formula $s = r\theta$ to find the length of the arc. The circle's radius is 10 inches: $r = 10$ inches. The measure of the central angle, in radians, is $\dfrac{2\pi}{3}$: $\theta = \dfrac{2\pi}{3}$. The length of the arc intercepted by this central angle is

$$s = r\theta = (10 \text{ inches})\left(\frac{2\pi}{3}\right) = \frac{20\pi}{3} \text{ inches} \approx 20.94 \text{ inches.} \quad \bullet\bullet\bullet$$

GREAT QUESTION!

What unit do I use when expressing the length of a circular arc?

The unit used to describe the length of a circular arc is the same unit that is given in the circle's radius.

⊘ Check Point **8** A circle has a radius of 6 inches. Find the length of the arc intercepted by a central angle of $45°$. Express arc length in terms of π. Then round your answer to two decimal places.

8 Find the area of a sector.

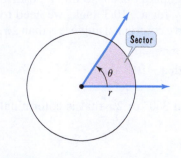

FIGURE 1.18 The shaded region is a sector of the circle.

Area of a Sector of a Circle

We now turn our attention from the length of a circular arc to the area of a sector. A **sector of a circle** is the portion of a circle's interior formed by a central angle θ. **Figure 1.18** shows a sector in a circle of radius r.

The central angle θ in **Figure 1.18** is measured in radians. If $\theta = 2\pi$, the central angle forms a sector that covers the entire interior of the circle. For a central angle θ, where $0 < \theta < 2\pi$, the area of the sector in **Figure 1.18** makes up the fraction $\dfrac{\theta}{2\pi}$ of the complete circle's area. Thus,

$$\text{Area of a sector} = \frac{\theta}{2\pi}(\pi r^2) = \frac{1}{2}r^2\theta, \ \theta \text{ in radians.}$$

Fraction of the complete circle   Area of the complete circle

The Area of a Sector

Let r be the radius of a circle and θ the nonnegative radian measure of a central angle of the circle. The area of the sector formed by the central angle is

$$A = \tfrac{1}{2}r^2\theta.$$

GREAT QUESTION!

What do the formulas for arc length, $s = r\theta$, and area of a sector, $A = \tfrac{1}{2}r^2\theta$, have in common?

The variables in both formulas involve the radius r of a circle and a central angle θ. When using either formula, the value of θ must be expressed in radians.

EXAMPLE 9   Finding the Area of a Sector of a Circle

A circle has a radius of 4 feet. Find the area of the sector formed by a central angle of $60°$.

SOLUTION

The formula $A = \tfrac{1}{2}r^2\theta$ can be used only when θ is expressed in radians. Thus, we begin by converting $60°$ to radians. Multiply by $\dfrac{\pi \text{ radians}}{180°}$.

$$60° = 60° \cdot \frac{\pi \text{ radians}}{180°} = \frac{60\pi}{180}\text{ radians} = \frac{\pi}{3}\text{ radians}$$

Now we can use the formula $A = \tfrac{1}{2}r^2\theta$ to find the area of the sector. The circle's radius is 4 feet: $r = 4$ feet. The measure of the central angle, in radians, is $\dfrac{\pi}{3}$: $\theta = \dfrac{\pi}{3}$. The area of the sector formed by this central angle is

$$A = \frac{1}{2}r^2\theta = \frac{1}{2}(4 \text{ feet})^2\left(\frac{\pi}{3}\right) = \frac{8\pi}{3}\text{ square feet} \approx 8.38 \text{ square feet}$$

The area of the sector, rounded to two decimal places, is 8.38 square feet.   •••

✓ **Check Point 9** A circle has a radius of 6 feet. Find the area of the sector formed by a central angle of $150°$. Express area in terms of π. Then round your answer to two decimal places.

⑨ Use linear and angular speed to describe motion on a circular path.

Linear and Angular Speed

A carousel contains four circular rows of animals. As the carousel revolves, the animals in the outer row travel a greater distance per unit of time than those in the inner rows. These animals have a greater *linear speed* than those in the inner rows. By contrast, all animals, regardless of the row, complete the same number of revolutions per unit of time. All animals in the four circular rows travel at the same *angular speed*.

Using v for linear speed and ω (omega) for angular speed, we define these two kinds of speed along a circular path as follows:

Definitions of Linear and Angular Speed

If a point is in motion on a circle of radius r through an angle of θ radians in time t, then its **linear speed** is

$$v = \frac{s}{t},$$

where s is the arc length given by $s = r\theta$, and its **angular speed** is

$$\omega = \frac{\theta}{t}.$$

The hard drive in a computer rotates at 3600 revolutions per minute. This angular speed, expressed in revolutions per minute, can also be expressed in revolutions per second, radians per minute, and radians per second. Using 2π radians = 1 revolution, we express the angular speed of a hard drive in radians per minute as follows:

$$3600 \text{ revolutions per minute}$$

$$= \frac{3600 \text{ revolutions}}{1 \text{ minute}} \cdot \frac{2\pi \text{ radians}}{1 \text{ revolution}} = \frac{7200\pi \text{ radians}}{1 \text{ minute}}$$

$$= 7200\pi \text{ radians per minute.}$$

We can establish a relationship between the two kinds of speed by dividing both sides of the arc length formula, $s = r\theta$, by t:

$$\frac{s}{t} = \frac{r\theta}{t} = r\frac{\theta}{t}.$$

<u>This expression defines linear speed.</u> <u>This expression defines angular speed.</u>

Thus, linear speed is the product of the radius and the angular speed.

Linear Speed in Terms of Angular Speed

The linear speed, v, of a point a distance r from the center of rotation is given by

$$v = r\omega,$$

where ω is the angular speed in radians per unit of time.

In words: Linear speed is the radius times the angular speed.

EXAMPLE 10   Finding Linear Speed

A wind machine used to generate electricity has blades that are 10 feet in length (see **Figure 1.19**). The propeller is rotating at four revolutions per second. Find the linear speed, in feet per second, of the tips of the blades.

SOLUTION

We are given ω, the angular speed.

$$\omega = 4 \text{ revolutions per second}$$

We use the formula $v = r\omega$ to find v, the linear speed. Before applying the formula, we must express ω in radians per second.

FIGURE 1.19

$$\omega = \frac{4 \text{ revolutions}}{1 \text{ second}} \cdot \frac{2\pi \text{ radians}}{1 \text{ revolution}} = \frac{8\pi \text{ radians}}{1 \text{ second}} \quad \text{or} \quad \frac{8\pi}{1 \text{ second}}$$

The angular speed of the propeller is 8π radians per second. The linear speed is

$$v = r\omega = 10 \text{ feet} \cdot \frac{8\pi}{1 \text{ second}} = \frac{80\pi \text{ feet}}{\text{second}}.$$

The linear speed of the tips of the blades is 80π feet per second, which is approximately 251 feet per second.   • • •

✓ Check Point **10** Long before iPods that hold thousands of songs and play them with superb audio quality, individual songs were delivered on 75-rpm and 45-rpm circular records. A 45-rpm record has an angular speed of 45 revolutions per minute. Find the linear speed, in inches per minute, at the point where the needle is 1.5 inches from the record's center.

ACHIEVING SUCCESS

It is impossible to learn trigonometry without knowing its special language. The exercises in the Concept and Vocabulary Check, mainly fill-in-the-blank and true/false items, test your understanding of the definitions and concepts presented in each section. **Work all of the exercises in the Concept and Vocabulary Check** regardless of which exercises your professor assigns in the Exercise Set that follows.

CONCEPT AND VOCABULARY CHECK

Fill in each blank so that the resulting statement is true.

1. An angle in a rectangular coordinate system is in standard position if its vertex is at the _____ and its initial side lies along the positive _____.
2. Positive angles are generated by _____ rotation. Negative angles are generated by _____ rotation.
3. If $0° < \theta < 90°$, θ is a/an _____ angle.
   If $\theta = 90°$, θ is a/an _____ angle.
   If $90° < \theta < 180°$, θ is a/an _____ angle.
   If $\theta = 180°$, θ is a/an _____ angle.

4. The radian measure of θ shown in the figure is $\theta =$ _____.
5. To convert degrees to radians, multiply degrees by _____.
6. To convert radians to degrees, multiply radians by _____.

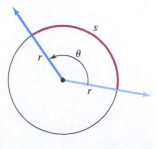

7. Two angles with the same initial and terminal sides but possibly different rotations are called _____ angles. Increasing or decreasing the degree measure of an angle in standard position by an integer multiple of _____ results in such an angle. Increasing or decreasing the radian measure of an angle in standard position by an integer multiple of _____ results in such an angle.

8. Using the figure shown, the length of the arc intercepted by the central angle θ is $s =$ _____.

9. Using the figure shown in Exercise 8, the area of the sector formed by the central angle θ is $A =$ _____.

10. True or false: If $r = 10$ centimeters and $\theta = 20°$, then $s = 10 \cdot 20 = 200$ centimeters. _____

11. The linear speed, v, of a point a distance r from the center of rotation is given by $v =$ _____, where ω is the _____ speed in radians per unit of time.

EXERCISE SET 1.1

Practice Exercises

In Exercises 1–6, the measure of an angle is given. Classify the angle as acute, right, obtuse, or straight.

1. 135°   **2.** 177°   **3.** 83.135°

4. 87.177°   **5.** π **6.** $\dfrac{\pi}{2}$

In Exercises 7–12, find the radian measure of the central angle of a circle of radius r that intercepts an arc of length s.

	Radius, r	Arc Length, s
7.	10 inches	40 inches
8.	5 feet	30 feet
9.	6 yards	8 yards
10.	8 yards	18 yards
11.	1 meter	400 centimeters
12.	1 meter	600 centimeters

In Exercises 13–20, convert each angle in degrees to radians. Express your answer as a multiple of π.

13. 45°   **14.** 18°   **15.** 135°

16. 150°   **17.** 300°   **18.** 330°

19. −225°   **20.** −270°

In Exercises 21–28, convert each angle in radians to degrees.

21. $\dfrac{\pi}{2}$ **22.** $\dfrac{\pi}{9}$ **23.** $\dfrac{2\pi}{3}$

24. $\dfrac{3\pi}{4}$ **25.** $\dfrac{7\pi}{6}$ **26.** $\dfrac{11\pi}{6}$

27. −3π **28.** −4π

In Exercises 29–34, convert each angle in degrees to radians. Round to two decimal places.

29. 18°   **30.** 76°   **31.** −40°

32. −50°   **33.** 200°   **34.** 250°

In Exercises 35–40, convert each angle in radians to degrees. Round to two decimal places.

35. 2 radians   **36.** 3 radians

37. $\dfrac{\pi}{13}$ radians   **38.** $\dfrac{\pi}{17}$ radians

39. −4.8 radians   **40.** −5.2 radians

In Exercises 41–56, use the circle shown in the rectangular coordinate system to draw each angle in standard position. State the quadrant in which the angle lies. When an angle's measure is given in radians, work the exercise without converting to degrees.

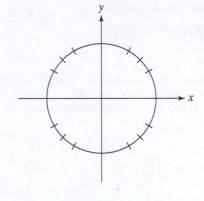

41. $\dfrac{7\pi}{6}$ **42.** $\dfrac{4\pi}{3}$ **43.** $\dfrac{3\pi}{4}$

44. $\dfrac{7\pi}{4}$ **45.** $-\dfrac{2\pi}{3}$ **46.** $-\dfrac{5\pi}{6}$

47. $-\dfrac{5\pi}{4}$ **48.** $-\dfrac{7\pi}{4}$ **49.** $\dfrac{16\pi}{3}$

50. $\dfrac{14\pi}{3}$ **51.** 120°   **52.** 150°

53. −210°   **54.** −240°   **55.** 420°

56. 405°

In Exercises 57–70, find a positive angle less than 360° or 2π that is coterminal with the given angle.

57. 395°   **58.** 415°   **59.** −150°

60. −160°   **61.** −765°   **62.** −760°

63. $\dfrac{19\pi}{6}$ **64.** $\dfrac{17\pi}{5}$ **65.** $\dfrac{23\pi}{5}$

66. $\dfrac{25\pi}{6}$ **67.** $-\dfrac{\pi}{50}$ **68.** $-\dfrac{\pi}{40}$

69. $-\dfrac{31\pi}{7}$ **70.** $-\dfrac{38\pi}{9}$

In Exercises 71–74, find the length of the arc on a circle of radius r intercepted by a central angle θ. Express arc length in terms of π. Then round your answer to two decimal places.

Radius, *r*	Central Angle, *θ*
71. 12 inches	$\theta = 45°$
72. 16 inches	$\theta = 60°$
73. 8 feet	$\theta = 225°$
74. 9 yards	$\theta = 315°$

In Exercises 75–78, find the area of the sector of a circle of radius r formed by a central angle θ. Express area in terms of π. Then round your answer to two decimal places.

Radius, *r*	Central Angle, *θ*
75. 10 meters	$\theta = 18°$
76. 6 yards	$\theta = 15°$
77. 4 inches	$\theta = 240°$
78. 3 inches	$\theta = 330°$

In Exercises 79–80, express each angular speed in radians per second.

79. 6 revolutions per second

80. 20 revolutions per second

Practice Plus

Use the circle shown in the rectangular coordinate system to solve Exercises 81–86. Find two angles, in radians, between -2π and 2π such that each angle's terminal side passes through the origin and the given point.

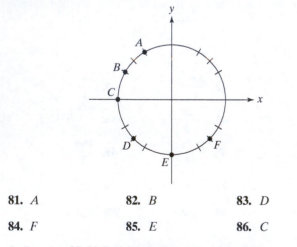

81. *A*	**82.** *B*	**83.** *D*
84. *F*	**85.** *E*	**86.** *C*

In Exercises 87–90, find the absolute value of the radian measure of the angle that the second hand of a clock moves through in the given time.

87. 55 seconds   **88.** 35 seconds

89. 3 minutes and 40 seconds

90. 4 minutes and 25 seconds

In Exercises 91–92, find the measure of the central angle on a circle of radius r that forms a sector with the given area.

Radius, *r*	Area of the Sector, *A*
91. 10 feet	25 square feet
92. 6 yards	36 square yards

Application Exercises

93. The minute hand of a clock moves from 12 to 2 o'clock, or $\frac{1}{6}$ of a complete revolution. Through how many degrees does it move? Through how many radians does it move?

94. The minute hand of a clock moves from 12 to 4 o'clock, or $\frac{1}{3}$ of a complete revolution. Through how many degrees does it move? Through how many radians does it move?

95. The minute hand of a clock is 8 inches long and moves from 12 to 2 o'clock. How far does the tip of the minute hand move? Express your answer in terms of π and then round to two decimal places.

96. The minute hand of a clock is 6 inches long and moves from 12 to 4 o'clock. How far does the tip of the minute hand move? Express your answer in terms of π and then round to two decimal places.

97. The figure shows a highway sign that warns of a railway crossing. The lines that form the cross pass through the circle's center and intersect at right angles. If the radius of the circle is 24 inches, find the length of each of the four arcs formed by the cross. Express your answer in terms of π and then round to two decimal places.

98. The radius of a wheel rolling on the ground is 80 centimeters. If the wheel rotates through an angle of 60°, how many centimeters does it move? Express your answer in terms of π and then round to two decimal places.

99. A lawn sprinkler rotates through an angle of 135° and projects water over a distance of 40 feet. What is the area of the lawn watered by the sprinkler? Express your answer in terms of π and then round to two decimal places.

100. A lawn sprinkler rotates through an angle of 120° and projects water over a distance of 30 feet. What is the area of the lawn watered by the sprinkler? Express your answer in terms of π and then round to two decimal places.

How do we measure the distance between two points, A and B, on Earth? We measure along a circle with a center, C, at the center of Earth. The radius of the circle is equal to the distance from C to the surface. Use the fact that Earth is a sphere of radius equal to approximately 4000 miles to solve Exercises 101–104.

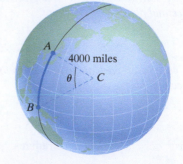

(In Exercises 101–104, be sure to refer to the figure at the bottom of the previous page.)

101. If two points, A and B, are 8000 miles apart, express angle θ in radians and in degrees.

102. If two points, A and B, are 10,000 miles apart, express angle θ in radians and in degrees.

103. If $\theta = 30°$, find the distance between A and B to the nearest mile.

104. If $\theta = 10°$, find the distance between A and B to the nearest mile.

105. The angular speed of a point on Earth is $\frac{\pi}{12}$ radian per hour. The Equator lies on a circle of radius approximately 4000 miles. Find the linear velocity, in miles per hour, of a point on the Equator.

106. A Ferris wheel has a radius of 25 feet. The wheel is rotating at two revolutions per minute. Find the linear speed, in feet per minute, of a seat on this Ferris wheel.

107. A water wheel has a radius of 12 feet. The wheel is rotating at 20 revolutions per minute. Find the linear speed, in feet per minute, of the water.

108. On a carousel, the outer row of animals is 20 feet from the center. The inner row of animals is 10 feet from the center. The carousel is rotating at 2.5 revolutions per minute. What is the difference, in feet per minute, in the linear speeds of the animals in the outer and inner rows? Round to the nearest foot per minute.

Explaining the Concepts

ACHIEVING SUCCESS

An effective way to understand something is to explain it to someone else. You can do this by using the Explaining the Concepts exercises that ask you to respond with verbal or written explanations. Speaking or writing about a new concept uses a different part of your brain than thinking about the concept. Explaining new ideas verbally will quickly reveal any gaps in your understanding. It will also help you to remember new concepts for longer periods of time.

109. What is an angle?

110. What determines the size of an angle?

111. Describe an angle in standard position.

112. Explain the difference between positive and negative angles. What are coterminal angles?

113. Explain what is meant by one radian.

114. Explain how to find the radian measure of a central angle.

115. Describe how to convert an angle in degrees to radians.

116. Explain how to convert an angle in radians to degrees.

117. Explain how to find the length of a circular arc.

118. Explain how to find the area of a sector.

119. If a carousel is rotating at 2.5 revolutions per minute, explain how to find the linear speed of a child seated on one of the animals.

120. The angular velocity of a point on Earth is $\frac{\pi}{12}$ radian per hour. Describe what happens every 24 hours.

Technology Exercises

In Exercises 121–124, use the keys on your calculator or graphing utility for converting an angle in degrees, minutes, and seconds $(D°M'S'')$ into decimal form, and vice versa.

In Exercises 121–122, convert each angle to a decimal in degrees. Round your answer to two decimal places.

121. $30°15'10''$ **122.** $65°45'20''$

In Exercises 123–124, convert each angle to $D°M'S''$ form. Round your answer to the nearest second.

123. $30.42°$ **124.** $50.42°$

Critical Thinking Exercises

Make Sense?   *In Exercises 125–128, determine whether each statement makes sense or does not make sense, and explain your reasoning.*

125. I made an error because the angle I drew in standard position exceeded a straight angle.

126. When an angle's measure is given in terms of π, I know that it's measured using radians.

127. When I convert degrees to radians, I multiply by 1, choosing $\frac{\pi}{180°}$ for 1.

128. Using radian measure, I can always find a positive angle less than 2π coterminal with a given angle by adding or subtracting 2π.

129. If $\theta = \frac{3}{2}$, is this angle larger or smaller than a right angle?

130. A railroad curve is laid out on a circle. What radius should be used if the track is to change direction by $20°$ in a distance of 100 miles? Round your answer to the nearest mile.

131. A lawn sprinkler projects water over a distance of 30 feet. Through how many degrees should the sprinkler rotate to water 700 square feet of lawn? Round to the nearest degree.

132. Assuming Earth to be a sphere of radius 4000 miles, how many miles north of the Equator is Miami, Florida, if it is $26°$ north from the Equator? Round your answer to the nearest mile.

Preview Exercises

Exercises 133–135 will help you prepare for the material covered in the next section. In each exercise, let θ be an acute angle in a right triangle, as shown in the figure. These exercises require the use of the Pythagorean Theorem.

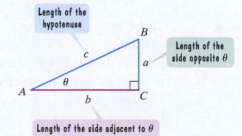

133. If $a = 5$ and $b = 12$, find the ratio of the length of the side opposite θ to the length of the hypotenuse.

134. If $a = 1$ and $b = 1$, find the ratio of the length of the side opposite θ to the length of the hypotenuse. Simplify the ratio by rationalizing the denominator.

135. Simplify: $\left(\dfrac{a}{c}\right)^2 + \left(\dfrac{b}{c}\right)^2$.

A Brief Review • Triangles

- The sum of the measures of the three angles of any triangle is 180°.

EXAMPLE Find the measure of angle A.

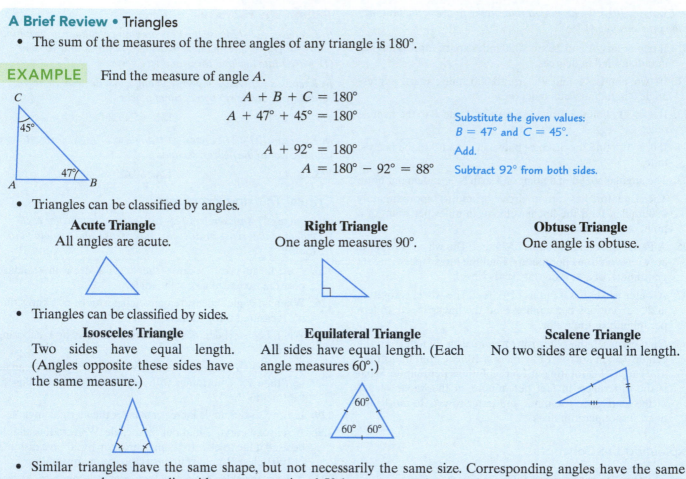

$$A + B + C = 180°$$
$$A + 47° + 45° = 180°$$

Substitute the given values: $B = 47°$ and $C = 45°$.

$$A + 92° = 180°$$

Add.

$$A = 180° - 92° = 88°$$

Subtract 92° from both sides.

- Triangles can be classified by angles.

Acute Triangle	**Right Triangle**	**Obtuse Triangle**
All angles are acute.	One angle measures 90°.	One angle is obtuse.

- Triangles can be classified by sides.

Isosceles Triangle
Two sides have equal length. (Angles opposite these sides have the same measure.)

Equilateral Triangle
All sides have equal length. (Each angle measures 60°.)

Scalene Triangle
No two sides are equal in length.

- Similar triangles have the same shape, but not necessarily the same size. Corresponding angles have the same measure and corresponding sides are proportional. If the measures of two angles of one triangle are equal to those of two angles of a second triangle, then the two triangles are similar.

EXAMPLE The triangles are similar. Find the missing length, x.

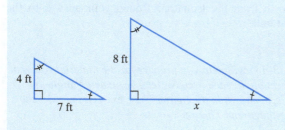

$$\frac{4}{8} = \frac{7}{x}$$

Corresponding sides of similar triangles are proportional.

$$4x = 8 \cdot 7$$

If $\frac{a}{b} = \frac{c}{d}$, then $ad = bc$.

$$4x = 56$$

Simplify.

$$\frac{4x}{4} = \frac{56}{4}$$

Divide both sides by 4.

$$x = 14$$

Simplify.

The missing length, x, is 14 feet.

- **The Pythagorean Theorem**
    The sum of the squares of the legs of a right triangle equals the square of the length of the hypotenuse.
    If the legs have lengths a and b and the hypotenuse has length c, then
    $$a^2 + b^2 = c^2.$$

EXAMPLE Find the missing length, a.

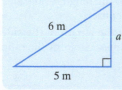

$$a^2 + 5^2 = 6^2$$
$$a^2 + 25 = 36$$
$$a^2 = 11$$
$$a = \sqrt{11} \approx 3.3 \text{ m}$$

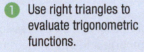

Section 1.2 — Right Triangle Trigonometry

What am I supposed to learn?

After studying this section, you should be able to:

1 Use right triangles to evaluate trigonometric functions.

2 Find function values for $30° \left(\dfrac{\pi}{6}\right)$, $45° \left(\dfrac{\pi}{4}\right)$, and $60° \left(\dfrac{\pi}{3}\right)$.

3 Recognize and use fundamental identities.

4 Use equal cofunctions of complements.

5 Evaluate trigonometric functions with a calculator.

6 Use right triangle trigonometry to solve applied problems.

In the last century, Ang Rita Sherpa climbed Mount Everest ten times, all without the use of bottled oxygen.

Mountain climbers have forever been fascinated by reaching the top of Mount Everest, sometimes with tragic results. The mountain, on Asia's Tibet–Nepal border, is Earth's highest, peaking at an incredible 29,035 feet. The heights of mountains can be found using **trigonometry**. The word *trigonometry* means "*measurement of triangles.*" Trigonometry is used in navigation, building, and engineering. For centuries, Muslims used trigonometry and the stars to navigate across the Arabian desert to Mecca, the birthplace of the prophet Muhammad, the founder of Islam. The ancient Greeks used trigonometry to record the locations of thousands of stars and worked out the motion of the Moon relative to Earth. Today, trigonometry is used to study the structure of DNA, the master molecule that determines how we grow from a single cell to a complex, fully developed adult.

A Brief Review • Functions

- A relation is any set of ordered pairs. The set of first components of the ordered pairs is the domain and the set of second components is the range. A function is a relation in which each member of the domain corresponds to exactly one member of the range. No two ordered pairs of a function can have the same first component and different second components.

EXAMPLE

The domain of the relation $\{(1, 2), (3, 4), (3, 7)\}$ is $\{1, 3\}$. The range is $\{2, 4, 7\}$. The relation is not a function because 3 in the domain corresponds to both 4 and 7 in the range.

- If a function is defined by an equation, the notation $f(x)$, read "f of x" or "f at x" describes the value of the function at the number, or input, x.

EXAMPLE   If $f(x) = x^2 + 3x + 5$, find $f(2)$.

Substitute 2 for x in the function's equation.

$$f(2) = 2^2 + 3 \cdot 2 + 5 = 4 + 6 + 5 = 15$$

Thus, $f(2) = 15$.   *f of 2 is 15.*

1 Use right triangles to evaluate trigonometric functions.

The Six Trigonometric Functions

We begin the study of trigonometry by defining six functions, the six *trigonometric functions*. The inputs for these functions are measures of acute angles in right triangles. The outputs are the ratios of the lengths of the sides of right triangles.

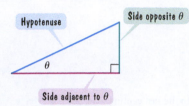

FIGURE 1.20 Naming a right triangle's sides from the point of view of an acute angle θ

Figure 1.20 shows a right triangle with one of its acute angles labeled θ. The side opposite the right angle is known as the **hypotenuse**. The other sides of the triangle are described by their position relative to the acute angle θ. One side is opposite θ and one is adjacent to θ.

The trigonometric functions have names that are words, rather than single letters such as f, g, and h. For example, the **sine of θ** is the length of the side opposite θ divided by the length of the hypotenuse:

$$\sin \theta = \frac{\text{length of side opposite } \theta}{\text{length of hypotenuse}}.$$

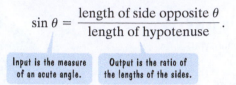

The ratio of lengths depends on angle θ and thus is a function of θ. The expression $\sin \theta$ really means $\sin(\theta)$, where sine is the name of the function and θ, the measure of an acute angle, is the input.

Here are the names of the six trigonometric functions, along with their abbreviations:

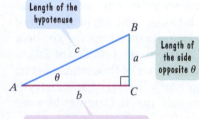

FIGURE 1.21

Name	Abbreviation	Name	Abbreviation
sine	sin	cosecant	csc
cosine	cos	secant	sec
tangent	tan	cotangent	cot

Now, let θ be an acute angle in a right triangle, as shown in **Figure 1.21**. The length of the side opposite θ is a, the length of the side adjacent to θ is b, and the length of the hypotenuse is c.

Right Triangle Definitions of Trigonometric Functions

See **Figure 1.21**. The six **trigonometric functions of the acute angle θ** are defined as follows:

$$\sin \theta = \frac{\text{length of side opposite angle } \theta}{\text{length of hypotenuse}} = \frac{a}{c} \qquad \csc \theta = \frac{\text{length of hypotenuse}}{\text{length of side opposite angle } \theta} = \frac{c}{a}$$

$$\cos \theta = \frac{\text{length of side adjacent to angle } \theta}{\text{length of hypotenuse}} = \frac{b}{c} \qquad \sec \theta = \frac{\text{length of hypotenuse}}{\text{length of side adjacent to angle } \theta} = \frac{c}{b}$$

$$\tan \theta = \frac{\text{length of side opposite angle } \theta}{\text{length of side adjacent to angle } \theta} = \frac{a}{b} \qquad \cot \theta = \frac{\text{length of side adjacent to angle } \theta}{\text{length of side opposite angle } \theta} = \frac{b}{a}.$$

Each of the trigonometric functions of the acute angle θ is positive. Observe that the ratios in the second column in the box are the reciprocals of the corresponding ratios in the first column.

GREAT QUESTION!

Is there a way to help me remember the right triangle definitions of any of the trigonometric functions?

The word

$$\text{SOHCAHTOA (pronounced: so-cah-tow-ah)}$$

may be helpful in remembering the definitions for sine, cosine, and tangent.

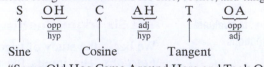

"Some Old Hog Came Around Here and Took Our Apples."

Figure 1.22 shows four right triangles of varying sizes. In each of the triangles, θ is the same acute angle, measuring approximately $56.3°$. All four of these similar triangles have the same shape and the lengths of corresponding sides are in the same ratio. In each triangle, the tangent function has the same value for the angle θ: $\tan \theta = \frac{3}{2}$.

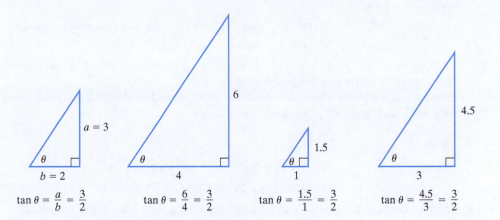

$$\tan \theta = \frac{a}{b} = \frac{3}{2} \qquad \tan \theta = \frac{6}{4} = \frac{3}{2} \qquad \tan \theta = \frac{1.5}{1} = \frac{3}{2} \qquad \tan \theta = \frac{4.5}{3} = \frac{3}{2}$$

FIGURE 1.22 A particular acute angle always gives the same ratio of opposite to adjacent sides.

In general, **the trigonometric function values of θ depend only on the size of angle θ and not on the size of the triangle**.

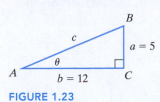

FIGURE 1.23

EXAMPLE 1 Evaluating Trigonometric Functions

Find the value of each of the six trigonometric functions of θ in **Figure 1.23**.

SOLUTION

We need to find the values of the six trigonometric functions of θ. However, we must know the lengths of all three sides of the triangle (a, b, and c) to evaluate all six functions. The values of a and b are given. We can use the Pythagorean Theorem, $c^2 = a^2 + b^2$, to find c.

$a = 5$ $\quad$ $b = 12$

$$c^2 = a^2 + b^2 = 5^2 + 12^2 = 25 + 144 = 169$$
$$c = \sqrt{169} = 13$$

Now that we know the lengths of the three sides of the triangle, we apply the definitions of the six trigonometric functions of θ. Referring to these lengths as opposite, adjacent, and hypotenuse, we have

$$\sin \theta = \frac{\text{opposite}}{\text{hypotenuse}} = \frac{5}{13} \qquad \csc \theta = \frac{\text{hypotenuse}}{\text{opposite}} = \frac{13}{5}$$

$$\cos \theta = \frac{\text{adjacent}}{\text{hypotenuse}} = \frac{12}{13} \qquad \sec \theta = \frac{\text{hypotenuse}}{\text{adjacent}} = \frac{13}{12}$$

$$\tan \theta = \frac{\text{opposite}}{\text{adjacent}} = \frac{5}{12} \qquad \cot \theta = \frac{\text{adjacent}}{\text{opposite}} = \frac{12}{5}.$$

$\bullet\bullet\bullet$

GREAT QUESTION!

Do I have to use the definitions of the trigonometric functions to get the function values shown in the second column?

No. The function values in the second column are reciprocals of those in the first column. You can obtain each of these values by interchanging the numerator and denominator of the corresponding ratio in the first column.

✓ **Check Point 1** Find the value of each of the six trigonometric functions of θ in the figure.

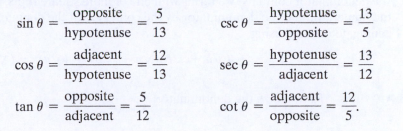

A Brief Review • Operations with Square Roots

- A square root is simplified when its radicand (the expression under the square root sign) has no factors other than 1 that are perfect squares. Simplification is accomplished using $\sqrt{ab} = \sqrt{a} \cdot \sqrt{b}$: The square root of a product is the product of the square roots.

EXAMPLE Simplify: $\sqrt{500}$.

$$\sqrt{500} = \sqrt{100 \cdot 5}$$ Factor 500: 100 is the greatest perfect square factor.

$$= \sqrt{100}\sqrt{5}$$ $\sqrt{ab} = \sqrt{a} \cdot \sqrt{b}$

$$= 10\sqrt{5}$$ Write $\sqrt{100}$ as 10. We read $10\sqrt{5}$ as "ten times the square root of 5."

- Adding and Subtracting Square Roots
  Two or more square roots can be combined using the distributive property provided that they have the same radicand. Such radicals are called like radicals.

EXAMPLES

$$7\sqrt{2} + 5\sqrt{2} = (7 + 5)\sqrt{2} = 12\sqrt{2}$$
$$\sqrt{5x} - 7\sqrt{5x} = 1\sqrt{5x} - 7\sqrt{5x} = (1 - 7)\sqrt{5x} = -6\sqrt{5x}$$

- In some cases, square roots can be combined once they have been simplified.

EXAMPLE

$$\sqrt{2} + \sqrt{8} = \sqrt{2} + \sqrt{4 \cdot 2} = 1\sqrt{2} + 2\sqrt{2} = (1 + 2)\sqrt{2} = 3\sqrt{2}$$

- The process of rewriting a square root expression as an equivalent expression in which the denominator no longer contains any square roots is called rationalizing the denominator. If the denominator consists of the square root of a natural number that is not a perfect square, multiply the numerator and the denominator by the smallest number that produces the square root of a perfect square in the denominator.

EXAMPLE Rationalize the denominator: $\dfrac{1}{\sqrt{2}}$.

$$\frac{1}{\sqrt{2}} = \frac{1}{\sqrt{2}} \cdot \frac{\sqrt{2}}{\sqrt{2}} = \frac{\sqrt{2}}{\sqrt{4}} = \frac{\sqrt{2}}{2}.$$

Multiply by 1.

- Square root expressions that involve the sum and difference of the same two terms are called conjugates.

EXAMPLE

The conjugate of $5 + \sqrt{3}$ is $5 - \sqrt{3}$.

- If a denominator contains two terms with one or more square roots, we rationalize the denominator by multiplying the numerator and the denominator by the conjugate of the denominator. The product of the denominator and its conjugate is found using

$$(\sqrt{a} + \sqrt{b})(\sqrt{a} - \sqrt{b}) = (\sqrt{a})^2 - (\sqrt{b})^2 = a - b.$$

EXAMPLE Rationalize the denominator: $\dfrac{7}{5 + \sqrt{3}}$.

$$\frac{7}{5 + \sqrt{3}} = \frac{7}{5 + \sqrt{3}} \cdot \frac{5 - \sqrt{3}}{5 - \sqrt{3}} = \frac{7(5 - \sqrt{3})}{5^2 - (\sqrt{3})^2} = \frac{7(5 - \sqrt{3})}{25 - 3} = \frac{7(5 - \sqrt{3})}{22} \text{ or } \frac{35 - 7\sqrt{3}}{22}$$

Multiply by 1. $(\sqrt{a} + \sqrt{b})(\sqrt{a} - \sqrt{b})$
$= (\sqrt{a})^2 - (\sqrt{b})^2$

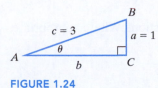

FIGURE 1.24

EXAMPLE 2   Evaluating Trigonometric Functions

Find the value of each of the six trigonometric functions of θ in **Figure 1.24**.

SOLUTION

We begin by finding b.

$$a^2 + b^2 = c^2 \qquad \text{Use the Pythagorean Theorem.}$$
$$1^2 + b^2 = 3^2 \qquad \text{Figure 1.24 shows that } a = 1 \text{ and } c = 3.$$
$$1 + b^2 = 9 \qquad 1^2 = 1 \text{ and } 3^2 = 9.$$
$$b^2 = 8 \qquad \text{Subtract 1 from both sides.}$$
$$b = \sqrt{8} = 2\sqrt{2} \qquad \text{Take the principal square root and simplify:}$$
$$\sqrt{8} = \sqrt{4 \cdot 2} = \sqrt{4}\sqrt{2} = 2\sqrt{2}.$$

Now that we know the lengths of the three sides of the triangle, we apply the definitions of the six trigonometric functions of θ.

$$\sin \theta = \frac{\text{opposite}}{\text{hypotenuse}} = \frac{1}{3} \qquad \csc \theta = \frac{\text{hypotenuse}}{\text{opposite}} = \frac{3}{1} = 3$$

$$\cos \theta = \frac{\text{adjacent}}{\text{hypotenuse}} = \frac{2\sqrt{2}}{3} \qquad \sec \theta = \frac{\text{hypotenuse}}{\text{adjacent}} = \frac{3}{2\sqrt{2}}$$

$$\tan \theta = \frac{\text{opposite}}{\text{adjacent}} = \frac{1}{2\sqrt{2}} \qquad \cot \theta = \frac{\text{adjacent}}{\text{opposite}} = \frac{2\sqrt{2}}{1} = 2\sqrt{2}$$

Because fractional expressions are usually written without radicals in the denominators, we simplify the values of $\tan \theta$ and $\sec \theta$ by rationalizing the denominators:

$$\tan \theta = \frac{1}{2\sqrt{2}} = \frac{1}{2\sqrt{2}} \cdot \frac{\sqrt{2}}{\sqrt{2}} = \frac{\sqrt{2}}{2 \cdot 2} = \frac{\sqrt{2}}{4} \qquad \sec \theta = \frac{3}{2\sqrt{2}} = \frac{3}{2\sqrt{2}} \cdot \frac{\sqrt{2}}{\sqrt{2}} = \frac{3\sqrt{2}}{2 \cdot 2} = \frac{3\sqrt{2}}{4}.$$

We are multiplying by 1 and not changing the value of $\frac{1}{2\sqrt{2}}$.

We are multiplying by 1 and not changing the value of $\frac{3}{2\sqrt{2}}$.

• • •

✓ **Check Point 2** Find the value of each of the six trigonometric functions of θ in the figure. Express each value in simplified form.

Function Values for Some Special Angles

② Find function values for $30° \left(\dfrac{\pi}{6} \right)$, $45° \left(\dfrac{\pi}{4} \right)$, and $60° \left(\dfrac{\pi}{3} \right)$.

A 45°, or $\dfrac{\pi}{4}$ radian, angle occurs frequently in trigonometry. How do we find the values of the trigonometric functions of 45°? We construct a right triangle with a 45° angle, as shown in **Figure 1.25**. The triangle actually has two 45° angles. Thus, the triangle is isosceles—that is, it has two sides of the same length. Assume that each leg of the triangle has a length equal to 1. We can find the length of the hypotenuse using the Pythagorean Theorem.

$$(\text{length of hypotenuse})^2 = 1^2 + 1^2 = 2$$
$$\text{length of hypotenuse} = \sqrt{2}$$

With **Figure 1.25**, we can determine the trigonometric function values for 45°.

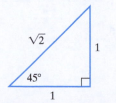

FIGURE 1.25 An isosceles right triangle

EXAMPLE 3   Evaluating Trigonometric Functions of 45°

Use **Figure 1.25** to find sin 45°, cos 45°, and tan 45°.

SOLUTION

We apply the definitions of these three trigonometric functions. Where appropriate, we simplify by rationalizing denominators.

$$\sin 45° = \frac{\text{length of side opposite } 45°}{\text{length of hypotenuse}} = \frac{1}{\sqrt{2}} = \frac{1}{\sqrt{2}} \cdot \frac{\sqrt{2}}{\sqrt{2}} = \frac{\sqrt{2}}{2}$$

Rationalize denominators.

$$\cos 45° = \frac{\text{length of side adjacent to } 45°}{\text{length of hypotenuse}} = \frac{1}{\sqrt{2}} = \frac{1}{\sqrt{2}} \cdot \frac{\sqrt{2}}{\sqrt{2}} = \frac{\sqrt{2}}{2}$$

$$\tan 45° = \frac{\text{length of side opposite } 45°}{\text{length of side adjacent to } 45°} = \frac{1}{1} = 1$$

FIGURE 1.25
(repeated)

● ● ●

✓ **Check Point 3** Use **Figure 1.25** to find csc 45°, sec 45°, and cot 45°.

When you worked Check Point 3, did you actually use **Figure 1.25** or did you use reciprocals to find the values?

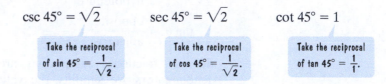

$$\csc 45° = \sqrt{2} \qquad \sec 45° = \sqrt{2} \qquad \cot 45° = 1$$

Take the reciprocal of $\sin 45° = \frac{1}{\sqrt{2}}$. Take the reciprocal of $\cos 45° = \frac{1}{\sqrt{2}}$. Take the reciprocal of $\tan 45° = \frac{1}{1}$.

Notice that if you use reciprocals, you should take the reciprocal of a function value before the denominator is rationalized. In this way, the reciprocal value will not contain a radical in the denominator.

Two other angles that occur frequently in trigonometry are 30°, or $\frac{\pi}{6}$ radian, and 60°, or $\frac{\pi}{3}$ radian, angles. We can find the values of the trigonometric functions of 30° and 60° by using a right triangle. To form this right triangle, draw an equilateral triangle—that is, a triangle with all sides the same length. Assume that each side has a length equal to 2. Now take half of the equilateral triangle. We obtain the right triangle in **Figure 1.26**. This right triangle has a hypotenuse of length 2 and a leg of length 1. The other leg has length a, which can be found using the Pythagorean Theorem.

$$a^2 + 1^2 = 2^2$$
$$a^2 + 1 = 4$$
$$a^2 = 3$$
$$a = \sqrt{3}$$

With the right triangle in **Figure 1.26**, we can determine the trigonometric functions for 30° and 60°.

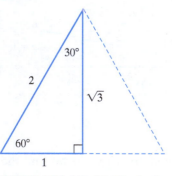

FIGURE 1.26 30°–60°–90° triangle

EXAMPLE 4 Evaluating Trigonometric Functions of 30° and 60°

Use **Figure 1.26** to find sin 60°, cos 60°, sin 30°, and cos 30°.

SOLUTION

We begin with 60°. Use the angle on the lower left in **Figure 1.26**.

$$\sin 60° = \frac{\text{length of side opposite } 60°}{\text{length of hypotenuse}} = \frac{\sqrt{3}}{2}$$

$$\cos 60° = \frac{\text{length of side adjacent to } 60°}{\text{length of hypotenuse}} = \frac{1}{2}$$

To find sin 30° and cos 30°, use the angle on the upper right in **Figure 1.26**.

$$\sin 30° = \frac{\text{length of side opposite } 30°}{\text{length of hypotenuse}} = \frac{1}{2}$$

$$\cos 30° = \frac{\text{length of side adjacent to } 30°}{\text{length of hypotenuse}} = \frac{\sqrt{3}}{2}$$

• • •

☑ Check Point **4** Use **Figure 1.26** to find tan 60° and tan 30°. If a radical appears in a denominator, rationalize the denominator.

Because we will often use the function values of 30°, 45°, and 60°, you should learn to construct the right triangles in **Figure 1.25** and **Figure 1.26**, repeated below. With sufficient practice, you will memorize the values in **Table 1.2**.

Table 1.2 Trigonometric Functions of Special Angles

θ	$30° = \frac{\pi}{6}$	$45° = \frac{\pi}{4}$	$60° = \frac{\pi}{3}$
$\sin \theta$	$\frac{1}{2}$	$\frac{\sqrt{2}}{2}$	$\frac{\sqrt{3}}{2}$
$\cos \theta$	$\frac{\sqrt{3}}{2}$	$\frac{\sqrt{2}}{2}$	$\frac{1}{2}$
$\tan \theta$	$\frac{\sqrt{3}}{3}$	1	$\sqrt{3}$

FIGURE 1.25
(repeated)

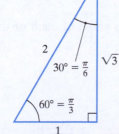

FIGURE 1.26
(repeated)

3 Recognize and use fundamental identities.

A Brief Review • Identities

An equation that is true for all values of the variable for which both sides are defined is an identity.

EXAMPLE

$$x + 3 = x + 2 + 1.$$

Fundamental Identities

Many relationships exist among the six trigonometric functions. These relationships are described using **trigonometric identities**. For example, csc θ is defined as the reciprocal of sin θ. This relationship can be expressed by the identity

$$\csc \theta = \frac{1}{\sin \theta}.$$

This identity is one of six **reciprocal identities**.

Reciprocal Identities

$$\sin \theta = \frac{1}{\csc \theta} \qquad \csc \theta = \frac{1}{\sin \theta}$$

$$\cos \theta = \frac{1}{\sec \theta} \qquad \sec \theta = \frac{1}{\cos \theta}$$

$$\tan \theta = \frac{1}{\cot \theta} \qquad \cot \theta = \frac{1}{\tan \theta}$$

Two other relationships that follow from the definitions of the trigonometric functions are called the **quotient identities**.

Quotient Identities

$$\tan \theta = \frac{\sin \theta}{\cos \theta} \qquad \cot \theta = \frac{\cos \theta}{\sin \theta}$$

If sin θ and cos θ are known, a quotient identity and three reciprocal identities make it possible to find the value of each of the four remaining trigonometric functions.

EXAMPLE 5   Using Quotient and Reciprocal Identities

Given $\sin\theta = \dfrac{2}{5}$ and $\cos\theta = \dfrac{\sqrt{21}}{5}$, find the value of each of the four remaining trigonometric functions.

SOLUTION

We can find $\tan\theta$ by using the quotient identity that describes $\tan\theta$ as the quotient of $\sin\theta$ and $\cos\theta$.

$$\tan\theta = \frac{\sin\theta}{\cos\theta} = \frac{\dfrac{2}{5}}{\dfrac{\sqrt{21}}{5}} = \frac{2}{5}\cdot\frac{5}{\sqrt{21}} = \frac{2}{\sqrt{21}} = \frac{2}{\sqrt{21}}\cdot\frac{\sqrt{21}}{\sqrt{21}} = \frac{2\sqrt{21}}{21}$$

> Rationalize the denominator.

We use the reciprocal identities to find the value of each of the remaining three functions.

$$\csc\theta = \frac{1}{\sin\theta} = \frac{1}{\dfrac{2}{5}} = \frac{5}{2}$$

$$\sec\theta = \frac{1}{\cos\theta} = \frac{1}{\dfrac{\sqrt{21}}{5}} = \frac{5}{\sqrt{21}} = \frac{5}{\sqrt{21}}\cdot\frac{\sqrt{21}}{\sqrt{21}} = \frac{5\sqrt{21}}{21}$$

> Rationalize the denominator.

$$\cot\theta = \frac{1}{\tan\theta} = \frac{1}{\dfrac{2}{\sqrt{21}}} = \frac{\sqrt{21}}{2}$$

> We found $\tan\theta = \dfrac{2}{\sqrt{21}}$. We could use $\tan\theta = \dfrac{2\sqrt{21}}{21}$, but then we would have to rationalize the denominator.

• • •

✓ **Check Point 5** Given $\sin\theta = \dfrac{2}{3}$ and $\cos\theta = \dfrac{\sqrt{5}}{3}$, find the value of each of the four remaining trigonometric functions.

Other relationships among trigonometric functions follow from the Pythagorean Theorem. Using **Figure 1.27**, the Pythagorean Theorem states that

$$a^2 + b^2 = c^2.$$

To obtain ratios that correspond to trigonometric functions, divide both sides of this equation by c^2.

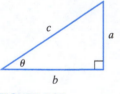

FIGURE 1.27

$$\frac{a^2}{c^2} + \frac{b^2}{c^2} = 1 \quad\text{or}\quad \left(\frac{a}{c}\right)^2 + \left(\frac{b}{c}\right)^2 = 1$$

> In **Figure 1.27**, $\sin\theta = \dfrac{a}{c}$, so this is $(\sin\theta)^2$.

> In **Figure 1.27**, $\cos\theta = \dfrac{b}{c}$, so this is $(\cos\theta)^2$.

Based on the observations in the voice balloons, we see that

$$(\sin\theta)^2 + (\cos\theta)^2 = 1.$$

We will use the notation $\sin^2\theta$ for $(\sin\theta)^2$ and $\cos^2\theta$ for $(\cos\theta)^2$. With this notation, we can write the identity as

$$\sin^2\theta + \cos^2\theta = 1.$$

Two additional identities can be obtained from $a^2 + b^2 = c^2$ by dividing both sides by b^2 and a^2, respectively. The three identities are called the **Pythagorean identities**.

> **Pythagorean Identities**
>
> $$\sin^2 \theta + \cos^2 \theta = 1 \qquad 1 + \tan^2 \theta = \sec^2 \theta \qquad 1 + \cot^2 \theta = \csc^2 \theta$$

EXAMPLE 6 Using a Pythagorean Identity

Given that $\sin \theta = \frac{3}{5}$ and θ is an acute angle, find the value of $\cos \theta$ using a trigonometric identity.

SOLUTION

We can find the value of $\cos \theta$ by using the Pythagorean identity

$$\sin^2 \theta + \cos^2 \theta = 1.$$

$$\left(\frac{3}{5}\right)^2 + \cos^2 \theta = 1 \qquad \text{We are given that } \sin \theta = \frac{3}{5}.$$

$$\frac{9}{25} + \cos^2 \theta = 1 \qquad \text{Square } \frac{3}{5}: \left(\frac{3}{5}\right)^2 = \frac{3^2}{5^2} = \frac{9}{25}.$$

$$\cos^2 \theta = 1 - \frac{9}{25} \qquad \text{Subtract } \frac{9}{25} \text{ from both sides.}$$

$$\cos^2 \theta = \frac{16}{25} \qquad \text{Simplify: } 1 - \frac{9}{25} = \frac{25}{25} - \frac{9}{25} = \frac{16}{25}.$$

$$\cos \theta = \sqrt{\frac{16}{25}} = \frac{4}{5} \qquad \text{Because } \theta \text{ is an acute angle, } \cos \theta \text{ is positive.}$$

Thus, $\cos \theta = \frac{4}{5}$. $\bullet\bullet\bullet$

$\oslash$ **Check Point 6** Given that $\sin \theta = \frac{1}{2}$ and θ is an acute angle, find the value of $\cos \theta$ using a trigonometric identity.

④ Use equal cofunctions of complements.

Trigonometric Functions and Complements

Two positive angles are **complements** if their sum is 90° or $\frac{\pi}{2}$. For example, angles of 70° and 20° are complements because 70° + 20° = 90°.

Another relationship among trigonometric functions is based on angles that are complements. Refer to **Figure 1.28**. Because the sum of the angles of any triangle is 180°, in a right triangle the sum of the acute angles is 90°. Thus, the acute angles are complements. If the degree measure of one acute angle is θ, then the degree measure of the other acute angle is $(90° - \theta)$. This angle is shown on the upper right in **Figure 1.28**.

Let's use **Figure 1.28** to compare $\sin \theta$ and $\cos(90° - \theta)$.

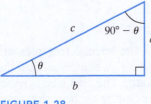

FIGURE 1.28

$$\sin \theta = \frac{\text{length of side opposite } \theta}{\text{length of hypotenuse}} = \frac{a}{c}$$

$$\cos(90° - \theta) = \frac{\text{length of side adjacent to } (90° - \theta)}{\text{length of hypotenuse}} = \frac{a}{c}$$

Thus, $\sin \theta = \cos(90° - \theta)$.

Because $\sin \theta = \cos(90° - \theta)$, if two angles are complements, the sine of one equals the cosine of the other. Because of this relationship, the sine and cosine are called *cofunctions* of each other. The name *cosine* is a shortened form of the phrase *complement's sine*.

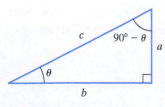

FIGURE 1.28 (repeated)

Any pair of trigonometric functions f and g for which

$$f(\theta) = g(90° - \theta) \quad \text{and} \quad g(\theta) = f(90° - \theta)$$

are called **cofunctions**. Using **Figure 1.28**, we can show that the tangent and cotangent are also cofunctions of each other. So are the secant and cosecant.

Cofunction Identities

The value of a trigonometric function of θ is equal to the cofunction of the complement of θ. Cofunctions of complementary angles are equal.

$$\sin \theta = \cos(90° - \theta) \qquad \cos \theta = \sin(90° - \theta)$$
$$\tan \theta = \cot(90° - \theta) \qquad \cot \theta = \tan(90° - \theta)$$
$$\sec \theta = \csc(90° - \theta) \qquad \csc \theta = \sec(90° - \theta)$$

If θ is in radians, replace $90°$ with $\dfrac{\pi}{2}$.

EXAMPLE 7   Using Cofunction Identities

Find a cofunction with the same value as the given expression:

a. $\sin 72°$ **b.** $\csc \dfrac{\pi}{3}$.

SOLUTION

Because the value of a trigonometric function of θ is equal to the cofunction of the complement of θ, we need to find the complement of each angle. We do this by subtracting the angle's measure from $90°$ or its radian equivalent, $\dfrac{\pi}{2}$.

a. $\sin 72° = \cos(90° - 72°) = \cos 18°$

> We have a function and its cofunction.

b. $\csc \dfrac{\pi}{3} = \sec\left(\dfrac{\pi}{2} - \dfrac{\pi}{3}\right) = \sec\left(\dfrac{3\pi}{6} - \dfrac{2\pi}{6}\right) = \sec \dfrac{\pi}{6}$

> We have a cofunction and its function.

> Perform the subtraction using the least common denominator, 6.

• • •

✍ **Check Point 7** Find a cofunction with the same value as the given expression:

a. $\sin 46°$ **b.** $\cot \dfrac{\pi}{12}$.

⑤ Evaluate trigonometric functions with a calculator.

Using a Calculator to Evaluate Trigonometric Functions

The values of the trigonometric functions obtained with the special triangles are exact values. For most acute angles other than $30°$, $45°$, and $60°$, we approximate the value of each of the trigonometric functions using a calculator. The first step is to set the calculator to the correct *mode*, degrees or radians, depending on how the acute angle is measured.

Most calculators have keys marked $\boxed{\text{SIN}}$, $\boxed{\text{COS}}$, and $\boxed{\text{TAN}}$. For example, to find the value of $\sin 30°$, set the calculator to the degree mode and enter 30 $\boxed{\text{SIN}}$ on most scientific calculators and $\boxed{\text{SIN}}$ 30 $\boxed{\text{ENTER}}$ on most graphing calculators. Consult the manual for your calculator.

To evaluate the cosecant, secant, and cotangent functions, use the key for the respective reciprocal function, $\boxed{\text{SIN}}$, $\boxed{\text{COS}}$, or $\boxed{\text{TAN}}$, and then use the reciprocal key. The reciprocal key is $\boxed{1/x}$ on many scientific calculators and $\boxed{x^{-1}}$ on many graphing calculators. For example, we can evaluate $\sec\dfrac{\pi}{12}$ using the following reciprocal relationship:

$$\sec\frac{\pi}{12} = \frac{1}{\cos\dfrac{\pi}{12}}.$$

Using the radian mode, enter one of the following keystroke sequences:

Many Scientific Calculators

$$\boxed{\pi}\ \boxed{\div}\ \boxed{12}\ \boxed{=}\ \boxed{\text{COS}}\ \boxed{1/x}$$

Many Graphing Calculators

$$\boxed{(}\ \boxed{\text{COS}}\ \boxed{(}\ \boxed{(}\ \boxed{\pi}\ \boxed{\div}\ \boxed{12}\ \boxed{)}\ \boxed{)}\ \boxed{x^{-1}}\ \boxed{\text{ENTER}}.$$

Rounding the display to four decimal places, we obtain $\sec\dfrac{\pi}{12} \approx 1.0353$.

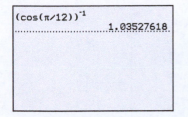

DISCOVERY

If you are using a graphing calculator, find the value of each of the following expressions, entering parentheses exactly as shown. (Your calculator may automatically insert the left parenthesis immediately following *cos*.) Use radian mode.

$$(\cos(\pi \div 12))^{-1}$$
$$\cos(\pi \div 12)^{-1}$$
$$\cos((\pi \div 12)^{-1})$$

Which two expression have the same value? What order of operations does the calculator follow when evaluating the second expression? Do you think that the second expression is ambiguous? Why do you think we used the outer set of parentheses in the keystrokes shown for the graphing calculator solution in 8(b)?

EXAMPLE 8 Evaluating Trigonometric Functions with a Calculator

Use a calculator to find the value to four decimal places:

    a. $\cos 48.2°$ **b.** $\cot 1.2$.

SOLUTION

Scientific Calculator Solution

Function	Mode	Keystrokes	Display, Rounded to Four Decimal Places
a. $\cos 48.2°$	Degree	48.2 $\boxed{\text{COS}}$	0.6665
b. $\cot 1.2$	Radian	1.2 $\boxed{\text{TAN}}$ $\boxed{1/x}$	0.3888

Graphing Calculator Solution

Function	Mode	Keystrokes	Display, Rounded to Four Decimal Places
a. $\cos 48.2°$	Degree	$\boxed{\text{COS}}$ 48.2 $\boxed{\text{ENTER}}$	0.6665
b. $\cot 1.2$	Radian	$\boxed{(}$ $\boxed{\text{TAN}}$ 1.2 $\boxed{)}$ $\boxed{x^{-1}}$ $\boxed{\text{ENTER}}$	0.3888

• • •

✓ **Check Point 8** Use a calculator to find the value to four decimal places:

    a. $\sin 72.8°$ **b.** $\csc 1.5$.

GREAT QUESTION!

Now that I can use my calculator to evaluate trigonometric functions, do I have to use pictures of right triangles to find function values for 30°, 45°, and 60°?

Yes. Don't become too calculator dependent. If you need to find *exact values* of trigonometric functions, leave the calculator alone. For most angles, the best that your calculator can do for you is to provide *approximate values*. When directions in trigonometry involve the word *exact*, put down the calculator and don't bother to ask Siri.

6 Use right triangle trigonometry to solve applied problems.

Applications

Many applications of right triangle trigonometry involve the angle made with an imaginary horizontal line. As shown in **Figure 1.29**, an angle formed by a horizontal line and the line of sight to an object that is above the horizontal line is called the **angle of elevation**. The angle formed by a horizontal line and the line of sight to an object that is below the horizontal line is called the **angle of depression**. Transits and sextants are instruments used to measure such angles.

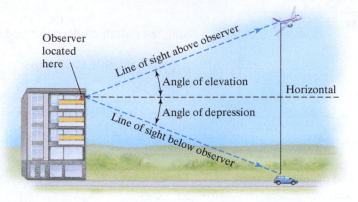

Observer located here
Line of sight above observer
Angle of elevation
Horizontal
Angle of depression
Line of sight below observer

FIGURE 1.29

EXAMPLE 9   Problem Solving Using an Angle of Elevation

Sighting the top of a building, a surveyor measured the angle of elevation to be 22°. The transit is 5 feet above the ground and 300 feet from the building. Find the building's height.

SOLUTION

The situation is illustrated in **Figure 1.30**. Let a be the height of the portion of the building that lies above the transit. The height of the building is the transit's height, 5 feet, plus a. Thus, we need to identify a trigonometric function that will make it possible to find a. In terms of the 22° angle, we are looking for the side opposite the angle. The transit is 300 feet from the building, so the side adjacent to the 22° angle is 300 feet. Because we have a known angle, an unknown opposite side, and a known adjacent side, we select the tangent function.

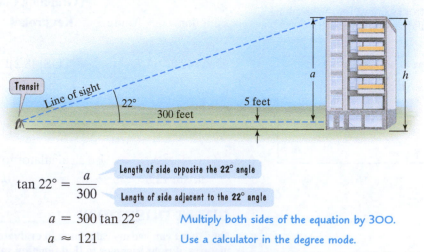

Transit
Line of sight
22°
300 feet
5 feet
a
h

FIGURE 1.30

$$\tan 22° = \frac{a}{300}$$

Length of side opposite the **22°** angle

Length of side adjacent to the **22°** angle

$a = 300 \tan 22°$ Multiply both sides of the equation by 300.

$a \approx 121$ Use a calculator in the degree mode.

The height of the part of the building above the transit is approximately 121 feet. Thus, the height of the building is determined by adding the transit's height, 5 feet, to 121 feet.

$$h \approx 5 + 121 = 126$$

The building's height is approximately 126 feet. $\bullet\bullet\bullet$

✓ Check Point **9** The irregular blue shape in **Figure 1.31** represents a lake. The distance across the lake, *a*, is unknown. To find this distance, a surveyor took the measurements shown in the figure. What is the distance across the lake?

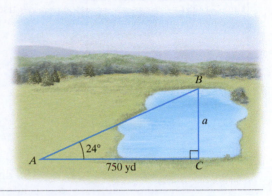

FIGURE 1.31

If two sides of a right triangle are known, an appropriate trigonometric function can be used to find an acute angle θ in the triangle. You will also need to use an inverse trigonometric key on a calculator. These keys use a function value to display the acute angle θ. For example, suppose that $\sin \theta = 0.866$. We can find θ in the degree mode by using the secondary *inverse sine* key, usually labeled $\boxed{\text{SIN}^{-1}}$. The key $\boxed{\text{SIN}^{-1}}$ is not a button you will actually press. It is the secondary function for the button labeled $\boxed{\text{SIN}}$.

```
sin⁻¹(.866)
          59.99708907
■
```

Many Scientific Calculators:

.866 $\boxed{\text{2nd}}$ $\boxed{\text{SIN}}$

Pressing $\boxed{\text{2nd}}$ $\boxed{\text{SIN}}$ accesses the inverse sine key, $\boxed{\text{SIN}^{-1}}$.

Many Graphing Calculators:

$\boxed{\text{2nd}}$ $\boxed{\text{SIN}}$ .866 $\boxed{\text{ENTER}}$

The display should show approximately 59.997, which can be rounded to 60. Thus, if $\sin \theta = 0.866$, then $\theta \approx 60°$.

EXAMPLE 10   Determining the Angle of Elevation

A building that is 21 meters tall casts a shadow 25 meters long. Find the angle of elevation of the Sun to the nearest degree.

SOLUTION

The situation is illustrated in **Figure 1.32**. We are asked to find θ.

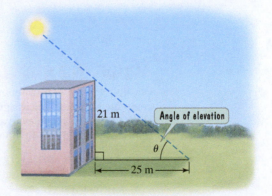

FIGURE 1.32

We begin with the tangent function.

$$\tan \theta = \frac{\text{side opposite } \theta}{\text{side adjacent to } \theta} = \frac{21}{25}$$

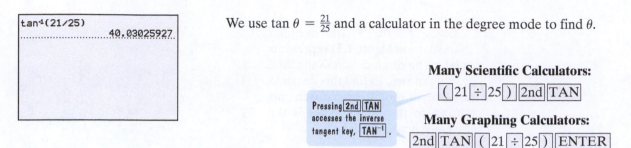

We use $\tan \theta = \frac{21}{25}$ and a calculator in the degree mode to find θ.

Many Scientific Calculators:

$(\ 21\ \div\ 25\ )\ \boxed{2\text{nd}}\ \boxed{\text{TAN}}$

Pressing $\boxed{2\text{nd}}$ $\boxed{\text{TAN}}$ accesses the inverse tangent key, $\boxed{\text{TAN}^{-1}}$.

Many Graphing Calculators:

$\boxed{2\text{nd}}\ \boxed{\text{TAN}}\ (\ 21\ \div\ 25\ )\ \boxed{\text{ENTER}}$

The display should show approximately 40. Thus, the angle of elevation of the sun is approximately 40°.

● ● ●

✓ Check Point **10** A flagpole that is 14 meters tall casts a shadow 10 meters long. Find the angle of elevation of the sun to the nearest degree.

Blitzer Bonus ‖ The Mountain Man

In the 1930s, a *National Geographic* team headed by Brad Washburn used trigonometry to create a map of the 5000-square-mile region of the Yukon, near the Canadian border. The team started with aerial photography. By drawing a network of angles on the photographs, the approximate locations of the major mountains and their rough heights were determined. The expedition then spent three months on foot to find the exact heights. Team members established two base points a known distance apart, one directly under the mountain's peak. By measuring the angle of elevation from one of the base points to the peak, the tangent function was used to determine the peak's height. The Yukon expedition was a major advance in the way maps are made.

ACHIEVING SUCCESS

Because concepts in trigonometry build on each other, **it is extremely important that you complete all homework assignments.** This requires more than attempting a few of the assigned exercises. When it comes to assigned homework, you need to do four things and to do these things consistently throughout any math course:

1. Attempt to work every assigned problem.
2. Check your answers.
3. Correct your errors.
4. Ask for help with the problems you have attempted, but do not understand.

Having said this, **don't panic at the length of the Exercise Sets.** You are not expected to work all, or even most, of the problems. Your professor will provide guidance on which exercises to work by assigning those problems that are consistent with the goals and objectives of your trig course.

CONCEPT AND VOCABULARY CHECK

Fill in each blank so that the resulting statement is true.

1. Using lengths a, b, and c in the right triangle shown, the trigonometric functions of θ are defined as follows:

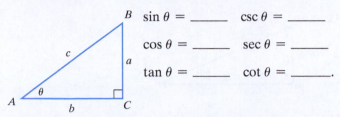

$\sin\theta =$ _____ $\csc\theta =$ _____

$\cos\theta =$ _____ $\sec\theta =$ _____

$\tan\theta =$ _____ $\cot\theta =$ _____.

2. Using the definitions in Exercise 1, we refer to a as the length of the side _____ angle θ, b as the length of the side _____ angle θ, and c as the length of the _____.

3. True or false: The trigonometric functions of θ in Exercise 1 depend only on the size of θ and not on the size of the triangle. _____

4. According to the reciprocal identities,
$$\frac{1}{\csc\theta} = \underline{\quad}, \frac{1}{\sec\theta} = \underline{\quad}, \text{ and } \frac{1}{\cot\theta} = \underline{\quad}.$$

5. According to the quotient identities,
$$\frac{\sin\theta}{\cos\theta} = \underline{\quad} \text{ and } \frac{\cos\theta}{\sin\theta} = \underline{\quad}.$$

6. According to the Pythagorean identities,
$\sin^2\theta + \cos^2\theta =$ _____, $1 + \tan^2\theta =$ _____, and $1 + \cot^2\theta =$ _____.

7. According to the cofunction identities,
$\cos(90° - \theta) =$ _____, $\cot(90° - \theta) =$ _____, and $\csc(90° - \theta) =$ _____.

EXERCISE SET 1.2

Practice Exercises

In Exercises 1–8, use the Pythagorean Theorem to find the length of the missing side of each right triangle. Then find the value of each of the six trigonometric functions of θ.

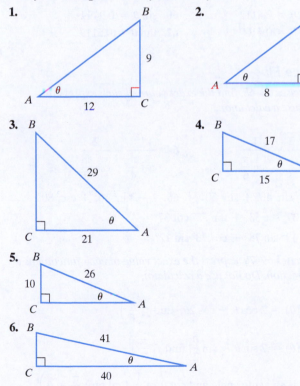

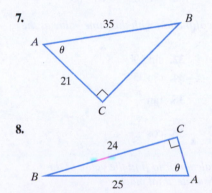

7.

8.

In Exercises 9–16, use the given triangles to evaluate each expression. If necessary, express the value without a square root in the denominator by rationalizing the denominator.

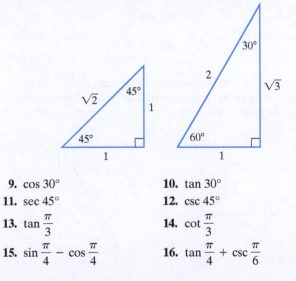

9. $\cos 30°$
10. $\tan 30°$
11. $\sec 45°$
12. $\csc 45°$
13. $\tan\frac{\pi}{3}$
14. $\cot\frac{\pi}{3}$
15. $\sin\frac{\pi}{4} - \cos\frac{\pi}{4}$
16. $\tan\frac{\pi}{4} + \csc\frac{\pi}{6}$

In Exercises 17–20, θ is an acute angle and sin θ and cos θ are given. Use identities to find tan θ, csc θ, sec θ, and cot θ. Where necessary, rationalize denominators.

17. $\sin\theta = \dfrac{8}{17}$, $\cos\theta = \dfrac{15}{17}$

18. $\sin\theta = \dfrac{3}{5}$, $\cos\theta = \dfrac{4}{5}$

19. $\sin\theta = \dfrac{1}{3}$, $\cos\theta = \dfrac{2\sqrt{2}}{3}$

20. $\sin\theta = \dfrac{6}{7}$, $\cos\theta = \dfrac{\sqrt{13}}{7}$

In Exercises 21–24, θ is an acute angle and sin θ is given. Use the Pythagorean identity $\sin^2\theta + \cos^2\theta = 1$ to find cos θ.

21. $\sin\theta = \dfrac{6}{7}$

22. $\sin\theta = \dfrac{7}{8}$

23. $\sin\theta = \dfrac{\sqrt{39}}{8}$

24. $\sin\theta = \dfrac{\sqrt{21}}{5}$

In Exercises 25–30, use an identity to find the value of each expression. Do not use a calculator.

25. $\sin 37° \csc 37°$

26. $\cos 53° \sec 53°$

27. $\sin^2\dfrac{\pi}{9} + \cos^2\dfrac{\pi}{9}$

28. $\sin^2\dfrac{\pi}{10} + \cos^2\dfrac{\pi}{10}$

29. $\sec^2 23° - \tan^2 23°$

30. $\csc^2 63° - \cot^2 63°$

In Exercises 31–38, find a cofunction with the same value as the given expression.

31. $\sin 7°$

32. $\sin 19°$

33. $\csc 25°$

34. $\csc 35°$

35. $\tan\dfrac{\pi}{9}$

36. $\tan\dfrac{\pi}{7}$

37. $\cos\dfrac{2\pi}{5}$

38. $\cos\dfrac{3\pi}{8}$

In Exercises 39–48, use a calculator to find the value of the trigonometric function to four decimal places.

39. $\sin 38°$

40. $\cos 21°$

41. $\tan 32.7°$

42. $\tan 52.6°$

43. $\csc 17°$

44. $\sec 55°$

45. $\cos\dfrac{\pi}{10}$

46. $\sin\dfrac{3\pi}{10}$

47. $\cot\dfrac{\pi}{12}$

48. $\cot\dfrac{\pi}{18}$

In Exercises 49–54, find the measure of the side of the right triangle whose length is designated by a lowercase letter. Round answers to the nearest whole number.

49. **50.**

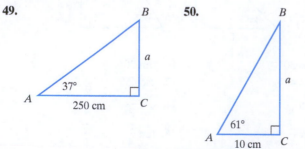

51. **52.**

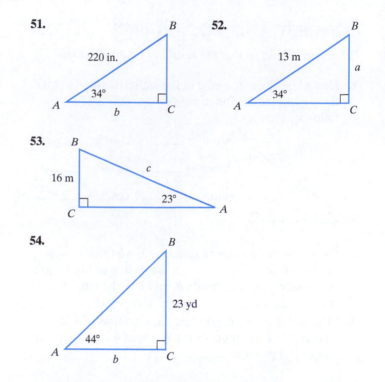

53.

54.

In Exercises 55–58, use a calculator to find the value of the acute angle θ to the nearest degree.

55. $\sin\theta = 0.2974$

56. $\cos\theta = 0.8771$

57. $\tan\theta = 4.6252$

58. $\tan\theta = 26.0307$

In Exercises 59–62, use a calculator to find the value of the acute angle θ in radians, rounded to three decimal places.

59. $\cos\theta = 0.4112$

60. $\sin\theta = 0.9499$

61. $\tan\theta = 0.4169$

62. $\tan\theta = 0.5117$

Practice Plus

In Exercises 63–68, find the exact value of each expression. Do not use a calculator.

63. $\dfrac{\tan\dfrac{\pi}{3}}{2} - \dfrac{1}{\sec\dfrac{\pi}{6}}$

64. $\dfrac{1}{\cot\dfrac{\pi}{4}} - \dfrac{2}{\csc\dfrac{\pi}{6}}$

65. $1 + \sin^2 40° + \sin^2 50°$

66. $1 - \tan^2 10° + \csc^2 80°$

67. $\csc 37° \sec 53° - \tan 53° \cot 37°$

68. $\cos 12° \sin 78° + \cos 78° \sin 12°$

In Exercises 69–70, express the exact value of each function as a single fraction. Do not use a calculator.

69. If $f(\theta) = 2\cos\theta - \cos 2\theta$, find $f\left(\dfrac{\pi}{6}\right)$.

70. If $f(\theta) = 2\sin\theta - \sin\dfrac{\theta}{2}$, find $f\left(\dfrac{\pi}{3}\right)$.

71. If θ is an acute angle and $\cot\theta = \dfrac{1}{4}$, find $\tan\left(\dfrac{\pi}{2} - \theta\right)$.

72. If θ is an acute angle and $\cos\theta = \dfrac{1}{3}$, find $\csc\left(\dfrac{\pi}{2} - \theta\right)$.

Application Exercises

73. To find the distance across a lake, a surveyor took the measurements shown in the figure. Use these measurements to determine how far it is across the lake. Round to the nearest yard.

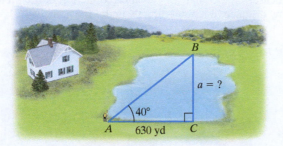

74. At a certain time of day, the angle of elevation of the Sun is 40°. To the nearest foot, find the height of a tree whose shadow is 35 feet long.

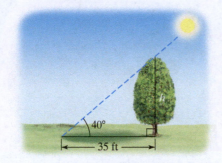

75. A tower that is 125 feet tall casts a shadow 172 feet long. Find the angle of elevation of the Sun to the nearest degree.

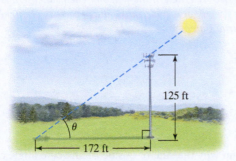

76. The Washington Monument is 555 feet high. If you are standing one quarter of a mile, or 1320 feet, from the base of the monument and looking to the top, find the angle of elevation to the nearest degree.

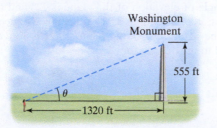

77. A plane rises from take-off and flies at an angle of 10° with the horizontal runway. When it has gained 500 feet, find the distance, to the nearest foot, the plane has flown.

78. A road is inclined at an angle of 5°. After driving 5000 feet along this road, find the driver's increase in altitude. Round to the nearest foot.

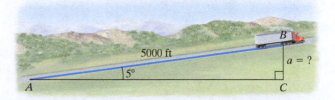

79. A telephone pole is 60 feet tall. A guy wire 75 feet long is attached from the ground to the top of the pole. Find the angle between the wire and the pole to the nearest degree.

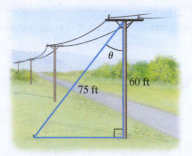

80. A telephone pole is 55 feet tall. A guy wire 80 feet long is attached from the ground to the top of the pole. Find the angle between the wire and the pole to the nearest degree.

Explaining the Concepts

81. If you are given the lengths of the sides of a right triangle, describe how to find the sine of either acute angle.

82. Describe one similarity and one difference between the definitions of $\sin \theta$ and $\cos \theta$, where θ is an acute angle of a right triangle.

83. Describe the triangle used to find the trigonometric functions of 45°.

84. Describe the triangle used to find the trigonometric functions of 30° and 60°.

85. What is a trigonometric identity?

86. Use words (not an equation) to describe one of the reciprocal identities.

87. Use words (not an equation) to describe one of the quotient identities.

88. Use words (not an equation) to describe one of the Pythagorean identities.

89. Describe a relationship among trigonometric functions that is based on angles that are complements.

90. Describe what is meant by an angle of elevation and an angle of depression.

91. Stonehenge, the famous "stone circle" in England, was built between 2750 B.C. and 1300 B.C. using solid stone blocks weighing over 99,000 pounds each. It required 550 people to pull a single stone up a ramp inclined at a 9° angle. Describe how right triangle trigonometry can be used to determine the distance the 550 workers had to drag a stone in order to raise it to a height of 30 feet.

Technology Exercises

92. Use a calculator in the radian mode to fill in the values in the following table. Then draw a conclusion about $\frac{\sin \theta}{\theta}$ as θ approaches 0.

θ	0.4	0.3	0.2	0.1	0.01	0.001	0.0001	0.00001
$\sin \theta$								
$\frac{\sin \theta}{\theta}$								

93. Use a calculator in the radian mode to fill in the values in the following table. Then draw a conclusion about $\frac{\cos \theta - 1}{\theta}$ as θ approaches 0.

θ	0.4	0.3	0.2	0.1	0.01	0.001	0.0001	0.00001
$\cos \theta$								
$\frac{\cos \theta - 1}{\theta}$								

Critical Thinking Exercises

Make Sense?   *In Exercises 94–97, determine whether each statement makes sense or does not make sense, and explain your reasoning.*

94. For a given angle θ, I found a slight increase in $\sin \theta$ as the size of the triangle increased.

95. Although I can use an isosceles right triangle to determine the exact value of $\sin \frac{\pi}{4}$, I can also use my calculator to obtain this value.

96. I can rewrite $\tan \theta$ as $\frac{1}{\cot \theta}$, as well as $\frac{\sin \theta}{\cos \theta}$.

97. Standing under this arch, I can determine its height by measuring the angle of elevation to the top of the arch and my distance to a point directly under the arch.

Delicate Arch in Arches National Park, Utah

In Exercises 98–101, determine whether each statement is true or false. If the statement is false, make the necessary change(s) to produce a true statement.

98. $\frac{\tan 45°}{\tan 15°} = \tan 3°$

99. $\tan^2 15° - \sec^2 15° = -1$

100. $\sin 45° + \cos 45° = 1$

101. $\tan^2 5° = \tan 25°$

102. Explain why the sine or cosine of an acute angle cannot be greater than or equal to 1.

103. Describe what happens to the tangent of an acute angle as the angle gets close to 90°.

104. From the top of a 250-foot lighthouse, a plane is sighted overhead and a ship is observed directly below the plane. The angle of elevation of the plane is 22° and the angle of depression of the ship is 35°. Find **a.** the distance of the ship from the lighthouse; **b.** the plane's height above the water. Round to the nearest foot.

Preview Exercises

Exercises 105–107 will help you prepare for the material covered in the next section.

Use these figures to solve Exercises 105–106.

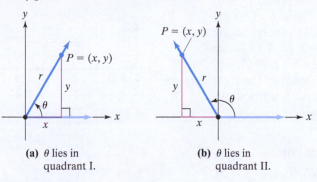

(a) θ lies in quadrant I. (b) θ lies in quadrant II.

105. a. Write a ratio that expresses $\sin \theta$ for the right triangle in **Figure (a)**.

b. Assuming that $r > 0$ in both figures, determine the ratio that you wrote in part (a) for **Figure (b)** with $x = -3$ and $y = 4$. Is this ratio positive or negative?

106. a. Write a ratio that expresses $\cos \theta$ for the right triangle in **Figure (a)**.

b. Assuming that $r > 0$ in both figures, determine the ratio that you wrote in part (a) for **Figure (b)** with $x = -3$ and $y = 5$. Is this ratio positive or negative?

107. Find the positive angle θ' formed by the terminal side of θ and the x-axis.

a.

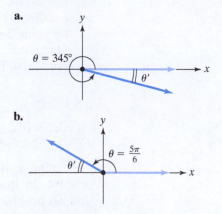

$\theta = 345°$

b.

$\theta = \frac{5\pi}{6}$

Section 1.3   Trigonometric Functions of Any Angle

What am I supposed to learn?

After studying this section, you should be able to:

1. Use the definitions of trigonometric functions of any angle.
2. Use the signs of the trigonometric functions.
3. Find reference angles.
4. Use reference angles to evaluate trigonometric functions.

There is something comforting in the repetition of some of nature's patterns. The ocean level at a beach varies between high and low tide approximately every 12 hours. The number of hours of daylight oscillates from a maximum on the summer solstice, June 21, to a minimum on the winter solstice, December 21. Then it increases to the same maximum the following June 21. Some believe that cycles, called biorhythms, represent physical, emotional, and intellectual aspects of our lives. Throughout the remainder of this chapter, we will see how the trigonometric functions are used to model phenomena that occur again and again. To do this, we need to move beyond right triangles.

A Brief Review • Plotting Points in the Rectangular Coordinate System

• The figure on the right shows how we **plot**, or locate, the points $(-5, 3)$ and $(3, -5)$. We plot $(-5, 3)$ by going 5 units from 0 to the left along the x-axis. Then we go 3 units up parallel to the y-axis. We plot $(3, -5)$ by going 3 units from 0 to the right along the x-axis and 5 units down parallel to the y-axis.

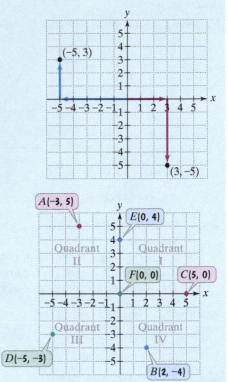

EXAMPLE   Plot the points:

$A(-3, 5)$, $B(2, -4)$, $C(5, 0)$, $D(-5, -3)$, $E(0, 4)$, and $F(0, 0)$.

Note that in quadrant I, x and y are positive. In quadrant II, x is negative and y is positive. In quadrant III, x and y are negative. In quadrant IV, x is positive and y is negative. All points along the x-axis have y-coordinates of 0. All points along the y-axis have x-coordinates of 0. The origin is represented by $(0, 0)$.

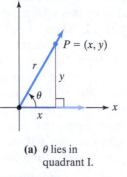

1 Use the definitions of trigonometric functions of any angle.

Trigonometric Functions of Any Angle

In the last section, we evaluated trigonometric functions of acute angles, such as that shown in **Figure 1.33(a)**. Note that this angle is in standard position. The point $P = (x, y)$ is a point r units from the origin on the terminal side of θ. A right triangle is formed by drawing a line segment from $P = (x, y)$ perpendicular to the x-axis. Note that y is the length of the side opposite θ and x is the length of the side adjacent to θ.

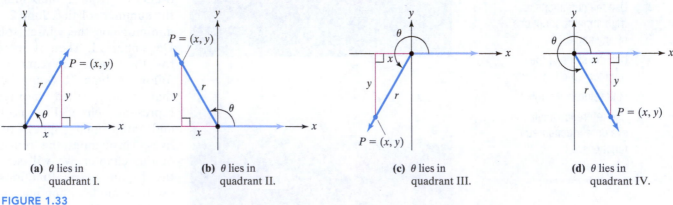

(a) θ lies in quadrant I.

(b) θ lies in quadrant II.

(c) θ lies in quadrant III.

(d) θ lies in quadrant IV.

FIGURE 1.33

Figures 1.33(b), **(c)**, and **(d)** show angles in standard position, but they are not acute. We can extend our definitions of the six trigonometric functions to include such angles, as well as quadrantal angles. (Recall that a quadrantal angle has its terminal side on the x-axis or y-axis; such angles are *not* shown in **Figure 1.33**.) The point $P = (x, y)$ may be any point on the terminal side of the angle θ other than the origin, $(0, 0)$.

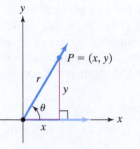

FIGURE 1.33(a) (repeated) θ lies in quadrant I.

GREAT QUESTION!

Is there a way to make it easier for me to remember the definitions of trigonometric functions of any angle?

Yes. If θ is acute, we have the right triangle shown in **Figure 1.33(a)**. In this situation, the definitions in the box are the right triangle definitions of the trigonometric functions. This should make it easier for you to remember the six definitions.

Definitions of Trigonometric Functions of Any Angle

Let θ be any angle in standard position and let $P = (x, y)$ be a point on the terminal side of θ. If $r = \sqrt{x^2 + y^2}$ is the distance from $(0, 0)$ to (x, y), as shown in **Figure 1.33** on the previous page, the **six trigonometric functions of θ** are defined by the following ratios:

$$\sin \theta = \frac{y}{r} \qquad\qquad \csc \theta = \frac{r}{y}, y \neq 0$$

$$\cos \theta = \frac{x}{r} \qquad\qquad \sec \theta = \frac{r}{x}, x \neq 0$$

$$\tan \theta = \frac{y}{x}, x \neq 0 \qquad\qquad \cot \theta = \frac{x}{y}, y \neq 0.$$

The ratios in the second column are the reciprocals of the corresponding ratios in the first column.

Because the point $P = (x, y)$ is any point on the terminal side of θ other than the origin, $(0, 0)$, $r = \sqrt{x^2 + y^2}$ cannot be zero. Examine the six trigonometric functions defined above. Note that the denominator of the sine and cosine functions is r. Because $r \neq 0$, the sine and cosine functions are defined for any angle θ. This is not true for the other four trigonometric functions. Note that the denominator of the tangent and secant functions is x: $\tan \theta = \frac{y}{x}$ and $\sec \theta = \frac{r}{x}$. These functions are not defined if $x = 0$. If the point $P = (x, y)$ is on the y-axis, then $x = 0$. Thus, the tangent and secant functions are undefined for all quadrantal angles with terminal sides on the positive or negative y-axis. Likewise, if $P = (x, y)$ is on the x-axis, then $y = 0$, and the cotangent and cosecant functions are undefined: $\cot \theta = \frac{x}{y}$ and $\csc \theta = \frac{r}{y}$. The cotangent and cosecant functions are undefined for all quadrantal angles with terminal sides on the positive or negative x-axis.

EXAMPLE 1 Evaluating Trigonometric Functions

Let $P = (-3, -5)$ be a point on the terminal side of θ. Find each of the six trigonometric functions of θ.

SOLUTION

The situation is shown in **Figure 1.34**. We need values for x, y, and r to evaluate all six trigonometric functions. We are given the values of x and y. Because $P = (-3, -5)$ is a point on the terminal side of θ, $x = -3$ and $y = -5$. Furthermore,

$$r = \sqrt{x^2 + y^2} = \sqrt{(-3)^2 + (-5)^2} = \sqrt{9 + 25} = \sqrt{34}.$$

Now that we know x, y, and r, we can find the six trigonometric functions of θ. Where appropriate, we will rationalize denominators.

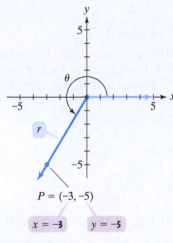

FIGURE 1.34

$$\sin \theta = \frac{y}{r} = \frac{-5}{\sqrt{34}} = -\frac{5}{\sqrt{34}} \cdot \frac{\sqrt{34}}{\sqrt{34}} = -\frac{5\sqrt{34}}{34} \qquad \csc \theta = \frac{r}{y} = \frac{\sqrt{34}}{-5} = -\frac{\sqrt{34}}{5}$$

$$\cos \theta = \frac{x}{r} = \frac{-3}{\sqrt{34}} = -\frac{3}{\sqrt{34}} \cdot \frac{\sqrt{34}}{\sqrt{34}} = -\frac{3\sqrt{34}}{34} \qquad \sec \theta = \frac{r}{x} = \frac{\sqrt{34}}{-3} = -\frac{\sqrt{34}}{3}$$

$$\tan \theta = \frac{y}{x} = \frac{-5}{-3} = \frac{5}{3} \qquad\qquad\qquad\qquad \cot \theta = \frac{x}{y} = \frac{-3}{-5} = \frac{3}{5} \quad \bullet\bullet\bullet$$

✓ **Check Point 1** Let $P = (1, -3)$ be a point on the terminal side of θ. Find each of the six trigonometric functions of θ.

How do we find the values of the trigonometric functions for a quadrantal angle? First, draw the angle in standard position. Second, choose a point P on the angle's terminal side. The trigonometric function values of θ depend only on the size of θ and not on the distance of point P from the origin. Thus, we will choose a point that is 1 unit from the origin. Finally, apply the definitions of the appropriate trigonometric functions.

EXAMPLE 2   Trigonometric Functions of Quadrantal Angles

Evaluate, if possible, the sine function and the tangent function at the following four quadrantal angles:

a. $\theta = 0° = 0$ **b.** $\theta = 90° = \dfrac{\pi}{2}$ **c.** $\theta = 180° = \pi$ **d.** $\theta = 270° = \dfrac{3\pi}{2}$.

SOLUTION

a. If $\theta = 0° = 0$ radians, then the terminal side of the angle is on the positive x-axis. Let us select the point $P = (1, 0)$ with $x = 1$ and $y = 0$. This point is 1 unit from the origin, so $r = 1$. **Figure 1.35** shows values of x, y, and r corresponding to $\theta = 0°$ or 0 radians. Now that we know x, y, and r, we can apply the definitions of the sine and tangent functions.

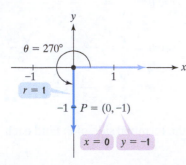

FIGURE 1.35

$$\sin 0° = \sin 0 = \frac{y}{r} = \frac{0}{1} = 0$$

$$\tan 0° = \tan 0 = \frac{y}{x} = \frac{0}{1} = 0$$

b. If $\theta = 90° = \dfrac{\pi}{2}$ radians, then the terminal side of the angle is on the positive y-axis. Let us select the point $P = (0, 1)$ with $x = 0$ and $y = 1$. This point is 1 unit from the origin, so $r = 1$. **Figure 1.36** shows values of x, y, and r corresponding to $\theta = 90°$ or $\dfrac{\pi}{2}$. Now that we know x, y, and r, we can apply the definitions of the sine and tangent functions.

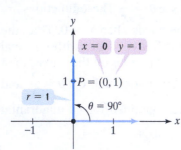

FIGURE 1.36

$$\sin 90° = \sin \frac{\pi}{2} = \frac{y}{r} = \frac{1}{1} = 1$$

$$\tan 90° = \tan \frac{\pi}{2} = \frac{y}{x} = \frac{1}{0}$$

Because division by 0 is undefined, tan 90° is undefined.

c. If $\theta = 180° = \pi$ radians, then the terminal side of the angle is on the negative x-axis. Let us select the point $P = (-1, 0)$ with $x = -1$ and $y = 0$. This point is 1 unit from the origin, so $r = 1$. **Figure 1.37** shows values of x, y, and r corresponding to $\theta = 180°$ or π. Now that we know x, y, and r, we can apply the definitions of the sine and tangent functions.

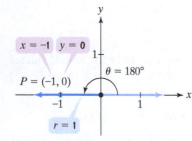

FIGURE 1.37

$$\sin 180° = \sin \pi = \frac{y}{r} = \frac{0}{1} = 0$$

$$\tan 180° = \tan \pi = \frac{y}{x} = \frac{0}{-1} = 0$$

d. If $\theta = 270° = \dfrac{3\pi}{2}$ radians, then the terminal side of the angle is on the negative y-axis. Let us select the point $P = (0, -1)$ with $x = 0$ and $y = -1$. This point is 1 unit from the origin, so $r = 1$. **Figure 1.38** shows values of x, y, and r corresponding to $\theta = 270°$ or $\dfrac{3\pi}{2}$. Now that we know x, y, and r, we can apply the definitions of the sine and tangent functions.

$$\sin 270° = \sin \frac{3\pi}{2} = \frac{y}{r} = \frac{-1}{1} = -1$$

$$\tan 270° = \tan \frac{3\pi}{2} = \frac{y}{x} = \frac{-1}{0}$$

FIGURE 1.38

DISCOVERY

Try finding tan 90° and tan 270° with your calculator. Describe what occurs.

Because division by 0 is undefined, tan 270° is undefined.   **• • •**

✓ **Check Point 2** Evaluate, if possible, the cosine function and the cosecant function at the following four quadrantal angles:

a. $\theta = 0° = 0$ **b.** $\theta = 90° = \dfrac{\pi}{2}$ **c.** $\theta = 180° = \pi$ **d.** $\theta = 270° = \dfrac{3\pi}{2}$.

② Use the signs of the trigonometric functions.

Quadrant II	Quadrant I
sine and cosecant positive	All functions positive
Quadrant III	**Quadrant IV**
tangent and cotangent positive	cosine and secant positive

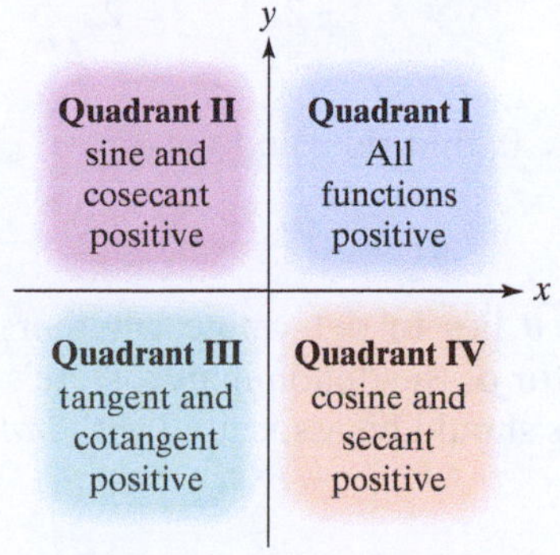

FIGURE 1.39 The signs of the trigonometric functions

The Signs of the Trigonometric Functions

In Example 2, we evaluated trigonometric functions of quadrantal angles. However, we will now return to the trigonometric functions of nonquadrantal angles. **If θ is not a quadrantal angle, the sign of a trigonometric function depends on the quadrant in which θ lies.** In all four quadrants, r is positive. However, x and y can be positive or negative. For example, if θ lies in quadrant II, x is negative and y is positive. Thus, the only positive ratios in this quadrant are $\dfrac{y}{r}$ and its reciprocal, $\dfrac{r}{y}$. These ratios are the function values for the sine and cosecant, respectively. In short, if θ lies in quadrant II, $\sin \theta$ and $\csc \theta$ are positive. The other four trigonometric functions are negative.

Figure 1.39 summarizes the signs of the trigonometric functions. If θ lies in quadrant I, all six functions are positive. If θ lies in quadrant II, only $\sin \theta$ and $\csc \theta$ are positive. If θ lies in quadrant III, only $\tan \theta$ and $\cot \theta$ are positive. Finally, if θ lies in quadrant IV, only $\cos \theta$ and $\sec \theta$ are positive. Observe that the positive functions in each quadrant occur in reciprocal pairs.

GREAT QUESTION!

Is there a way to remember the signs of the trigonometric functions?

Here's a phrase that may be helpful:

A	**S**mart	**T**rig	**C**lass.
All trig functions are positive in QI.	Sine and its reciprocal, cosecant, are positive in QII.	Tangent and its reciprocal, cotangent, are positive in QIII.	Cosine and its reciprocal, secant, are positive in QIV.

EXAMPLE 3   Finding the Quadrant in Which an Angle Lies

If $\tan \theta < 0$ and $\cos \theta > 0$, name the quadrant in which angle θ lies.

SOLUTION

When $\tan \theta < 0$, θ lies in quadrant II or IV. When $\cos \theta > 0$, θ lies in quadrant I or IV. When both conditions are met ($\tan \theta < 0$ and $\cos \theta > 0$), θ must lie in quadrant IV. ● ● ●

✓ **Check Point 3** If $\sin \theta < 0$ and $\cos \theta < 0$, name the quadrant in which angle θ lies.

EXAMPLE 4   Evaluating Trigonometric Functions

Given $\tan \theta = -\dfrac{2}{3}$ and $\cos \theta > 0$, find $\cos \theta$ and $\csc \theta$.

SOLUTION

Because the tangent is negative and the cosine is positive, θ lies in quadrant IV. This will help us to determine whether the negative sign in $\tan \theta = -\dfrac{2}{3}$ should be associated with the numerator or the denominator. Keep in mind that in quadrant IV, x is positive and y is negative. Thus,

In quadrant IV, y is negative.

$$\tan \theta = -\frac{2}{3} = \frac{y}{x} = \frac{-2}{3}.$$

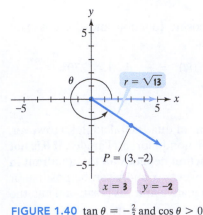

$r = \sqrt{13}$

$P = (3, -2)$

$x = 3$ $\quad$ $y = -2$

FIGURE 1.40 $\tan \theta = -\frac{2}{3}$ and $\cos \theta > 0$

Using $\tan \theta = \dfrac{y}{x} = \dfrac{-2}{3}$, we conclude that $x = 3$ and $y = -2$. (See **Figure 1.40.**) Furthermore,

$$r = \sqrt{x^2 + y^2} = \sqrt{3^2 + (-2)^2} = \sqrt{9 + 4} = \sqrt{13}.$$

Now that we know x, y, and r, we can find $\cos \theta$ and $\csc \theta$.

$$\cos \theta = \frac{x}{r} = \frac{3}{\sqrt{13}} = \frac{3}{\sqrt{13}} \cdot \frac{\sqrt{13}}{\sqrt{13}} = \frac{3\sqrt{13}}{13} \qquad \csc \theta = \frac{r}{y} = \frac{\sqrt{13}}{-2} = -\frac{\sqrt{13}}{2} \quad \bullet\bullet\bullet$$

✅ **Check Point 4** Given $\tan \theta = -\frac{1}{3}$ and $\cos \theta < 0$, find $\sin \theta$ and $\sec \theta$.

In Example 4, we used the quadrant in which θ lies to determine whether a negative sign should be associated with the numerator or the denominator. Here's a situation, similar to Example 4, where negative signs should be associated with *both* the numerator and the denominator:

$$\tan \theta = \frac{3}{5} \quad \text{and} \quad \cos \theta < 0.$$

Because the tangent is positive and the cosine is negative, θ lies in quadrant III. In quadrant III, x is negative and y is negative. Thus,

$$\tan \theta = \frac{3}{5} = \frac{y}{x} = \frac{-3}{-5}.$$

We see that $x = -5$ and $y = -3$.

③ Find reference angles.

Reference Angles

We will often evaluate trigonometric functions of positive angles greater than 90° and all negative angles by making use of a positive acute angle. This positive acute angle is called a *reference angle*.

> ### Definition of a Reference Angle
>
> Let θ be a nonacute angle in standard position that lies in a quadrant. Its **reference angle** is the positive acute angle θ' formed by the terminal side of θ and the x-axis.

Figure 1.41 shows the reference angle for θ lying in quadrants II, III, and IV. Notice that the formula used to find θ', the reference angle, varies according to the quadrant in which θ lies. You may find it easier to find the reference angle for a given angle by making a figure that shows the angle in standard position. The acute angle formed by the terminal side of this angle and the x-axis is the reference angle.

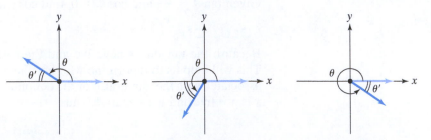

FIGURE 1.41 Reference angles, θ', for positive angles, θ, in quadrants II, III, and IV

If $90° < \theta < 180°$, then $\theta' = 180° - \theta$.

If $180° < \theta < 270°$, then $\theta' = \theta - 180°$.

If $270° < \theta < 360°$, then $\theta' = 360° - \theta$.

EXAMPLE 5 Finding Reference Angles

Find the reference angle, θ', for each of the following angles:

a. $\theta = 345°$ **b.** $\theta = \dfrac{5\pi}{6}$ **c.** $\theta = -135°$ **d.** $\theta = 2.5$.

SOLUTION

a. A 345° angle in standard position is shown in **Figure 1.42**. Because 345° lies in quadrant IV, the reference angle is

$$\theta' = 360° - 345° = 15°.$$

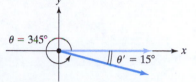

FIGURE 1.42

b. Because $\dfrac{5\pi}{6}$ lies between $\dfrac{\pi}{2} = \dfrac{3\pi}{6}$ and $\pi = \dfrac{6\pi}{6}$, $\theta = \dfrac{5\pi}{6}$ lies in quadrant II. The angle is shown in **Figure 1.43**. The reference angle is

$$\theta' = \pi - \frac{5\pi}{6} = \frac{6\pi}{6} - \frac{5\pi}{6} = \frac{\pi}{6}.$$

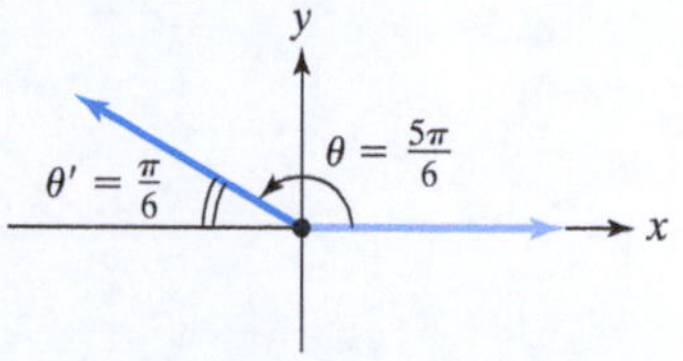

FIGURE 1.43

DISCOVERY

Solve part (c) by first finding a positive coterminal angle for −135° less than 360°. Use the positive coterminal angle to find the reference angle.

c. A −135° angle in standard position is shown in **Figure 1.44**. The figure indicates that the positive acute angle formed by the terminal side of θ and the x-axis is 45°. The reference angle is

$$\theta' = 45°.$$

FIGURE 1.44

d. The angle $\theta = 2.5$ lies between $\dfrac{\pi}{2} \approx 1.57$ and $\pi \approx 3.14$. This means that $\theta = 2.5$ is in quadrant II, shown in **Figure 1.45**. The reference angle is

$$\theta' = \pi - 2.5 \approx 0.64.$$

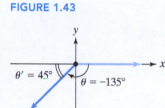

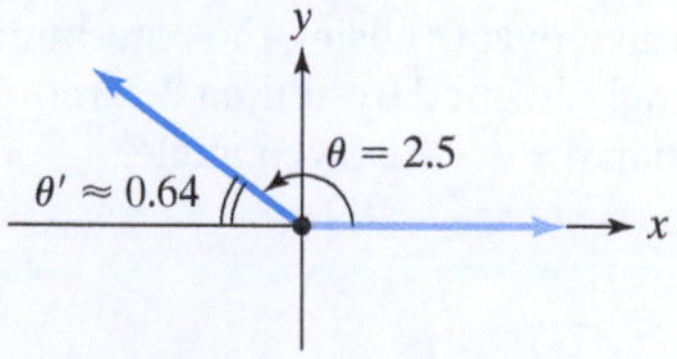

FIGURE 1.45   • • •

☑️ **Check Point 5** Find the reference angle, θ', for each of the following angles:

a. $\theta = 210°$ **b.** $\theta = \dfrac{7\pi}{4}$ **c.** $\theta = -240°$ **d.** $\theta = 3.6$.

Finding reference angles for angles that are greater than 360° (2π) or less than −360° (-2π) involves using coterminal angles. We have seen that coterminal angles have the same initial and terminal sides. Recall that coterminal angles can be obtained by increasing or decreasing an angle's measure by an integer multiple of 360° or 2π.

Finding Reference Angles for Angles Greater Than 360° (2π) or Less Than −360° (-2π)

1. Find a positive angle α less than 360° or 2π that is coterminal with the given angle.
2. Draw α in standard position.
3. Use the drawing to find the reference angle for the given angle. The positive acute angle formed by the terminal side of α and the x-axis is the reference angle.

EXAMPLE 6   Finding Reference Angles

Find the reference angle for each of the following angles:

a. $\theta = 580°$ **b.** $\theta = \dfrac{8\pi}{3}$ **c.** $\theta = -\dfrac{13\pi}{6}$.

SOLUTION

a. For a 580° angle, subtract 360° to find a positive coterminal angle less than 360°.

$$580° - 360° = 220°$$

Figure 1.46 shows $\alpha = 220°$ in standard position. Because 220° lies in quadrant III, the reference angle is

$$\alpha' = 220° - 180° = 40°.$$

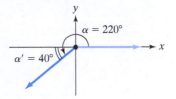

FIGURE 1.46

b. For an $\dfrac{8\pi}{3}$ angle, note that $\dfrac{8}{3} = 2\dfrac{2}{3}$, so subtract 2π to find a positive coterminal angle less than 2π.

$$\frac{8\pi}{3} - 2\pi = \frac{8\pi}{3} - \frac{6\pi}{3} = \frac{2\pi}{3}$$

Figure 1.47 shows $\alpha = \dfrac{2\pi}{3}$ in standard position. Because $\dfrac{2\pi}{3}$ lies in quadrant II, the reference angle is

$$\alpha' = \pi - \frac{2\pi}{3} = \frac{3\pi}{3} - \frac{2\pi}{3} = \frac{\pi}{3}.$$

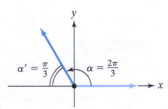

FIGURE 1.47

DISCOVERY

Solve part (c) using the coterminal angle formed by adding 2π, rather than 4π, to the given angle.

c. For a $-\dfrac{13\pi}{6}$ angle, note that $-\dfrac{13}{6} = -2\dfrac{1}{6}$, so add 4π to find a positive coterminal angle less than 2π.

$$-\frac{13\pi}{6} + 4\pi = -\frac{13\pi}{6} + \frac{24\pi}{6} = \frac{11\pi}{6}$$

Figure 1.48 shows $\alpha = \dfrac{11\pi}{6}$ in standard position. Because $\dfrac{11\pi}{6}$ lies in quadrant IV, the reference angle is

$$\alpha' = 2\pi - \frac{11\pi}{6} = \frac{12\pi}{6} - \frac{11\pi}{6} = \frac{\pi}{6}. \qquad \bullet\bullet\bullet$$

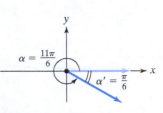

FIGURE 1.48

✓ **Check Point 6** Find the reference angle for each of the following angles:

a. $\theta = 665°$ **b.** $\theta = \dfrac{15\pi}{4}$ **c.** $\theta = -\dfrac{11\pi}{3}$.

④ Use reference angles to evaluate trigonometric functions.

Evaluating Trigonometric Functions Using Reference Angles

The way that reference angles are defined makes them useful in evaluating trigonometric functions.

> **Using Reference Angles to Evaluate Trigonometric Functions**
>
> The values of the trigonometric functions of a given angle, θ, are the same as the values of the trigonometric functions of the reference angle, θ', except possibly for the sign. A function value of the acute reference angle, θ', is always positive. However, the same function value for θ may be positive or negative.

For example, we can use a reference angle, θ', to obtain an exact value for tan 120°. The reference angle for $\theta = 120°$ is $\theta' = 180° - 120° = 60°$. We know the exact value of the tangent function of the reference angle: tan 60° = $\sqrt{3}$. We also know that the value of a trigonometric function of a given angle, θ, is the same as that of its reference angle, θ', except possibly for the sign. Thus, we can conclude that tan 120° equals $-\sqrt{3}$ or $\sqrt{3}$.

What sign should we attach to $\sqrt{3}$? A 120° angle lies in quadrant II, where only the sine and cosecant are positive. Thus, the tangent function is negative for a 120° angle. Therefore,

> Prefix by a negative sign to show tangent is negative in quadrant II.

$$\tan 120° = -\tan 60° = -\sqrt{3}.$$

> The reference angle for 120° is 60°.

In the previous section, we used two right triangles to find exact trigonometric values of 30°, 45°, and 60°. Using a procedure similar to finding tan 120°, we can now find the exact function values of all angles for which 30°, 45°, or 60° are reference angles.

A Procedure for Using Reference Angles to Evaluate Trigonometric Functions

The value of a trigonometric function of any angle θ is found as follows:

1. Find the associated reference angle, θ', and the function value for θ'.
2. Use the quadrant in which θ lies to prefix the appropriate sign to the function value in step 1.

DISCOVERY

Draw the two right triangles involving 30°, 45°, and 60°. Indicate the length of each side. Use these lengths to verify the function values for the reference angles in the solution to Example 7.

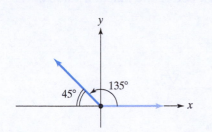

FIGURE 1.49 Reference angle for 135°

EXAMPLE 7 Using Reference Angles to Evaluate Trigonometric Functions

Use reference angles to find the exact value of each of the following trigonometric functions:

a. sin 135°　　**b.** $\cos\dfrac{4\pi}{3}$　　**c.** $\cot\left(-\dfrac{\pi}{3}\right)$.

SOLUTION

a. We use our two-step procedure to find sin 135°.

Step 1　Find the reference angle, θ', and sin θ'. Figure 1.49 shows 135° lies in quadrant II. The reference angle is

$$\theta' = 180° - 135° = 45°.$$

The function value for the reference angle is sin 45° = $\dfrac{\sqrt{2}}{2}$.

Step 2　Use the quadrant in which θ lies to prefix the appropriate sign to the function value in step 1. The angle $\theta = 135°$ lies in quadrant II. Because the sine is positive in quadrant II, we put a + sign before the function value of the reference angle. Thus,

> The sine is positive in quadrant II.

$$\sin 135° = +\sin 45° = \frac{\sqrt{2}}{2}.$$

> The reference angle for 135° is 45°.

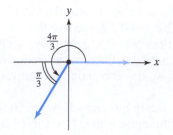

FIGURE 1.50 Reference angle for $\dfrac{4\pi}{3}$

b. We use our two-step procedure to find $\cos \dfrac{4\pi}{3}$.

Step 1 **Find the reference angle, θ', and $\cos \theta'$.** **Figure 1.50** shows that $\theta = \dfrac{4\pi}{3}$ lies in quadrant III. The reference angle is

$$\theta' = \dfrac{4\pi}{3} - \pi = \dfrac{4\pi}{3} - \dfrac{3\pi}{3} = \dfrac{\pi}{3}.$$

The function value for the reference angle is

$$\cos \dfrac{\pi}{3} = \dfrac{1}{2}.$$

Step 2 **Use the quadrant in which θ lies to prefix the appropriate sign to the function value in step 1.** The angle $\theta = \dfrac{4\pi}{3}$ lies in quadrant III. Because only the tangent and cotangent are positive in quadrant III, the cosine is negative in this quadrant. We put a $-$ sign before the function value of the reference angle. Thus,

The cosine is negative in quadrant III.

$$\cos \dfrac{4\pi}{3} = -\cos \dfrac{\pi}{3} = -\dfrac{1}{2}.$$

The reference angle for $\dfrac{4\pi}{3}$ is $\dfrac{\pi}{3}$.

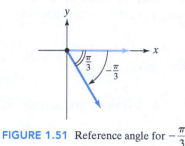

FIGURE 1.51 Reference angle for $-\dfrac{\pi}{3}$

c. We use our two-step procedure to find $\cot\left(-\dfrac{\pi}{3}\right)$.

Step 1 **Find the reference angle, θ', and $\cot \theta'$.** **Figure 1.51** shows that $\theta = -\dfrac{\pi}{3}$ lies in quadrant IV. The reference angle is $\theta' = \dfrac{\pi}{3}$. The function value for the reference angle is $\cot \dfrac{\pi}{3} = \dfrac{\sqrt{3}}{3}$.

Step 2 **Use the quadrant in which θ lies to prefix the appropriate sign to the function value in step 1.** The angle $\theta = -\dfrac{\pi}{3}$ lies in quadrant IV. Because only the cosine and secant are positive in quadrant IV, the cotangent is negative in this quadrant. We put a $-$ sign before the function value of the reference angle. Thus,

The cotangent is negative in quadrant IV.

$$\cot\left(-\dfrac{\pi}{3}\right) = -\cot \dfrac{\pi}{3} = -\dfrac{\sqrt{3}}{3}.$$

The reference angle for $-\dfrac{\pi}{3}$ is $\dfrac{\pi}{3}$.

• • •

⊘ **Check Point 7** Use reference angles to find the exact value of the following trigonometric functions:

a. $\sin 300°$ **b.** $\tan \dfrac{5\pi}{4}$ **c.** $\sec\left(-\dfrac{\pi}{6}\right)$.

In our final example, we use positive coterminal angles less than 2π to find the reference angles.

EXAMPLE 8   Using Reference Angles to Evaluate
Trigonometric Functions

Use reference angles to find the exact value of each of the following trigonometric
functions:

a. $\tan \dfrac{14\pi}{3}$ **b.** $\sec\left(-\dfrac{17\pi}{4}\right)$.

SOLUTION

a. We use our two-step procedure to find $\tan \dfrac{14\pi}{3}$.

Step 1   Find the reference angle, θ', and $\tan \theta'$. Because $\dfrac{14}{3} = 4\dfrac{2}{3}$,
subtract 4π from $\dfrac{14\pi}{3}$ to find a positive coterminal angle less than 2π.

$$\theta = \frac{14\pi}{3} - 4\pi = \frac{14\pi}{3} - \frac{12\pi}{3} = \frac{2\pi}{3}$$

Figure 1.52 shows $\theta = \dfrac{2\pi}{3}$ in standard position. The angle lies in quadrant II.
The reference angle is

$$\theta' = \pi - \frac{2\pi}{3} = \frac{3\pi}{3} - \frac{2\pi}{3} = \frac{\pi}{3}.$$

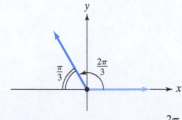

FIGURE 1.52 Reference angle for $\dfrac{2\pi}{3}$

The function value for the reference angle is $\tan \dfrac{\pi}{3} = \sqrt{3}$.

**Step 2   Use the quadrant in which θ lies to prefix the appropriate sign to the
function value in step 1.** The coterminal angle $\theta = \dfrac{2\pi}{3}$ lies in quadrant II.
Because the tangent is negative in quadrant II, we put a $-$ sign before the
function value of the reference angle. Thus,

> The tangent is negative
> in quadrant II.

$$\tan \frac{14\pi}{3} = \tan \frac{2\pi}{3} = -\tan \frac{\pi}{3} = -\sqrt{3}.$$

> The reference angle
> for $\dfrac{2\pi}{3}$ is $\dfrac{\pi}{3}$.

b. We use our two-step procedure to find $\sec\left(-\dfrac{17\pi}{4}\right)$.

Step 1   Find the reference angle, θ', and $\sec \theta'$. Because $-\dfrac{17}{4} = -4\dfrac{1}{4}$,
add 6π (three multiples of 2π) to $-\dfrac{17\pi}{4}$ to find a positive coterminal angle
less than 2π.

$$\theta = -\frac{17\pi}{4} + 6\pi = -\frac{17\pi}{4} + \frac{24\pi}{4} = \frac{7\pi}{4}$$

Figure 1.53 shows $\theta = \dfrac{7\pi}{4}$ in standard position. The angle lies in quadrant IV.
The reference angle is

$$\theta' = 2\pi - \frac{7\pi}{4} = \frac{8\pi}{4} - \frac{7\pi}{4} = \frac{\pi}{4}.$$

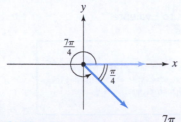

FIGURE 1.53 Reference angle for $\dfrac{7\pi}{4}$

The function value for the reference angle is $\sec \dfrac{\pi}{4} = \sqrt{2}$.

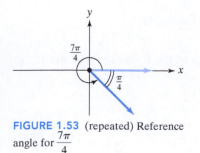

FIGURE 1.53 (repeated) Reference angle for $\dfrac{7\pi}{4}$

Step 2 Use the quadrant in which θ lies to prefix the appropriate sign to the function value in step 1. The coterminal angle $\theta = \dfrac{7\pi}{4}$ lies in quadrant IV. Because the secant is positive in quadrant IV, we put a + sign before the function value of the reference angle. Thus,

The secant is positive in quadrant IV.

$$\sec\left(-\frac{17\pi}{4}\right) = \sec\frac{7\pi}{4} = +\sec\frac{\pi}{4} = \sqrt{2}.$$

The reference angle for $\frac{7\pi}{4}$ is $\frac{\pi}{4}$.

• • •

Check Point 8 Use reference angles to find the exact value of each of the following trigonometric functions:

a. $\cos\dfrac{17\pi}{6}$ **b.** $\sin\left(-\dfrac{22\pi}{3}\right)$.

ACHIEVING SUCCESS

To be successful in trigonometry, it is often necessary to connect concepts.

For example, evaluating trigonometric functions like those in Example 8 and Check Point 8 involves using a number of concepts, including finding coterminal angles and reference angles, locating special angles, determining the signs of trigonometric functions in specific quadrants, and finding the trigonometric functions of special angles $\left(30° = \dfrac{\pi}{6}, 45° = \dfrac{\pi}{4}, \text{ and } 60° = \dfrac{\pi}{3}\right)$. Here's an early reference sheet showing some of the concepts you should have at your fingertips (or memorized).

Degree and Radian Measures of Special and Quadrantal Angles

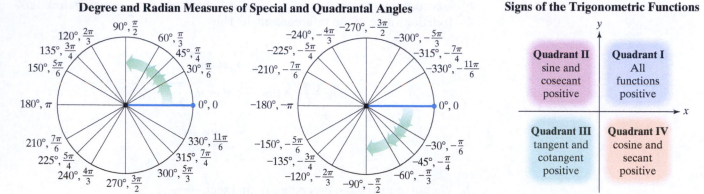

Signs of the Trigonometric Functions

Quadrant II sine and cosecant positive	**Quadrant I** All functions positive
Quadrant III tangent and cotangent positive	**Quadrant IV** cosine and secant positive

Special Right Triangles and Trigonometric Functions of Special Angles

θ	$30° = \dfrac{\pi}{6}$	$45° = \dfrac{\pi}{4}$	$60° = \dfrac{\pi}{3}$
$\sin\theta$	$\dfrac{1}{2}$	$\dfrac{\sqrt{2}}{2}$	$\dfrac{\sqrt{3}}{2}$
$\cos\theta$	$\dfrac{\sqrt{3}}{2}$	$\dfrac{\sqrt{2}}{2}$	$\dfrac{1}{2}$
$\tan\theta$	$\dfrac{\sqrt{3}}{3}$	1	$\sqrt{3}$

Trigonometric Functions of Quadrantal Angles

θ	$0° = 0$	$90° = \dfrac{\pi}{2}$	$180° = \pi$	$270° = \dfrac{3\pi}{2}$
$\sin\theta$	0	1	0	-1
$\cos\theta$	1	0	-1	0
$\tan\theta$	0	undefined	0	undefined

Using Reference Angles to Evaluate Trigonometric Functions

$$\sin\theta = \boxed{}\ \sin\theta'$$
$$\cos\theta = \boxed{}\ \cos\theta'$$
$$\tan\theta = \boxed{}\ \tan\theta'$$

+ or − in $\boxed{}$ is determined by the quadrant in which θ lies and the sign of the function in that quadrant.

CONCEPT AND VOCABULARY CHECK

Fill in each blank so that the resulting statement is true.

1. Let θ be any angle in standard position and let $P = (x, y)$ be any point besides the origin on the terminal side of θ. If $r = \sqrt{x^2 + y^2}$ is the distance from $(0, 0)$ to (x, y), the trigonometric functions of θ are defined as follows:

$\sin \theta =$ _____ $\csc \theta =$ _____

$\cos \theta =$ _____ $\sec \theta =$ _____

$\tan \theta =$ _____ $\cot \theta =$ _____.

2. Using the definitions in Exercise 1, the trigonometric functions that are undefined when $x = 0$ are _____ and _____. The trigonometric functions that are undefined when $y = 0$ are _____ and _____. The trigonometric functions that do not depend on the value of r are _____ and _____.

3. If θ lies in quadrant II, _____ and _____ are positive.

4. If θ lies in quadrant III, _____ and _____ are positive.

5. If θ lies in quadrant IV, _____ and _____ are positive.

6. Let θ be a nonacute angle in standard position that lies in a quadrant. Its reference angle is the positive acute angle formed by the _____ side of θ and the _____-axis.

7. Complete each statement for a positive angle θ and its reference angle θ'.
   a. If $90° < \theta < 180°$, then $\theta' =$ _____.
   b. If $180° < \theta < 270°$, then $\theta' =$ _____.
   c. If $270° < \theta < 360°$, then $\theta' =$ _____.

EXERCISE SET 1.3

Practice Exercises

In Exercises 1–8, a point on the terminal side of angle θ is given. Find the exact value of each of the six trigonometric functions of θ.

1. $(-4, 3)$ 2. $(-12, 5)$ 3. $(2, 3)$
4. $(3, 7)$ 5. $(3, -3)$ 6. $(5, -5)$
7. $(-2, -5)$ 8. $(-1, -3)$

In Exercises 9–16, evaluate the trigonometric function at the quadrantal angle, or state that the expression is undefined.

9. $\cos \pi$ 10. $\tan \pi$ 11. $\sec \pi$

12. $\csc \pi$ 13. $\tan \dfrac{3\pi}{2}$ 14. $\cos \dfrac{3\pi}{2}$

15. $\cot \dfrac{\pi}{2}$ 16. $\tan \dfrac{\pi}{2}$

In Exercises 17–22, let θ be an angle in standard position. Name the quadrant in which θ lies.

17. $\sin \theta > 0$, $\cos \theta > 0$ 18. $\sin \theta < 0$, $\cos \theta > 0$
19. $\sin \theta < 0$, $\cos \theta < 0$ 20. $\tan \theta < 0$, $\sin \theta < 0$
21. $\tan \theta < 0$, $\cos \theta < 0$ 22. $\cot \theta > 0$, $\sec \theta < 0$

In Exercises 23–34, find the exact value of each of the remaining trigonometric functions of θ.

23. $\cos \theta = -\frac{3}{5}$, θ in quadrant III
24. $\sin \theta = -\frac{12}{13}$, θ in quadrant III
25. $\sin \theta = \frac{5}{13}$, θ in quadrant II
26. $\cos \theta = \frac{4}{5}$, θ in quadrant IV
27. $\cos \theta = \frac{8}{17}$, $270° < \theta < 360°$
28. $\cos \theta = \frac{1}{3}$, $270° < \theta < 360°$
29. $\tan \theta = -\frac{2}{3}$, $\sin \theta > 0$ 30. $\tan \theta = -\frac{1}{3}$, $\sin \theta > 0$
31. $\tan \theta = \frac{4}{3}$, $\cos \theta < 0$ 32. $\tan \theta = \frac{5}{12}$, $\cos \theta < 0$
33. $\sec \theta = -3$, $\tan \theta > 0$ 34. $\csc \theta = -4$, $\tan \theta > 0$

In Exercises 35–60, find the reference angle for each angle.

35. $160°$ 36. $170°$ 37. $205°$
38. $210°$ 39. $355°$ 40. $351°$

41. $\dfrac{7\pi}{4}$ 42. $\dfrac{5\pi}{4}$ 43. $\dfrac{5\pi}{6}$

44. $\dfrac{5\pi}{7}$ 45. $-150°$ 46. $-250°$

47. $-335°$ 48. $-359°$ 49. 4.7
50. 5.5 51. $565°$ 52. $553°$

53. $\dfrac{17\pi}{6}$ 54. $\dfrac{11\pi}{4}$ 55. $\dfrac{23\pi}{4}$

56. $\dfrac{17\pi}{3}$ 57. $-\dfrac{11\pi}{4}$ 58. $-\dfrac{17\pi}{6}$

59. $-\dfrac{25\pi}{6}$ 60. $-\dfrac{13\pi}{3}$

In Exercises 61–86, use reference angles to find the exact value of each expression. Do not use a calculator.

61. $\cos 225°$ 62. $\sin 300°$ 63. $\tan 210°$
64. $\sec 240°$ 65. $\tan 420°$ 66. $\tan 405°$

67. $\sin \dfrac{2\pi}{3}$ 68. $\cos \dfrac{3\pi}{4}$ 69. $\csc \dfrac{7\pi}{6}$

70. $\cot \dfrac{7\pi}{4}$ 71. $\tan \dfrac{9\pi}{4}$ 72. $\tan \dfrac{9\pi}{2}$

73. $\sin(-240°)$ 74. $\sin(-225°)$ 75. $\tan\left(-\dfrac{\pi}{4}\right)$

76. $\tan\left(-\dfrac{\pi}{6}\right)$ 77. $\sec 495°$ 78. $\sec 510°$

79. $\cot \dfrac{19\pi}{6}$ 80. $\cot \dfrac{13\pi}{3}$ 81. $\cos \dfrac{23\pi}{4}$

82. $\cos \dfrac{35\pi}{6}$ 83. $\tan\left(-\dfrac{17\pi}{6}\right)$ 84. $\tan\left(-\dfrac{11\pi}{4}\right)$

85. $\sin\left(-\dfrac{17\pi}{3}\right)$ 86. $\sin\left(-\dfrac{35\pi}{6}\right)$

Practice Plus

In Exercises 87–92, find the exact value of each expression. Write the answer as a single fraction. Do not use a calculator.

87. $\sin \dfrac{\pi}{3} \cos \pi - \cos \dfrac{\pi}{3} \sin \dfrac{3\pi}{2}$

88. $\sin \dfrac{\pi}{4} \cos 0 - \sin \dfrac{\pi}{6} \cos \pi$

89. $\sin \dfrac{11\pi}{4} \cos \dfrac{5\pi}{6} + \cos \dfrac{11\pi}{4} \sin \dfrac{5\pi}{6}$

90. $\sin \dfrac{17\pi}{3} \cos \dfrac{5\pi}{4} + \cos \dfrac{17\pi}{3} \sin \dfrac{5\pi}{4}$

91. $\sin \dfrac{3\pi}{2} \tan\left(-\dfrac{15\pi}{4}\right) - \cos\left(-\dfrac{5\pi}{3}\right)$

92. $\sin \dfrac{3\pi}{2} \tan\left(-\dfrac{8\pi}{3}\right) + \cos\left(-\dfrac{5\pi}{6}\right)$

In Exercises 93–98, let

$$f(x) = \sin x, \ g(x) = \cos x, \text{ and } h(x) = 2x.$$

Find the exact value of each expression. Do not use a calculator.

93. $f\left(\dfrac{4\pi}{3} + \dfrac{\pi}{6}\right) + f\left(\dfrac{4\pi}{3}\right) + f\left(\dfrac{\pi}{6}\right)$

94. $g\left(\dfrac{5\pi}{6} + \dfrac{\pi}{6}\right) + g\left(\dfrac{5\pi}{6}\right) + g\left(\dfrac{\pi}{6}\right)$

95. $(h \circ g)\left(\dfrac{17\pi}{3}\right)$

96. $(h \circ f)\left(\dfrac{11\pi}{4}\right)$

97. the average rate of change of f from

$x_1 = \dfrac{5\pi}{4}$ to $x_2 = \dfrac{3\pi}{2}$

$\left(\text{Hint: The average rate of change of } f \text{ from } x_1 \text{ to } x_2 \text{ is} \right.$

$\left. \dfrac{f(x_2) - f(x_1)}{x_2 - x_1}.\right)$

98. the average rate of change of g from

$x_1 = \dfrac{3\pi}{4}$ to $x_2 = \pi$

(See the hint for Exercise 97.)

In Exercises 99–104, find two values of $\theta, 0 \le \theta < 2\pi$, that satisfy each equation.

99. $\sin \theta = \dfrac{\sqrt{2}}{2}$

100. $\cos \theta = \dfrac{1}{2}$

101. $\sin \theta = -\dfrac{\sqrt{2}}{2}$

102. $\cos \theta = -\dfrac{1}{2}$

103. $\tan \theta = -\sqrt{3}$

104. $\tan \theta = -\dfrac{\sqrt{3}}{3}$

Explaining the Concepts

105. If you are given a point on the terminal side of angle θ, explain how to find $\sin \theta$.

106. Explain why tan 90° is undefined.

107. If $\cos \theta > 0$ and $\tan \theta < 0$, explain how to find the quadrant in which θ lies.

108. What is a reference angle? Give an example with your description.

109. Explain how reference angles are used to evaluate trigonometric functions. Give an example with your description.

Critical Thinking Exercises

Make Sense? *In Exercises 110–113, determine whether each statement makes sense or does not make sense, and explain your reasoning.*

110. I'm working with a quadrantal angle θ for which $\sin \theta$ is undefined.

111. This angle θ is in a quadrant in which $\sin \theta < 0$ and $\csc \theta > 0$.

112. I am given that $\tan \theta = \dfrac{3}{5}$, so I can conclude that $y = 3$ and $x = 5$.

113. When I found the exact value of $\cos \dfrac{14\pi}{3}$, I used a number of concepts, including coterminal angles, reference angles, finding the cosine of a special angle, and knowing the cosine's sign in various quadrants.

Preview Exercises

Exercises 114–116 will help you prepare for the material covered in the next section.

114. Graph: $x^2 + y^2 = 1$. Then locate the point $\left(-\dfrac{1}{2}, \dfrac{\sqrt{3}}{2}\right)$ on the graph.

115. Use your graph of $x^2 + y^2 = 1$ from Exercise 114 to determine the relation's domain and range.

116. a. Find the exact value of $\sin\left(\dfrac{\pi}{4}\right)$, $\sin\left(-\dfrac{\pi}{4}\right)$, $\sin\left(\dfrac{\pi}{3}\right)$, and $\sin\left(-\dfrac{\pi}{3}\right)$. Based on your results, can the sine function be an even function? Explain your answer.

b. Find the exact value of $\cos\left(\dfrac{\pi}{4}\right)$, $\cos\left(-\dfrac{\pi}{4}\right)$, $\cos\left(\dfrac{\pi}{3}\right)$, and $\cos\left(-\dfrac{\pi}{3}\right)$. Based on your results, can the cosine function be an odd function? Explain your answer.

WHAT YOU KNOW: We learned to use radians to measure angles: One radian (approximately 57°) is the measure of the central angle that intercepts an arc equal in length to the radius of the circle. Using $180° = \pi$ radians, we converted degrees to radians $\left(\text{multiply by } \dfrac{\pi}{180°}\right)$ and radians to degrees $\left(\text{multiply by } \dfrac{180°}{\pi}\right)$. We developed formulas for arc length, $s = r\theta$, and area of a sector, $A = \frac{1}{2}r^2\theta$, for a circle of radius r and a central angle θ, measured in radians. We expressed linear speed, v, in terms of angular speed, $v = rw$, where w is the angular speed in radians per unit of time. We defined the six trigonometric functions using right triangles and angles in standard position. Evaluating trigonometric functions using reference angles involved connecting a number of concepts, including finding coterminal and reference angles, locating special angles, determining the signs of the trigonometric functions in specific quadrants, and finding the function values at special angles. Use the important Achieving Success box on page 50 as a reference sheet to help connect these concepts. Relationships among the six trigonometric functions include the reciprocal identities (see the box on page 27), the quotient identities $\left(\tan \theta = \dfrac{\sin \theta}{\cos \theta}, \cot \theta = \dfrac{\cos \theta}{\sin \theta}\right)$, the Pythagorean identities $(\sin^2 \theta + \cos^2 \theta = 1, 1 + \tan^2 \theta = \sec^2 \theta, 1 + \cot^2 \theta = \csc^2 \theta)$, and the cofunction identities (see the box on page 30).

In Exercises 1–2, convert each angle in degrees to radians. Express your answer as a multiple of π.

1. $10°$

2. $-105°$

In Exercises 3–4, convert each angle in radians to degrees.

3. $\dfrac{5\pi}{12}$

4. $-\dfrac{13\pi}{20}$

In Exercises 5–7,

   a. *Find a positive angle less than 360° or 2π that is coterminal with the given angle.*

   b. *Draw the given angle in standard position.*

   c. *Find the reference angle for the given angle.*

5. $\dfrac{11\pi}{3}$

6. $-\dfrac{19\pi}{4}$

7. $510°$

8. Use the triangle to find each of the six trigonometric functions of θ.

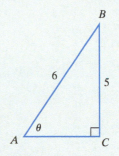

9. Use the point on the terminal side of θ to find each of the six trigonometric functions of θ.

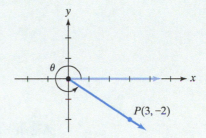

In Exercises 10–11, find the exact value of the remaining trigonometric functions of θ.

10. $\tan \theta = -\dfrac{3}{4}$, $\cos \theta < 0$ **11.** $\cos \theta = \dfrac{3}{7}$, $\sin \theta < 0$

In Exercises 12–13, find the measure of the side of the right triangle whose length is designated by a lowercase letter. Round the answer to the nearest whole number.

12.

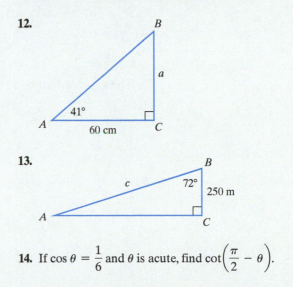

13.

14. If $\cos \theta = \dfrac{1}{6}$ and θ is acute, find $\cot\left(\dfrac{\pi}{2} - \theta\right)$.

In Exercises 15–24, find the exact value of each expression.
Do not use a calculator.

15. $\tan 30°$

16. $\cot 120°$

17. $\cos 240°$

18. $\sec \dfrac{11\pi}{6}$

19. $\sin^2 \dfrac{\pi}{7} + \cos^2 \dfrac{\pi}{7}$

20. $\sin\left(-\dfrac{2\pi}{3}\right)$

21. $\csc\left(\dfrac{22\pi}{3}\right)$

22. $\cos 495°$

23. $\tan\left(-\dfrac{17\pi}{6}\right)$

24. $\sin^2 \dfrac{\pi}{2} - \cos \pi$

25. A circle has a radius of 40 centimeters. Find the length of the arc intercepted by a central angle of 36°. Express the answer in terms of π. Then round to two decimal places.

26. A circle has a radius of 8 yards. Find the area of the sector formed by a central angle of 210°. Express the answer in terms of π. Then round to two decimal places.

27. A merry-go-round makes 8 revolutions per minute. Find the linear speed, in feet per minute, of a horse 10 feet from the center. Express the answer in terms of π. Then round to one decimal place.

28. A plane takes off at an angle of 6°. After traveling for one mile, or 5280 feet, along this flight path, find the plane's height, to the nearest tenth of a foot, above the ground.

29. A tree that is 50 feet tall casts a shadow that is 60 feet long. Find the angle of elevation, to the nearest degree, of the Sun.

A Brief Review • The Standard Form of the Equation of a Circle

- Let (x, y) represent the coordinates of any point on a circle in the rectangular coordinate system. The standard form of the equation of a circle with center (h, k) and radius r is

$$(x - h)^2 + (y - k)^2 = r^2.$$

EXAMPLE

The standard form of the equation of the circle with center $(0, 0)$ and radius 1 is

$$(x - 0)^2 + (y - 0)^2 = 1^2, \quad \text{or} \quad x^2 + y^2 = 1.$$

EXAMPLE

The standard form of the equation of the circle with center $(-2, 3)$ and radius 4 is

$$[x - (-2)]^2 + (y - 3)^2 = 4^2, \quad \text{or} \quad (x + 2)^2 + (y - 3)^2 = 16.$$

Section 1.4

Trigonometric Functions: The Unit Circle

What am I supposed to learn?

After studying this section, you should be able to:

① Use a unit circle to define trigonometric functions of real numbers.

② Recognize the domain and range of sine and cosine functions.

③ Use even and odd trigonometric functions.

④ Use periodic properties.

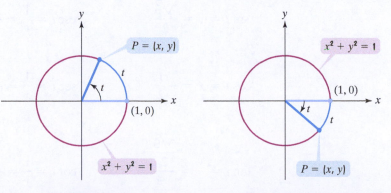

Cycles govern many aspects of life—heartbeats, sleep patterns, seasons, and tides all follow regular, predictable cycles. In this section, we will see why trigonometric functions are used to model phenomena that occur in cycles. To do this, we need to move beyond angles and consider trigonometric functions of real numbers.

Trigonometric Functions of Real Numbers

Thus far, we have considered trigonometric functions of angles measured in degrees or radians. To define trigonometric functions of real numbers, rather than angles, we use a unit circle. A **unit circle** is a circle of radius 1, with its center at the origin of a rectangular coordinate system. The equation of this

① Use a unit circle to define trigonometric functions of real numbers.

unit circle is $x^2 + y^2 = 1$. **Figure 1.54** shows a unit circle in which the central angle measures t radians. We can use the formula for the length of a circular arc, $s = r\theta$, to find the length of the intercepted arc.

$$s = r\theta = 1 \cdot t = t$$

| The radius of a unit circle is 1. | The radian measure of the central angle is t. |

Thus, the length of the intercepted arc is t. This is also the radian measure of the central angle. Thus, **in a unit circle, the radian measure of the central angle is equal to the length of the intercepted arc.** Both are given by the same *real number t*.

In **Figure 1.55**, the radian measure of the angle and the length of the intercepted arc are both shown by t. Let $P = (x, y)$ denote the point on the unit circle that has arc length t from $(1, 0)$. **Figure 1.55(a)** shows that if t is positive, point P is reached by moving counterclockwise along the unit circle from $(1, 0)$. **Figure 1.55(b)** shows that if t is negative, point P is reached by moving clockwise along the unit circle from $(1, 0)$. For each real number t, there corresponds a point $P = (x, y)$ on the unit circle.

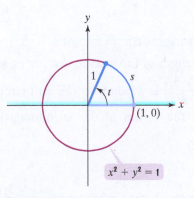

FIGURE 1.54 Unit circle with a central angle measuring t radians

FIGURE 1.55

(a) t is positive.

(b) t is negative.

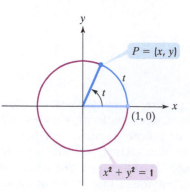

FIGURE 1.55(a) (repeated) *t* is positive.

Using **Figure 1.55**, we define the cosine function at *t* as the *x*-coordinate of *P* and the sine function at *t* as the *y*-coordinate of *P*. Thus,

$$x = \cos t \quad \text{and} \quad y = \sin t.$$

For example, a point $P = (x, y)$ on the unit circle corresponding to a real number *t* is shown in **Figure 1.56** for $\pi < t < \dfrac{3\pi}{2}$. We see that the coordinates of $P = (x, y)$ are $x = -\frac{3}{5}$ and $y = -\frac{4}{5}$. Because the cosine function is the *x*-coordinate of *P* and the sine function is the *y*-coordinate of *P*, the values of these trigonometric functions at the real number *t* are

$$\cos t = -\frac{3}{5} \quad \text{and} \quad \sin t = -\frac{4}{5}.$$

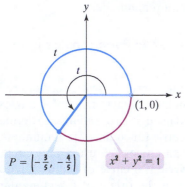

FIGURE 1.56

> ### Definitions of the Trigonometric Functions in Terms of a Unit Circle
>
> If *t* is a real number and $P = (x, y)$ is the point on the unit circle that corresponds to *t*, then
>
> $$\sin t = y \qquad\qquad \csc t = \frac{1}{y}, y \neq 0$$
>
> $$\cos t = x \qquad\qquad \sec t = \frac{1}{x}, x \neq 0$$
>
> $$\tan t = \frac{y}{x}, x \neq 0 \qquad\qquad \cot t = \frac{x}{y}, y \neq 0.$$

Because this definition expresses function values in terms of coordinates of a point on a unit circle, the trigonometric functions are sometimes called the **circular functions**.

EXAMPLE 1   Finding Values of the Trigonometric Functions

In **Figure 1.57**, *t* is a real number equal to the length of the intercepted arc of an angle that measures *t* radians and $P = \left(-\dfrac{1}{2}, \dfrac{\sqrt{3}}{2}\right)$ is a point on the unit circle that corresponds to *t*. Use the figure to find the values of the trigonometric functions at *t*.

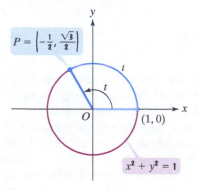

FIGURE 1.57

SOLUTION

The point *P* on the unit circle that corresponds to *t* has coordinates $\left(-\dfrac{1}{2}, \dfrac{\sqrt{3}}{2}\right)$. We use $x = -\dfrac{1}{2}$ and $y = \dfrac{\sqrt{3}}{2}$ to find the values of the trigonometric functions.

$$\sin t = y = \frac{\sqrt{3}}{2} \qquad\qquad \csc t = \frac{1}{y} = \frac{1}{\dfrac{\sqrt{3}}{2}} = \frac{2}{\sqrt{3}} = \frac{2}{\sqrt{3}} \cdot \frac{\sqrt{3}}{\sqrt{3}} = \frac{2\sqrt{3}}{3}$$

$$\cos t = x = -\frac{1}{2} \qquad\qquad \sec t = \frac{1}{x} = \frac{1}{-\dfrac{1}{2}} = -2$$

$$\tan t = \frac{y}{x} = \frac{\dfrac{\sqrt{3}}{2}}{-\dfrac{1}{2}} = -\sqrt{3} \qquad \cot t = \frac{x}{y} = \frac{-\dfrac{1}{2}}{\dfrac{\sqrt{3}}{2}} = -\frac{1}{\sqrt{3}} = -\frac{1}{\sqrt{3}} \cdot \frac{\sqrt{3}}{\sqrt{3}} = -\frac{\sqrt{3}}{3}$$

• • •

✅ **Check Point 1** Use the figure on the right to find the values of the trigonometric functions at t.

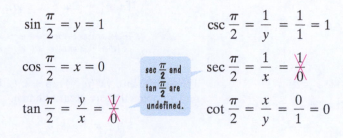

EXAMPLE 2 Finding Values of the Trigonometric Functions

Use **Figure 1.58** to find the values of the trigonometric functions at $t = \dfrac{\pi}{2}$.

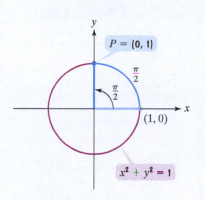

FIGURE 1.58

SOLUTION

The point P on the unit circle that corresponds to $t = \dfrac{\pi}{2}$ has coordinates $(0, 1)$. We use $x = 0$ and $y = 1$ to find the values of the trigonometric functions at $\dfrac{\pi}{2}$.

$$\sin \frac{\pi}{2} = y = 1 \qquad\qquad \csc \frac{\pi}{2} = \frac{1}{y} = \frac{1}{1} = 1$$

$$\cos \frac{\pi}{2} = x = 0 \qquad\qquad \sec \frac{\pi}{2} = \frac{1}{x} = \frac{1}{0}$$

$$\tan \frac{\pi}{2} = \frac{y}{x} = \frac{1}{0} \qquad\qquad \cot \frac{\pi}{2} = \frac{x}{y} = \frac{0}{1} = 0$$

sec $\dfrac{\pi}{2}$ and tan $\dfrac{\pi}{2}$ are undefined.

• • •

✅ **Check Point 2** Use the figure on the right to find the values of the trigonometric functions at $t = \pi$.

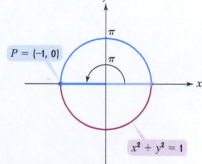

② Recognize the domain and range of sine and cosine functions.

Domain and Range of Sine and Cosine Functions

The value of a trigonometric function at the real number t is its value at an angle of t radians. However, using real number domains, we can observe properties of trigonometric functions that are not as apparent using the angle approach. For example, the domain and range of each trigonometric function can be found from the unit circle definition. At this point, let's look only at the sine and cosine functions,

$$\sin t = y \quad \text{and} \quad \cos t = x.$$

Figure 1.59 shows the sine function at t as the y-coordinate of a point along the unit circle:

$$y = \sin t.$$

The range is associated with y, the point's second coordinate.

The domain is associated with t, the angle's radian measure and the intercepted arc's length.

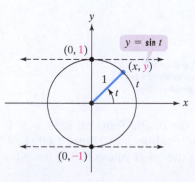

FIGURE 1.59

Because t can be any real number, the domain of the sine function is $(-\infty, \infty)$, the set of all real numbers. The radius of the unit circle is 1 and the dashed horizontal lines in **Figure 1.59** show that y cannot be less than -1 or greater than 1. Thus, the range of the sine function is $[-1, 1]$, the set of all real numbers from -1 to 1, inclusive.

A Brief Review • Interval Notation

- Some sets of real numbers can be represented using interval notation. Suppose that a and b are two real numbers such that $a < b$.

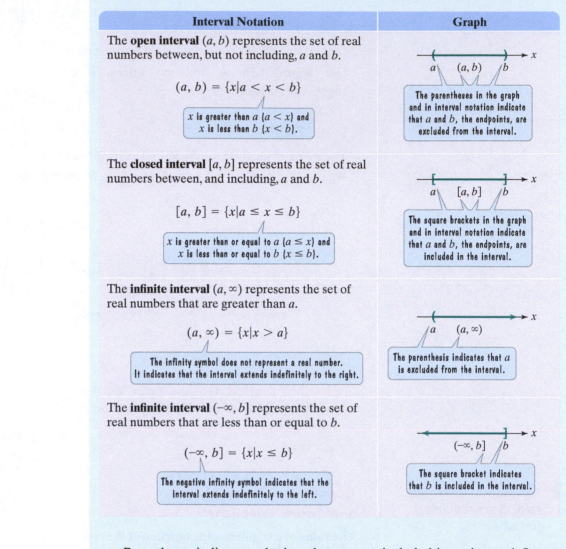

Interval Notation	Graph
The **open interval** (a, b) represents the set of real numbers between, but not including, a and b. $(a, b) = \{x \mid a < x < b\}$ *x is greater than a ($a < x$) and x is less than b ($x < b$).*	*The parentheses in the graph and in interval notation indicate that a and b, the endpoints, are excluded from the interval.*
The **closed interval** $[a, b]$ represents the set of real numbers between, and including, a and b. $[a, b] = \{x \mid a \leq x \leq b\}$ *x is greater than or equal to a ($a \leq x$) and x is less than or equal to b ($x \leq b$).*	*The square brackets in the graph and in interval notation indicate that a and b, the endpoints, are included in the interval.*
The **infinite interval** (a, ∞) represents the set of real numbers that are greater than a. $(a, \infty) = \{x \mid x > a\}$ *The infinity symbol does not represent a real number. It indicates that the interval extends indefinitely to the right.*	*The parenthesis indicates that a is excluded from the interval.*
The **infinite interval** $(-\infty, b]$ represents the set of real numbers that are less than or equal to b. $(-\infty, b] = \{x \mid x \leq b\}$ *The negative infinity symbol indicates that the interval extends indefinitely to the left.*	*The square bracket indicates that b is included in the interval.*

- Parentheses indicate endpoints that are not included in an interval. Square brackets indicate endpoints that are included in an interval. Parentheses are always used with ∞ or $-\infty$.

Figure 1.60 shows the cosine function at t as the x-coordinate of a point along the unit circle:

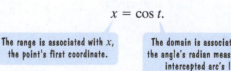

$$x = \cos t.$$

The range is associated with x, the point's first coordinate.   *The domain is associated with t, the angle's radian measure and the intercepted arc's length.*

Because t can be any real number, the domain of the cosine function is $(-\infty, \infty)$. The radius of the unit circle is 1 and the dashed vertical lines in **Figure 1.60** show that x cannot be less than -1 or greater than 1. Thus, the range of the cosine function is $[-1, 1]$.

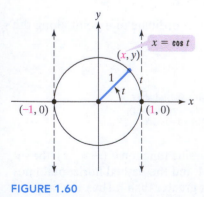

FIGURE 1.60

> ### The Domain and Range of the Sine and Cosine Functions
>
> The domain of the sine function and the cosine function is $(-\infty, \infty)$, the set of all real numbers. The range of these functions is $[-1, 1]$, the set of all real numbers from -1 to 1, inclusive.

③ Use even and odd trigonometric functions.

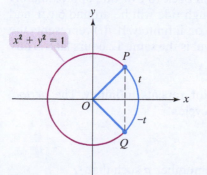

FIGURE 1.61

Even and Odd Trigonometric Functions

A function is even if $f(-t) = f(t)$ and odd if $f(-t) = -f(t)$. We can use **Figure 1.61** to show that the cosine function is an even function and the sine function is an odd function. By definition, the coordinates of the points P and Q in **Figure 1.61** are as follows:

$$P\text{: } (\cos t, \sin t)$$
$$Q\text{: } (\cos(-t), \ \sin(-t)).$$

In **Figure 1.61**, the x-coordinates of P and Q are the same. Thus,

$$\cos(-t) = \cos t.$$

This shows that the cosine function is an even function. By contrast, the y-coordinates of P and Q are negatives of each other. Thus,

$$\sin(-t) = -\sin t.$$

This shows that the sine function is an odd function.

This argument is valid regardless of the length of t. Thus, the arc may terminate in any of the four quadrants or on any axis. Using the unit circle definition of the trigonometric functions, we obtain the following results:

> ### Even and Odd Trigonometric Functions
>
> The cosine and secant functions are *even*.
>
> $$\cos(-t) = \cos t \qquad\qquad \sec(-t) = \sec t$$
>
> The sine, cosecant, tangent, and cotangent functions are *odd*.
>
> $$\sin(-t) = -\sin t \qquad\qquad \csc(-t) = -\csc t$$
>
> $$\tan(-t) = -\tan t \qquad\qquad \cot(-t) = -\cot t$$

EXAMPLE 3   Using Even and Odd Functions to Find Exact Values

Find the exact value of each trigonometric function:

    a. $\cos(-45°)$ **b.** $\tan\left(-\dfrac{\pi}{3}\right)$.

SOLUTION

    a. $\cos(-45°) = \cos 45° = \dfrac{\sqrt{2}}{2}$

    b. $\tan\left(-\dfrac{\pi}{3}\right) = -\tan\dfrac{\pi}{3} = -\sqrt{3}$ • • •

✓ **Check Point 3** Find the exact value of each trigonometric function:

    a. $\cos(-60°)$ **b.** $\tan\left(-\dfrac{\pi}{6}\right)$.

> **A Brief Review • Mathematical Modeling**
> - The process of finding formulas to describe real-world phenomena is called mathematical modeling. Such formulas, together with the meaning assigned to the variables, are called mathematical models. We often say that these formulas model, or describe, the relationships among the variables.

④ Use periodic properties.

Periodic Functions

Certain patterns in nature repeat again and again. For example, the ocean level at a beach varies from low tide to high tide and then back to low tide approximately every 12 hours. If low tide occurs at noon, then high tide will be around 6 P.M. and low tide will occur again around midnight, and so on infinitely. If $f(t)$ represents the ocean level at the beach at any time t, then the level is the same 12 hours later. Thus,

$$f(t + 12) = f(t).$$

The word *periodic* means that this tidal behavior repeats infinitely. The *period*, 12 hours, is the time it takes to complete one full cycle.

> **Definition of a Periodic Function**
>
> A function f is **periodic** if there exists a positive number p such that
> $$f(t + p) = f(t)$$
> for all t in the domain of f. The smallest positive number p for which f is periodic is called the **period** of f.

The trigonometric functions are used to model periodic phenomena. Why? If we begin at any point P on the unit circle and travel a distance of 2π units along the perimeter, we will return to the same point P. Because the trigonometric functions are defined in terms of the coordinates of that point P, we obtain the following results:

> **Periodic Properties of the Sine and Cosine Functions**
>
> $$\sin(t + 2\pi) = \sin t \quad \text{and} \quad \cos(t + 2\pi) = \cos t$$
>
> The sine and cosine functions are periodic functions and have period 2π.

EXAMPLE 4 Using Periodic Properties to Find Exact Values

Find the exact value of each trigonometric function:

a. $\cos 420°$ **b.** $\sin \dfrac{9\pi}{4}$.

SOLUTION

a. $\cos 420° = \cos(60° + 360°) = \cos 60° = \dfrac{1}{2}$

b. $\sin \dfrac{9\pi}{4} = \sin\left(\dfrac{\pi}{4} + 2\pi\right) = \sin \dfrac{\pi}{4} = \dfrac{\sqrt{2}}{2}$ • • •

✓ **Check Point 4** Find the exact value of each trigonometric function:

a. $\cos 405°$ **b.** $\sin \dfrac{7\pi}{3}$.

Like the sine and cosine functions, the secant and cosecant functions have period 2π. However, the tangent and cotangent functions have a smaller period.

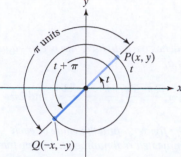

FIGURE 1.62 Tangent at P = tangent at Q

Figure 1.62 shows that if we begin at any point $P(x, y)$ on the unit circle and travel a distance of π units along the perimeter, we arrive at the point $Q(-x, -y)$. The tangent function, defined in terms of the coordinates of a point, is the same at (x, y) and $(-x, -y)$.

$$\boxed{\text{Tangent function at } (x, y)} \quad \frac{y}{x} = \frac{-y}{-x} \quad \boxed{\text{Tangent function } \pi \text{ radians later}}$$

We see that $\tan(t + \pi) = \tan t$. The same observations apply to the cotangent function.

> **Periodic Properties of the Tangent and Cotangent Functions**
>
> $$\tan(t + \pi) = \tan t \quad \text{and} \quad \cot(t + \pi) = \cot t$$
>
> The tangent and cotangent functions are periodic functions and have period π.

Why do the trigonometric functions model phenomena that repeat *indefinitely*? By starting at point P on the unit circle and traveling a distance of 2π units, 4π units, 6π units, and so on, we return to the starting point P. Because the trigonometric functions are defined in terms of the coordinates of that point P, if we add (or subtract) multiples of 2π to t, the values of the trigonometric functions of t do not change. Furthermore, the values for the tangent and cotangent functions of t do not change if we add (or subtract) multiples of π to t.

> **Repetitive Behavior of the Sine, Cosine, and Tangent Functions**
>
> For any integer n and real number t,
>
> $$\sin(t + 2\pi n) = \sin t, \quad \cos(t + 2\pi n) = \cos t, \quad \text{and} \quad \tan(t + \pi n) = \tan t.$$

ACHIEVING SUCCESS

Organizing and creating your own compact chapter summaries can reinforce what you know and help with the retention of this information. Imagine that your professor will permit two index cards of notes (3 by 5; front and back) on all exams. Organize and create such a two-card summary for the test on this chapter. Begin by determining what information you would find most helpful to include on the cards. Take as long as you need to create the summary. Based on how effective you find this strategy, you may decide to use the technique to help prepare for future exams.

CONCEPT AND VOCABULARY CHECK

Fill in each blank so that the resulting statement is true.

1. In a unit circle, the radian measure of the central angle is equal to the length of the _____.

2. If t is a real number and $P = (x, y)$ is a point on the unit circle that corresponds to t, then x is the _____ of t and y is the _____ of t.

3. The two trigonometric functions defined for all real numbers are the _____ function and the _____ function. The domain of each of these functions is _____.

4. The largest possible value for the sine function and the cosine function is _____ and the smallest possible value is _____. The range for each of these functions is _____.

5. $\cos(-t) = $ _____ and $\sec(-t) = $ _____, so the cosine and secant are _____ functions.

6. $\sin(-t) = $ _____, $\csc(-t) = $ _____, $\tan(-t) = $ _____, and $\cot(-t) = $ _____, so the sine, cosecant, tangent, and cotangent are _____ functions.

7. If there exists a positive number p such that $f(t + p) = f(t)$, function f is _____. The smallest positive number p for which $f(t + p) = f(t)$ is called the _____ of t.

8. $\sin(t + 2\pi) = $ _____ and $\cos(t + 2\pi) = $ _____, so the sine and cosine functions are _____ functions. The period of each of these functions is _____.

9. $\tan(t + \pi) = $ _____ and $\cot(t + \pi) = $ _____, so the tangent and cotangent functions are _____ functions. The period of each of these functions is _____.

EXERCISE SET 1.4

Practice Exercises

In Exercises 1–4, a point P(x, y) is shown on the unit circle corresponding to a real number t. Find the values of the trigonometric functions at t.

1.

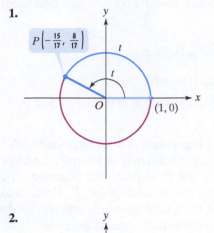

In Exercises 5–18, the unit circle has been divided into twelve equal arcs, corresponding to t-values of

$$0, \frac{\pi}{6}, \frac{\pi}{3}, \frac{\pi}{2}, \frac{2\pi}{3}, \frac{5\pi}{6}, \pi, \frac{7\pi}{6}, \frac{4\pi}{3}, \frac{3\pi}{2}, \frac{5\pi}{3}, \frac{11\pi}{6}, \text{and } 2\pi.$$

Use the (x, y) coordinates in the figure to find the value of each trigonometric function at the indicated real number, t, or state that the expression is undefined.

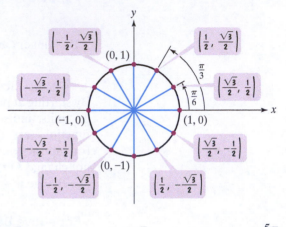

2.

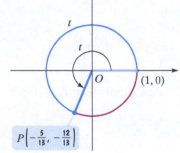

5. $\sin\frac{\pi}{6}$	**6.** $\sin\frac{\pi}{3}$	**7.** $\cos\frac{5\pi}{6}$
8. $\cos\frac{2\pi}{3}$	**9.** $\tan\pi$	**10.** $\tan 0$
11. $\csc\frac{7\pi}{6}$	**12.** $\csc\frac{4\pi}{3}$	**13.** $\sec\frac{11\pi}{6}$
14. $\sec\frac{5\pi}{3}$	**15.** $\sin\frac{3\pi}{2}$	**16.** $\cos\frac{3\pi}{2}$
17. $\sec\frac{3\pi}{2}$	**18.** $\tan\frac{3\pi}{2}$	

3.

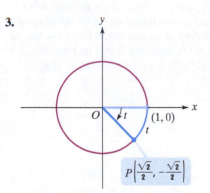

In Exercises 19–24,

   a. Use the unit circle shown for Exercises 5–18 to find the value of the trigonometric function.

   b. Use even and odd properties of trigonometric functions and your answer from part (a) to find the value of the same trigonometric function at the indicated real number.

4.

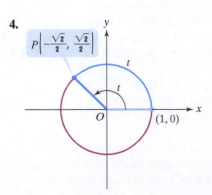

19. a. $\cos\frac{\pi}{6}$	**20. a.** $\cos\frac{\pi}{3}$
b. $\cos\left(-\frac{\pi}{6}\right)$	**b.** $\cos\left(-\frac{\pi}{3}\right)$
21. a. $\sin\frac{5\pi}{6}$	**22. a.** $\sin\frac{2\pi}{3}$
b. $\sin\left(-\frac{5\pi}{6}\right)$	**b.** $\sin\left(-\frac{2\pi}{3}\right)$
23. a. $\tan\frac{5\pi}{3}$	**24. a.** $\tan\frac{11\pi}{6}$
b. $\tan\left(-\frac{5\pi}{3}\right)$	**b.** $\tan\left(-\frac{11\pi}{6}\right)$

In Exercises 25–32, the unit circle has been divided into eight equal arcs, corresponding to t-values of

$$0, \frac{\pi}{4}, \frac{\pi}{2}, \frac{3\pi}{4}, \pi, \frac{5\pi}{4}, \frac{3\pi}{2}, \frac{7\pi}{4}, \text{ and } 2\pi.$$

a. Use the (x, y) coordinates in the figure to find the value of the trigonometric function.

b. Use periodic properties and your answer from part (a) to find the value of the same trigonometric function at the indicated real number.

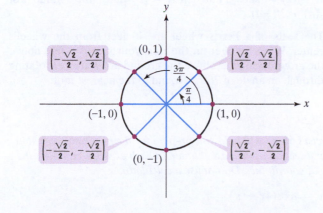

25. a. $\sin \dfrac{3\pi}{4}$

   b. $\sin \dfrac{11\pi}{4}$

26. a. $\cos \dfrac{3\pi}{4}$

   b. $\cos \dfrac{11\pi}{4}$

27. a. $\cos \dfrac{\pi}{2}$

   b. $\cos \dfrac{9\pi}{2}$

28. a. $\sin \dfrac{\pi}{2}$

   b. $\sin \dfrac{9\pi}{2}$

29. a. $\tan \pi$

   b. $\tan 17\pi$

30. a. $\cot \dfrac{\pi}{2}$

   b. $\cot \dfrac{15\pi}{2}$

31. a. $\sin \dfrac{7\pi}{4}$

   b. $\sin \dfrac{47\pi}{4}$

32. a. $\cos \dfrac{7\pi}{4}$

   b. $\cos \dfrac{47\pi}{4}$

Practice Plus

In Exercises 33–42, let

$$\sin t = a, \cos t = b, \text{ and } \tan t = c.$$

Write each expression in terms of a, b, and c.

33. $\sin(-t) - \sin t$

34. $\tan(-t) - \tan t$

35. $4 \cos(-t) - \cos t$

36. $3 \cos(-t) - \cos t$

37. $\sin(t + 2\pi) - \cos(t + 4\pi) + \tan(t + \pi)$

38. $\sin(t + 2\pi) + \cos(t + 4\pi) - \tan(t + \pi)$

39. $\sin(-t - 2\pi) - \cos(-t - 4\pi) - \tan(-t - \pi)$

40. $\sin(-t - 2\pi) + \cos(-t - 4\pi) - \tan(-t - \pi)$

41. $\cos t + \cos(t + 1000\pi) - \tan t - \tan(t + 999\pi) - \sin t +$

$$4 \sin(t - 1000\pi)$$

42. $-\cos t + 7 \cos(t + 1000\pi) + \tan t + \tan(t + 999\pi) +$

$$\sin t + \sin(t - 1000\pi)$$

Application Exercises

In Exercises 43–44, use a calculator in radian mode in parts (b) and (c).

43. The number of hours of daylight, H, on day t of any given year (on January 1, $t = 1$) in Fairbanks, Alaska, can be modeled by the function

$$H(t) = 12 + 8.3 \sin\left[\frac{2\pi}{365}(t - 80)\right].$$

a. March 21, the 80th day of the year, is the spring equinox. Find the number of hours of daylight in Fairbanks on this day.

b. June 21, the 172nd day of the year, is the summer solstice, the day with the maximum number of hours of daylight. To the nearest tenth of an hour, find the number of hours of daylight in Fairbanks on this day.

c. December 21, the 355th day of the year, is the winter solstice, the day with the minimum number of hours of daylight. Find, to the nearest tenth of an hour, the number of hours of daylight in Fairbanks on this day.

44. The number of hours of daylight, H, on day t of any given year (on January 1, $t = 1$) in San Diego, California, can be modeled by the function

$$H(t) = 12 + 2.4 \sin\left[\frac{2\pi}{365}(t - 80)\right].$$

a. March 21, the 80th day of the year, is the spring equinox. Find the number of hours of daylight in San Diego on this day.

b. June 21, the 172nd day of the year, is the summer solstice, the day with the maximum number of hours of daylight. Find, to the nearest tenth of an hour, the number of hours of daylight in San Diego on this day.

c. December 21, the 355th day of the year, is the winter solstice, the day with the minimum number of hours of daylight. To the nearest tenth of an hour, find the number of hours of daylight in San Diego on this day.

45. People who believe in biorhythms claim that there are three cycles that rule our behavior—the physical, emotional, and mental. Each is a sine function of a certain period. The function for our emotional fluctuations is

$$E = \sin \frac{\pi}{14} t,$$

where t is measured in days starting at birth. Emotional fluctuations, E, are measured from −1 to 1, inclusive, with 1 representing peak emotional well-being, −1 representing the low for emotional well-being, and 0 representing feeling neither emotionally high nor low.

a. Find E corresponding to $t = 7, 14, 21, 28,$ and 35. Describe what you observe.

b. What is the period of the emotional cycle?

46. The height of the water, H, in feet, at a boat dock t hours after 6 A.M. is given by

$$H = 10 + 4 \sin \frac{\pi}{6} t.$$

a. Find the height of the water at the dock at 6 A.M., 9 A.M., noon, 6 P.M., midnight, and 3 A.M.

b. When is low tide and when is high tide?

c. What is the period of this function and what does this mean about the tides?

Explaining the Concepts

47. Why are the trigonometric functions sometimes called circular functions?

48. What is the range of the sine function? Use the unit circle to explain where this range comes from.

49. What do we mean by even trigonometric functions? Which of the six functions fall into this category?

50. What is a periodic function? Why are the sine and cosine functions periodic?

51. Explain how you can use the function for emotional fluctuations in Exercise 45 to determine good days for having dinner with your moody boss.

52. Describe a phenomenon that repeats indefinitely. What is its period?

Critical Thinking Exercises

Make Sense?   *In Exercises 53–56, determine whether each statement makes sense or does not make sense, and explain your reasoning.*

53. Assuming that the innermost circle on this Navajo sand painting is a unit circle, as *A* moves around the circle, its coordinates define the cosine and sine functions, respectively.

54. I'm using a value for t and a point on the unit circle corresponding to t for which $\sin t = -\dfrac{\sqrt{10}}{2}$.

55. Because $\cos \dfrac{\pi}{6} = \dfrac{\sqrt{3}}{2}$, I can conclude that
$$\cos\left(-\dfrac{\pi}{6}\right) = -\dfrac{\sqrt{3}}{2}.$$

56. I can find the exact value of $\sin \dfrac{7\pi}{3}$ using periodic properties of the sine function, or using a coterminal angle and a reference angle.

57. Find the exact value of
$$\cos 0° + \cos 1° + \cos 2° + \cos 3° + \cdots + \cos 179° + \cos 180°.$$

58. If $f(x) = \sin x$ and $f(a) = \frac{1}{4}$, find the value of
$$f(a) + f(a + 2\pi) + f(a + 4\pi) + f(a + 6\pi).$$

59. If $f(x) = \sin x$ and $f(a) = \frac{1}{4}$, find the value of $f(a) + 2f(-a)$.

60. The seats of a Ferris wheel are 40 feet from the wheel's center. When you get on the ride, your seat is 5 feet above the ground. How far above the ground are you after rotating through an angle of 765°? Round to the nearest foot.

Preview Exercises

Exercises 61–63 will help you prepare for the material covered in the first section of the next chapter. In each exercise, complete the table of coordinates. Do not use a calculator.

61. $y = \frac{1}{2}\cos(4x + \pi)$

x	$-\dfrac{\pi}{4}$	$-\dfrac{\pi}{8}$	0	$\dfrac{\pi}{8}$	$\dfrac{\pi}{4}$
y					

62. $y = 4\sin\left(2x - \dfrac{2\pi}{3}\right)$

x	$\dfrac{\pi}{3}$	$\dfrac{7\pi}{12}$	$\dfrac{5\pi}{6}$	$\dfrac{13\pi}{12}$	$\dfrac{4\pi}{3}$
y					

63. $y = 3\sin\dfrac{\pi}{2}x$

x	0	$\dfrac{1}{3}$	1	$\dfrac{5}{3}$	2	$\dfrac{7}{3}$	3	$\dfrac{11}{3}$	4
y									

After completing this table of coordinates, plot the nine ordered pairs as points in a rectangular coordinate system. Then connect the points with a smooth curve.

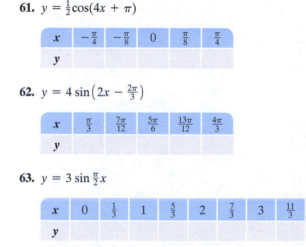

CHAPTER 1   Summary, Review, and Test

SUMMARY

DEFINITIONS AND CONCEPTS   EXAMPLES

1.1 Angles and Radian Measure

a. An angle consists of two rays with a common endpoint, the vertex.

b. An angle is in standard position if its vertex is at the origin and its initial side lies along the positive *x*-axis. Figure 1.3 on page 3 shows positive and negative angles in standard position.

DEFINITIONS AND CONCEPTS	**EXAMPLES**

c. A quadrantal angle is an angle with its terminal side on the *x*-axis or the *y*-axis.

---

d. Angles can be measured in degrees. $1°$ is $\frac{1}{360}$ of a complete rotation.

---

e. Acute angles measure more than $0°$ but less than $90°$, right angles $90°$, obtuse angles more than $90°$ but less than $180°$, and straight angles $180°$.

Figure 1.5, p. 3

---

f. Angles can be measured in radians. One radian is the measure of the central angle when the intercepted arc and radius have the same length. In general, the radian measure of a central angle is the length of the intercepted arc divided by the circle's radius: $\theta = \frac{s}{r}$.

Ex. 1, p. 5

---

g. To convert from degrees to radians, multiply degrees by $\frac{\pi \text{ radians}}{180°}$. To convert from radians to degrees, multiply radians by $\frac{180°}{\pi \text{ radians}}$.

Ex. 2, p. 6;
Ex. 3, p. 6

---

h. To draw angles measured in radians in standard position, it is helpful to "think in radians" without having to convert to degrees. See Figure 1.15 on page 9.

Ex. 4, p. 7

---

i. Two angles with the same initial and terminal sides are called coterminal angles. Increasing or decreasing an angle's measure by integer multiples of $360°$ or 2π produces coterminal angles.

Ex. 5, p. 10;
Ex. 6, p. 11;
Ex. 7, p. 12

---

j. The arc length formula is $s = r\theta$, as described in the box on page 13, where θ is measured in radians.

Ex. 8, p. 13

---

k. The formula for the area of a sector is $A = \frac{1}{2}r^2\theta$, as described in the box on page 14, where θ is measured in radians.

Ex. 9, p. 14

---

l. The definition of linear speed is $v = \frac{s}{t}$; angular speed is $\omega = \frac{\theta}{t}$, as given in the upper box on page 15.

---

m. Linear speed is expressed in terms of angular speed by $v = r\omega$, where v is the linear speed of a point a distance r from the center of rotation and ω is the angular speed in radians per unit of time.

Ex. 10, p. 16

1.2 Right Triangle Trigonometry

a. The right triangle definitions of the six trigonometric functions are given in the box on page 22:

$$\sin \theta = \frac{\text{opp}}{\text{hyp}}; \csc \theta = \frac{\text{hyp}}{\text{opp}}; \cos \theta = \frac{\text{adj}}{\text{hyp}}; \sec \theta = \frac{\text{hyp}}{\text{adj}}; \tan \theta = \frac{\text{opp}}{\text{adj}}; \cot \theta = \frac{\text{adj}}{\text{opp}}$$

Ex. 1, p. 23;
Ex. 2, p. 25

---

b. Function values for $30°$, $45°$, and $60°$ can be obtained using these special triangles.

Ex. 3, p. 25;
Ex. 4, p. 26

DEFINITIONS AND CONCEPTS

c. Fundamental Identities

   1. Reciprocal Identities

$$\sin\theta = \frac{1}{\csc\theta} \text{ and } \csc\theta = \frac{1}{\sin\theta}; \cos\theta = \frac{1}{\sec\theta} \text{ and } \sec\theta = \frac{1}{\cos\theta}; \tan\theta = \frac{1}{\cot\theta} \text{ and } \cot\theta = \frac{1}{\tan\theta}$$

   2. Quotient Identities

$$\tan\theta = \frac{\sin\theta}{\cos\theta}; \cot\theta = \frac{\cos\theta}{\sin\theta}$$

   3. Pythagorean Identities

$$\sin^2\theta + \cos^2\theta = 1; 1 + \tan^2\theta = \sec^2\theta; 1 + \cot^2\theta = \csc^2\theta$$

Ex. 5, p. 28;
Ex. 6, p. 29

---

d. Two angles are complements if their sum is 90° or $\frac{\pi}{2}$. The value of a trigonometric function of θ is equal to the cofunction of the complement of θ. Cofunction identities are listed in the box on page 30.

Ex. 7, p. 30

1.3 Trigonometric Functions of Any Angle

a. Definitions of the trigonometric functions of any angle are given in the box on page 41:

$$\sin\theta = \frac{y}{r}; \csc\theta = \frac{r}{y}; \cos\theta = \frac{x}{r}; \sec\theta = \frac{r}{x}; \tan\theta = \frac{y}{x}; \cot\theta = \frac{x}{y}; r = \sqrt{x^2 + y^2}$$

Ex. 1, p. 41;
Ex. 2, p. 42

---

b. Signs of the trigonometric functions: All functions are positive in quadrant I. If θ lies in quadrant II, $\sin\theta$ and $\csc\theta$ are positive. If θ lies in quadrant III, $\tan\theta$ and $\cot\theta$ are positive. If θ lies in quadrant IV, $\cos\theta$ and $\sec\theta$ are positive.

Ex. 3, p. 43;
Ex. 4, p. 43

---

c. If θ is a nonacute angle in standard position that lies in a quadrant, its reference angle is the positive acute angle θ' formed by the terminal side of θ and the x-axis. The reference angle for a given angle can be found by making a sketch that shows the angle in standard position. Figure 1.41 on page 44 shows reference angles for θ in quadrants II, III, and IV.

Ex. 5, p. 45;
Ex. 6, p. 46

---

d. The values of the trigonometric functions of a given angle are the same as the values of the functions of the reference angle, except possibly for the sign. A procedure for using reference angles to evaluate trigonometric functions is given in the box on page 47.

Ex. 7, p. 47;
Ex. 8, p. 49

1.4 Trigonometric Functions: The Unit Circle

a. Definitions of the trigonometric functions in terms of a unit circle are given in the box on page 56.

Ex. 1, p. 56;
Ex. 2, p. 57

---

b. The cosine and secant functions are even:

$$\cos(-t) = \cos t, \quad \sec(-t) = \sec t.$$

The other trigonometric functions are odd:

$$\sin(-t) = -\sin t, \quad \csc(-t) = -\csc t,$$

$$\tan(-t) = -\tan t, \quad \cot(-t) = -\cot t.$$

Ex. 3, p. 59

---

c. If $f(t + p) = f(t)$, the function f is periodic. The smallest positive value of p for which $f(t + p) = f(t)$ is the period of f. The tangent and cotangent functions have period π. The other four trigonometric functions have period 2π.

Ex. 4, p. 60

REVIEW EXERCISES

1.1

1. Find the radian measure of the central angle of a circle of radius 6 centimeters that intercepts an arc of length 27 centimeters.

In Exercises 2–4, convert each angle in degrees to radians. Express your answer as a multiple of π.

2. 15°   **3.** 120°   **4.** 315°

In Exercises 5–7, convert each angle in radians to degrees.

5. $\dfrac{5\pi}{3}$ **6.** $\dfrac{7\pi}{5}$ **7.** $-\dfrac{5\pi}{6}$

In Exercises 8–12, draw each angle in standard position.

8. $\dfrac{5\pi}{6}$ **9.** $-\dfrac{2\pi}{3}$ **10.** $\dfrac{8\pi}{3}$

11. 190°   **12.** −135°

In Exercises 13–17, find a positive angle less than 360° or 2π that is coterminal with the given angle.

13. 400°   **14.** −445°   **15.** $\dfrac{13\pi}{4}$

16. $\dfrac{31\pi}{6}$ **17.** $-\dfrac{8\pi}{3}$

18. Find the length of the arc on a circle of radius 10 feet intercepted by a 135° central angle. Express arc length in terms of π. Then round your answer to two decimal places.

19. Find the area of the sector of a circle with radius 6 yards formed by a 30° central angle. Express area in terms of π. Then round your answer to two decimal places.

20. The angular speed of a propeller on a wind generator is 10.3 revolutions per minute. Express this angular speed in radians per minute.

21. The propeller of an airplane has a radius of 3 feet. The propeller is rotating at 2250 revolutions per minute. Find the linear speed, in feet per minute, of the tip of the propeller.

1.2

22. Use the triangle to find each of the six trigonometric functions of θ.

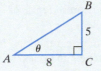

In Exercises 23–26, find the exact value of each expression. Do not use a calculator.

23. $\sin\dfrac{\pi}{6} + \tan^2\dfrac{\pi}{3}$ **24.** $\cos^2\dfrac{\pi}{4} - \tan^2\dfrac{\pi}{4}$

25. $\sec^2\dfrac{\pi}{5} - \tan^2\dfrac{\pi}{5}$ **26.** $\cos\dfrac{2\pi}{9}\sec\dfrac{2\pi}{9}$

27. If θ is an acute angle and $\sin\theta = \dfrac{2\sqrt{7}}{7}$, use the identity $\sin^2\theta + \cos^2\theta = 1$ to find $\cos\theta$.

In Exercises 28–29, find a cofunction with the same value as the given expression.

28. $\sin 70°$ **29.** $\cos\dfrac{\pi}{3}$

In Exercises 30–32, find the measure of the side of the right triangle whose length is designated by a lowercase letter. Round answers to the nearest whole number.

30.

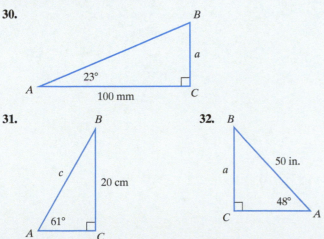

31.

32.

33. If $\sin\theta = \dfrac{1}{4}$ and θ is acute, find $\tan\left(\dfrac{\pi}{2} - \theta\right)$.

34. A hiker climbs for a half mile up a slope whose inclination is 17°. How many feet of altitude, to the nearest foot, does the hiker gain?

35. To find the distance across a lake, a surveyor took the measurements in the figure shown. What is the distance across the lake? Round to the nearest meter.

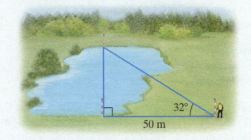

36. When a six-foot pole casts a four-foot shadow, what is the angle of elevation of the Sun? Round to the nearest whole degree.

1.3 and 1.4

In Exercises 37–38, a point on the terminal side of angle θ is given. Find the exact value of each of the six trigonometric functions of θ, or state that the function is undefined.

37. $(-1, -5)$ **38.** $(0, -1)$

In Exercises 39–40, let θ be an angle in standard position. Name the quadrant in which θ lies.

39. $\tan\theta > 0$ and $\sec\theta > 0$ **40.** $\tan\theta > 0$ and $\cos\theta < 0$

In Exercises 41–43, find the exact value of each of the remaining trigonometric functions of θ.

41. $\cos\theta = \dfrac{2}{5}$, $\sin\theta < 0$ **42.** $\tan\theta = -\dfrac{1}{3}$, $\sin\theta > 0$

43. $\cot\theta = 3$, $\cos\theta < 0$

In Exercises 44–48, find the reference angle for each angle.

44. 265°   **45.** $\dfrac{5\pi}{8}$ **46.** −410°

47. $\dfrac{17\pi}{6}$ **48.** $-\dfrac{11\pi}{3}$

In Exercises 49–59, find the exact value of each expression. Do not use a calculator.

49. $\sin 240°$ **50.** $\tan 120°$ **51.** $\sec\dfrac{7\pi}{4}$

52. $\cos\dfrac{11\pi}{6}$ **53.** $\cot(-210°)$ **54.** $\csc\left(-\dfrac{2\pi}{3}\right)$

55. $\sin\left(-\dfrac{\pi}{3}\right)$ **56.** $\sin 495°$ **57.** $\tan\dfrac{13\pi}{4}$

58. $\sin\dfrac{22\pi}{3}$ **59.** $\cos\left(-\dfrac{35\pi}{6}\right)$

CHAPTER 1 TEST

1. Convert 135° to an exact radian measure.

2. Find the length of the arc on a circle of radius 20 feet intercepted by a 75° central angle. Express arc length in terms of π. Then round your answer to two decimal places.

3. Find the area of the sector of a circle with radius 12 yards formed by a 150° central angle. Express area in terms of π. Then round your answer to two decimal places.

4. **a.** Find a positive angle less than 2π that is coterminal with $\frac{16\pi}{3}$.

   b. Find the reference angle for $\frac{16\pi}{3}$.

5. If $(-2, 5)$ is a point on the terminal side of angle θ, find the exact value of each of the six trigonometric functions of θ.

6. Determine the quadrant in which θ lies if $\cos \theta < 0$ and $\cot \theta > 0$.

7. If $\cos \theta = \frac{1}{3}$ and $\tan \theta < 0$, find the exact value of each of the remaining trigonometric functions of θ.

In Exercises 8–13, find the exact value of each expression. Do not use a calculator.

8. $\tan \dfrac{\pi}{6} \cos \dfrac{\pi}{3} - \cos \dfrac{\pi}{2}$

9. $\tan 300°$

10. $\sin \dfrac{7\pi}{4}$

11. $\sec \dfrac{22\pi}{3}$

12. $\cot\left(-\dfrac{8\pi}{3}\right)$

13. $\tan\left(\dfrac{7\pi}{3} + n\pi\right)$, n is an integer.

14. If $\sin \theta = a$ and $\cos \theta = b$, represent each of the following in terms of a and b.

    a. $\sin(-\theta) + \cos(-\theta)$

    b. $\tan \theta - \sec \theta$

15. The angle of elevation to the top of a building from a point on the ground 30 yards from its base is 37°. Find the height of the building to the nearest yard.

16. A 73-foot rope from the top of a circus tent pole is anchored to the flat ground 43 feet from the bottom of the pole. Find the angle, to the nearest tenth of a degree, that the rope makes with the pole.

17. Why are trigonometric functions ideally suited to model phenomena that repeat in cycles?

Graphs of the Trigonometric Functions; Inverse Trigonometric Functions

Music is all around us. A mere snippet of a song from the past can trigger vivid memories, including emotions ranging from unabashed joy to deep sorrow. Trigonometric functions describe the pitch, loudness, and quality of musical notes. In this chapter and the next, you will learn how trigonometry models the sound of music.

HERE'S WHERE YOU'LL FIND APPLICATIONS RELATED TO MUSIC:

- Synthesizers that electronically reproduce musical sounds: Section 2.4 opener
- Modeling music: Blitzer Bonus on page 133
- Modeling notes that have different tones: Section 3.2 opener
- Sound quality: Blitzer Bonus on page 160
- Sinusoidal sounds: Blitzer Bonus on page 187
- Using the sum of sines to describe the sounds of simple melodies: Exercise Set 3.4, Exercises 37, 38, and 62.

A Brief Review • Graphs of Functions

- The graph of a function is the graph of its ordered pairs. The graph can be used to determine the function's domain and its range. To find the domain, look for all the inputs on the x-axis that correspond to points on the graph. To find the range, look for all the outputs on the y-axis that correspond to points on the graph.

EXAMPLE

Use the graph on the left to identify the function's domain and its range.

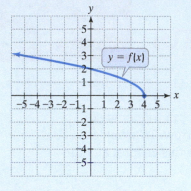

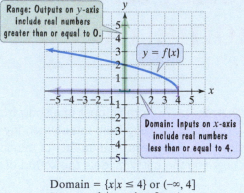

Range: Outputs on y-axis include real numbers greater than or equal to 0.

Domain: Inputs on x-axis include real numbers less than or equal to 4.

Domain = $\{x \mid x \le 4\}$ or $(-\infty, 4]$
Range = $\{y \mid y \ge 0\}$ or $[0, \infty)$

- The Vertical Line Test for Functions
  If any vertical line intersects a graph in more than one point, the graph does not define y as a function of x.

EXAMPLE

a.

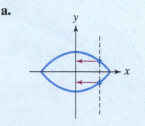

b.

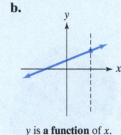

c.

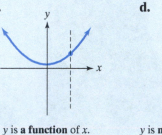

d.

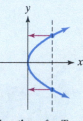

y is **not a function** of x. Two values of y correspond to an x-value.

y is **a function** of x.

y is **a function** of x.

y is **not a function** of x. Two values of y correspond to an x-value.

- The graph of an even function in which $f(-x) = f(x)$ is symmetric with respect to the y-axis. The graph of an odd function in which $f(-x) = -f(x)$ is symmetric with respect to the origin.

EXAMPLE

$f(x) = x^2 - 4$
$f(-x) = (-x)^2 - 4 = x^2 - 4 = f(x)$
Function f is even. The graph of f has y-axis symmetry.

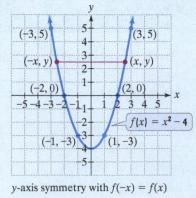

y-axis symmetry with $f(-x) = f(x)$

EXAMPLE

$f(x) = x^3$
$f(-x) = (-x)^3 = -x^3 = -f(x)$
Function f is odd. The graph of f has origin symmetry.

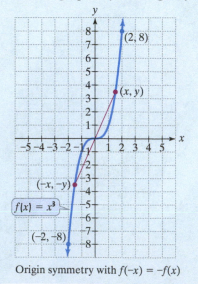

Origin symmetry with $f(-x) = -f(x)$

Section 2.1

Graphs of Sine and Cosine Functions

What am I supposed to learn?

After studying this section, you should be able to:

1. Understand the graph of $y = \sin x$.
2. Graph variations of $y = \sin x$.
3. Understand the graph of $y = \cos x$.
4. Graph variations of $y = \cos x$.
5. Use vertical shifts of sine and cosine curves.
6. Graph the sum of two trigonometric functions.
7. Model periodic behavior.

Take a deep breath and relax. Many relaxation exercises involve slowing down our breathing. Some people suggest that the way we breathe affects every part of our lives. Did you know that graphs of trigonometric functions can be used to analyze the breathing cycle, which is our closest link to both life and death?

In this section, we use graphs of sine and cosine functions to visualize their properties. We use the traditional symbol x, rather than θ or t, to represent the independent variable. We use the symbol y for the dependent variable, or the function's value at x. Thus, we will be graphing $y = \sin x$ and $y = \cos x$ in rectangular coordinates. In all graphs of trigonometric functions, the independent variable, x, is measured in radians.

The Graph of $y = \sin x$

1 Understand the graph of $y = \sin x$.

The trigonometric functions can be graphed in a rectangular coordinate system by plotting points whose coordinates satisfy the function. Thus, we graph $y = \sin x$ by listing some points on the graph. Because the period of the sine function is 2π, we will graph the function on the interval $[0, 2\pi]$. The rest of the graph is made up of repetitions of this portion.

Table 2.1 lists some values of (x, y) on the graph of $y = \sin x, 0 \le x \le 2\pi$.

Table 2.1 Values of (x, y) on the Graph of $y = \sin x$

x	0	$\frac{\pi}{6}$	$\frac{\pi}{3}$	$\frac{\pi}{2}$	$\frac{2\pi}{3}$	$\frac{5\pi}{6}$	π	$\frac{7\pi}{6}$	$\frac{4\pi}{3}$	$\frac{3\pi}{2}$	$\frac{5\pi}{3}$	$\frac{11\pi}{6}$	2π
$y = \sin x$	0	$\frac{1}{2}$	$\frac{\sqrt{3}}{2}$	1	$\frac{\sqrt{3}}{2}$	$\frac{1}{2}$	0	$-\frac{1}{2}$	$-\frac{\sqrt{3}}{2}$	-1	$-\frac{\sqrt{3}}{2}$	$-\frac{1}{2}$	0

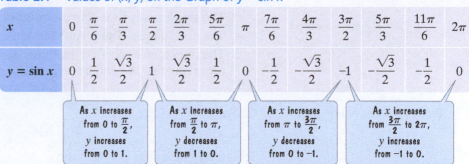

As x increases from 0 to $\frac{\pi}{2}$, y increases from 0 to 1.

As x increases from $\frac{\pi}{2}$ to π, y decreases from 1 to 0.

As x increases from π to $\frac{3\pi}{2}$, y decreases from 0 to -1.

As x increases from $\frac{3\pi}{2}$ to 2π, y increases from -1 to 0.

In plotting the points obtained in **Table 2.1**, we will use the approximation $\frac{\sqrt{3}}{2} \approx 0.87$. Rather than approximating π, we will mark off units on the x-axis in terms of π. If we connect these points with a smooth curve, we obtain the graph shown in **Figure 2.1** on the next page. The figure shows one period of the graph of $y = \sin x$.

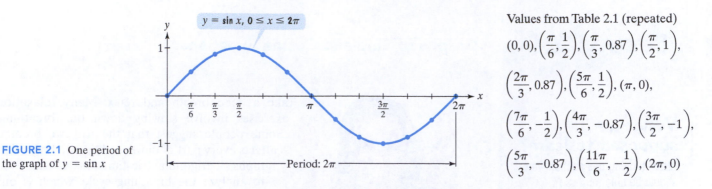

FIGURE 2.1 One period of the graph of $y = \sin x$

Values from Table 2.1 (repeated)

$(0, 0), \left(\dfrac{\pi}{6}, \dfrac{1}{2}\right), \left(\dfrac{\pi}{3}, 0.87\right), \left(\dfrac{\pi}{2}, 1\right),$

$\left(\dfrac{2\pi}{3}, 0.87\right), \left(\dfrac{5\pi}{6}, \dfrac{1}{2}\right), (\pi, 0),$

$\left(\dfrac{7\pi}{6}, -\dfrac{1}{2}\right), \left(\dfrac{4\pi}{3}, -0.87\right), \left(\dfrac{3\pi}{2}, -1\right),$

$\left(\dfrac{5\pi}{3}, -0.87\right), \left(\dfrac{11\pi}{6}, -\dfrac{1}{2}\right), (2\pi, 0)$

We can obtain a more complete graph of $y = \sin x$ by continuing the portion shown in **Figure 2.1** to the left and to the right. The graph of the sine function, called a **sine curve**, is shown in **Figure 2.2**. Any part of the graph that corresponds to one period (2π) is one cycle of the graph of $y = \sin x$.

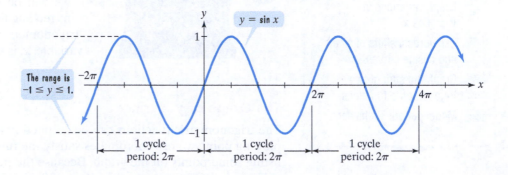

FIGURE 2.2 The graph of $y = \sin x$

The graph of $y = \sin x$ allows us to visualize some of the properties of the sine function.

- The domain is $(-\infty, \infty)$, the set of all real numbers. The graph extends indefinitely to the left and to the right with no gaps or holes.
- The range is $[-1, 1]$, the set of all real numbers between -1 and 1, inclusive. The graph never rises above 1 or falls below -1.
- The period is 2π. The graph's pattern repeats in every interval of length 2π.
- The function is an odd function: $\sin(-x) = -\sin x$. This can be seen by observing that the graph is symmetric with respect to the origin.

2 Graph variations of $y = \sin x$.

Graphing Variations of $y = \sin x$

To graph variations of $y = \sin x$ by hand, it is helpful to find x-intercepts, maximum points, and minimum points. One complete cycle of the sine curve includes three x-intercepts, one maximum point, and one minimum point. The graph of $y = \sin x$ has x-intercepts at the beginning, middle, and end of its full period, shown in **Figure 2.3**. The

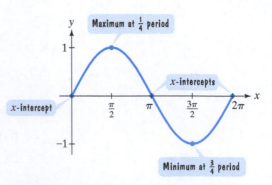

FIGURE 2.3 Key points in graphing the sine function

curve reaches its maximum point $\frac{1}{4}$ of the way through the period. It reaches its minimum point $\frac{3}{4}$ of the way through the period. Thus, key points in graphing sine functions are obtained by dividing the period into four equal parts. The x-coordinates of the five key points are as follows:

$$x_1 = \text{value of } x \text{ where the cycle begins}$$

$$x_2 = x_1 + \frac{\text{period}}{4}$$

$$x_3 = x_2 + \frac{\text{period}}{4}$$

$$x_4 = x_3 + \frac{\text{period}}{4}$$

$$x_5 = x_4 + \frac{\text{period}}{4}.$$

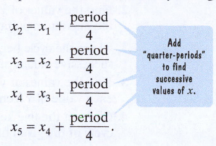

Add "quarter-periods" to find successive values of x.

The y-coordinates of the five key points are obtained by evaluating the given function at each of these values of x.

The graph of $y = \sin x$ forms the basis for graphing functions of the form

$$y = A \sin x.$$

For example, consider $y = 2 \sin x$, in which $A = 2$. We can obtain the graph of $y = 2 \sin x$ from that of $y = \sin x$ if we multiply each y-coordinate on the graph of $y = \sin x$ by 2. **Figure 2.4** shows the graphs. The basic sine curve is *stretched* and ranges between -2 and 2, rather than between -1 and 1. However, both $y = \sin x$ and $y = 2 \sin x$ have a period of 2π.

In general, the graph of $y = A \sin x$ ranges between $-|A|$ and $|A|$. Thus, the range of the function is $-|A| \leq y \leq |A|$. If $|A| > 1$, the basic sine curve is *stretched*, as in **Figure 2.4**. If $|A| < 1$, the basic sine curve is *shrunk*. We call $|A|$ the **amplitude** of $y = A \sin x$. The maximum value of y on the graph of $y = A \sin x$ is $|A|$, the amplitude.

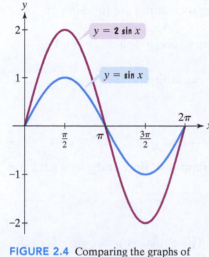

FIGURE 2.4 Comparing the graphs of $y = \sin x$ and $y = 2 \sin x$

A Brief Review • Transformations of Functions

- The graph of one function can be turned into the graph of a different function. The table summarizes the procedures for transforming the graph of $y = f(x)$. In each case, c represents a positive real number.

Summary of Transformations

To Graph:	Draw the Graph of f and:	Changes in the Equation of $y = f(x)$
Vertical shifts $y = f(x) + c$ $y = f(x) - c$	Raise the graph of f by c units. Lower the graph of f by c units.	c is added to $f(x)$. c is subtracted from $f(x)$.
Horizontal shifts $y = f(x + c)$ $y = f(x - c)$	Shift the graph of f to the left c units. Shift the graph of f to the right c units.	x is replaced with $x + c$. x is replaced with $x - c$.
Reflection about the x-axis $y = -f(x)$	Reflect the graph of f about the x-axis.	$f(x)$ is multiplied by -1.
Reflection about the y-axis $y = f(-x)$	Reflect the graph of f about the y-axis.	x is replaced with $-x$.
Vertical stretching or shrinking $y = cf(x), c > 1$ $y = cf(x), 0 < c < 1$	Multiply each y-coordinate of $y = f(x)$ by c, vertically stretching the graph of f. Multiply each y-coordinate of $y = f(x)$ by c, vertically shrinking the graph of f.	$f(x)$ is multiplied by $c, c > 1$. $f(x)$ is multiplied by $c, 0 < c < 1$.
Horizontal stretching or shrinking $y = f(cx), c > 1$ $y = f(cx), 0 < c < 1$	Divide each x-coordinate of $y = f(x)$ by c, horizontally shrinking the graph of f. Divide each x-coordinate of $y = f(x)$ by c, horizontally stretching the graph of f.	x is replaced with $cx, c > 1$. x is replaced with $cx, 0 < c < 1$.

Graphing Variations of $y = \sin x$

1. Identify the amplitude and the period.
2. Find the values of x for the five key points—the three x-intercepts, the maximum point, and the minimum point. Start with the value of x where the cycle begins and add quarter-periods—that is, $\dfrac{\text{period}}{4}$—to find successive values of x.
3. Find the values of y for the five key points by evaluating the function at each value of x from step 2.
4. Connect the five key points with a smooth curve and graph one complete cycle of the given function.
5. Extend the graph in step 4 to the left or right as desired.

EXAMPLE 1   Graphing a Variation of $y = \sin x$

Determine the amplitude of $y = \frac{1}{2} \sin x$. Then graph $y = \sin x$ and $y = \frac{1}{2} \sin x$ for $0 \le x \le 2\pi$.

SOLUTION

Step 1   Identify the amplitude and the period. The equation $y = \frac{1}{2} \sin x$ is of the form $y = A \sin x$ with $A = \frac{1}{2}$. Thus, the amplitude is $|A| = \frac{1}{2}$. This means that the maximum value of y is $\frac{1}{2}$ and the minimum value of y is $-\frac{1}{2}$. The period for both $y = \frac{1}{2} \sin x$ and $y = \sin x$ is 2π.

Step 2   Find the values of x for the five key points. We need to find the three x-intercepts, the maximum point, and the minimum point on the interval $[0, 2\pi]$. To do so, we begin by dividing the period, 2π, by 4.

$$\frac{\text{period}}{4} = \frac{2\pi}{4} = \frac{\pi}{2}$$

We start with the value of x where the cycle begins: $x_1 = 0$. Now we add quarter-periods, $\frac{\pi}{2}$, to generate x-values for each of the key points. The five x-values are

$$x_1 = 0, \quad x_2 = 0 + \frac{\pi}{2} = \frac{\pi}{2}, \quad x_3 = \frac{\pi}{2} + \frac{\pi}{2} = \pi,$$

$$x_4 = \pi + \frac{\pi}{2} = \frac{3\pi}{2}, \quad x_5 = \frac{3\pi}{2} + \frac{\pi}{2} = 2\pi.$$

Step 3   Find the values of y for the five key points. We evaluate the function at each value of x from step 2.

Value of x	Value of y: $y = \dfrac{1}{2} \sin x$	Coordinates of key point	
0	$y = \dfrac{1}{2} \sin 0 = \dfrac{1}{2} \cdot 0 = 0$	$(0, 0)$	
$\dfrac{\pi}{2}$	$y = \dfrac{1}{2} \sin \dfrac{\pi}{2} = \dfrac{1}{2} \cdot 1 = \dfrac{1}{2}$	$\left(\dfrac{\pi}{2}, \dfrac{1}{2}\right)$	maximum point
π	$y = \dfrac{1}{2} \sin \pi = \dfrac{1}{2} \cdot 0 = 0$	$(\pi, 0)$	
$\dfrac{3\pi}{2}$	$y = \dfrac{1}{2} \sin \dfrac{3\pi}{2} = \dfrac{1}{2}(-1) = -\dfrac{1}{2}$	$\left(\dfrac{3\pi}{2}, -\dfrac{1}{2}\right)$	minimum point
2π	$y = \dfrac{1}{2} \sin 2\pi = \dfrac{1}{2} \cdot 0 = 0$	$(2\pi, 0)$	

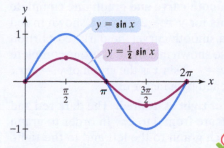

FIGURE 2.5 The graphs of $y = \sin x$ and $y = \frac{1}{2} \sin x$, $0 \le x \le 2\pi$

There are x-intercepts at 0, π, and 2π. The maximum and minimum points are indicated by the voice balloons.

Step 4 Connect the five key points with a smooth curve and graph one complete cycle of the given function. The five key points for $y = \frac{1}{2} \sin x$ are shown in red in **Figure 2.5**. By connecting the points with a smooth curve, the figure shows one complete cycle of $y = \frac{1}{2} \sin x$. Also shown is the graph of $y = \sin x$. The graph of $y = \frac{1}{2} \sin x$ is the graph of $y = \sin x$ vertically shrunk by a factor of $\frac{1}{2}$. •••

⊘ **Check Point 1** Determine the amplitude of $y = 3 \sin x$. Then graph $y = \sin x$ and $y = 3 \sin x$ for $0 \le x \le 2\pi$.

EXAMPLE 2   Graphing a Variation of $y = \sin x$

Determine the amplitude of $y = -2 \sin x$. Then graph $y = \sin x$ and $y = -2 \sin x$ for $-\pi \le x \le 3\pi$.

SOLUTION

Step 1   Identify the amplitude and the period. The equation $y = -2 \sin x$ is of the form $y = A \sin x$ with $A = -2$. Thus, the amplitude is $|A| = |-2| = 2$. This means that the maximum value of y is 2 and the minimum value of y is -2. Both $y = \sin x$ and $y = -2 \sin x$ have a period of 2π.

Step 2   Find the values of x for the five key points. Begin by dividing the period, 2π, by 4.

$$\frac{\text{period}}{4} = \frac{2\pi}{4} = \frac{\pi}{2}$$

Start with the value of x where the cycle begins: $x_1 = 0$. Adding quarter-periods, $\frac{\pi}{2}$, the five x-values for the key points are

$$x_1 = 0, \quad x_2 = 0 + \frac{\pi}{2} = \frac{\pi}{2}, \quad x_3 = \frac{\pi}{2} + \frac{\pi}{2} = \pi,$$

$$x_4 = \pi + \frac{\pi}{2} = \frac{3\pi}{2}, \quad x_5 = \frac{3\pi}{2} + \frac{\pi}{2} = 2\pi.$$

Although we will be graphing on $[-\pi, 3\pi]$, we select $x_1 = 0$ rather than $x_1 = -\pi$. Knowing the graph's shape on $[0, 2\pi]$ will enable us to continue the pattern and extend it to the left to $-\pi$ and to the right to 3π.

Step 3   Find the values of y for the five key points. We evaluate the function at each value of x from step 2.

Value of x	Value of y: $y = -2 \sin x$	Coordinates of key point	
0	$y = -2 \sin 0 = -2 \cdot 0 = 0$	$(0, 0)$	
$\dfrac{\pi}{2}$	$y = -2 \sin \dfrac{\pi}{2} = -2 \cdot 1 = -2$	$\left(\dfrac{\pi}{2}, -2\right)$	minimum point
π	$y = -2 \sin \pi = -2 \cdot 0 = 0$	$(\pi, 0)$	
$\dfrac{3\pi}{2}$	$y = -2 \sin \dfrac{3\pi}{2} = -2(-1) = 2$	$\left(\dfrac{3\pi}{2}, 2\right)$	maximum point
2π	$y = -2 \sin 2\pi = -2 \cdot 0 = 0$	$(2\pi, 0)$	

There are x-intercepts at 0, π, and 2π. The minimum and maximum points are indicated by the voice balloons.

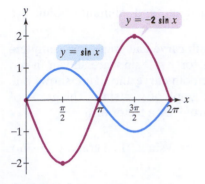

FIGURE 2.6 The graphs of $y = \sin x$ and $y = -2 \sin x$, $0 \leq x \leq 2\pi$. Key points for $y = -2 \sin x$ are $(0, 0)$, $\left(\frac{\pi}{2}, -2\right)$, $(\pi, 0)$, $\left(\frac{3\pi}{2}, 2\right)$, and $(2\pi, 0)$.

FIGURE 2.7 The graphs of $y = \sin x$ and $y = -2 \sin x$, $-\pi \leq x \leq 3\pi$.

Step 4   Connect the five key points with a smooth curve and graph one complete cycle of the given function. The five key points for $y = -2 \sin x$ are shown in red in **Figure 2.6**. By connecting the points with a smooth curve, the dark red portion shows one complete cycle of $y = -2 \sin x$. Also shown in dark blue is one complete cycle of the graph of $y = \sin x$. The graph of $y = -2 \sin x$ is the graph of $y = \sin x$ reflected about the x-axis and vertically stretched by a factor of 2.

Step 5   Extend the graph in step 4 to the left or right as desired. The dark red and dark blue portions of the graphs in **Figure 2.6** are from 0 to 2π. In order to graph for $-\pi \leq x \leq 3\pi$, continue the pattern of each graph to the left and to the right. These extensions are shown by the lighter colors in **Figure 2.7**.

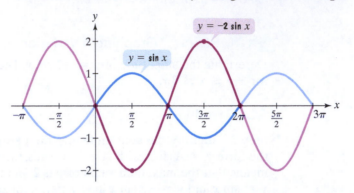

•••

✅ **Check Point 2** Determine the amplitude of $y = -\frac{1}{2} \sin x$. Then graph $y = \sin x$ and $y = -\frac{1}{2} \sin x$ for $-\pi \leq x \leq 3\pi$.

GREAT QUESTION!

What should I do to graph functions of the form $y = A \sin Bx$ if B is negative?

If $B < 0$ in $y = A \sin Bx$, use $\sin(-\theta) = -\sin \theta$ to rewrite the equation before obtaining its graph.

Now let us examine the graphs of functions of the form $y = A \sin Bx$, where B is the coefficient of x and $B > 0$. How do such graphs compare to those of functions of the form $y = A \sin x$? We know that $y = A \sin x$ completes one cycle from $x = 0$ to $x = 2\pi$. Thus, $y = A \sin Bx$ completes one cycle as Bx increases from 0 to 2π. Set up an inequality to represent this and solve for x to determine the values of x for which $y = \sin Bx$ completes one cycle.

$$0 \leq Bx \leq 2\pi \quad \text{\textcolor{blue}{$y = \sin Bx$ completes one cycle as Bx increases from 0 to 2π.}}$$

$$0 \leq x \leq \frac{2\pi}{B} \quad \text{\textcolor{blue}{Divide by B, where $B > 0$, and solve for x.}}$$

The inequality $0 \leq x \leq \dfrac{2\pi}{B}$ means that $y = A \sin Bx$ completes one cycle from 0 to $\dfrac{2\pi}{B}$. The period is $\dfrac{2\pi}{B}$. The graph of $y = A \sin Bx$ is the graph of $y = A \sin x$ horizontally shrunk by a factor of $\dfrac{1}{B}$ if $B > 1$ and horizontally stretched by a factor of $\dfrac{1}{B}$ if $0 < B < 1$.

Amplitudes and Periods

The graph of $y = A \sin Bx$, $B > 0$, has

$$\text{amplitude} = |A|$$

$$\text{period} = \frac{2\pi}{B}.$$

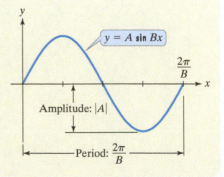

EXAMPLE 3 Graphing a Function of the Form $y = A \sin Bx$

Determine the amplitude and period of $y = 3 \sin 2x$. Then graph the function for $0 \le x \le 2\pi$.

SOLUTION

Step 1 **Identify the amplitude and the period.** The equation $y = 3 \sin 2x$ is of the form $y = A \sin Bx$ with $A = 3$ and $B = 2$.

$$\text{amplitude:} \quad |A| = |3| = 3$$

$$\text{period:} \quad \frac{2\pi}{B} = \frac{2\pi}{2} = \pi.$$

The amplitude, 3, tells us that the maximum value of y is 3 and the minimum value of y is -3. The period, π, tells us that the graph completes one cycle from 0 to π.

Step 2 **Find the values of x for the five key points.** Begin by dividing the period of $y = 3 \sin 2x$, π, by 4.

$$\frac{\text{period}}{4} = \frac{\pi}{4}$$

Start with the value of x where the cycle begins: $x_1 = 0$. Adding quarter-periods, $\frac{\pi}{4}$, the five x-values for the key points are

$$x_1 = 0, \quad x_2 = 0 + \frac{\pi}{4} = \frac{\pi}{4}, \quad x_3 = \frac{\pi}{4} + \frac{\pi}{4} = \frac{\pi}{2},$$

$$x_4 = \frac{\pi}{2} + \frac{\pi}{4} = \frac{3\pi}{4}, \quad x_5 = \frac{3\pi}{4} + \frac{\pi}{4} = \pi.$$

Step 3 **Find the values of y for the five key points.** We evaluate the function at each value of x from step 2.

Value of x	Value of y: $y = 3 \sin 2x$	Coordinates of key point	
0	$y = 3 \sin (2 \cdot 0)$ $= 3 \sin 0 = 3 \cdot 0 = 0$	$(0, 0)$	
$\frac{\pi}{4}$	$y = 3 \sin\left(2 \cdot \frac{\pi}{4}\right)$ $= 3 \sin \frac{\pi}{2} = 3 \cdot 1 = 3$	$\left(\frac{\pi}{4}, 3\right)$	maximum point
$\frac{\pi}{2}$	$y = 3 \sin\left(2 \cdot \frac{\pi}{2}\right)$ $= 3 \sin \pi = 3 \cdot 0 = 0$	$\left(\frac{\pi}{2}, 0\right)$	
$\frac{3\pi}{4}$	$y = 3 \sin\left(2 \cdot \frac{3\pi}{4}\right)$ $= 3 \sin \frac{3\pi}{2} = 3(-1) = -3$	$\left(\frac{3\pi}{4}, -3\right)$	minimum point
π	$y = 3 \sin (2 \cdot \pi)$ $= 3 \sin 2\pi = 3 \cdot 0 = 0$	$(\pi, 0)$	

In the interval $[0, \pi]$, there are x-intercepts at $0, \frac{\pi}{2}$, and π. The maximum and minimum points are indicated by the voice balloons.

Step 4 **Connect the five key points with a smooth curve and graph one complete cycle of the given function.** The five key points for $y = 3 \sin 2x$ are shown in **Figure 2.8**. By connecting the points with a smooth curve, the blue portion shows one complete cycle of $y = 3 \sin 2x$ from 0 to π. The graph of $y = 3 \sin 2x$ is the graph of $y = \sin x$ vertically stretched by a factor of 3 and horizontally shrunk by a factor of $\frac{1}{2}$.

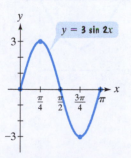

FIGURE 2.8 The graph of $y = 3 \sin 2x, 0 \le x \le \pi$

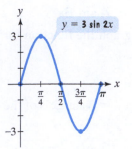

FIGURE 2.8 (repeated)

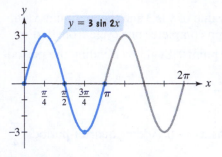

FIGURE 2.9

TECHNOLOGY

The graph of $y = 3 \sin 2x$ in a $\left[0, 2\pi, \dfrac{\pi}{2}\right]$ by $[-4, 4, 1]$ viewing rectangle verifies our hand-drawn graph in **Figure 2.9**.

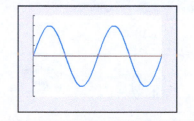

Step 5 Extend the graph in step 4 to the left or right as desired. The blue portion of the graph in **Figure 2.8** is from 0 to π. In order to graph for $0 \le x \le 2\pi$, we continue this portion and extend the graph another full period to the right. This extension is shown in gray in **Figure 2.9**.   •••

⊘ Check Point **3** Determine the amplitude and period of $y = 2 \sin \frac{1}{2}x$. Then graph the function for $0 \le x \le 8\pi$.

Now let us examine the graphs of functions of the form $y = A \sin(Bx - C)$, where $B > 0$. How do such graphs compare to those of functions of the form $y = A \sin Bx$? In both cases, the amplitude is $|A|$ and the period is $\dfrac{2\pi}{B}$. One complete cycle occurs as $Bx - C$ increases from 0 to 2π. This means that we can find an interval containing one cycle by solving the following inequality:

$$0 \le Bx - C \le 2\pi.$$

$y = A \sin(Bx - C)$ completes one cycle as $Bx - C$ increases from 0 to 2π.

$$C \le Bx \le C + 2\pi$$

Add C to all three parts.

$$\frac{C}{B} \le x \le \frac{C}{B} + \frac{2\pi}{B}$$

Divide by B, where $B > 0$, and solve for x.

This is the x-coordinate on the left where the cycle begins.

This is the x-coordinate on the right where the cycle ends. $\dfrac{2\pi}{B}$ is the period.

The voice balloon on the left indicates that the graph of $y = A \sin(Bx - C)$ is the graph of $y = A \sin Bx$ shifted horizontally by $\dfrac{C}{B}$. Thus, the number $\dfrac{C}{B}$ is the **phase shift** associated with the graph.

The Graph of $y = A \sin(Bx - C)$

The graph of $y = A \sin(Bx - C), B > 0$, is obtained by horizontally shifting the graph of $y = A \sin Bx$ so that the starting point of the cycle is shifted from $x = 0$ to $x = \dfrac{C}{B}$. If $\dfrac{C}{B} > 0$, the shift is to the right. If $\dfrac{C}{B} < 0$, the shift is to the left.

The number $\dfrac{C}{B}$ is called the **phase shift**.

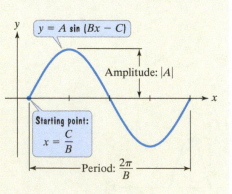

$$\text{amplitude} = |A|$$

$$\text{period} = \frac{2\pi}{B}$$

Graphing a Function of the Form $y = A \sin(Bx - C)$

Determine the amplitude, period, and phase shift of $y = 4 \sin\left(2x - \dfrac{2\pi}{3}\right)$. Then graph one period of the function.

SOLUTION

Step 1 Identify the amplitude, the period, and the phase shift. We must first identify values for A, B, and C.

> The equation is of the form
> $y = A \sin(Bx - C)$.

$$y = 4 \sin\left(2x - \frac{2\pi}{3}\right)$$

Using the voice balloon, we see that $A = 4$, $B = 2$, and $C = \dfrac{2\pi}{3}$.

amplitude: $|A| = |4| = 4$ The maximum value of y is 4 and the minimum is −4.

period: $\dfrac{2\pi}{B} = \dfrac{2\pi}{2} = \pi$ Each cycle is of length π.

phase shift: $\dfrac{C}{B} = \dfrac{\frac{2\pi}{3}}{2} = \dfrac{2\pi}{3} \cdot \dfrac{1}{2} = \dfrac{\pi}{3}$ A cycle starts at $x = \frac{\pi}{3}$.

Step 2 Find the values of x for the five key points. Begin by dividing the period, π, by 4.

$$\frac{\text{period}}{4} = \frac{\pi}{4}$$

Start with the value of x where the cycle begins: $x_1 = \dfrac{\pi}{3}$. Adding quarter-periods, $\dfrac{\pi}{4}$, the five x-values for the key points are

$$x_1 = \frac{\pi}{3}, \quad x_2 = \frac{\pi}{3} + \frac{\pi}{4} = \frac{4\pi}{12} + \frac{3\pi}{12} = \frac{7\pi}{12},$$

$$x_3 = \frac{7\pi}{12} + \frac{\pi}{4} = \frac{7\pi}{12} + \frac{3\pi}{12} = \frac{10\pi}{12} = \frac{5\pi}{6},$$

$$x_4 = \frac{5\pi}{6} + \frac{\pi}{4} = \frac{10\pi}{12} + \frac{3\pi}{12} = \frac{13\pi}{12},$$

$$x_5 = \frac{13\pi}{12} + \frac{\pi}{4} = \frac{13\pi}{12} + \frac{3\pi}{12} = \frac{16\pi}{12} = \frac{4\pi}{3}.$$

GREAT QUESTION!

Is there a way I can speed up the additions shown on the right?

Yes. First write the starting point, $\frac{\pi}{3}$, and the quarter-period, $\frac{\pi}{4}$, with a common denominator, 12.

$$\text{starting point} = \frac{\pi}{3} = \frac{4\pi}{12}$$

$$\text{quarter-period} = \frac{\pi}{4} = \frac{3\pi}{12}$$

GREAT QUESTION!

Is there a way to check my computations for the x-values for the five key points?

Yes. The difference between x_5 and x_1, or $x_5 - x_1$, should equal the period.

$$x_5 - x_1 = \frac{4\pi}{3} - \frac{\pi}{3} = \frac{3\pi}{3} = \pi$$

Because the period is π, this verifies that our five x-values are correct.

Step 3   Find the values of y for the five key points. We evaluate the function at each value of x from step 2.

Value of x	Value of y: $y = 4\sin\left(2x - \dfrac{2\pi}{3}\right)$	Coordinates of key point
$\dfrac{\pi}{3}$	$y = 4\sin\left(2 \cdot \dfrac{\pi}{3} - \dfrac{2\pi}{3}\right)$ $= 4\sin 0 = 4 \cdot 0 = 0$	$\left(\dfrac{\pi}{3}, 0\right)$
$\dfrac{7\pi}{12}$	$y = 4\sin\left(2 \cdot \dfrac{7\pi}{12} - \dfrac{2\pi}{3}\right)$ $= 4\sin\left(\dfrac{7\pi}{6} - \dfrac{4\pi}{6}\right)$ $= 4\sin\dfrac{3\pi}{6} = 4\sin\dfrac{\pi}{2} = 4 \cdot 1 = 4$	$\left(\dfrac{7\pi}{12}, 4\right)$ maximum point
$\dfrac{5\pi}{6}$	$y = 4\sin\left(2 \cdot \dfrac{5\pi}{6} - \dfrac{2\pi}{3}\right)$ $= 4\sin\left(\dfrac{5\pi}{3} - \dfrac{2\pi}{3}\right)$ $= 4\sin\dfrac{3\pi}{3} = 4\sin \pi = 4 \cdot 0 = 0$	$\left(\dfrac{5\pi}{6}, 0\right)$
$\dfrac{13\pi}{12}$	$y = 4\sin\left(2 \cdot \dfrac{13\pi}{12} - \dfrac{2\pi}{3}\right)$ $= 4\sin\left(\dfrac{13\pi}{6} - \dfrac{4\pi}{6}\right)$ $= 4\sin\dfrac{9\pi}{6} = 4\sin\dfrac{3\pi}{2} = 4(-1) = -4$	$\left(\dfrac{13\pi}{12}, -4\right)$ minimum point
$\dfrac{4\pi}{3}$	$y = 4\sin\left(2 \cdot \dfrac{4\pi}{3} - \dfrac{2\pi}{3}\right)$ $= 4\sin\dfrac{6\pi}{3} = 4\sin 2\pi = 4 \cdot 0 = 0$	$\left(\dfrac{4\pi}{3}, 0\right)$

In the interval $\left[\dfrac{\pi}{3}, \dfrac{4\pi}{3}\right]$, there are x-intercepts at $\dfrac{\pi}{3}, \dfrac{5\pi}{6},$ and $\dfrac{4\pi}{3}$. The maximum and minimum points are indicated by the voice balloons.

Step 4   Connect the five key points with a smooth curve and graph one complete cycle of the given function. The five key points are shown on the graph of $y = 4\sin\left(2x - \dfrac{2\pi}{3}\right)$ in **Figure 2.10**.

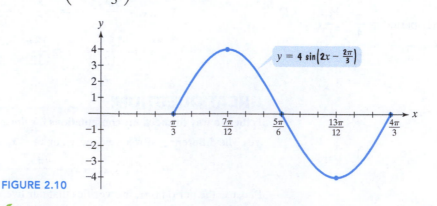

FIGURE 2.10

✓ Check Point **4** Determine the amplitude, period, and phase shift of $y = 3\sin\left(2x - \dfrac{\pi}{3}\right)$. Then graph one period of the function.

3 Understand the graph
of $y = \cos x$.

The Graph of $y = \cos x$

We graph $y = \cos x$ by listing some points on the graph. Because the period of the cosine function is 2π, we will concentrate on the graph of the basic cosine curve on the interval $[0, 2\pi]$. The rest of the graph is made up of repetitions of this portion. **Table 2.2** lists some values of (x, y) on the graph of $y = \cos x$.

Table 2.2   Values of (x, y) on the Graph of $y = \cos x$

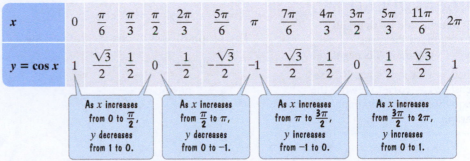

x	0	$\frac{\pi}{6}$	$\frac{\pi}{3}$	$\frac{\pi}{2}$	$\frac{2\pi}{3}$	$\frac{5\pi}{6}$	π	$\frac{7\pi}{6}$	$\frac{4\pi}{3}$	$\frac{3\pi}{2}$	$\frac{5\pi}{3}$	$\frac{11\pi}{6}$	2π
$y = \cos x$	1	$\frac{\sqrt{3}}{2}$	$\frac{1}{2}$	0	$-\frac{1}{2}$	$-\frac{\sqrt{3}}{2}$	-1	$-\frac{\sqrt{3}}{2}$	$-\frac{1}{2}$	0	$\frac{1}{2}$	$\frac{\sqrt{3}}{2}$	1

As x increases from 0 to $\frac{\pi}{2}$, y decreases from 1 to 0.

As x increases from $\frac{\pi}{2}$ to π, y decreases from 0 to -1.

As x increases from π to $\frac{3\pi}{2}$, y increases from -1 to 0.

As x increases from $\frac{3\pi}{2}$ to 2π, y increases from 0 to 1.

Plotting the points in **Table 2.2** and connecting them with a smooth curve, we obtain the graph shown in **Figure 2.11**. The portion of the graph in dark blue shows one complete period. We can obtain a more complete graph of $y = \cos x$ by extending this dark blue portion to the left and to the right.

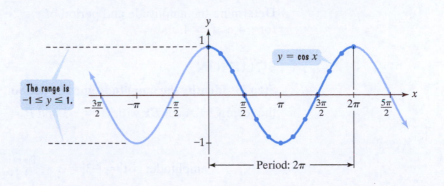

The range is $-1 \le y \le 1$.

$y = \cos x$

Period: 2π

FIGURE 2.11 The graph of $y = \cos x$

The graph of $y = \cos x$ allows us to visualize some of the properties of the cosine function.

- The domain is $(-\infty, \infty)$, the set of all real numbers. The graph extends indefinitely to the left and to the right with no gaps or holes.
- The range is $[-1, 1]$, the set of all real numbers between -1 and 1, inclusive. The graph never rises above 1 or falls below -1.
- The period is 2π. The graph's pattern repeats in every interval of length 2π.
- The function is an even function: $\cos(-x) = \cos x$. This can be seen by observing that the graph is symmetric with respect to the y-axis.

Take a second look at **Figure 2.11**. Can you see that the graph of $y = \cos x$ is the graph of $y = \sin x$ with a phase shift of $-\frac{\pi}{2}$? If you trace along the curve from $x = -\frac{\pi}{2}$ to $x = \frac{3\pi}{2}$, you are tracing one complete cycle of the sine curve. This can be expressed as an identity:

$$\cos x = \sin\left(x + \frac{\pi}{2}\right).$$

Because of this similarity, the graphs of sine functions and cosine functions are called **sinusoidal graphs**.

④ Graph variations of $y = \cos x$.

Graphing Variations of $y = \cos x$

We use the same steps to graph variations of $y = \cos x$ as we did for graphing variations of $y = \sin x$. We will continue finding key points by dividing the period into four equal parts. Amplitudes, periods, and phase shifts play an important role when graphing by hand.

GREAT QUESTION!

What should I do to graph functions of the form $y = A \cos Bx$ if B is negative?

If $B < 0$ in $y = A \cos Bx$, use $\cos(-\theta) = \cos \theta$ to rewrite the equation before obtaining its graph.

The Graph of $y = A \cos Bx$

The graph of $y = A \cos Bx$, $B > 0$, has

$$\text{amplitude} = |A|$$

$$\text{period} = \frac{2\pi}{B}.$$

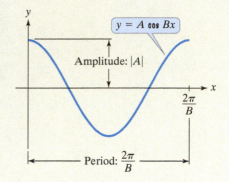

EXAMPLE 5   Graphing a Function of the Form $y = A \cos Bx$

Determine the amplitude and period of $y = -3 \cos \frac{\pi}{2} x$. Then graph the function for $-4 \le x \le 4$.

SOLUTION

Step 1 Identify the amplitude and the period. The equation $y = -3 \cos \frac{\pi}{2} x$ is of the form $y = A \cos Bx$ with $A = -3$ and $B = \frac{\pi}{2}$.

$$\text{amplitude:}\ \ |A| = |-3| = 3 \quad \boxed{\text{The maximum value of } y \text{ is 3 and the minimum is } -3.}$$

$$\text{period:}\ \ \frac{2\pi}{B} = \frac{2\pi}{\dfrac{\pi}{2}} = 2\pi \cdot \frac{2}{\pi} = 4 \quad \boxed{\text{Each cycle is of length 4.}}$$

Step 2 Find the values of x for the five key points. Begin by dividing the period, 4, by 4.

$$\frac{\text{period}}{4} = \frac{4}{4} = 1$$

Start with the value of x where the cycle begins: $x_1 = 0$. Adding quarter-periods, 1, the five x-values for the key points are

$$x_1 = 0, \quad x_2 = 0 + 1 = 1, \quad x_3 = 1 + 1 = 2, \quad x_4 = 2 + 1 = 3, \quad x_5 = 3 + 1 = 4.$$

Step 3 Find the values of y for the five key points. We evaluate the function at each value of x from step 2.

Value of x	Value of y: $y = -3 \cos \dfrac{\pi}{2}x$	Coordinates of key point	
0	$y = -3 \cos\left(\dfrac{\pi}{2} \cdot 0\right)$ $= -3 \cos 0 = -3 \cdot 1 = -3$	$(0, -3)$	minimum point
1	$y = -3 \cos\left(\dfrac{\pi}{2} \cdot 1\right)$ $= -3 \cos \dfrac{\pi}{2} = -3 \cdot 0 = 0$	$(1, 0)$	
2	$y = -3 \cos\left(\dfrac{\pi}{2} \cdot 2\right)$ $= -3 \cos \pi = -3(-1) = 3$	$(2, 3)$	maximum point
3	$y = -3 \cos\left(\dfrac{\pi}{2} \cdot 3\right)$ $= -3 \cos \dfrac{3\pi}{2} = -3 \cdot 0 = 0$	$(3, 0)$	
4	$y = -3 \cos\left(\dfrac{\pi}{2} \cdot 4\right)$ $= -3 \cos 2\pi = -3 \cdot 1 = -3$	$(4, -3)$	minimum point

In the interval $[0, 4]$, there are x-intercepts at 1 and 3. The minimum and maximum points are indicated by the voice balloons.

Step 4 Connect the five key points with a smooth curve and graph one complete cycle of the given function. The five key points for $y = -3 \cos \dfrac{\pi}{2}x$ are shown in **Figure 2.12**. By connecting the points with a smooth curve, the blue portion shows one complete cycle of $y = -3 \cos \dfrac{\pi}{2}x$ from 0 to 4.

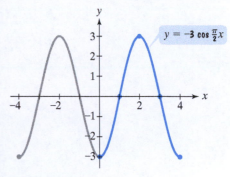

FIGURE 2.12

TECHNOLOGY

The graph of $y = -3 \cos \dfrac{\pi}{2}x$ in a $[-4, 4, 1]$ by $[-4, 4, 1]$ viewing rectangle verifies our hand-drawn graph in **Figure 2.12**.

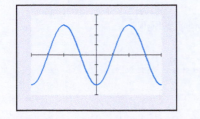

Step 5 Extend the graph in step 4 to the left or right as desired. The blue portion of the graph in **Figure 2.12** is for x from 0 to 4. In order to graph for $-4 \leq x \leq 4$, we continue this portion and extend the graph another full period to the left. This extension is shown in gray in **Figure 2.12**. •••

⊘ Check Point **5** Determine the amplitude and period of $y = -4 \cos \pi x$. Then graph the function for $-2 \leq x \leq 2$.

Finally, let us examine the graphs of functions of the form $y = A \cos(Bx - C)$. Graphs of these functions shift the graph of $y = A \cos Bx$ horizontally by $\dfrac{C}{B}$.

The Graph of $y = A \cos(Bx - C)$

The graph of $y = A \cos(Bx - C), B > 0$, is obtained by horizontally shifting the graph of $y = A \cos Bx$ so that the starting point of the cycle is shifted from $x = 0$ to $x = \dfrac{C}{B}$. If $\dfrac{C}{B} > 0$, the shift is to the right. If $\dfrac{C}{B} < 0$, the shift is to the left.

The number $\dfrac{C}{B}$ is called the **phase shift**.

$$\text{amplitude} = |A|$$

$$\text{period} \quad = \frac{2\pi}{B}$$

EXAMPLE 6   Graphing a Function of the Form $y = A \cos(Bx - C)$

Determine the amplitude, period, and phase shift of $y = \frac{1}{2} \cos(4x + \pi)$. Then graph one period of the function.

SOLUTION

Step 1 Identify the amplitude, the period, and the phase shift. We must first identify values for $A, B,$ and C. To do this, we need to express the equation in the form $y = A \cos(Bx - C)$. Thus, we write $y = \frac{1}{2} \cos(4x + \pi)$ as $y = \frac{1}{2} \cos[4x - (-\pi)]$. Now we can identify values for $A, B,$ and C.

The equation is of the form
$y = A \sin (Bx - C)$.

$$y = \frac{1}{2} \cos\big[4x - (-\pi)\big]$$

Using the voice balloon, we see that $A = \frac{1}{2}, B = 4,$ and $C = -\pi$.

amplitude: $|A| = \left|\dfrac{1}{2}\right| = \dfrac{1}{2}$ The maximum value of y is $\frac{1}{2}$ and the minimum is $-\frac{1}{2}$.

period: $\dfrac{2\pi}{B} = \dfrac{2\pi}{4} = \dfrac{\pi}{2}$ Each cycle is of length $\frac{\pi}{2}$.

phase shift: $\dfrac{C}{B} = -\dfrac{\pi}{4}$ A cycle starts at $x = -\frac{\pi}{4}$.

Step 2 Find the values of x for the five key points. Begin by dividing the period, $\dfrac{\pi}{2}$, by 4.

$$\frac{\text{period}}{4} = \frac{\dfrac{\pi}{2}}{4} = \frac{\pi}{8}$$

Start with the value of x where the cycle begins: $x_1 = -\dfrac{\pi}{4}$. Adding quarter-periods, $\dfrac{\pi}{8}$, the five x-values for the key points are

$$x_1 = -\frac{\pi}{4}, \quad x_2 = -\frac{\pi}{4} + \frac{\pi}{8} = -\frac{2\pi}{8} + \frac{\pi}{8} = -\frac{\pi}{8}, \quad x_3 = -\frac{\pi}{8} + \frac{\pi}{8} = 0,$$

$$x_4 = 0 + \frac{\pi}{8} = \frac{\pi}{8}, \quad x_5 = \frac{\pi}{8} + \frac{\pi}{8} = \frac{2\pi}{8} = \frac{\pi}{4}.$$

Step 3 Find the values of y for the five key points. Take a few minutes and use your calculator to evaluate the function at each value of x from step 2. Show that the key points are

$$\left(-\frac{\pi}{4}, \frac{1}{2}\right), \left(-\frac{\pi}{8}, 0\right), \left(0, -\frac{1}{2}\right), \left(\frac{\pi}{8}, 0\right), \text{and} \left(\frac{\pi}{4}, \frac{1}{2}\right).$$

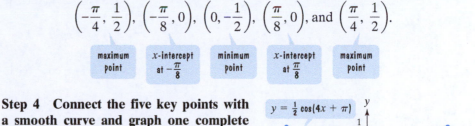

| maximum point | x-intercept at $-\frac{\pi}{8}$ | minimum point | x-intercept at $\frac{\pi}{8}$ | maximum point |

Step 4 Connect the five key points with a smooth curve and graph one complete cycle of the given function. The key points and the graph of $y = \frac{1}{2}\cos(4x + \pi)$ are shown in **Figure 2.13**.

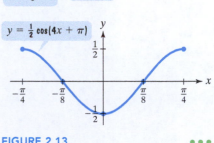

FIGURE 2.13 • • •

TECHNOLOGY

The graph of
$$y = \frac{1}{2}\cos(4x + \pi)$$
in a $\left[-\dfrac{\pi}{4}, \dfrac{\pi}{4}, \dfrac{\pi}{8}\right]$ by $[-1, 1, 1]$ viewing rectangle verifies our hand-drawn graph in **Figure 2.13**.

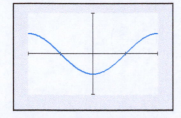

✓ **Check Point 6** Determine the amplitude, period, and phase shift of $y = \frac{3}{2}\cos(2x + \pi)$. Then graph one period of the function.

⑤ Use vertical shifts of sine and cosine curves.

Vertical Shifts of Sinusoidal Graphs

We now look at sinusoidal graphs of functions of the form

$$y = A\sin(Bx - C) + D \quad \text{and} \quad y = A\cos(Bx - C) + D.$$

The constant D causes a vertical shift in each of the graphs of $y = A\sin(Bx - C)$ and $y = A\cos(Bx - C)$. If D is positive, the shift is D units upward. If D is negative, the shift is $|D|$ units downward. These vertical shifts result in sinusoidal graphs oscillating about the horizontal line $y = D$ rather than about the x-axis. Thus, the maximum value of y is $D + |A|$ and the minimum value of y is $D - |A|$.

EXAMPLE 7 A Vertical Shift

Graph one period of the function $y = \frac{1}{2}\cos x - 1$.

SOLUTION

The graph of $y = \frac{1}{2}\cos x - 1$ is the graph of $y = \frac{1}{2}\cos x$ shifted one unit downward. The period of $y = \frac{1}{2}\cos x$ is 2π, which is also the period for the vertically shifted graph. The key points on the interval $[0, 2\pi]$ for $y = \frac{1}{2}\cos x - 1$ are found by first determining their x-coordinates. The quarter-period is $\dfrac{2\pi}{4}$, or $\dfrac{\pi}{2}$.

The cycle begins at $x = 0$. As always, we add quarter-periods to generate x-values for each of the key points. The five x-values are

$$x_1 = 0, \quad x_2 = 0 + \frac{\pi}{2} = \frac{\pi}{2}, \quad x_3 = \frac{\pi}{2} + \frac{\pi}{2} = \pi,$$

$$x_4 = \pi + \frac{\pi}{2} = \frac{3\pi}{2}, \quad x_5 = \frac{3\pi}{2} + \frac{\pi}{2} = 2\pi.$$

The values of y for the five key points and their coordinates are determined as follows.

Value of x	Value of y: $y = \dfrac{1}{2}\cos x - 1$	Coordinates of key point
0	$y = \dfrac{1}{2}\cos 0 - 1$ $= \dfrac{1}{2} \cdot 1 - 1 = -\dfrac{1}{2}$	$\left(0, -\dfrac{1}{2}\right)$
$\dfrac{\pi}{2}$	$y = \dfrac{1}{2}\cos\dfrac{\pi}{2} - 1$ $= \dfrac{1}{2} \cdot 0 - 1 = -1$	$\left(\dfrac{\pi}{2}, -1\right)$
π	$y = \dfrac{1}{2}\cos \pi - 1$ $= \dfrac{1}{2}(-1) - 1 = -\dfrac{3}{2}$	$\left(\pi, -\dfrac{3}{2}\right)$
$\dfrac{3\pi}{2}$	$y = \dfrac{1}{2}\cos\dfrac{3\pi}{2} - 1$ $= \dfrac{1}{2} \cdot 0 - 1 = -1$	$\left(\dfrac{3\pi}{2}, -1\right)$
2π	$y = \dfrac{1}{2}\cos 2\pi - 1$ $= \dfrac{1}{2} \cdot 1 - 1 = -\dfrac{1}{2}$	$\left(2\pi, -\dfrac{1}{2}\right)$

The five key points for $y = \frac{1}{2}\cos x - 1$ are shown in **Figure 2.14**. By connecting the points with a smooth curve, we obtain one period of the graph.

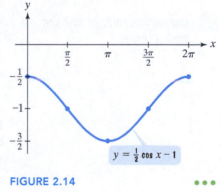

FIGURE 2.14   • • •

✅ **Check Point 7** Graph one period of the function $y = 2\cos x + 1$.

---

⑥ Graph the sum of two trigonometric functions.

Graphing the Sum of Two Trigonometric Functions

We can graph the sum of two (or more) trigonometric functions using a method that involves adding y-coordinates. For example, to graph $y = \sin x + \cos x$, we begin by graphing $y_1 = \sin x$ and $y_2 = \cos x$ in the same rectangular coordinate system. Then we select several values of x at which we add the y-coordinates of $\sin x$ and $\cos x$. Plotting the points $(x, \sin x + \cos x)$ and joining them with a smooth curve gives the graph of the desired function. This method is illustrated in Example 8.

EXAMPLE 8   Graphing the Sum of Two Trigonometric Functions

Use the method of adding y-coordinates to graph $y = \sin x + \cos x$ for $0 \le x \le 2\pi$.

SOLUTION

We begin by graphing $y_1 = \sin x$ and $y_2 = \cos x$, $0 \le x \le 2\pi$, in the same rectangular coordinate system, shown in **Figure 2.15**.

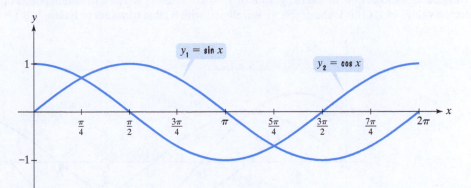

FIGURE 2.15 The graphs of $y_1 = \sin x$ and $y_2 = \cos x$, $0 \le x \le 2\pi$

Now we select several values of x at which we determine $y = \sin x + \cos x$ by adding the y-coordinates of $\sin x$ and $\cos x$.

x	0	$\dfrac{\pi}{4}$	$\dfrac{\pi}{2}$	$\dfrac{3\pi}{4}$	π	$\dfrac{5\pi}{4}$	$\dfrac{3\pi}{2}$	$\dfrac{7\pi}{4}$	2π
$y_1 = \sin x$	0	$0.7\,*$	1	$0.7\,*$	0	$-0.7\,*$	-1	$-0.7\,*$	0
$y_2 = \cos x$	1	$0.7\,*$	0	$-0.7\,*$	-1	$-0.7\,*$	0	$0.7\,*$	1
$y = \sin x + \cos x$	$0 + 1$ $= 1$	$0.7 + 0.7$ $= 1.4$	$1 + 0$ $= 1$	$0.7 + (-0.7)$ $= 0$	$0 + (-1)$ $= -1$	$-0.7 + (-0.7)$ $= -1.4$	$-1 + 0$ $= -1$	$-0.7 + 0.7$ $= 0$	$0 + 1$ $= 1$
Points on the graph of $y = \sin x + \cos x$	$(0, 1)$	$\left(\dfrac{\pi}{4}, 1.4\right)$	$\left(\dfrac{\pi}{2}, 1\right)$	$\left(\dfrac{3\pi}{4}, 0\right)$	$(\pi, -1)$	$\left(\dfrac{5\pi}{4}, -1.4\right)$	$\left(\dfrac{3\pi}{2}, -1\right)$	$\left(\dfrac{7\pi}{4}, 0\right)$	$(2\pi, 1)$

*These values of y are approximate. Exact values are $\dfrac{\sqrt{2}}{2} \approx 0.7$ and $-\dfrac{\sqrt{2}}{2} \approx -0.7$.

We plot the nine points in the table and connect them with a smooth curve. This results in the graph of $y = \sin x + \cos x$, $0 \le x \le 2\pi$, shown in red in **Figure 2.16**. Also shown in the coordinate system are the graphs of $y_1 = \sin x$ and $y_2 = \cos x$.

DISCOVERY

It can be shown that

$$\sin x + \cos x = \sqrt{2}\sin\left(x + \frac{\pi}{4}\right).$$

Based on this identity, determine the amplitude, period, and phase shift of $y = \sin x + \cos x$. Are these values consistent with the graph of $y = \sin x + \cos x$ shown in **Figure 2.16**?

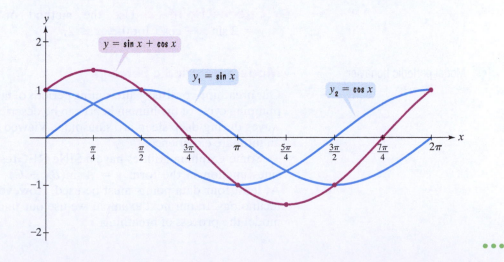

FIGURE 2.16

• • •

GREAT QUESTION!

When graphing $y = y_1 + y_2$, I begin by graphing y_1 and y_2 in the same rectangular coordinate system. Is there a way to use the graphs of y_1 and y_2 to obtain the graph of y without making a table of values?

Yes. Select several values of x and observe the graphs of y_1 and y_2 at each of these selected values.

For each choice of x, graphically add the value of y from the graph of y_2 to the corresponding value of y from the graph of y_1. If $y_2 > 0$, then $y_1 + y_2$ will be above y_1 by a distance equal to y_2. If $y_2 < 0$, then $y_1 + y_2$ will be below y_1 by a distance equal to $|y_2|$. If $y_2 = 0$, then $y_1 + y_2$ will lie on the graph of y_1.

Figure 2.17 shows how to use the graphs of $y_1 = \sin x$ and $y_2 = \cos x$ to obtain points on the graph of $y = \sin x + \cos x$ for selected values of x. (This is the graph we developed with a table of values in Example 8.)

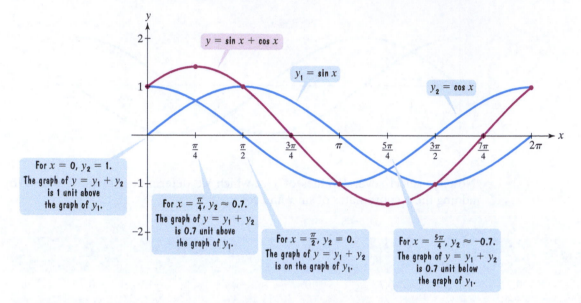

FIGURE 2.17 Using the graphs of y_1 and y_2 to obtain the graph of $y = y_1 + y_2$

TECHNOLOGY

The graphs of $y_1 = \sin x$, $y_2 = \cos x$, both in blue, and $y_3 = y_1 + y_2$, in red, in a $\left[0, 2\pi, \dfrac{\pi}{4}\right]$ by $[-2, 2, 1]$ viewing rectangle, reinforce our hand-drawn graphs in **Figure 2.17**.

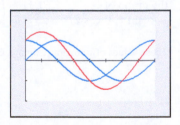

✔ Check Point **8** Use the method of adding y-coordinates to graph $y = 2 \sin x + \cos x$ for $0 \le x \le 2\pi$.

7 Model periodic behavior.

Modeling Periodic Behavior

Our breathing consists of alternating periods of inhaling and exhaling. Each complete pumping cycle of the human heart can be described using a sine function. Our brain waves during deep sleep are sinusoidal. Viewed in this way, trigonometry becomes an intimate experience.

Some graphing utilities have a SINe REGression feature. This feature gives the sine function in the form $y = A \sin(Bx + C) + D$ of best fit for wavelike data. At least four data points must be used. However, it is not always necessary to use technology. In our next example, we use our understanding of sinusoidal graphs to model the process of breathing.

EXAMPLE 9 A Trigonometric Breath of Life

The graph in **Figure 2.18** shows one complete normal breathing cycle. The cycle consists of inhaling and exhaling. It takes place every 5 seconds. Velocity of air flow is positive when we inhale and negative when we exhale. It is measured in liters per second. If y represents velocity of air flow after x seconds, find a function of the form $y = A \sin Bx$ that models air flow in a normal breathing cycle.

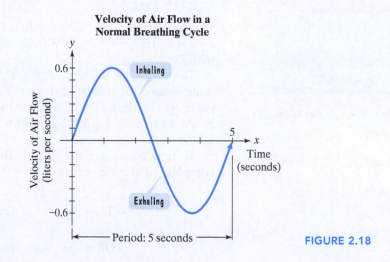

FIGURE 2.18

SOLUTION

We need to determine values for A and B in the equation $y = A \sin Bx$. The amplitude, A, is the maximum value of y. **Figure 2.18** shows that this maximum value is 0.6. Thus, $A = 0.6$.

The value of B in $y = A \sin Bx$ can be found using the formula for the period: period $= \dfrac{2\pi}{B}$. The period of our breathing cycle is 5 seconds. Thus,

$$5 = \frac{2\pi}{B} \qquad \text{\color{blue}Our goal is to solve this equation for } B.$$

$$5B = 2\pi \qquad \text{\color{blue}Multiply both sides of the equation by } B.$$

$$B = \frac{2\pi}{5}. \qquad \text{\color{blue}Divide both sides of the equation by 5.}$$

We see that $A = 0.6$ and $B = \dfrac{2\pi}{5}$. Substitute these values into $y = A \sin Bx$. The breathing cycle is modeled by

$$y = 0.6 \sin \frac{2\pi}{5}x. \qquad \bullet\bullet\bullet$$

✓ **Check Point 9** Find an equation of the form $y = A \sin Bx$ that produces the graph shown in the figure on the right.

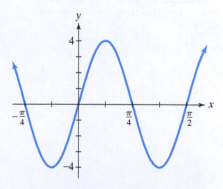

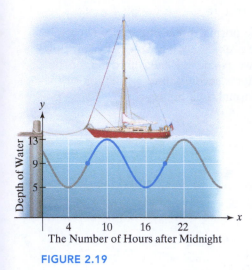

Depth of Water

FIGURE 2.19

EXAMPLE 10 Modeling a Tidal Cycle

Figure 2.19 shows that the depth of water at a boat dock varies with the tides. The depth is 5 feet at low tide and 13 feet at high tide. On a certain day, low tide occurs at 4 A.M. and high tide at 10 A.M. If y represents the depth of the water, in feet, x hours after midnight, use a sine function of the form $y = A \sin(Bx - C) + D$ to model the water's depth.

SOLUTION

We need to determine values for $A, B, C,$ and D in the equation $y = A \sin(Bx - C) + D$. We can find these values using **Figure 2.19**. We begin with D.

To find D, we use the vertical shift. Because the water's depth ranges from a minimum of 5 feet to a maximum of 13 feet, the curve oscillates about the middle value, 9 feet. Thus, $D = 9$, which is the vertical shift.

At maximum depth, the water is 4 feet above 9 feet. Thus, A, the amplitude, is 4: $A = 4$.

To find B, we use the period. The blue portion of the graph shows that one complete tidal cycle occurs in $19 - 7$, or 12 hours. The period is 12. Thus,

$$12 = \frac{2\pi}{B} \qquad \text{Our goal is to solve this equation for } B.$$

$$12B = 2\pi \qquad \text{Multiply both sides by } B.$$

$$B = \frac{2\pi}{12} = \frac{\pi}{6}. \qquad \text{Divide both sides by 12.}$$

TECHNOLOGY

Graphic Connections

We can use a graphing utility to verify that the model in Example 10,

$$y = 4 \sin\left(\frac{\pi}{6}x - \frac{7\pi}{6}\right) + 9,$$

is correct. The graph of the function is shown in a $[0, 28, 4]$ by $[0, 15, 5]$ viewing rectangle.

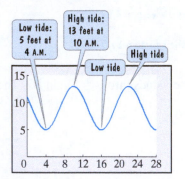

To find C, we use the phase shift. The blue portion of the graph shows that the starting point of the cycle is shifted from 0 to 7. The phase shift, $\dfrac{C}{B}$, is 7.

$$7 = \frac{C}{B} \qquad \text{The phase shift of } y = A \sin(Bx - C) \text{ is } \frac{C}{B}.$$

$$7 = \frac{C}{\frac{\pi}{6}} \qquad \text{From above, we have } B = \frac{\pi}{6}.$$

$$\frac{7\pi}{6} = C \qquad \text{Multiply both sides of the equation by } \frac{\pi}{6}.$$

We see that $A = 4$, $B = \dfrac{\pi}{6}$, $C = \dfrac{7\pi}{6}$, and $D = 9$. Substitute these values into $y = A \sin(Bx - C) + D$. The water's depth, in feet, x hours after midnight is modeled by

$$y = 4 \sin\left(\frac{\pi}{6}x - \frac{7\pi}{6}\right) + 9. \qquad \bullet\bullet\bullet$$

⟡ Check Point **10** A region that is 30° north of the Equator averages a minimum of 10 hours of daylight in December. Hours of daylight are at a maximum of 14 hours in June. Let x represent the month of the year, with 1 for January, 2 for February, 3 for March, and 12 for December. If y represents the number of hours of daylight in month x, use a sine function of the form $y = A \sin(Bx - C) + D$ to model the hours of daylight.

ACHIEVING SUCCESS

Check out the *Learning Guide* that accompanies this textbook.

Benefits of using the *Learning Guide* include:

- It will help you become better organized. This includes organizing your class notes, assigned homework, quizzes, and tests.
- It will enable you to use your textbook more efficiently.
- It will help increase your study skills.
- It will help you prepare for the chapter tests.

Ask your professor about the availability of this textbook supplement.

CONCEPT AND VOCABULARY CHECK

Fill in each blank so that the resulting statement is true.

1. The graph of $y = A \sin Bx$ has amplitude = _____ and period = _____.

2. The amplitude of $y = 3 \sin \frac{1}{2}x$ is _____ and the period is _____.

3. The period of $y = 4 \sin 2x$ is _____, so the x-values for the five key points are $x_1 =$ _____, $x_2 =$ _____, $x_3 =$ _____, $x_4 =$ _____, and $x_5 =$ _____.

4. The graph of $y = A \sin (Bx - C)$ has phase shift _____. If this phase shift is positive, the graph of $y = A \sin Bx$ is shifted to the _____. If this phase shift is negative, the graph of $y = A \sin Bx$ is shifted to the _____.

5. The graph of $y = A \cos Bx$ has amplitude = _____ and period = _____.

6. The amplitude of $y = \frac{1}{2} \cos 3x$ is _____ and the period is _____.

7. True or false: The graph of $y = \cos\left(x + \frac{\pi}{4}\right)$ lies $\frac{\pi}{4}$ units to the right of the graph of $y = \cos x$. _____

8. True or false: The graph of $y = \cos\left(2x - \frac{\pi}{2}\right)$ has phase shift $\frac{\pi}{4}$. _____

9. True or false: The maximum value of the function $y = -2 \cos x + 5$ is 7. _____

10. True or false: The minimum value of the function $y = 2 \sin x + 1$ is -1. _____

EXERCISE SET 2.1

Practice Exercises

In Exercises 1–6, determine the amplitude of each function. Then graph the function and $y = \sin x$ in the same rectangular coordinate system for $0 \le x \le 2\pi$.

1. $y = 4 \sin x$
2. $y = 5 \sin x$
3. $y = \frac{1}{3} \sin x$
4. $y = \frac{1}{4} \sin x$
5. $y = -3 \sin x$
6. $y = -4 \sin x$

In Exercises 7–16, determine the amplitude and period of each function. Then graph one period of the function.

7. $y = \sin 2x$
8. $y = \sin 4x$
9. $y = 3 \sin \frac{1}{2}x$
10. $y = 2 \sin \frac{1}{4}x$
11. $y = 4 \sin \pi x$
12. $y = 3 \sin 2\pi x$
13. $y = -3 \sin 2\pi x$
14. $y = -2 \sin \pi x$
15. $y = -\sin \frac{2}{3}x$
16. $y = -\sin \frac{4}{3}x$

In Exercises 17–30, determine the amplitude, period, and phase shift of each function. Then graph one period of the function.

17. $y = \sin(x - \pi)$
18. $y = \sin\left(x - \frac{\pi}{2}\right)$
19. $y = \sin(2x - \pi)$
20. $y = \sin\left(2x - \frac{\pi}{2}\right)$
21. $y = 3 \sin(2x - \pi)$
22. $y = 3 \sin\left(2x - \frac{\pi}{2}\right)$
23. $y = \frac{1}{2} \sin\left(x + \frac{\pi}{2}\right)$
24. $y = \frac{1}{2} \sin(x + \pi)$
25. $y = -2 \sin\left(2x + \frac{\pi}{2}\right)$
26. $y = -3 \sin\left(2x + \frac{\pi}{2}\right)$
27. $y = 3 \sin(\pi x + 2)$
28. $y = 3 \sin(2\pi x + 4)$
29. $y = -2 \sin(2\pi x + 4\pi)$
30. $y = -3 \sin(2\pi x + 4\pi)$

In Exercises 31–34, determine the amplitude of each function. Then graph the function and $y = \cos x$ in the same rectangular coordinate system for $0 \le x \le 2\pi$.

31. $y = 2 \cos x$ **32.** $y = 3 \cos x$

33. $y = -2 \cos x$ **34.** $y = -3 \cos x$

In Exercises 35–42, determine the amplitude and period of each function. Then graph one period of the function.

35. $y = \cos 2x$ **36.** $y = \cos 4x$

37. $y = 4 \cos 2\pi x$ **38.** $y = 5 \cos 2\pi x$

39. $y = -4 \cos \frac{1}{2} x$ **40.** $y = -3 \cos \frac{1}{3} x$

41. $y = -\frac{1}{2} \cos \frac{\pi}{3} x$ **42.** $y = -\frac{1}{2} \cos \frac{\pi}{4} x$

In Exercises 43–52, determine the amplitude, period, and phase shift of each function. Then graph one period of the function.

43. $y = \cos\left(x - \frac{\pi}{2}\right)$ **44.** $y = \cos\left(x + \frac{\pi}{2}\right)$

45. $y = 3 \cos(2x - \pi)$ **46.** $y = 4 \cos(2x - \pi)$

47. $y = \frac{1}{2} \cos\left(3x + \frac{\pi}{2}\right)$ **48.** $y = \frac{1}{2} \cos(2x + \pi)$

49. $y = -3 \cos\left(2x - \frac{\pi}{2}\right)$ **50.** $y = -4 \cos\left(2x - \frac{\pi}{2}\right)$

51. $y = 2 \cos(2\pi x + 8\pi)$ **52.** $y = 3 \cos(2\pi x + 4\pi)$

In Exercises 53–60, use a vertical shift to graph one period of the function.

53. $y = \sin x + 2$ **54.** $y = \sin x - 2$

55. $y = \cos x - 3$ **56.** $y = \cos x + 3$

57. $y = 2 \sin \frac{1}{2} x + 1$ **58.** $y = 2 \cos \frac{1}{2} x + 1$

59. $y = -3 \cos 2\pi x + 2$ **60.** $y = -3 \sin 2\pi x + 2$

In Exercises 61–66, use the method of adding y-coordinates to graph each function for $0 \le x \le 2\pi$.

61. $y = 2 \cos x + \sin x$ **62.** $y = 3 \cos x + \sin x$

63. $y = \sin x + \sin 2x$ **64.** $y = \cos x + \cos 2x$

65. $y = \sin x + \cos 2x$ **66.** $y = \cos x + \sin 2x$

In Exercises 67–68, use the method of adding y-coordinates to graph each function for $0 \le x \le 4$.

67. $y = \sin \pi x + \cos \frac{\pi}{2} x$ **68.** $y = \cos \pi x + \sin \frac{\pi}{2} x$

Practice Plus

In Exercises 69–74, find an equation for each graph.

69.

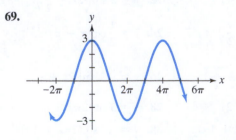

70.

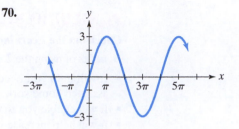

71.

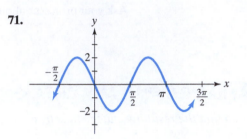

72.

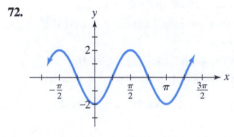

73.

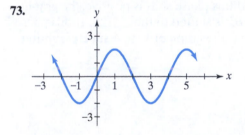

74.

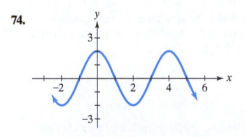

In Exercises 75–78, graph one period of each function.

75. $y = \left|2 \cos \frac{x}{2}\right|$ **76.** $y = \left|3 \cos \frac{2x}{3}\right|$

77. $y = -|3 \sin \pi x|$ **78.** $y = -\left|2 \sin \frac{\pi x}{2}\right|$

In Exercises 79–82, graph f, g, and h in the same rectangular coordinate system for $0 \leq x \leq 2\pi$. Obtain the graph of h by adding or subtracting the corresponding y-coordinates on the graphs of f and g.

79. $f(x) = -2 \sin x, g(x) = \sin 2x, h(x) = (f + g)(x)$

80. $f(x) = 2 \cos x, g(x) = \cos 2x, h(x) = (f + g)(x)$

81. $f(x) = \sin x, g(x) = \cos 2x, h(x) = (f - g)(x)$

82. $f(x) = \cos x, g(x) = \sin 2x, h(x) = (f - g)(x)$

Application Exercises

In the theory of biorhythms, sine functions are used to measure a person's potential. You can obtain your biorhythm chart online by simply entering your date of birth, the date you want your biorhythm chart to begin, and the number of months you wish to have included in the plot. Shown below is your author's chart, beginning January 25, 2015, when he was 25,473 days old. We all have cycles with the same amplitudes and periods as those shown here. Each of our three basic cycles begins at birth. Use the biorhythm chart shown to solve Exercises 83–90. The longer tick marks correspond to the dates shown.

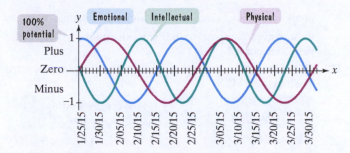

83. What is the period of the physical cycle?

84. What is the period of the emotional cycle?

85. What is the period of the intellectual cycle?

86. For the period shown, what is the worst day in February for your author to run in a marathon?

87. For the period shown, what is the best day in March for your author to meet an online friend for the first time?

88. For the period shown, what is the best day in February for your author to begin writing this trigonometry chapter?

89. If you extend these sinusoidal graphs to the end of the year, is there a day when your author should not even bother getting out of bed?

90. If you extend these sinusoidal graphs to the end of the year, are there any days where your author is at near-peak physical, emotional, and intellectual potential?

91. Rounded to the nearest hour, Los Angeles averages 14 hours of daylight in June, 10 hours in December, and 12 hours in March and September. Let x represent the number of months after June and let y represent the number of hours of daylight in month x. Make a graph that displays the information from June of one year to June of the following year.

92. A clock with an hour hand that is 15 inches long is hanging on a wall. At noon, the distance between the tip of the hour hand and the ceiling is 23 inches. At 3 P.M., the distance is 38 inches; at 6 P.M., 53 inches; at 9 P.M., 38 inches; and at midnight the distance is again 23 inches. If y represents the distance between the tip of the hour hand and the ceiling x hours after noon, make a graph that displays the information for $0 \leq x \leq 24$.

93. The number of hours of daylight in Boston is given by

$$y = 3 \sin \frac{2\pi}{365}(x - 79) + 12,$$

where x is the number of days after January 1.

a. What is the amplitude of this function?

b. What is the period of this function?

c. How many hours of daylight are there on the longest day of the year?

d. How many hours of daylight are there on the shortest day of the year?

e. Graph the function for one period, starting on January 1.

94. The average monthly temperature, y, in degrees Fahrenheit, for Juneau, Alaska, can be modeled by $y = 16 \sin\left(\frac{\pi}{6}x - \frac{2\pi}{3}\right) + 40$, where x is the month of the year (January = 1, February = 2, ... December = 12). Graph the function for $1 \leq x \leq 12$. What is the highest average monthly temperature? In which month does this occur?

95. The following figure shows the depth of water at the end of a boat dock. The depth is 6 feet at low tide and 12 feet at high tide. On a certain day, low tide occurs at 6 A.M. and high tide at noon. If y represents the depth of the water x hours after midnight, use a cosine function of the form $y = A \cos Bx + D$ to model the water's depth.

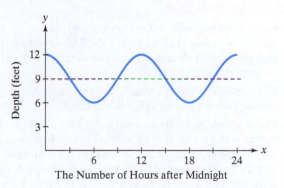

The Number of Hours after Midnight

96. The following figure shows the depth of water at the end of a boat dock. The depth is 5 feet at high tide and 3 feet at low tide. On a certain day, high tide occurs at noon and low tide at 6 P.M. If y represents the depth of the water x hours after noon, use a cosine function of the form $y = A \cos Bx + D$ to model the water's depth.

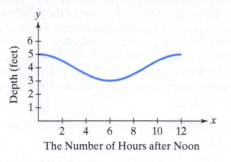

The Number of Hours after Noon

Explaining the Concepts

97. Without drawing a graph, describe the behavior of the basic sine curve.

98. What is the amplitude of the sine function? What does this tell you about the graph?

99. If you are given the equation of a sine function, how do you determine the period?

100. What does a phase shift indicate about the graph of a sine function? How do you determine the phase shift from the function's equation?

101. Describe a general procedure for obtaining the graph of $y = A \sin(Bx - C)$.

102. Without drawing a graph, describe the behavior of the basic cosine curve.

103. Describe a relationship between the graphs of $y = \sin x$ and $y = \cos x$.

104. Describe the relationship between the graphs of $y = A \cos(Bx - C)$ and $y = A \cos(Bx - C) + D$.

105. Describe how to graph the sum of two functions. Use $y = \sin x + \cos x$ in your description.

106. Biorhythm cycles provide interesting applications of sinusoidal graphs. But do you believe in the validity of biorhythms? Write a few sentences explaining why or why not.

Technology Exercises

107. Use a graphing utility to verify any five of the sine curves that you drew by hand in Exercises 7–30. The amplitude, period, and phase shift should help you to determine appropriate viewing rectangle settings.

108. Use a graphing utility to verify any five of the cosine curves that you drew by hand in Exercises 35–52.

109. Use a graphing utility to verify any two of the sinusoidal curves with vertical shifts that you drew in Exercises 53–60.

110. Use a graphing utility to verify any two of your hand-drawn graphs in Exercises 61–68. For each selected exercise, show the graphs of y_1, y_2, and $y_1 + y_2$ in the same viewing rectangle.

In Exercises 111–114, use a graphing utility to graph two periods of the function.

111. $y = 3 \sin(2x + \pi)$

112. $y = -2 \cos\left(2\pi x - \dfrac{\pi}{2}\right)$

113. $y = 0.2 \sin\left(\dfrac{\pi}{10}x + \pi\right)$

114. $y = 3 \sin(2x - \pi) + 5$

115. Use a graphing utility to graph $y = \sin x$ and $y = x - \dfrac{x^3}{6} + \dfrac{x^5}{120}$ in a $\left[-\pi, \pi, \dfrac{\pi}{2}\right]$ by $[-2, 2, 1]$ viewing rectangle. How do the graphs compare?

116. Use a graphing utility to graph $y = \cos x$ and $y = 1 - \dfrac{x^2}{2} + \dfrac{x^4}{24}$ in a $\left[-\pi, \pi, \dfrac{\pi}{2}\right]$ by $[-2, 2, 1]$ viewing rectangle. How do the graphs compare?

117. Use a graphing utility to graph

$$y = \sin x + \frac{\sin 2x}{2} + \frac{\sin 3x}{3} + \frac{\sin 4x}{4}$$

in a $\left[-2\pi, 2\pi, \dfrac{\pi}{2}\right]$ by $[-2, 2, 1]$ viewing rectangle. How do these waves compare to the smooth rolling waves of the basic sine curve?

118. Use a graphing utility to graph

$$y = \sin x - \frac{\sin 3x}{9} + \frac{\sin 5x}{25}$$

in a $\left[-2\pi, 2\pi, \dfrac{\pi}{2}\right]$ by $[-2, 2, 1]$ viewing rectangle. How do these waves compare to the smooth rolling waves of the basic sine curve?

119. The data show the average monthly temperatures for Washington, D.C.

x (Month)	Average Monthly Temperature, °F
1 (January)	34.6
2 (February)	37.5
3 (March)	47.2
4 (April)	56.5
5 (May)	66.4
6 (June)	75.6
7 (July)	80.0
8 (August)	78.5
9 (September)	71.3
10 (October)	59.7
11 (November)	49.8
12 (December)	39.4

Source: U.S. National Oceanic and Atmospheric Administration

a. Use your graphing utility to draw a scatter plot of the data from $x = 1$ through $x = 12$.

b. Use the SINe REGression feature to find the sinusoidal function of the form $y = A \sin(Bx + C) + D$ that best fits the data.

c. Use your graphing utility to draw the sinusoidal function of best fit on the scatter plot.

120. Repeat Exercise 119 for data of your choice. The data can involve the average monthly temperatures for the region where you live or any data whose scatter plot takes the form of a sinusoidal function.

Critical Thinking Exercises

Make Sense?   *In Exercises 121–124, determine whether each statement makes sense or does not make sense, and explain your reasoning.*

121. When graphing one complete cycle of $y = A \sin(Bx - C)$, I find it easiest to begin my graph on the x-axis.

122. When graphing one complete cycle of $y = A \cos(Bx - C)$, I find it easiest to begin my graph on the x-axis.

123. Using the equation $y = A \sin Bx$, if I replace either A or B with its opposite, the graph of the resulting equation is a reflection of the graph of the original equation about the x-axis.

124. A ride on a circular Ferris wheel is like riding sinusoidal graphs.

125. Determine the range of each of the following functions. Then give a viewing rectangle, or window, that shows two periods of the function's graph.

    a. $f(x) = 3 \sin\left(x + \dfrac{\pi}{6}\right) - 2$

    b. $g(x) = \sin 3\left(x + \dfrac{\pi}{6}\right) - 2$

126. Write the equation for a cosine function with amplitude π, period 1, and phase shift -2.

In Chapter 3, we will prove the following identities:

$$\sin^2 x = \frac{1}{2} - \frac{1}{2}\cos 2x$$

$$\cos^2 x = \frac{1}{2} + \frac{1}{2}\cos 2x.$$

Use these identities to solve Exercises 127–128.

127. Use the identity for $\sin^2 x$ to graph one period of $y = \sin^2 x$.

128. Use the identity for $\cos^2 x$ to graph one period of $y = \cos^2 x$.

Group Exercise

129. This exercise is intended to provide some fun with biorhythms, regardless of whether you believe they have any validity. We will use each member's chart to determine biorhythmic compatibility. Before meeting, each group member should go online and obtain his or her biorhythm chart. The date of the group meeting is the date on which your chart should begin. Include 12 months in the plot. At the meeting, compare differences and similarities among the intellectual sinusoidal curves. Using these comparisons, each person should find the one other person with whom he or she would be most intellectually compatible.

Retaining the Concepts

ACHIEVING SUCCESS

According to the Ebbinghaus retention model, you forget 50% of what you learn within one hour. You lose 60% within 24 hours. After 30 days, 70% is gone. Reviewing previously covered topics is an effective way to counteract this phenomenon. From here on, each Exercise Set will contain three review exercises. It is essential to review previously covered topics to improve your understanding of the topics and to help maintain your mastery of the material. If you are not certain how to solve a review exercise, turn to the section and the worked example given in parentheses at the end of each exercise. The more you review the material, the more you retain. Answers to all Retaining the Concepts Exercises are given in the answer section.

130. Use an identity and not a calculator to find the value of each expression.

    a. $\cos 47° \sec 47°$

    b. $\sin^2 \dfrac{\pi}{5} + \cos^2 \dfrac{\pi}{5}$

    (Section 1.2, Examples 5 and 6)

131. Find the value of each of the six trigonometric functions of θ in the right triangle. Where necessary, rationalize denominators.

    (Section 1.2, Example 2)

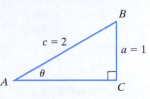

132. Use a reference angle to find the exact value of $\tan \dfrac{4\pi}{3}$. (Section 1.3, Example 7)

Preview Exercises

Exercises 133–135 will help you prepare for the material covered in the next section.

133. Solve: $-\dfrac{\pi}{2} < x + \dfrac{\pi}{4} < \dfrac{\pi}{2}$.

134. Simplify: $\dfrac{-\dfrac{3\pi}{4} + \dfrac{\pi}{4}}{2}$.

135. a. Graph $y = -3 \cos \dfrac{x}{2}$ for $-\pi \le x \le 5\pi$.

    b. Consider the reciprocal function of $y = -3 \cos \frac{x}{2}$, namely, $y = -3 \sec \frac{x}{2}$. What does your graph from part (a) indicate about this reciprocal function for $x = -\pi, \pi, 3\pi$, and 5π?

Graphs of Other Trigonometric Functions

What am I supposed to learn?

After studying this section, you should be able to:

1. Understand the graph of $y = \tan x$.
2. Graph variations of $y = \tan x$.
3. Understand the graph of $y = \cot x$.
4. Graph variations of $y = \cot x$.
5. Understand the graphs of $y = \csc x$ and $y = \sec x$.
6. Graph variations of $y = \csc x$ and $y = \sec x$.

The debate over whether Earth is warming up is over: Humankind's reliance on fossil fuels—coal, fuel oil, and natural gas—is to blame for global warming. In this section's Exercise Set (Exercise 87), you will see how trigonometric graphs reveal interesting patterns in carbon dioxide concentration from 1990 through 2008. In this section, trigonometric graphs will reveal patterns involving the tangent, cotangent, secant, and cosecant functions.

The Graph of $y = \tan x$

1. Understand the graph of $y = \tan x$.

The properties of the tangent function discussed in Section 1.4 will help us determine its graph. Because the tangent function has properties that are different from sinusoidal functions, its graph differs significantly from those of sine and cosine. Properties of the tangent function include the following:

- The period is π. It is only necessary to graph $y = \tan x$ over an interval of length π. The remainder of the graph consists of repetitions of that graph at intervals of π.
- The tangent function is an odd function: $\tan(-x) = -\tan x$. The graph is symmetric with respect to the origin.
- The tangent function is undefined at $\frac{\pi}{2}$. The graph of $y = \tan x$ has a vertical asymptote at $x = \frac{\pi}{2}$.

We obtain the graph of $y = \tan x$ using some points on the graph and origin symmetry. **Table 2.3** lists some values of (x, y) on the graph of $y = \tan x$ on the interval $\left[0, \frac{\pi}{2}\right)$.

Table 2.3 Values of (x, y) on the Graph of $y = \tan x$

x	0	$\frac{\pi}{6}$	$\frac{\pi}{4}$	$\frac{\pi}{3}$	$\frac{5\pi}{12}$ (75°)	$\frac{17\pi}{36}$ (85°)	$\frac{89\pi}{180}$ (89°)	1.57	$\frac{\pi}{2}$
$y = \tan x$	0	$\frac{\sqrt{3}}{3} \approx 0.6$	1	$\sqrt{3} \approx 1.7$	3.7	11.4	57.3	1255.8	undefined

As x increases from 0 toward $\frac{\pi}{2}$, y increases slowly at first, then more and more rapidly.

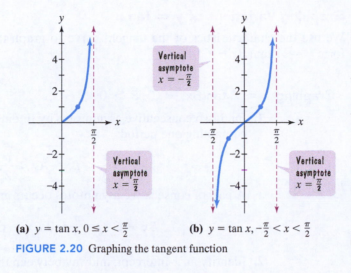

(a) $y = \tan x, 0 \le x < \frac{\pi}{2}$ **(b)** $y = \tan x, -\frac{\pi}{2} < x < \frac{\pi}{2}$

FIGURE 2.20 Graphing the tangent function

The graph in **Figure 2.20(a)** is based on our observation that as x increases from 0 toward $\frac{\pi}{2}$, y increases slowly at first, then more and more rapidly. Notice that y increases without bound as x approaches $\frac{\pi}{2}$. As the figure shows, the graph of $y = \tan x$ has a vertical asymptote at $x = \frac{\pi}{2}$.

The graph of $y = \tan x$ can be completed on the interval $\left(-\frac{\pi}{2}, \frac{\pi}{2}\right)$ by using origin symmetry. **Figure 2.20(b)** shows the result of reflecting the graph in **Figure 2.20(a)** about the origin. The graph of $y = \tan x$ has another vertical asymptote at $x = -\frac{\pi}{2}$. Notice that y decreases without bound as x approaches $-\frac{\pi}{2}$.

Because the period of the tangent function is π, the graph in **Figure 2.20(b)** shows one complete period of $y = \tan x$. We obtain the complete graph of $y = \tan x$ by repeating the graph in **Figure 2.20(b)** to the left and right over intervals of π. The resulting graph and its main characteristics are shown in the following box:

The Tangent Curve: The Graph of $y = \tan x$ and Its Characteristics

Characteristics

- **Period:** π
- **Domain:** All real numbers except odd multiples of $\frac{\pi}{2}$
- **Range:** All real numbers
- **Vertical asymptotes** at odd multiples of $\frac{\pi}{2}$
- **An x-intercept** occurs midway between each pair of consecutive asymptotes.
- **Odd function** with origin symmetry
- Points on the graph $\frac{1}{4}$ and $\frac{3}{4}$ of the way between consecutive asymptotes have y-coordinates of -1 and 1, respectively.

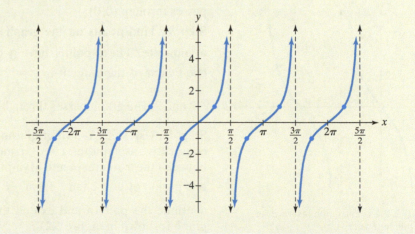

② Graph variations of $y = \tan x$.

Graphing Variations of $y = \tan x$

We use the characteristics of the tangent curve to graph tangent functions of the form $y = A \tan(Bx - C)$.

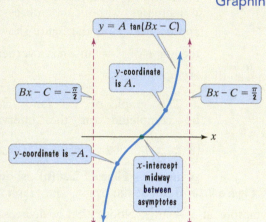

Graphing $y = A \tan(Bx - C)$, $B > 0$

1. Find two consecutive asymptotes by finding an interval containing one period:

$$-\frac{\pi}{2} < Bx - C < \frac{\pi}{2}.$$

A pair of consecutive asymptotes occurs at

$$Bx - C = -\frac{\pi}{2} \text{ and } Bx - C = \frac{\pi}{2}.$$

2. Identify an x-intercept, midway between the consecutive asymptotes.

3. Find the points on the graph $\frac{1}{4}$ and $\frac{3}{4}$ of the way between the consecutive asymptotes. These points have y-coordinates of $-A$ and A, respectively.

4. Use steps 1–3 to graph one full period of the function. Add additional cycles to the left or right as needed.

EXAMPLE 1   Graphing a Tangent Function

Graph $y = 2 \tan \dfrac{x}{2}$ for $-\pi < x < 3\pi$.

SOLUTION

Refer to **Figure 2.21** as you read each step.

Step 1   Find two consecutive asymptotes. We do this by finding an interval containing one period.

$$-\frac{\pi}{2} < \frac{x}{2} < \frac{\pi}{2} \qquad \text{Set up the inequality } -\frac{\pi}{2} < \text{ variable expression in tangent } < \frac{\pi}{2}.$$

$$-\pi < x < \pi \qquad \text{Multiply all parts by 2 and solve for } x.$$

An interval containing one period is $(-\pi, \pi)$. Thus, two consecutive asymptotes occur at $x = -\pi$ and $x = \pi$.

Step 2   Identify an x-intercept, midway between the consecutive asymptotes. Midway between $x = -\pi$ and $x = \pi$ is $x = 0$. An x-intercept is 0 and the graph passes through $(0, 0)$.

Step 3   Find points on the graph $\dfrac{1}{4}$ and $\dfrac{3}{4}$ of the way between the consecutive asymptotes. These points have y-coordinates of $-A$ and A. Because A, the coefficient of the tangent in $y = 2 \tan \dfrac{x}{2}$, is 2, these points have y-coordinates of -2 and 2. The graph passes through $\left(-\dfrac{\pi}{2}, -2\right)$ and $\left(\dfrac{\pi}{2}, 2\right)$.

Step 4   Use steps 1–3 to graph one full period of the function. We use the two consecutive asymptotes, $x = -\pi$ and $x = \pi$, an x-intercept of 0, and points midway between the x-intercept and asymptotes with y-coordinates of -2 and 2. We graph one period of $y = 2 \tan \dfrac{x}{2}$ from $-\pi$ to π. In order to graph for $-\pi < x < 3\pi$, we continue the pattern and extend the graph another full period to the right. The graph is shown in **Figure 2.21**.

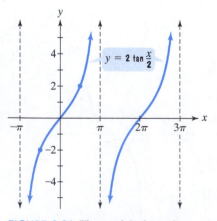

FIGURE 2.21 The graph is shown for two full periods.

● ● ●

✅ **Check Point 1** Graph $y = 3 \tan 2x$ for $-\dfrac{\pi}{4} < x < \dfrac{3\pi}{4}$.

EXAMPLE 2   Graphing a Tangent Function

Graph two full periods of $y = \tan\left(x + \dfrac{\pi}{4}\right)$.

SOLUTION

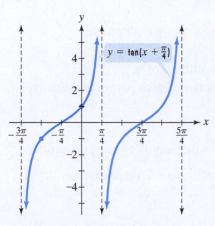

FIGURE 2.22 The graph is shown for two full periods.

The graph of $y = \tan\left(x + \dfrac{\pi}{4}\right)$ is the graph of $y = \tan x$ shifted horizontally to the left $\dfrac{\pi}{4}$ units. Refer to **Figure 2.22** as you read each step.

Step 1 Find two consecutive asymptotes. We do this by finding an interval containing one period.

$$-\frac{\pi}{2} < x + \frac{\pi}{4} < \frac{\pi}{2} \qquad \text{\color{blue}{Set up the inequality } } -\frac{\pi}{2} < \text{\color{blue}{variable expression in tangent}} < \frac{\pi}{2}.$$

$$-\frac{\pi}{2} - \frac{\pi}{4} < x < \frac{\pi}{2} - \frac{\pi}{4} \qquad \text{\color{blue}{Subtract } } \frac{\pi}{4} \text{ \color{blue}{from all parts and solve for } } x.$$

$$-\frac{3\pi}{4} < x < \frac{\pi}{4} \qquad \text{\color{blue}{Simplify: } } -\frac{\pi}{2} - \frac{\pi}{4} = -\frac{2\pi}{4} - \frac{\pi}{4} = -\frac{3\pi}{4}$$

$$\text{\color{blue}{and } } \frac{\pi}{2} - \frac{\pi}{4} = \frac{2\pi}{4} - \frac{\pi}{4} = \frac{\pi}{4}.$$

An interval containing one period is $\left(-\dfrac{3\pi}{4}, \dfrac{\pi}{4}\right)$. Thus, two consecutive asymptotes occur at $x = -\dfrac{3\pi}{4}$ and $x = \dfrac{\pi}{4}$.

Step 2 Identify an x-intercept, midway between the consecutive asymptotes.

$$x\text{-intercept} = \frac{-\dfrac{3\pi}{4} + \dfrac{\pi}{4}}{2} = \frac{-\dfrac{2\pi}{4}}{2} = -\frac{2\pi}{8} = -\frac{\pi}{4}$$

An x-intercept is $-\dfrac{\pi}{4}$ and the graph passes through $\left(-\dfrac{\pi}{4}, 0\right)$.

Step 3 Find points on the graph $\dfrac{1}{4}$ and $\dfrac{3}{4}$ of the way between the consecutive asymptotes. These points have y-coordinates of $-A$ and A. Because A, the coefficient of the tangent in $y = \tan\left(x + \dfrac{\pi}{4}\right)$, is 1, these points have y-coordinates of -1 and 1. They are shown as blue dots in **Figure 2.22**.

Step 4 Use steps 1–3 to graph one full period of the function. We use the two consecutive asymptotes, $x = -\dfrac{3\pi}{4}$ and $x = \dfrac{\pi}{4}$, to graph one full period of $y = \tan\left(x + \dfrac{\pi}{4}\right)$ from $-\dfrac{3\pi}{4}$ to $\dfrac{\pi}{4}$. We graph two full periods by continuing the pattern and extending the graph another full period to the right. The graph is shown in **Figure 2.22**.   • • •

✅ **Check Point 2** Graph two full periods of $y = \tan\left(x - \dfrac{\pi}{2}\right)$.

③ Understand the graph of $y = \cot x$.

The Graph of $y = \cot x$

Like the tangent function, the cotangent function, $y = \cot x$, has a period of π. The graph and its main characteristics are shown in the following box.

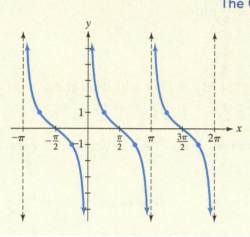

The Cotangent Curve: The Graph of $y = \cot x$ and Its Characteristics

Characteristics

- **Period:** π
- **Domain:** All real numbers except integral multiples of π
- **Range:** All real numbers
- **Vertical asymptotes** at integral multiples of π
- **An x-intercept** occurs midway between each pair of consecutive asymptotes.
- **Odd function** with origin symmetry
- Points on the graph $\frac{1}{4}$ and $\frac{3}{4}$ of the way between consecutive asymptotes have y-coordinates of 1 and -1, respectively.

④ Graph variations of $y = \cot x$.

Graphing Variations of $y = \cot x$

We use the characteristics of the cotangent curve to graph cotangent functions of the form $y = A \cot(Bx - C)$.

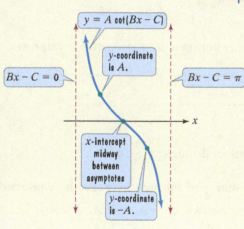

Graphing $y = A \cot(Bx - C)$, $B > 0$

1. Find two consecutive asymptotes by finding an interval containing one full period:
$$0 < Bx - C < \pi.$$
A pair of consecutive asymptotes occurs at
$$Bx - C = 0 \text{ and } Bx - C = \pi.$$

2. Identify an x-intercept, midway between the consecutive asymptotes.

3. Find the points on the graph $\frac{1}{4}$ and $\frac{3}{4}$ of the way between the consecutive asymptotes. These points have y-coordinates of A and $-A$, respectively.

4. Use steps 1–3 to graph one full period of the function. Add additional cycles to the left or right as needed.

EXAMPLE 3　Graphing a Cotangent Function

Graph $y = 3 \cot 2x$.

SOLUTION

Step 1　Find two consecutive asymptotes. We do this by finding an interval containing one period.

$0 < 2x < \pi$　Set up the inequality $0 <$ variable expression in cotangent $< \pi$.

$0 < x < \dfrac{\pi}{2}$　Divide all parts by 2 and solve for x.

An interval containing one period is $\left(0, \dfrac{\pi}{2}\right)$. Thus, two consecutive asymptotes occur at $x = 0$ and $x = \dfrac{\pi}{2}$, shown in **Figure 2.23**.

Step 2   Identify an x-intercept, midway between the consecutive asymptotes. Midway between $x = 0$ and $x = \dfrac{\pi}{2}$ is $x = \dfrac{\pi}{4}$. An x-intercept is $\dfrac{\pi}{4}$ and the graph passes through $\left(\dfrac{\pi}{4}, 0\right)$.

Step 3   Find points on the graph $\dfrac{1}{4}$ and $\dfrac{3}{4}$ of the way between consecutive asymptotes. These points have y-coordinates of A and $-A$. Because A, the coefficient of the cotangent in $y = 3 \cot 2x$, is 3, these points have y-coordinates of 3 and -3. They are shown as blue dots in **Figure 2.23**.

Step 4   Use steps 1–3 to graph one full period of the function. We use the two consecutive asymptotes, $x = 0$ and $x = \dfrac{\pi}{2}$, to graph one full period of $y = 3 \cot 2x$. This curve is repeated to the left and right, as shown in **Figure 2.23**.   •••

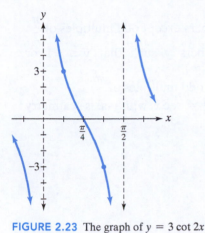

FIGURE 2.23 The graph of $y = 3 \cot 2x$

✓ **Check Point 3** Graph $y = \dfrac{1}{2} \cot \dfrac{\pi}{2} x$.

⑤ Understand the graphs of $y = \csc x$ and $y = \sec x$.

The Graphs of $y = \csc x$ and $y = \sec x$

We obtain the graphs of the cosecant and secant curves by using the reciprocal identities

$$\csc x = \frac{1}{\sin x} \quad \text{and} \quad \sec x = \frac{1}{\cos x}.$$

The identity $\csc x = \dfrac{1}{\sin x}$ tells us that the value of the cosecant function $y = \csc x$ at a given value of x equals the reciprocal of the corresponding value of the sine function, provided that the value of the sine function is not 0. If the value of $\sin x$ is 0, then at each of these values of x, the cosecant function is not defined. A vertical asymptote is associated with each of these values on the graph of $y = \csc x$.

We obtain the graph of $y = \csc x$ by taking reciprocals of the y-values in the graph of $y = \sin x$. Vertical asymptotes of $y = \csc x$ occur at the x-intercepts of $y = \sin x$. Likewise, we obtain the graph of $y = \sec x$ by taking the reciprocal of $y = \cos x$. Vertical asymptotes of $y = \sec x$ occur at the x-intercepts of $y = \cos x$. The graphs of $y = \csc x$ and $y = \sec x$ and their key characteristics are shown in the following boxes. We have used dashed red curves to graph $y = \sin x$ and $y = \cos x$ first, drawing vertical asymptotes through the x-intercepts.

The Cosecant Curve: The Graph of $y = \csc x$ and Its Characteristics

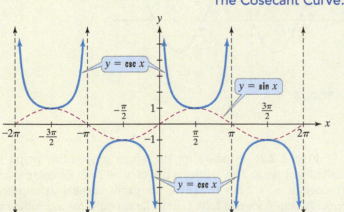

Characteristics

- **Period:** 2π
- **Domain:** All real numbers except integral multiples of π
- **Range:** All real numbers y such that $y \leq -1$ or $y \geq 1$: $(-\infty, -1] \cup [1, \infty)$
- **Vertical asymptotes** at integral multiples of π
- **Odd function,** $\csc(-x) = -\csc x$, with origin symmetry

The Secant Curve: The Graph of $y = \sec x$ and Its Characteristics

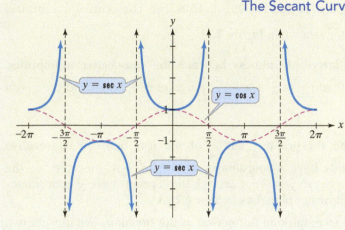

Characteristics

- **Period:** 2π
- **Domain:** All real numbers except odd multiples of $\dfrac{\pi}{2}$
- **Range:** All real numbers y such that $y \leq -1$ or $y \geq 1$: $(-\infty, -1] \cup [1, \infty)$
- **Vertical asymptotes** at odd multiples of $\dfrac{\pi}{2}$
- **Even function,** $\sec(-x) = \sec x$, with y-axis symmetry

⑥ Graph variations of $y = \csc x$ and $y = \sec x$.

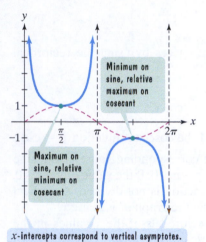

FIGURE 2.24

Graphing Variations of $y = \csc x$ and $y = \sec x$

We use graphs of functions involving the corresponding reciprocal functions to obtain graphs of cosecant and secant functions. To graph a cosecant or secant curve, begin by graphing the function where cosecant or secant is replaced by its reciprocal function. For example, to graph $y = 2 \csc 2x$, we use the graph of $y = 2 \sin 2x$.

Likewise, to graph $y = -3 \sec \dfrac{x}{2}$, we use the graph of $y = -3 \cos \dfrac{x}{2}$.

Figure 2.24 illustrates how we use a sine curve to obtain a cosecant curve. Notice that

- x-intercepts on the red sine curve correspond to vertical asymptotes of the blue cosecant curve.
- A maximum point on the red sine curve corresponds to a minimum point on a continuous portion of the blue cosecant curve.
- A minimum point on the red sine curve corresponds to a maximum point on a continuous portion of the blue cosecant curve.

EXAMPLE 4   Using a Sine Curve to Obtain a Cosecant Curve

Use the graph of $y = 2 \sin 2x$ in **Figure 2.25** to obtain the graph of $y = 2 \csc 2x$.

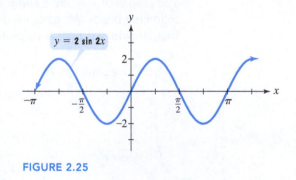

FIGURE 2.25

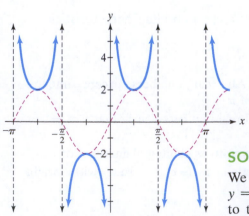

FIGURE 2.26 Using a sine curve to graph $y = 2 \csc 2x$

SOLUTION

We begin our work in **Figure 2.26** by showing the given graph, the graph of $y = 2 \sin 2x$, using dashed red lines. The x-intercepts of $y = 2 \sin 2x$ correspond to the vertical asymptotes of $y = 2 \csc 2x$. Thus, we draw vertical asymptotes through the x-intercepts, shown in **Figure 2.26**. Using the asymptotes as guides, we sketch the graph of $y = 2 \csc 2x$ in **Figure 2.26**.

•••

✓ **Check Point 4** Use the graph of $y = \sin\left(x + \dfrac{\pi}{4}\right)$, shown on the right, to obtain the graph of $y = \csc\left(x + \dfrac{\pi}{4}\right)$.

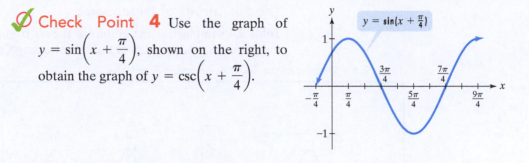

We use a cosine curve to obtain a secant curve in exactly the same way we used a sine curve to obtain a cosecant curve. Thus,

- x-intercepts on the cosine curve correspond to vertical asymptotes on the secant curve.
- A maximum point on the cosine curve corresponds to a minimum point on a continuous portion of the secant curve.
- A minimum point on the cosine curve corresponds to a maximum point on a continuous portion of the secant curve.

EXAMPLE 5 Graphing a Secant Function

Graph $y = -3 \sec \dfrac{x}{2}$ for $-\pi < x < 5\pi$.

SOLUTION

We begin by graphing the function $y = -3 \cos \dfrac{x}{2}$, where secant has been replaced by cosine, its reciprocal function. This equation is of the form $y = A \cos Bx$ with $A = -3$ and $B = \frac{1}{2}$.

amplitude: $|A| = |-3| = 3$ 〔The maximum value of y is 3 and the minimum is −3.〕

period: $\dfrac{2\pi}{B} = \dfrac{2\pi}{\frac{1}{2}} = 4\pi$ 〔Each cycle is of length 4π.〕

We use quarter-periods, $\dfrac{4\pi}{4}$, or π, to find the x-values for the five key points. Starting with $x = 0$, the x-values are $0, \pi, 2\pi, 3\pi,$ and 4π. Evaluating the function $y = -3 \cos \dfrac{x}{2}$ at each of these values of x, the key points are

$$(0, -3), (\pi, 0), (2\pi, 3), (3\pi, 0), \text{ and } (4\pi, -3).$$

We use these key points to graph $y = -3 \cos \dfrac{x}{2}$ from 0 to 4π, shown using a dashed red line in **Figure 2.27**. In order to graph $y = -3 \sec \dfrac{x}{2}$ for $-\pi < x < 5\pi$, extend the dashed red graph of the cosine function π units to the left and π units to the right. Now use this dashed red graph to obtain the graph of the corresponding secant function, its reciprocal function. Draw vertical asymptotes through the x-intercepts. Using these asymptotes as guides, the graph of $y = -3 \sec \dfrac{x}{2}$ is shown in blue in **Figure 2.27**. •••

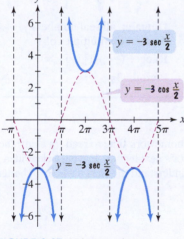

FIGURE 2.27 Using a cosine curve to graph $y = -3 \sec \dfrac{x}{2}$

✓ **Check Point 5** Graph $y = 2 \sec 2x$ for $-\dfrac{3\pi}{4} < x < \dfrac{3\pi}{4}$.

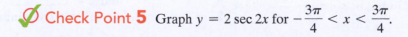

The Six Curves of Trigonometry

Table 2.4 summarizes the graphs of the six trigonometric functions. Below each of the graphs is a description of the domain, range, and period of the function.

Table 2.4   Graphs of the Six Trigonometric Functions

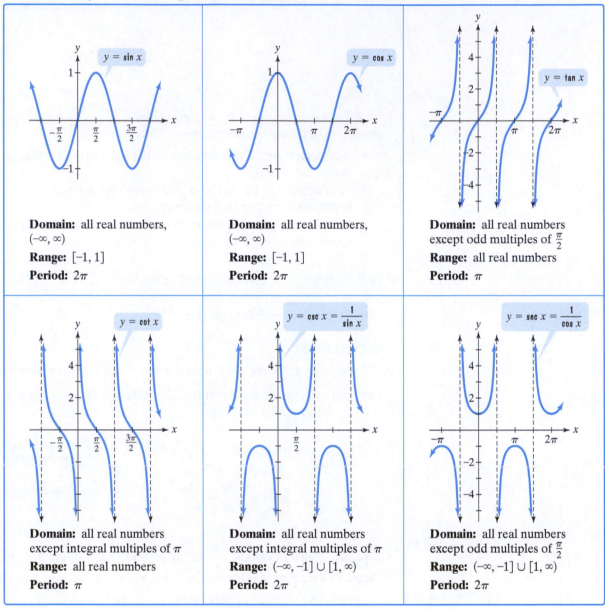

Domain: all real numbers, $(-\infty, \infty)$

Range: $[-1, 1]$

Period: 2π

Domain: all real numbers, $(-\infty, \infty)$

Range: $[-1, 1]$

Period: 2π

Domain: all real numbers except odd multiples of $\frac{\pi}{2}$

Range: all real numbers

Period: π

Domain: all real numbers except integral multiples of π

Range: all real numbers

Period: π

Domain: all real numbers except integral multiples of π

Range: $(-\infty, -1] \cup [1, \infty)$

Period: 2π

Domain: all real numbers except odd multiples of $\frac{\pi}{2}$

Range: $(-\infty, -1] \cup [1, \infty)$

Period: 2π

ACHIEVING SUCCESS

Read ahead. You might find it helpful to use some of your homework time to read (or skim) the section in the textbook that will be covered in your professor's next lecture. Having a clear idea of the new material that will be discussed will help you to understand the class a whole lot better.

CONCEPT AND VOCABULARY CHECK

Fill in each blank so that the resulting statement is true.

1. In order to graph $y = \frac{1}{2}\tan 2x$, an interval containing one period is found by solving $-\frac{\pi}{2} < 2x < \frac{\pi}{2}$. An interval containing one period is _____. Thus, two consecutive asymptotes occur at $x =$ _____ and $x =$ _____.

2. An interval containing one period of $y = \tan\left(x - \frac{\pi}{2}\right)$ is _____. Thus, two consecutive asymptotes occur at $x =$ _____ and $x =$ _____.

3. In order to graph $y = 3 \cot \dfrac{\pi}{2} x$, an interval containing one period is found by solving $0 < \dfrac{\pi}{2} x < \pi$. An interval containing one period is _____. Thus, two consecutive asymptotes occur at $x = $ _____ and $x = $ _____.

4. An interval containing one period of $y = 4 \cot\left(x + \dfrac{\pi}{4}\right)$ is _____. Thus, two consecutive asymptotes occur at $x = $ _____ and $x = $ _____.

5. It is easiest to graph $y = 3 \csc 2x$ by first graphing $y = $ _____.

6. It is easiest to graph $y = 2 \sec \pi x$ by first graphing _____.

7. True or false: The graphs of $y = \sec \dfrac{x}{2}$ and $y = \cos \dfrac{x}{2}$ are identical. _____

8. True or false: The graph of $y = 2 \sin 2x$ has an x-intercept at $\dfrac{\pi}{2}$, so $x = \dfrac{\pi}{2}$ is a vertical asymptote of $y = 2 \csc 2x$. _____

EXERCISE SET 2.2

Practice Exercises

In Exercises 1–4, the graph of a tangent function is given. Select the equation for each graph from the following options:

$$y = \tan\left(x + \frac{\pi}{2}\right), \quad y = \tan(x + \pi), \quad y = -\tan x, \quad y = -\tan\left(x - \frac{\pi}{2}\right).$$

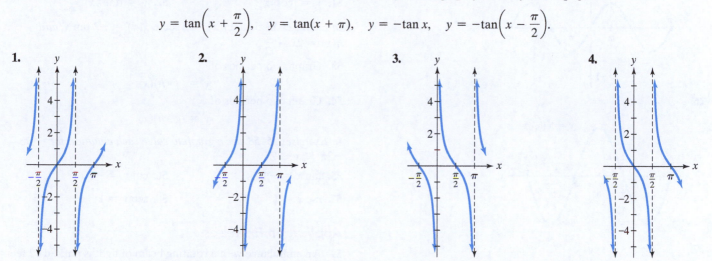

1.   **2.**   **3.**   **4.**

In Exercises 5–12, graph two periods of the given tangent function.

5. $y = 3 \tan \dfrac{x}{4}$ **6.** $y = 2 \tan \dfrac{x}{4}$ **7.** $y = \dfrac{1}{2} \tan 2x$ **8.** $y = 2 \tan 2x$

9. $y = -2 \tan \dfrac{1}{2}x$ **10.** $y = -3 \tan \dfrac{1}{2}x$ **11.** $y = \tan(x - \pi)$ **12.** $y = \tan\left(x - \dfrac{\pi}{4}\right)$

In Exercises 13–16, the graph of a cotangent function is given. Select the equation for each graph from the following options:

$$y = \cot\left(x + \frac{\pi}{2}\right), \quad y = \cot(x + \pi), \quad y = -\cot x, \quad y = -\cot\left(x - \frac{\pi}{2}\right).$$

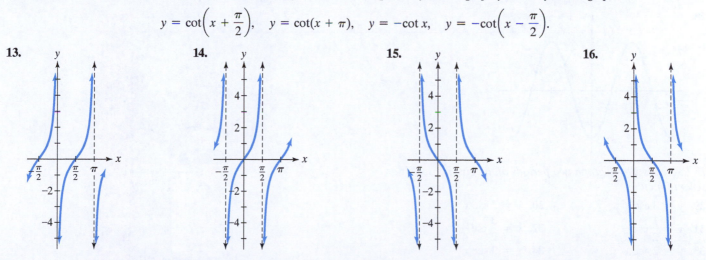

13.   **14.**   **15.**   **16.**

In Exercises 17–24, graph two periods of the given cotangent function.

17. $y = 2 \cot x$

18. $y = \dfrac{1}{2} \cot x$

19. $y = \dfrac{1}{2} \cot 2x$

20. $y = 2 \cot 2x$

21. $y = -3 \cot \dfrac{\pi}{2} x$

22. $y = -2 \cot \dfrac{\pi}{4} x$

23. $y = 3 \cot\left(x + \dfrac{\pi}{2}\right)$

24. $y = 3 \cot\left(x + \dfrac{\pi}{4}\right)$

In Exercises 25–28, use each graph to obtain the graph of the corresponding reciprocal function, cosecant or secant. Give the equation of the function for the graph that you obtain.

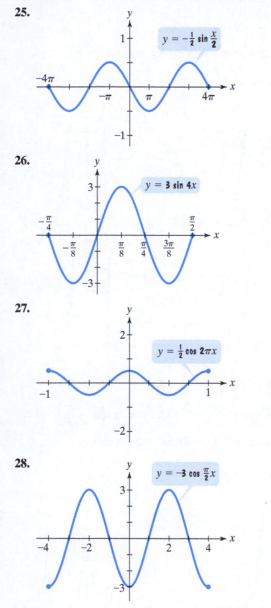

25.

$y = -\dfrac{1}{2} \sin \dfrac{x}{2}$

26.

$y = 3 \sin 4x$

27.

$y = \dfrac{1}{2} \cos 2\pi x$

28.

$y = -3 \cos \dfrac{\pi}{2} x$

In Exercises 29–44, graph two periods of the given cosecant or secant function.

29. $y = 3 \csc x$

30. $y = 2 \csc x$

31. $y = \dfrac{1}{2} \csc \dfrac{x}{2}$

32. $y = \dfrac{3}{2} \csc \dfrac{x}{4}$

33. $y = 2 \sec x$

34. $y = 3 \sec x$

35. $y = \sec \dfrac{x}{3}$

36. $y = \sec \dfrac{x}{2}$

37. $y = -2 \csc \pi x$

38. $y = -\dfrac{1}{2} \csc \pi x$

39. $y = -\dfrac{1}{2} \sec \pi x$

40. $y = -\dfrac{3}{2} \sec \pi x$

41. $y = \csc(x - \pi)$

42. $y = \csc\left(x - \dfrac{\pi}{2}\right)$

43. $y = 2 \sec(x + \pi)$

44. $y = 2 \sec\left(x + \dfrac{\pi}{2}\right)$

Practice Plus

In Exercises 45–52, graph two periods of each function.

45. $y = 2 \tan\left(x - \dfrac{\pi}{6}\right) + 1$

46. $y = 2 \cot\left(x + \dfrac{\pi}{6}\right) - 1$

47. $y = \sec\left(2x + \dfrac{\pi}{2}\right) - 1$

48. $y = \csc\left(2x - \dfrac{\pi}{2}\right) + 1$

49. $y = \csc|x|$

50. $y = \sec|x|$

51. $y = \left|\cot \dfrac{1}{2}x\right|$

52. $y = \left|\tan \dfrac{1}{2}x\right|$

In Exercises 53–54, let $f(x) = 2 \sec x$, $g(x) = -2 \tan x$, and $h(x) = 2x - \dfrac{\pi}{2}$.

53. Graph two periods of

$$y = (f \circ h)(x).$$

54. Graph two periods of

$$y = (g \circ h)(x).$$

In Exercises 55–58, use a graph to solve each equation for $-2\pi \le x \le 2\pi$.

55. $\tan x = -1$

56. $\cot x = -1$

57. $\csc x = 1$

58. $\sec x = 1$

Application Exercises

59. An ambulance with a rotating beam of light is parked 12 feet from a building. The function

$$d = 12 \tan 2\pi t$$

describes the distance, d, in feet, of the rotating beam of light from point C after t seconds.

a. Graph the function on the interval $[0, 2]$.

b. For what values of t in $[0, 2]$ is the function undefined? What does this mean in terms of the rotating beam of light in the figure shown?

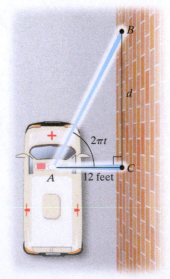

60. The angle of elevation from the top of a house to a jet flying 2 miles above the house is x radians. If d represents the horizontal distance, in miles, of the jet from the house, express d in terms of a trigonometric function of x. Then graph the function for $0 < x < \pi$.

61. Your best friend is marching with a band and has asked you to film him. The figure below shows that you have set yourself up 10 feet from the street where your friend will be passing from left to right. If d represents your distance, in feet, from your friend and x is the radian measure of the angle shown, express d in terms of a trigonometric function of x. Then graph the function for $-\dfrac{\pi}{2} < x < \dfrac{\pi}{2}$. Negative angles indicate that your marching buddy is on your left.

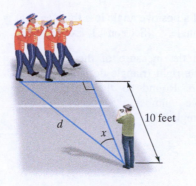

In Exercises 62–64, sketch a reasonable graph that models the given situation.

62. The number of hours of daylight per day in your hometown over a two-year period

63. The motion of a diving board vibrating 10 inches in each direction per second just after someone has dived off

64. The distance of a rotating beam of light from a point on a wall (See the figure for Exercise 59.)

Explaining the Concepts

65. Without drawing a graph, describe the behavior of the basic tangent curve.

66. If you are given the equation of a tangent function, how do you find a pair of consecutive asymptotes?

67. If you are given the equation of a tangent function, how do you identify an x-intercept?

68. Without drawing a graph, describe the behavior of the basic cotangent curve.

69. If you are given the equation of a cotangent function, how do you find a pair of consecutive asymptotes?

70. Explain how to determine the range of $y = \csc x$ from the graph. What is the range?

71. Explain how to use a sine curve to obtain a cosecant curve. Why can the same procedure be used to obtain a secant curve from a cosine curve?

72. Scientists record brain activity by attaching electrodes to the scalp and then connecting these electrodes to a machine. The brain activity recorded with this machine is shown in the three graphs. Which trigonometric functions would be most appropriate for describing the oscillations in brain activity? Describe similarities and differences among these functions when modeling brain activity when awake, during dreaming sleep, and during non-dreaming sleep.

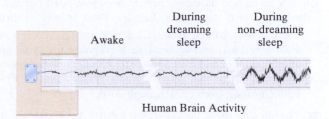

Human Brain Activity

Technology Exercises

73. Use a graphing utility to verify any two of the tangent curves that you drew by hand in Exercises 5–12.

74. Use a graphing utility to verify any two of the cotangent curves that you drew by hand in Exercises 17–24.

75. Use a graphing utility to verify any two of the cosecant curves that you drew by hand in Exercises 29–44.

76. Use a graphing utility to verify any two of the secant curves that you drew by hand in Exercises 29–44.

In Exercises 77–82, use a graphing utility to graph each function. Use a viewing rectangle that shows the graph for at least two periods.

77. $y = \tan \dfrac{x}{4}$ **78.** $y = \tan 4x$

79. $y = \cot 2x$ **80.** $y = \cot \dfrac{x}{2}$

81. $y = \dfrac{1}{2}\tan \pi x$ **82.** $y = \dfrac{1}{2}\tan(\pi x + 1)$

In Exercises 83–86, use a graphing utility to graph each pair of functions in the same viewing rectangle. Use a viewing rectangle that shows the graphs for at least two periods.

83. $y = 0.8 \sin \dfrac{x}{2}$ and $y = 0.8 \csc \dfrac{x}{2}$

84. $y = -2.5 \sin \dfrac{\pi}{3}x$ and $y = -2.5 \csc \dfrac{\pi}{3}x$

85. $y = 4 \cos\left(2x - \dfrac{\pi}{6}\right)$ and $y = 4 \sec\left(2x - \dfrac{\pi}{6}\right)$

86. $y = -3.5 \cos\left(\pi x - \dfrac{\pi}{6}\right)$ and $y = -3.5 \sec\left(\pi x - \dfrac{\pi}{6}\right)$

87. Carbon dioxide particles in our atmosphere trap heat and raise the planet's temperature. Even if all greenhouse-gas emissions miraculously ended today, the planet would continue to warm through the rest of the century because of the amount of carbon we have already added to the atmosphere. Carbon dioxide accounts for about half of global warming. The function

$$y = 2.5 \sin 2\pi x + 0.0216x^2 + 0.654x + 316$$

models carbon dioxide concentration, y, in parts per million, where $x = 0$ represents January 1960; $x = \frac{1}{12}$, February 1960; $x = \frac{2}{12}$, March 1960; ..., $x = 1$, January 1961; $x = \frac{13}{12}$, February 1961; and so on. Use a graphing utility to graph the function in a $[30, 48, 5]$ by $[310, 420, 5]$ viewing rectangle. Describe what the graph reveals about carbon dioxide concentration from 1990 through 2008.

88. Graph $y = \sin \dfrac{1}{x}$ in a $[-0.2, 0.2, 0.01]$ by $[-1.2, 1.2, 0.01]$ viewing rectangle. What is happening as x approaches 0 from the left or the right? Explain this behavior.

Critical Thinking Exercises

Make Sense? *In Exercises 89–92, determine whether each statement makes sense or does not make sense, and explain your reasoning.*

89. I use the pattern

asymptote, $-A$, x-intercept, A, asymptote

to graph one full period of $y = A \tan(Bx - C)$.

90. After using the four-step procedure to graph $y = -\cot\left(x + \dfrac{\pi}{4}\right)$, I checked my graph by verifying it was the graph of $y = \cot x$ shifted left $\dfrac{\pi}{4}$ unit and reflected about the x-axis.

91. I used the graph of $y = 3 \cos 2x$ to obtain the graph of $y = 3 \csc 2x$.

92. I used a tangent function to model the average monthly temperature of New York City, where $x = 1$ represents January, $x = 2$ represents February, and so on.

In Exercises 93–94, write an equation for each blue graph.

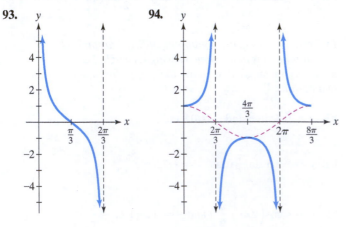

In Exercises 95–96, write the equation for a cosecant function satisfying the given conditions.

95. period: 3π; range: $(-\infty, -2] \cup [2, \infty)$

96. period: 2; range: $(-\infty, -\pi] \cup [\pi, \infty)$

97. Determine the range of the following functions. Then give a viewing rectangle, or window, that shows two periods of the function's graph.

a. $f(x) = \sec\left(3x + \dfrac{\pi}{2}\right)$

b. $g(x) = 3 \sec \pi\left(x + \dfrac{1}{2}\right)$

98. For $x > 0$, what effect does 2^{-x} in $y = 2^{-x} \sin x$ have on the graph of $y = \sin x$? What kind of behavior can be modeled by a function such as $y = 2^{-x} \sin x$?

Retaining the Concepts

99. A circle has a radius of 8 inches. Find the length of the arc intercepted by a central angle of 150°. Express arc length in terms of π. Then round your answer to two decimal places. (Section 1.1, Example 8)

100. Find a positive angle less than 2π that is coterminal with a $\frac{25\pi}{3}$ angle.   (Section 1.1, Example 7)

101. Find the measure of the side of the right triangle designated by a. Round to the nearest whole number. (Section 1.2, Example 9)

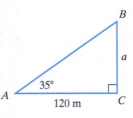

Preview Exercises

Exercises 102–104 will help you prepare for the material covered in the next section.

102. a. Graph $y = \sin x$ for $-\dfrac{\pi}{2} \le x \le \dfrac{\pi}{2}$.

b. Based on your graph in part (a), does $y = \sin x$ have an inverse function if the domain is restricted to $\left[-\dfrac{\pi}{2}, \dfrac{\pi}{2}\right]$? Explain your answer.

c. Determine the angle in the interval $\left[-\dfrac{\pi}{2}, \dfrac{\pi}{2}\right]$ whose sine is $-\dfrac{1}{2}$. Identify this information as a point on your graph in part (a).

103. a. Graph $y = \cos x$ for $0 \le x \le \pi$.

b. Based on your graph in part (a), does $y = \cos x$ have an inverse function if the domain is restricted to $[0, \pi]$? Explain your answer.

c. Determine the angle in the interval $[0, \pi]$ whose cosine is $-\dfrac{\sqrt{3}}{2}$. Identify this information as a point on your graph in part (a).

104. a. Graph $y = \tan x$ for $-\dfrac{\pi}{2} < x < \dfrac{\pi}{2}$.

b. Based on your graph in part (a), does $y = \tan x$ have an inverse function if the domain is restricted to $\left(-\dfrac{\pi}{2}, \dfrac{\pi}{2}\right)$? Explain your answer.

c. Determine the angle in the interval $\left(-\dfrac{\pi}{2}, \dfrac{\pi}{2}\right)$ whose tangent is $-\sqrt{3}$. Identify this information as a point on your graph in part (a).

CHAPTER 2   Mid-Chapter Check Point

WHAT YOU KNOW: We used the graphs of the six trigonometric functions, shown in Table 2.4 on page 104, to graph variations of these functions. The graphs of $y = A \sin(Bx - C)$ and $y = A \cos(Bx - C), B > 0$, were graphed using amplitude $= |A|$, period $= \frac{2\pi}{B}$, and phase shift $= \frac{C}{B}$. The constant D in $y = A \sin(Bx - C) + D$ and $y = A \cos(Bx - C) + D$ caused vertical shifts with oscillation about the horizontal line $y = D$. We used the method of adding y-coordinates to graph the sum of two trigonometric functions. We graphed $y = A \tan(Bx - C), B > 0$, using consecutive asymptotes (solve $-\frac{\pi}{2} < Bx - C < \frac{\pi}{2}$; consecutive asymptotes occur at $Bx - C = -\frac{\pi}{2}$ and $Bx - C = \frac{\pi}{2}$) and an x-intercept midway between them. We graphed $y = A \cot(Bx - C), B > 0$, using consecutive asymptotes (solve $0 < Bx - C < \pi$; consecutive asymptotes occur at $Bx - C = 0$ and $Bx - C = \pi$) and an x-intercept midway between them. To graph a cosecant curve, we began by graphing the corresponding sine curve. We drew vertical asymptotes through x-intercepts, using asymptotes as guides to sketch the graph. We graphed secant curves by first graphing corresponding cosine curves and using the same procedure.

In Exercises 1–2, determine the amplitude and period of each function. Then graph one period of the function.

1. $y = 4 \sin 2x$ **2.** $y = \frac{1}{2} \cos \frac{\pi}{3} x$

In Exercises 3–4, determine the amplitude, period, and phase shift of each function. Then graph one period of the function.

3. $y = 3 \sin(x - \pi)$ **4.** $y = 2 \cos\left(2x - \frac{\pi}{4}\right)$

5. Use a vertical shift to graph one period of $y = \cos 2x + 1$.

6. Use the method of adding y-coordinates to graph $y = 2 \sin x + 2 \cos x$ for $0 \le x \le 2\pi$.

In Exercises 7–8, graph two full periods of the given tangent or cotangent function.

7. $y = 2 \tan \frac{\pi}{4} x$ **8.** $y = 4 \cot 2x$

In Exercises 9–10, graph two full periods of the given secant or cosecant function.

9. $y = -2 \sec \pi x$ **10.** $y = 3 \csc 2\pi x$

A Brief Review • Composite and Inverse Functions

- The composition of functions f and g, $f \circ g$, is defined by $(f \circ g)(x) = f(g(x))$. This composite function is obtained by replacing each occurrence of x in the equation for f with $g(x)$.

EXAMPLE

$$f(x) = x^2 + x \quad \text{and} \quad g(x) = 2x + 1$$

$$(f \circ g)(x) = f(g(x)) = (g(x))^2 + g(x)$$

Replace x with $g(x)$.

$$= (2x + 1)^2 + (2x + 1) = 4x^2 + 4x + 1 + 2x + 1$$
$$= 4x^2 + 6x + 2$$

- The composition of functions g and f, $g \circ f$, is defined by $(g \circ f)(x) = g(f(x))$. This composite function is obtained by replacing each occurrence of x in the equation for g with $f(x)$.

EXAMPLE

$$f(x) = x^2 + x \quad \text{and} \quad g(x) = 2x + 1$$

$$(g \circ f)(x) = g(f(x)) = 2f(x) + 1$$

Replace x with $f(x)$.

$$= 2(x^2 + x) + 1 = 2x^2 + 2x + 1$$

- If $f(g(x)) = x$ and $g(f(x)) = x$, function g is the inverse of function f, denoted f^{-1} and read "f inverse." The procedure for finding a function's inverse uses a switch-and-solve strategy: Switch x and y, then solve for y.

EXAMPLE

If $f(x) = 2x - 5$, find $f^{-1}(x)$.

$$y = 2x - 5 \quad \text{Replace } f(x) \text{ with } y.$$
$$x = 2y - 5 \quad \text{Exchange } x \text{ and } y.$$
$$x + 5 = 2y \quad \text{Solve for } y.$$
$$\frac{x + 5}{2} = y$$
$$f^{-1}(x) = \frac{x + 5}{2} \quad \text{Replace } y \text{ with } f^{-1}(x).$$

- If $f(a) = b$, then $f^{-1}(b) = a$. The domain of f is the range of f^{-1}. The range of f is the domain of f^{-1}.
- $f(f^{-1}(x)) = x$ and $f^{-1}(f(x)) = x$.
- The Horizontal Line Test for Inverse Functions

  A function, f, has an inverse that is a function, f^{-1}, if there is no horizontal line that intersects the graph of f at more than one point. A one-to-one function is one in which no two different ordered pairs have the same second component. Only one-to-one functions have inverse functions.

EXAMPLE

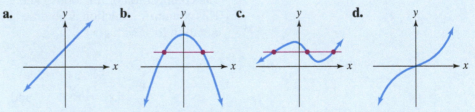

a. Has an inverse function
b. No inverse function
c. No inverse function
d. Has an inverse function

- If the point (a, b) is on the graph of f, then the point (b, a) is on the graph of f^{-1}. The graph of f^{-1} is a reflection of the graph of f about the line $y = x$.

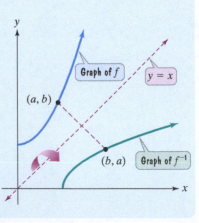

Section 2.3

Inverse Trigonometric Functions

Star Wars Episode VII: The Force Awakens

What am I supposed to learn?

After studying this section, you should be able to:

❶ Understand and use the inverse sine function.

❷ Understand and use the inverse cosine function.

❸ Understand and use the inverse tangent function.

❹ Use a calculator to evaluate inverse trigonometric functions.

❺ Find exact values of composite functions with inverse trigonometric functions.

Movies are very much a visual medium. Though music and sound effects are important to the experience, the power of film is captured by the phrase "watching the movie." Where in the theater should you sit to maximize the visual impact of the astonishing worlds created by film? In this section's Exercise Set (Exercises 93 and 94), you will see how an inverse trigonometric function can enhance your movie-going experience.

GREAT QUESTION!

What are the most important things from the Brief Review on pages 109–110 that I should know about inverse functions?

Here are some helpful things to remember from the Brief Review.

- If no horizontal line intersects the graph of a function more than once, the function is one-to-one and has an inverse function.
- If the point (a, b) is on the graph of f, then the point (b, a) is on the graph of the inverse function, denoted f^{-1}. The graph of f^{-1} is a reflection of the graph of f about the line $y = x$.

❶ Understand and use the inverse sine function.

The Inverse Sine Function

Figure 2.28 shows the graph of $y = \sin x$. Can you see that every horizontal line that can be drawn between -1 and 1 intersects the graph infinitely many times? Thus, the sine function is not one-to-one and has no inverse function.

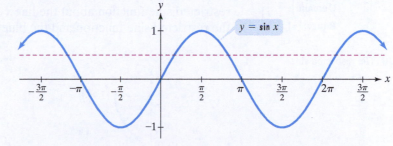

FIGURE 2.28 The horizontal line test shows that the sine function is not one-to-one and has no inverse function.

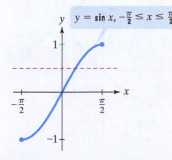

FIGURE 2.29 The restricted sine function passes the horizontal line test. It is one-to-one and has an inverse function.

In **Figure 2.29**, we have taken a portion of the sine curve, restricting the domain of the sine function to $-\dfrac{\pi}{2} \le x \le \dfrac{\pi}{2}$. With this restricted domain, every horizontal line that can be drawn between -1 and 1 intersects the graph exactly once. Thus, the restricted function passes the horizontal line test and is one-to-one.

On the restricted domain $-\frac{\pi}{2} \le x \le \frac{\pi}{2}, y = \sin x$ has an inverse function. The inverse of the restricted sine function is called the **inverse sine function**. Two notations are commonly used to denote the inverse sine function:

$$y = \sin^{-1} x \quad \text{or} \quad y = \arcsin x.$$

In this text, we will use $y = \sin^{-1} x$. This notation has the same symbol as the inverse function notation $f^{-1}(x)$.

> ### The Inverse Sine Function
>
> The **inverse sine function**, denoted by $\sin^{-1}$, is the inverse of the restricted sine function $y = \sin x, -\frac{\pi}{2} \le x \le \frac{\pi}{2}$. Thus,
>
> $$y = \sin^{-1} x \quad \text{means} \quad \sin y = x,$$
>
> where $-\frac{\pi}{2} \le y \le \frac{\pi}{2}$ and $-1 \le x \le 1$. We read $y = \sin^{-1} x$ as "y equals the inverse sine at x."

GREAT QUESTION!

Is $\sin^{-1} x$ the same thing as $\frac{1}{\sin x}$?

No. The notation $y = \sin^{-1} x$ does not mean $y = \frac{1}{\sin x}$. The notation $y = \frac{1}{\sin x}$, or the reciprocal of the sine function, is written $y = (\sin x)^{-1}$ and means $y = \csc x$.

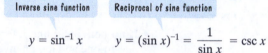

Inverse sine function   Reciprocal of sine function

$$y = \sin^{-1} x \qquad y = (\sin x)^{-1} = \frac{1}{\sin x} = \csc x$$

One way to graph $y = \sin^{-1} x$ is to take points on the graph of the restricted sine function and reverse the order of the coordinates. For example, **Figure 2.30** shows that $\left(-\frac{\pi}{2}, -1\right)$, $(0, 0)$, and $\left(\frac{\pi}{2}, 1\right)$ are on the graph of the restricted sine function. Reversing the order of the coordinates gives $\left(-1, -\frac{\pi}{2}\right)$, $(0, 0)$, and $\left(1, \frac{\pi}{2}\right)$. We now use these three points to sketch the inverse sine function. The graph of $y = \sin^{-1} x$ is shown in **Figure 2.31**.

Another way to obtain the graph of $y = \sin^{-1} x$ is to reflect the graph of the restricted sine function about the line $y = x$, shown in **Figure 2.32**. The red graph is the restricted sine function and the blue graph is the graph of $y = \sin^{-1} x$.

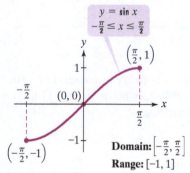

Domain: $\left[-\frac{\pi}{2}, \frac{\pi}{2}\right]$
Range: $[-1, 1]$

FIGURE 2.30 The restricted sine function

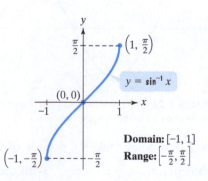

FIGURE 2.31 The graph of the inverse sine function

Domain: $[-1, 1]$
Range: $\left[-\frac{\pi}{2}, \frac{\pi}{2}\right]$

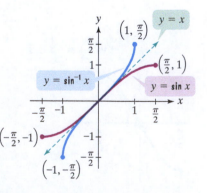

FIGURE 2.32 Using a reflection to obtain the graph of the inverse sine function

Exact values of $\sin^{-1} x$ can be found by thinking of **$\sin^{-1} x$ as the angle in the interval $\left[-\dfrac{\pi}{2}, \dfrac{\pi}{2} \right]$ whose sine is x.** For example, we can use the two endpoints on the blue graph of the inverse sine function in **Figure 2.31** to write

$$\sin^{-1}(-1) = -\frac{\pi}{2} \quad \text{and} \quad \sin^{-1} 1 = \frac{\pi}{2}.$$

The angle whose sine is -1 is $-\dfrac{\pi}{2}$.

The angle whose sine is 1 is $\dfrac{\pi}{2}$.

Because we are thinking of $\sin^{-1} x$ in terms of an angle, we will represent such an angle by θ.

Finding Exact Values of $\sin^{-1} x$

1. Let $\theta = \sin^{-1} x$.

2. Rewrite $\theta = \sin^{-1} x$ as $\sin \theta = x$, where $-\dfrac{\pi}{2} \le \theta \le \dfrac{\pi}{2}$.

3. Use the exact values in **Table 2.5** to find the value of θ in $\left[-\dfrac{\pi}{2}, \dfrac{\pi}{2} \right]$ that satisfies $\sin \theta = x$.

Table 2.5   Exact Values for $\sin \theta, -\dfrac{\pi}{2} \le \theta \le \dfrac{\pi}{2}$

θ	$-\dfrac{\pi}{2}$	$-\dfrac{\pi}{3}$	$-\dfrac{\pi}{4}$	$-\dfrac{\pi}{6}$	0	$\dfrac{\pi}{6}$	$\dfrac{\pi}{4}$	$\dfrac{\pi}{3}$	$\dfrac{\pi}{2}$
$\sin \theta$	-1	$-\dfrac{\sqrt{3}}{2}$	$-\dfrac{\sqrt{2}}{2}$	$-\dfrac{1}{2}$	0	$\dfrac{1}{2}$	$\dfrac{\sqrt{2}}{2}$	$\dfrac{\sqrt{3}}{2}$	1

EXAMPLE 1   **Finding the Exact Value of an Inverse Sine Function**

Find the exact value of $\sin^{-1} \dfrac{\sqrt{2}}{2}$.

SOLUTION

Step 1   Let $\theta = \sin^{-1} x$. Thus,

$$\theta = \sin^{-1} \frac{\sqrt{2}}{2}.$$

We must find the angle θ, $-\dfrac{\pi}{2} \le \theta \le \dfrac{\pi}{2}$, whose sine equals $\dfrac{\sqrt{2}}{2}$.

Step 2   Rewrite $\theta = \sin^{-1} x$ as $\sin \theta = x$, where $-\dfrac{\pi}{2} \le \theta \le \dfrac{\pi}{2}$. Using the definition of the inverse sine function, we rewrite $\theta = \sin^{-1} \dfrac{\sqrt{2}}{2}$ as

$$\sin \theta = \frac{\sqrt{2}}{2}, \text{ where } -\frac{\pi}{2} \le \theta \le \frac{\pi}{2}.$$

Step 3   Use the exact values in Table 2.5 to find the value of θ in $\left[-\dfrac{\pi}{2}, \dfrac{\pi}{2} \right]$ that satisfies $\sin \theta = x$. Table 2.5 shows that the only angle in the interval $\left[-\dfrac{\pi}{2}, \dfrac{\pi}{2} \right]$ that satisfies $\sin \theta = \dfrac{\sqrt{2}}{2}$ is $\dfrac{\pi}{4}$. Thus, $\theta = \dfrac{\pi}{4}$. Because θ, in step 1, represents $\sin^{-1} \dfrac{\sqrt{2}}{2}$, we conclude that

$$\sin^{-1} \frac{\sqrt{2}}{2} = \frac{\pi}{4}. \quad \text{The angle in } \left[-\frac{\pi}{2}, \frac{\pi}{2} \right] \text{ whose sine is } \frac{\sqrt{2}}{2} \text{ is } \frac{\pi}{4}. \quad \bullet\bullet\bullet$$

✅ **Check Point 1** Find the exact value of $\sin^{-1}\dfrac{\sqrt{3}}{2}$.

EXAMPLE 2 Finding the Exact Value of an Inverse Sine Function

Find the exact value of $\sin^{-1}\left(-\dfrac{1}{2}\right)$.

SOLUTION

Step 1 Let $\theta = \sin^{-1}x$. Thus,

$$\theta = \sin^{-1}\left(-\dfrac{1}{2}\right).$$

We must find the angle θ, $-\dfrac{\pi}{2} \le \theta \le \dfrac{\pi}{2}$, whose sine equals $-\dfrac{1}{2}$.

Step 2 Rewrite $\theta = \sin^{-1}x$ as $\sin\theta = x$, where $-\dfrac{\pi}{2} \le \theta \le \dfrac{\pi}{2}$. We rewrite $\theta = \sin^{-1}\left(-\dfrac{1}{2}\right)$ and obtain

$$\sin\theta = -\dfrac{1}{2}, \text{ where } -\dfrac{\pi}{2} \le \theta \le \dfrac{\pi}{2}.$$

Step 3 Use the exact values in Table 2.5 to find the value of θ in $\left[-\dfrac{\pi}{2}, \dfrac{\pi}{2}\right]$ that satisfies $\sin\theta = x$. Table 2.5 on the previous page shows that the only angle in the interval $\left[-\dfrac{\pi}{2}, \dfrac{\pi}{2}\right]$ that satisfies $\sin\theta = -\dfrac{1}{2}$ is $-\dfrac{\pi}{6}$. Thus,

$$\sin^{-1}\left(-\dfrac{1}{2}\right) = -\dfrac{\pi}{6}. \qquad \bullet\bullet\bullet$$

✅ **Check Point 2** Find the exact value of $\sin^{-1}\left(-\dfrac{\sqrt{2}}{2}\right)$.

Some inverse sine expressions cannot be evaluated. Because the domain of the inverse sine function is $[-1, 1]$, it is only possible to evaluate $\sin^{-1}x$ for values of x in this domain. Thus, $\sin^{-1}3$ cannot be evaluated. There is no angle whose sine is 3.

② Understand and use the inverse cosine function.

The Inverse Cosine Function

Figure 2.33 shows how we restrict the domain of the cosine function so that it becomes one-to-one and has an inverse function. Restrict the domain to the interval $[0, \pi]$, shown by the dark blue graph. Over this interval, the restricted cosine function passes the horizontal line test and has an inverse function.

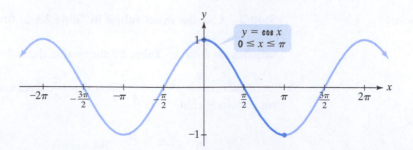

FIGURE 2.33 $y = \cos x$ is one-to-one on the interval $[0, \pi]$.

The Inverse Cosine Function

The **inverse cosine function**, denoted by $\cos^{-1}$, is the inverse of the restricted cosine function $y = \cos x, 0 \le x \le \pi$. Thus,

$$y = \cos^{-1} x \quad \text{means} \quad \cos y = x,$$

where $0 \le y \le \pi$ and $-1 \le x \le 1$.

One way to graph $y = \cos^{-1} x$ is to take points on the graph of the restricted cosine function and reverse the order of the coordinates. For example, **Figure 2.34** shows that $(0, 1)$, $\left(\dfrac{\pi}{2}, 0\right)$, and $(\pi, -1)$ are on the graph of the restricted cosine function. Reversing the order of the coordinates gives $(1, 0)$, $\left(0, \dfrac{\pi}{2}\right)$, and $(-1, \pi)$.

We now use these three points to sketch the inverse cosine function. The graph of $y = \cos^{-1} x$ is shown in **Figure 2.35**. You can also obtain this graph by reflecting the graph of the restricted cosine function about the line $y = x$.

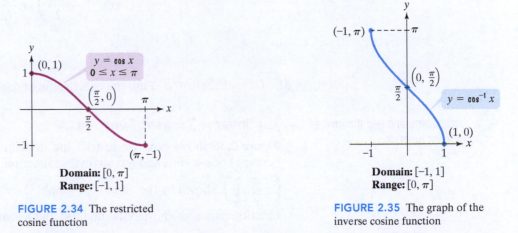

Domain: $[0, \pi]$
Range: $[-1, 1]$

FIGURE 2.34 The restricted cosine function

Domain: $[-1, 1]$
Range: $[0, \pi]$

FIGURE 2.35 The graph of the inverse cosine function

Exact values of $\cos^{-1} x$ can be found by thinking of $\textbf{cos}^{-1} x$ **as the angle in the interval** $[0, \pi]$ **whose cosine is** x.

Finding Exact Values of $\cos^{-1} x$

1. Let $\theta = \cos^{-1} x$.

2. Rewrite $\theta = \cos^{-1} x$ as $\cos \theta = x$, where $0 \le \theta \le \pi$.

3. Use the exact values in **Table 2.6** to find the value of θ in $[0, \pi]$ that satisfies $\cos \theta = x$.

Table 2.6 Exact Values for $\cos \theta, 0 \le \theta \le \pi$

θ	0	$\dfrac{\pi}{6}$	$\dfrac{\pi}{4}$	$\dfrac{\pi}{3}$	$\dfrac{\pi}{2}$	$\dfrac{2\pi}{3}$	$\dfrac{3\pi}{4}$	$\dfrac{5\pi}{6}$	π
$\cos \theta$	1	$\dfrac{\sqrt{3}}{2}$	$\dfrac{\sqrt{2}}{2}$	$\dfrac{1}{2}$	0	$-\dfrac{1}{2}$	$-\dfrac{\sqrt{2}}{2}$	$-\dfrac{\sqrt{3}}{2}$	-1

EXAMPLE 3 Finding the Exact Value of an Inverse Cosine Function

Find the exact value of $\cos^{-1}\left(-\dfrac{\sqrt{3}}{2}\right)$.

SOLUTION

Step 1   **Let $\theta = \cos^{-1}x$. Thus,**

$$\theta = \cos^{-1}\left(-\frac{\sqrt{3}}{2}\right).$$

We must find the angle θ, $0 \leq \theta \leq \pi$, whose cosine equals $-\dfrac{\sqrt{3}}{2}$.

Step 2   **Rewrite $\theta = \cos^{-1}x$ as $\cos\theta = x$, where $0 \leq \theta \leq \pi$.** We obtain

$$\cos\theta = -\frac{\sqrt{3}}{2}, \text{ where } 0 \leq \theta \leq \pi.$$

Step 3   **Use the exact values in Table 2.6 to find the value of θ in $[0, \pi]$ that satisfies $\cos\theta = x$.** Table 2.6 on the previous page shows that the only angle in the interval $[0, \pi]$ that satisfies $\cos\theta = -\dfrac{\sqrt{3}}{2}$ is $\dfrac{5\pi}{6}$. Thus, $\theta = \dfrac{5\pi}{6}$ and

$$\cos^{-1}\left(-\frac{\sqrt{3}}{2}\right) = \frac{5\pi}{6}. \qquad \text{\color{blue}The angle in } [O, \pi] \text{ whose cosine is}$$

$$\color{blue}-\frac{\sqrt{3}}{2} \text{ is } \frac{5\pi}{6}.$$

•••

✅ **Check Point 3**   Find the exact value of $\cos^{-1}\left(-\dfrac{1}{2}\right)$.

③ Understand and use the inverse tangent function.

The Inverse Tangent Function

Figure 2.36 shows how we restrict the domain of the tangent function so that it becomes one-to-one and has an inverse function. Restrict the domain to the interval $\left(-\dfrac{\pi}{2}, \dfrac{\pi}{2}\right)$, shown by the solid blue graph. Over this interval, the restricted tangent function passes the horizontal line test and has an inverse function.

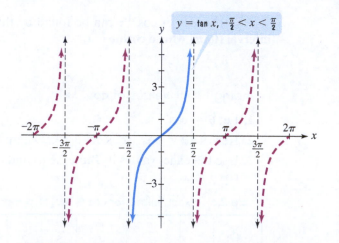

$y = \tan x,\ -\dfrac{\pi}{2} < x < \dfrac{\pi}{2}$

FIGURE 2.36 $y = \tan x$ is one-to-one on the interval $\left(-\dfrac{\pi}{2}, \dfrac{\pi}{2}\right)$.

The Inverse Tangent Function

The **inverse tangent function**, denoted by $\tan^{-1}$, is the inverse of the restricted tangent function $y = \tan x$, $-\dfrac{\pi}{2} < x < \dfrac{\pi}{2}$. Thus,

$$y = \tan^{-1}x \quad \text{means} \quad \tan y = x,$$

where $-\dfrac{\pi}{2} < y < \dfrac{\pi}{2}$ and $-\infty < x < \infty$.

We graph $y = \tan^{-1} x$ by taking points on the graph of the restricted function and reversing the order of the coordinates. **Figure 2.37** shows that $\left(-\frac{\pi}{4}, -1\right)$, $(0, 0)$, and $\left(\frac{\pi}{4}, 1\right)$ are on the graph of the restricted tangent function. Reversing the order gives $\left(-1, -\frac{\pi}{4}\right)$, $(0, 0)$, and $\left(1, \frac{\pi}{4}\right)$. We now use these three points to graph the inverse tangent function. The graph of $y = \tan^{-1} x$ is shown in **Figure 2.38**. Notice that the vertical asymptotes become horizontal asymptotes for the graph of the inverse function.

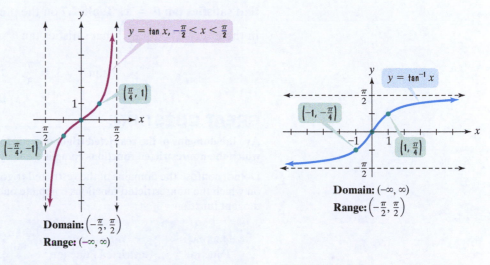

$y = \tan x, -\frac{\pi}{2} < x < \frac{\pi}{2}$
$\left(\frac{\pi}{4}, 1\right)$
$\left(-\frac{\pi}{4}, -1\right)$
1
$-\frac{\pi}{2}$
$\frac{\pi}{2}$
x
y
Domain: $\left(-\frac{\pi}{2}, \frac{\pi}{2}\right)$
Range: $(-\infty, \infty)$

FIGURE 2.37 The restricted tangent function

$y = \tan^{-1} x$
$\left(-1, -\frac{\pi}{4}\right)$
$\left(1, \frac{\pi}{4}\right)$
$\frac{\pi}{2}$
$-\frac{\pi}{2}$
−1
1
x
y
Domain: $(-\infty, \infty)$
Range: $\left(-\frac{\pi}{2}, \frac{\pi}{2}\right)$

FIGURE 2.38 The graph of the inverse tangent function

Exact values of $\tan^{-1} x$ can be found by thinking of **$\tan^{-1} x$ as the angle in the interval $\left(-\frac{\pi}{2}, \frac{\pi}{2}\right)$ whose tangent is x.**

Finding Exact Values of $\tan^{-1} x$

1. Let $\theta = \tan^{-1} x$.
2. Rewrite $\theta = \tan^{-1} x$ as $\tan \theta = x$, where $-\frac{\pi}{2} < \theta < \frac{\pi}{2}$.
3. Use the exact values in **Table 2.7** to find the value of θ in $\left(-\frac{\pi}{2}, \frac{\pi}{2}\right)$ that satisfies $\tan \theta = x$.

Table 2.7 Exact Values for $\tan \theta, -\frac{\pi}{2} < \theta < \frac{\pi}{2}$

θ	$-\frac{\pi}{3}$	$-\frac{\pi}{4}$	$-\frac{\pi}{6}$	0	$\frac{\pi}{6}$	$\frac{\pi}{4}$	$\frac{\pi}{3}$
$\tan \theta$	$-\sqrt{3}$	-1	$-\frac{\sqrt{3}}{3}$	0	$\frac{\sqrt{3}}{3}$	1	$\sqrt{3}$

EXAMPLE 4 Finding the Exact Value of an Inverse Tangent Function

Find the exact value of $\tan^{-1}\sqrt{3}$.

SOLUTION

Step 1 Let $\theta = \tan^{-1} x$. Thus,

$$\theta = \tan^{-1}\sqrt{3}.$$

We must find the angle θ, $-\dfrac{\pi}{2} < \theta < \dfrac{\pi}{2}$, whose tangent equals $\sqrt{3}$.

Step 2 Rewrite $\theta = \tan^{-1} x$ as $\tan \theta = x$, where $-\dfrac{\pi}{2} < \theta < \dfrac{\pi}{2}$. We obtain

$$\tan \theta = \sqrt{3}, \text{ where } -\frac{\pi}{2} < \theta < \frac{\pi}{2}.$$

Step 3 Use the exact values in Table 2.7 to find the value of θ in $\left(-\dfrac{\pi}{2}, \dfrac{\pi}{2}\right)$ that satisfies $\tan \theta = x$. Table 2.7 on the previous page shows that the only angle in the interval $\left(-\dfrac{\pi}{2}, \dfrac{\pi}{2}\right)$ that satisfies $\tan \theta = \sqrt{3}$ is $\dfrac{\pi}{3}$. Thus, $\theta = \dfrac{\pi}{3}$ and

$$\tan^{-1}\sqrt{3} = \frac{\pi}{3}. \qquad \text{The angle in } \left(-\frac{\pi}{2}, \frac{\pi}{2}\right) \text{ whose tangent is } \sqrt{3} \text{ is } \frac{\pi}{3}.$$

• • •

GREAT QUESTION!

Are the domains of the restricted trigonometric functions the same as the intervals on which the nonrestricted functions complete one cycle?

Do not confuse the domains of the restricted trigonometric functions with the intervals on which the nonrestricted functions complete one cycle. They are only the same for the tangent function.

Trigonometric Function	Domain of Restricted Function	Interval on Which Nonrestricted Function's Graph Completes One Period	
$y = \sin x$	$\left[-\dfrac{\pi}{2}, \dfrac{\pi}{2}\right]$	$[0, 2\pi]$	Period: 2π
$y = \cos x$	$[0, \pi]$	$[0, 2\pi]$	Period: 2π
$y = \tan x$	$\left(-\dfrac{\pi}{2}, \dfrac{\pi}{2}\right)$	$\left(-\dfrac{\pi}{2}, \dfrac{\pi}{2}\right)$	Period: π

These domain restrictions are the range for $y = \sin^{-1} x$, $y = \cos^{-1} x$, and $y = \tan^{-1} x$, respectively.

✓ **Check Point 4** Find the exact value of $\tan^{-1}(-1)$.

Table 2.8 summarizes the graphs of the three basic inverse trigonometric functions. Below each of the graphs is a description of the function's domain and range.

Table 2.8 Graphs of the Three Basic Inverse Trigonometric Functions

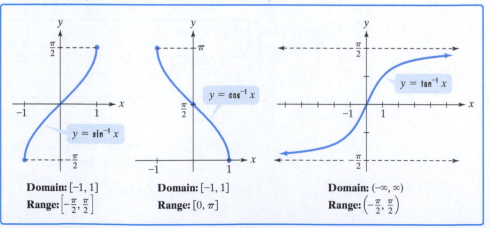

Domain: $[-1, 1]$
Range: $\left[-\dfrac{\pi}{2}, \dfrac{\pi}{2}\right]$

Domain: $[-1, 1]$
Range: $[0, \pi]$

Domain: $(-\infty, \infty)$
Range: $\left(-\dfrac{\pi}{2}, \dfrac{\pi}{2}\right)$

4 Use a calculator to evaluate inverse trigonometric functions.

Using a Calculator to Evaluate Inverse Trigonometric Functions

Calculators give approximate values of inverse trigonometric functions. Use the secondary keys marked $\boxed{\text{SIN}^{-1}}$, $\boxed{\text{COS}^{-1}}$, and $\boxed{\text{TAN}^{-1}}$. These keys are not buttons that you actually press. They are the secondary functions for the buttons labeled $\boxed{\text{SIN}}$, $\boxed{\text{COS}}$, and $\boxed{\text{TAN}}$, respectively. Consult your manual for the location of this feature.

EXAMPLE 5   Calculators and Inverse Trigonometric Functions

Use a calculator to find the value to four decimal places of each function:

a. $\sin^{-1}\dfrac{1}{4}$ **b.** $\tan^{-1}(-9.65)$.

SOLUTION

Scientific Calculator Solution

Function	Mode	Keystrokes	Display, Rounded to Four Places
a. $\sin^{-1}\dfrac{1}{4}$	Radian	1 $\boxed{\div}$ 4 $\boxed{=}$ $\boxed{\text{2nd}}$ $\boxed{\text{SIN}}$	0.2527
b. $\tan^{-1}(-9.65)$	Radian	9.65 $\boxed{+/-}$ $\boxed{\text{2nd}}$ $\boxed{\text{TAN}}$	-1.4675

Graphing Calculator Solution

Function	Mode	Keystrokes	Display, Rounded to Four Places
a. $\sin^{-1}\dfrac{1}{4}$	Radian	$\boxed{\text{2nd}}$ $\boxed{\text{SIN}}$ $\boxed{(}$ $\boxed{1}$ $\boxed{\div}$ $\boxed{4}$ $\boxed{)}$ $\boxed{\text{ENTER}}$	0.2527
b. $\tan^{-1}(-9.65)$	Radian	$\boxed{\text{2nd}}$ $\boxed{\text{TAN}}$ $\boxed{(-)}$ 9.65 $\boxed{\text{ENTER}}$	-1.4675

•••

```
sin⁻¹(1/4)
              .2526802551
tan⁻¹(-9.65)
             -1.467537946
```

✓ **Check Point 5** Use a calculator to find the value to four decimal places of each function:

a. $\cos^{-1}\dfrac{1}{3}$ **b.** $\tan^{-1}(-35.85)$.

GREAT QUESTION!

What happens if I attempt to evaluate an inverse trigonometric function at a value that is not in its domain?

In real number mode, most calculators will display an error message. For example, an error message can result if you attempt to approximate $\cos^{-1}3$. There is no angle whose cosine is 3. The domain of the inverse cosine function is $[-1, 1]$ and 3 does not belong to this domain.

5 Find exact values of composite functions with inverse trigonometric functions.

Composition of Functions Involving Inverse Trigonometric Functions

In our Brief Review of composite and inverse functions, we saw that

$$f(f^{-1}(x)) = x \quad \text{and} \quad f^{-1}(f(x)) = x.$$

x must be in the domain of f^{-1}. x must be in the domain of f.

We apply these properties to the sine, cosine, tangent, and their inverse functions to obtain the following properties:

> **Inverse Properties**
>
> **The Sine Function and Its Inverse**
>
> $$\sin(\sin^{-1} x) = x \qquad \text{for every } x \text{ in the interval } [-1, 1]$$
>
> $$\sin^{-1}(\sin x) = x \qquad \text{for every } x \text{ in the interval } \left[-\frac{\pi}{2}, \frac{\pi}{2}\right]$$
>
> **The Cosine Function and Its Inverse**
>
> $$\cos(\cos^{-1} x) = x \qquad \text{for every } x \text{ in the interval } [-1, 1]$$
>
> $$\cos^{-1}(\cos x) = x \qquad \text{for every } x \text{ in the interval } [0, \pi]$$
>
> **The Tangent Function and Its Inverse**
>
> $$\tan(\tan^{-1} x) = x \qquad \text{for every real number } x$$
>
> $$\tan^{-1}(\tan x) = x \qquad \text{for every } x \text{ in the interval } \left(-\frac{\pi}{2}, \frac{\pi}{2}\right)$$

The restrictions on x in the inverse properties are a bit tricky. For example,

$$\sin^{-1}\left(\sin \frac{\pi}{4}\right) = \frac{\pi}{4}.$$

> $\sin^{-1}(\sin x) = x$ for x in $\left[-\frac{\pi}{2}, \frac{\pi}{2}\right]$.
> Observe that $\frac{\pi}{4}$ is in this interval.

Can we use $\sin^{-1}(\sin x) = x$ to find the exact value of $\sin^{-1}\left(\sin \frac{5\pi}{4}\right)$? Is $\frac{5\pi}{4}$ in the interval $\left[-\frac{\pi}{2}, \frac{\pi}{2}\right]$? No. Thus, to evaluate $\sin^{-1}\left(\sin \frac{5\pi}{4}\right)$, we must first find $\sin \frac{5\pi}{4}$.

> $\frac{5\pi}{4}$ is in quadrant III, where the sine is negative.

$$\sin \frac{5\pi}{4} = -\sin \frac{\pi}{4} = -\frac{\sqrt{2}}{2}$$

> The reference angle for $\frac{5\pi}{4}$ is $\frac{\pi}{4}$.

We evaluate $\sin^{-1}\left(\sin \frac{5\pi}{4}\right)$ as follows:

$$\sin^{-1}\left(\sin \frac{5\pi}{4}\right) = \sin^{-1}\left(-\frac{\sqrt{2}}{2}\right) = -\frac{\pi}{4}.$$ If necessary, see Table 2.5 on page 113.

To determine how to evaluate the composition of functions involving inverse trigonometric functions, first examine the value of x. **You can use the inverse properties in the box only if x is in the specified interval.**

EXAMPLE 6 Evaluating Compositions of Functions and Their Inverses

Find the exact value, if possible:

a. $\cos(\cos^{-1} 0.6)$ **b.** $\sin^{-1}\left(\sin \frac{3\pi}{2}\right)$ **c.** $\cos(\cos^{-1} 1.5)$.

SOLUTION

a. The inverse property $\cos(\cos^{-1} x) = x$ applies for every x in $[-1, 1]$. To evaluate $\cos(\cos^{-1} 0.6)$, observe that $x = 0.6$. This value of x lies in $[-1, 1]$, which is the domain of the inverse cosine function. This means that we can use the inverse property $\cos(\cos^{-1} x) = x$. Thus,

$$\cos(\cos^{-1} 0.6) = 0.6.$$

b. The inverse property $\sin^{-1}(\sin x) = x$ applies for every x in $\left[-\dfrac{\pi}{2}, \dfrac{\pi}{2}\right]$. To evaluate $\sin^{-1}\left(\sin \dfrac{3\pi}{2}\right)$, observe that $x = \dfrac{3\pi}{2}$. This value of x does not lie in $\left[-\dfrac{\pi}{2}, \dfrac{\pi}{2}\right]$. To evaluate this expression, we first find $\sin \dfrac{3\pi}{2}$.

$$\sin^{-1}\left(\sin \frac{3\pi}{2}\right) = \sin^{-1}(-1) = -\frac{\pi}{2} \quad \text{\color{blue}{The angle in } } \left[-\frac{\pi}{2}, \frac{\pi}{2}\right] \text{\color{blue}{ whose sine is } } {-1} \text{\color{blue}{ is }} -\frac{\pi}{2}.$$

c. The inverse property $\cos(\cos^{-1} x) = x$ applies for every x in $[-1, 1]$. To attempt to evaluate $\cos(\cos^{-1} 1.5)$, observe that $x = 1.5$. This value of x does not lie in $[-1, 1]$, which is the domain of the inverse cosine function. Thus, the expression $\cos(\cos^{-1} 1.5)$ is not defined because $\cos^{-1} 1.5$ is not defined. • • •

✓ **Check Point 6** Find the exact value, if possible:

a. $\cos(\cos^{-1} 0.7)$ **b.** $\sin^{-1}(\sin \pi)$ **c.** $\cos[\cos^{-1}(-1.2)]$.

We can use points on terminal sides of angles in standard position to find exact values of expressions involving the composition of a function and a different inverse function. Here are two examples:

$$\cos\left(\tan^{-1} \frac{5}{12}\right) \qquad \cot\left[\sin^{-1}\left(-\frac{1}{3}\right)\right].$$

> Inner part involves the angle in $\left(-\frac{\pi}{2}, \frac{\pi}{2}\right)$ whose tangent is $\frac{5}{12}$.

> Inner part involves the angle in $\left[-\frac{\pi}{2}, \frac{\pi}{2}\right]$ whose sine is $-\frac{1}{3}$.

The inner part of each expression involves an angle. To evaluate such expressions, we represent such angles by θ. Then we use a sketch that illustrates our representation. Examples 7 and 8 show how to carry out such evaluations.

EXAMPLE 7   Evaluating a Composite Trigonometric Expression

Find the exact value of $\cos\left(\tan^{-1} \dfrac{5}{12}\right)$.

SOLUTION

We let θ represent the angle in $\left(-\dfrac{\pi}{2}, \dfrac{\pi}{2}\right)$ whose tangent is $\dfrac{5}{12}$. Thus,

$$\theta = \tan^{-1} \frac{5}{12}.$$

We are looking for the exact value of $\cos\left(\tan^{-1} \dfrac{5}{12}\right)$, with $\theta = \tan^{-1} \dfrac{5}{12}$. Using the definition of the inverse tangent function, we can rewrite $\theta = \tan^{-1} \dfrac{5}{12}$ as

$$\tan \theta = \frac{5}{12}, \quad \text{where} \quad -\frac{\pi}{2} < \theta < \frac{\pi}{2}.$$

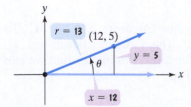

FIGURE 2.39 Representing $\tan \theta = \frac{5}{12}$

Because $\tan \theta$ is positive, θ must be an angle in $\left(0, \frac{\pi}{2}\right)$. Thus, θ is a first-quadrant angle. **Figure 2.39** shows a right triangle in quadrant I with

$$\tan \theta = \frac{5}{12}.$$

Side opposite θ, or y
Side adjacent to θ, or x

The hypotenuse of the triangle, r, or the distance from the origin to $(12, 5)$, is found using $r = \sqrt{x^2 + y^2}$.

$$r = \sqrt{x^2 + y^2} = \sqrt{12^2 + 5^2} = \sqrt{144 + 25} = \sqrt{169} = 13$$

We use the values for x and r to find the exact value of $\cos\left(\tan^{-1}\frac{5}{12}\right)$.

$$\cos\left(\tan^{-1}\frac{5}{12}\right) = \cos\theta = \frac{\text{side adjacent to } \theta, \text{ or } x}{\text{hypotenuse, or } r} = \frac{12}{13}$$

• • •

✓ **Check Point 7** Find the exact value of $\sin\left(\tan^{-1}\frac{3}{4}\right)$.

EXAMPLE 8 Evaluating a Composite Trigonometric Expression

Find the exact value of $\cot\left[\sin^{-1}\left(-\frac{1}{3}\right)\right]$.

SOLUTION

We let θ represent the angle in $\left[-\frac{\pi}{2}, \frac{\pi}{2}\right]$ whose sine is $-\frac{1}{3}$. Thus,

$$\theta = \sin^{-1}\left(-\frac{1}{3}\right) \quad \text{and} \quad \sin\theta = -\frac{1}{3}, \quad \text{where} \quad -\frac{\pi}{2} \le \theta \le \frac{\pi}{2}.$$

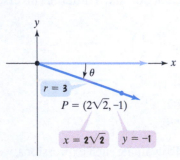

FIGURE 2.40 Representing $\sin\theta = -\frac{1}{3}$

Because $\sin\theta$ is negative in $\sin\theta = -\frac{1}{3}$, θ must be an angle in $\left[-\frac{\pi}{2}, 0\right)$. Thus, θ is a negative angle that lies in quadrant IV. **Figure 2.40** shows angle θ in quadrant IV with

In quadrant IV, y is negative.

$$\sin\theta = -\frac{1}{3} = \frac{y}{r} = \frac{-1}{3}.$$

Thus, $y = -1$ and $r = 3$. The value of x can be found using $r = \sqrt{x^2 + y^2}$ or $x^2 + y^2 = r^2$.

$$x^2 + (-1)^2 = 3^2 \qquad \text{Use } x^2 + y^2 = r^2 \text{ with } y = -1 \text{ and } r = 3.$$
$$x^2 + 1 = 9 \qquad \text{Square } -1 \text{ and square 3.}$$
$$x^2 = 8 \qquad \text{Subtract 1 from both sides.}$$
$$x = \sqrt{8} = \sqrt{4 \cdot 2} = 2\sqrt{2} \qquad \text{Use the square root property. Remember that } x \text{ is positive in quadrant IV.}$$

We use values for x and y to find the exact value of $\cot\left[\sin^{-1}\left(-\frac{1}{3}\right)\right]$.

$$\cot\left[\sin^{-1}\left(-\frac{1}{3}\right)\right] = \cot\theta = \frac{x}{y} = \frac{2\sqrt{2}}{-1} = -2\sqrt{2}$$

• • •

✓ **Check Point 8** Find the exact value of $\cos\left[\sin^{-1}\left(-\frac{1}{2}\right)\right]$.

Some composite functions with inverse trigonometric functions can be simplified to algebraic expressions. To simplify such an expression, we represent the inverse trigonometric function in the expression by θ. Then we use a right triangle.

> ### EXAMPLE 9   Simplifying an Expression Involving $\sin^{-1} x$

If $0 < x \le 1$, write $\cos(\sin^{-1} x)$ as an algebraic expression in x.

SOLUTION

We let θ represent the angle in $\left[-\dfrac{\pi}{2}, \dfrac{\pi}{2} \right]$ whose sine is x. Thus,

$$\theta = \sin^{-1} x \quad \text{and} \quad \sin \theta = x, \quad \text{where} \quad -\frac{\pi}{2} \le \theta \le \frac{\pi}{2}.$$

Because $0 < x \le 1$, $\sin \theta$ is positive. Thus, θ is a first-quadrant angle and can be represented as an acute angle of a right triangle. **Figure 2.41** shows a right triangle with

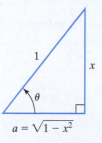

$$\sin \theta = x = \frac{x}{1}. \quad \boxed{\text{Side opposite } \theta} \quad \boxed{\text{Hypotenuse}}$$

FIGURE 2.41 Representing $\sin \theta = x$

The third side, a in **Figure 2.41**, can be found using the Pythagorean Theorem.

$$a^2 + x^2 = 1^2 \qquad \text{Apply the Pythagorean Theorem to the right triangle in Figure 2.41.}$$

$$a^2 = 1 - x^2 \qquad \text{Subtract } x^2 \text{ from both sides.}$$

$$a = \sqrt{1 - x^2} \qquad \text{Use the square root property and solve for } a. \text{ Remember that side } a \text{ is positive.}$$

We use the right triangle in **Figure 2.41** to write $\cos(\sin^{-1} x)$ as an algebraic expression.

$$\cos(\sin^{-1} x) = \cos \theta = \frac{\text{side adjacent to } \theta}{\text{hypotenuse}} = \frac{\sqrt{1 - x^2}}{1} = \sqrt{1 - x^2} \qquad \bullet\bullet\bullet$$

✓ **Check Point 9** If $x > 0$, write $\sec(\tan^{-1} x)$ as an algebraic expression in x.

The inverse secant function, $y = \sec^{-1} x$, is used in calculus. However, inverse cotangent and inverse cosecant functions are rarely used. Two of these remaining inverse trigonometric functions are briefly developed in the Exercise Set (Exercises 73 and 74) that follows.

ACHIEVING SUCCESS

Write down all the steps. In this textbook, examples are written that provide step-by-step solutions. No steps are omitted and each step is explained to the right of the mathematics. Some professors are careful to write down every step of a problem as they are doing it in class; others aren't so fastidious. In either case, write down what the professor puts up and when you get home, fill in whatever steps have been omitted (if any). In your math work, including homework and tests, show clear step-by-step solutions. Detailed solutions help organize your thoughts and enhance understanding. Doing too many steps mentally often results in preventable mistakes.

CONCEPT AND VOCABULARY CHECK

Fill in each blank so that the resulting statement is true.

1. By restricting the domain of $y = \sin x$ to

   _____, the restricted sine function has an inverse function. The inverse sine function is denoted by $y =$ _____.

2. By restricting the domain of $y = \cos x$ to _____, the restricted cosine function has an inverse function. The inverse cosine function is denoted by $y =$ _____.

3. By restricting the domain of $y = \tan x$ to

   _____, the restricted tangent function has an inverse function. The inverse of the tangent function is denoted by $y =$ _____.

4. The domain of $y = \sin^{-1} x$ is _____ and the range

   is _____.

5. The domain of $y = \cos^{-1} x$ is _____ and the range is _____.

6. The domain of $y = \tan^{-1} x$ is _____ and the

   range is _____.

7. $\sin^{-1}(\sin x) = x$ for every x in the interval

   _____.

8. $\cos^{-1}(\cos x) = x$ for every x in the interval _____.

9. $\tan^{-1}(\tan x) = x$ for every x in the interval

   _____.

10. True or false: $\cos^{-1} x = \dfrac{1}{\cos x}$ _____

EXERCISE SET 2.3

Practice Exercises

In Exercises 1–18, find the exact value of each expression.

1. $\sin^{-1}\dfrac{1}{2}$
2. $\sin^{-1} 0$
3. $\sin^{-1}\dfrac{\sqrt{2}}{2}$
4. $\sin^{-1}\dfrac{\sqrt{3}}{2}$
5. $\sin^{-1}\left(-\dfrac{1}{2}\right)$
6. $\sin^{-1}\left(-\dfrac{\sqrt{3}}{2}\right)$
7. $\cos^{-1}\dfrac{\sqrt{3}}{2}$
8. $\cos^{-1}\dfrac{\sqrt{2}}{2}$
9. $\cos^{-1}\left(-\dfrac{\sqrt{2}}{2}\right)$
10. $\cos^{-1}\left(-\dfrac{\sqrt{3}}{2}\right)$
11. $\cos^{-1} 0$
12. $\cos^{-1} 1$
13. $\tan^{-1}\dfrac{\sqrt{3}}{3}$
14. $\tan^{-1} 1$
15. $\tan^{-1} 0$
16. $\tan^{-1}(-1)$
17. $\tan^{-1}(-\sqrt{3})$
18. $\tan^{-1}\left(-\dfrac{\sqrt{3}}{3}\right)$

In Exercises 19–30, use a calculator to find the value of each expression rounded to two decimal places.

19. $\sin^{-1} 0.3$
20. $\sin^{-1} 0.47$
21. $\sin^{-1}(-0.32)$
22. $\sin^{-1}(-0.625)$
23. $\cos^{-1}\dfrac{3}{8}$
24. $\cos^{-1}\dfrac{4}{9}$
25. $\cos^{-1}\dfrac{\sqrt{5}}{7}$
26. $\cos^{-1}\dfrac{\sqrt{7}}{10}$
27. $\tan^{-1}(-20)$
28. $\tan^{-1}(-30)$
29. $\tan^{-1}(-\sqrt{473})$
30. $\tan^{-1}(-\sqrt{5061})$

In Exercises 31–46, find the exact value of each expression, if possible. Do not use a calculator.

31. $\sin(\sin^{-1} 0.9)$
32. $\cos(\cos^{-1} 0.57)$
33. $\sin^{-1}\left(\sin\dfrac{\pi}{3}\right)$
34. $\cos^{-1}\left(\cos\dfrac{2\pi}{3}\right)$
35. $\sin^{-1}\left(\sin\dfrac{5\pi}{6}\right)$
36. $\cos^{-1}\left(\cos\dfrac{4\pi}{3}\right)$
37. $\tan(\tan^{-1} 125)$
38. $\tan(\tan^{-1} 380)$
39. $\tan^{-1}\left[\tan\left(-\dfrac{\pi}{6}\right)\right]$
40. $\tan^{-1}\left[\tan\left(-\dfrac{\pi}{3}\right)\right]$
41. $\tan^{-1}\left(\tan\dfrac{2\pi}{3}\right)$
42. $\tan^{-1}\left(\tan\dfrac{3\pi}{4}\right)$
43. $\sin^{-1}(\sin \pi)$
44. $\cos^{-1}(\cos 2\pi)$
45. $\sin(\sin^{-1}\pi)$
46. $\cos(\cos^{-1} 3\pi)$

In Exercises 47–62, use a sketch to find the exact value of each expression.

47. $\cos\left(\sin^{-1}\dfrac{4}{5}\right)$
48. $\sin\left(\tan^{-1}\dfrac{7}{24}\right)$
49. $\tan\left(\cos^{-1}\dfrac{5}{13}\right)$
50. $\cot\left(\sin^{-1}\dfrac{5}{13}\right)$
51. $\tan\left[\sin^{-1}\left(-\dfrac{3}{5}\right)\right]$
52. $\cos\left[\sin^{-1}\left(-\dfrac{4}{5}\right)\right]$
53. $\sin\left(\cos^{-1}\dfrac{\sqrt{2}}{2}\right)$
54. $\cos\left(\sin^{-1}\dfrac{1}{2}\right)$
55. $\sec\left[\sin^{-1}\left(-\dfrac{1}{4}\right)\right]$
56. $\sec\left[\sin^{-1}\left(-\dfrac{1}{2}\right)\right]$
57. $\tan\left[\cos^{-1}\left(-\dfrac{1}{3}\right)\right]$
58. $\tan\left[\cos^{-1}\left(-\dfrac{1}{4}\right)\right]$
59. $\csc\left[\cos^{-1}\left(-\dfrac{\sqrt{3}}{2}\right)\right]$
60. $\sec\left[\sin^{-1}\left(-\dfrac{\sqrt{2}}{2}\right)\right]$
61. $\cos\left[\tan^{-1}\left(-\dfrac{2}{3}\right)\right]$
62. $\sin\left[\tan^{-1}\left(-\dfrac{3}{4}\right)\right]$

In Exercises 63–72, use a right triangle to write each expression as an algebraic expression. Assume that x is positive and that the given inverse trigonometric function is defined for the expression in x.

63. $\tan(\cos^{-1} x)$

64. $\sin(\tan^{-1} x)$

65. $\cos(\sin^{-1} 2x)$

66. $\sin(\cos^{-1} 2x)$

67. $\cos\left(\sin^{-1}\dfrac{1}{x}\right)$

68. $\sec\left(\cos^{-1}\dfrac{1}{x}\right)$

69. $\cot\left(\tan^{-1}\dfrac{x}{\sqrt{3}}\right)$

70. $\cot\left(\tan^{-1}\dfrac{x}{\sqrt{2}}\right)$

71. $\sec\left(\sin^{-1}\dfrac{x}{\sqrt{x^2+4}}\right)$

72. $\cot\left(\sin^{-1}\dfrac{\sqrt{x^2-9}}{x}\right)$

73. a. Graph the restricted secant function, $y = \sec x$, by restricting x to the intervals $\left[0, \dfrac{\pi}{2}\right)$ and $\left(\dfrac{\pi}{2}, \pi\right]$.

  b. Use the horizontal line test to explain why the restricted secant function has an inverse function.

  c. Use the graph of the restricted secant function to graph $y = \sec^{-1} x$.

74. a. Graph the restricted cotangent function, $y = \cot x$, by restricting x to the interval $(0, \pi)$.

  b. Use the horizontal line test to explain why the restricted cotangent function has an inverse function.

  c. Use the graph of the restricted cotangent function to graph $y = \cot^{-1} x$.

Practice Plus

*The graphs of $y = \sin^{-1} x$, $y = \cos^{-1} x$, and $y = \tan^{-1} x$ are shown in **Table 2.8** on page 118. In Exercises 75–84, use transformations (vertical shifts, horizontal shifts, reflections, stretching, or shrinking) of these graphs to graph each function. Then use interval notation to give the function's domain and range.*

75. $f(x) = \sin^{-1} x + \dfrac{\pi}{2}$

76. $f(x) = \cos^{-1} x + \dfrac{\pi}{2}$

77. $g(x) = \cos^{-1}(x + 1)$

78. $g(x) = \sin^{-1}(x + 1)$

79. $h(x) = -2\tan^{-1} x$

80. $h(x) = -3\tan^{-1} x$

81. $f(x) = \sin^{-1}(x - 2) - \dfrac{\pi}{2}$

82. $f(x) = \cos^{-1}(x - 2) - \dfrac{\pi}{2}$

83. $g(x) = \cos^{-1}\dfrac{x}{2}$

84. $g(x) = \sin^{-1}\dfrac{x}{2}$

In Exercises 85–92, determine the domain and the range of each function.

85. $f(x) = \sin(\sin^{-1} x)$

86. $f(x) = \cos(\cos^{-1} x)$

87. $f(x) = \cos^{-1}(\cos x)$

88. $f(x) = \sin^{-1}(\sin x)$

89. $f(x) = \sin^{-1}(\cos x)$

90. $f(x) = \cos^{-1}(\sin x)$

91. $f(x) = \sin^{-1} x + \cos^{-1} x$

92. $f(x) = \cos^{-1} x - \sin^{-1} x$

Application Exercises

93. Your neighborhood movie theater has a 25-foot-high screen located 8 feet above your eye level. If you sit too close to the screen, your viewing angle is too small, resulting in a distorted picture. By contrast, if you sit too far back, the image is quite small, diminishing the movie's visual impact. If you sit x feet back from the screen, your viewing angle, θ, is given by

$$\theta = \tan^{-1}\dfrac{33}{x} - \tan^{-1}\dfrac{8}{x}.$$

25 feet

8 feet

Find the viewing angle, in radians, at distances of 5 feet, 10 feet, 15 feet, 20 feet, and 25 feet.

94. The function $\theta = \tan^{-1}\dfrac{33}{x} - \tan^{-1}\dfrac{8}{x}$, described in Exercise 93, is graphed below in a $[0, 50, 10]$ by $[0, 1, 0.1]$ viewing rectangle. Use the graph to describe what happens to your viewing angle as you move farther back from the screen. How far back from the screen, to the nearest foot, should you sit to maximize your viewing angle? Verify this observation by finding the viewing angle one foot closer to the screen and one foot farther from the screen for this ideal viewing distance.

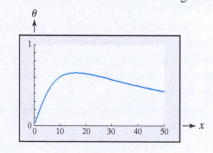

The formula

$$\theta = 2\tan^{-1}\dfrac{21.634}{x}$$

gives the viewing angle, θ, in radians, for a camera whose lens is x millimeters wide. Use this formula to solve Exercises 95–96.

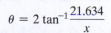

95. Find the viewing angle, in radians and in degrees (to the nearest tenth of a degree), of a 28-millimeter lens.

96. Find the viewing angle, in radians and in degrees (to the nearest tenth of a degree), of a 300-millimeter telephoto lens.

For years, mathematicians were challenged by the following problem: What is the area of a region under a curve between two values of x? The problem was solved in the seventeenth century with the development of integral calculus. Using calculus, the area of the region under $y = \dfrac{1}{x^2 + 1}$, above the x-axis, and between $x = a$ and $x = b$ is $\tan^{-1} b - \tan^{-1} a$. Use this result, shown in the figure, to find the area of the region under $y = \dfrac{1}{x^2 + 1}$, above the x-axis, and between the values of a and b given in Exercises 97–98.

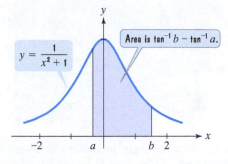

$y = \dfrac{1}{x^2 + 1}$

Area is $\tan^{-1} b - \tan^{-1} a$.

97. $a = 0$ and $b = 2$

98. $a = -2$ and $b = 1$

Explaining the Concepts

99. Explain why, without restrictions, no trigonometric function has an inverse function.

100. Describe the restriction on the sine function so that it has an inverse function.

101. How can the graph of $y = \sin^{-1} x$ be obtained from the graph of the restricted sine function?

102. Without drawing a graph, describe the behavior of the graph of $y = \sin^{-1} x$. Mention the function's domain and range in your description.

103. Describe the restriction on the cosine function so that it has an inverse function.

104. Without drawing a graph, describe the behavior of the graph of $y = \cos^{-1} x$. Mention the function's domain and range in your description.

105. Describe the restriction on the tangent function so that it has an inverse function.

106. Without drawing a graph, describe the behavior of the graph of $y = \tan^{-1} x$. Mention the function's domain and range in your description.

107. If $\sin^{-1}\left(\sin\dfrac{\pi}{3}\right) = \dfrac{\pi}{3}$, is $\sin^{-1}\left(\sin\dfrac{5\pi}{6}\right) = \dfrac{5\pi}{6}$? Explain your answer.

108. Explain how a right triangle can be used to find the exact value of $\sec\left(\sin^{-1}\dfrac{4}{5}\right)$.

109. Find the height of the screen and the number of feet that it is located above eye level in your favorite movie theater. Modify the formula given in Exercise 93 so that it applies to your theater. Then describe where in the theater you should sit so that a movie creates the greatest visual impact.

Technology Exercises

In Exercises 110–113, graph each pair of functions in the same viewing rectangle. Use your knowledge of the domain and range for the inverse trigonometric function to select an appropriate viewing rectangle. How is the graph of the second equation in each exercise related to the graph of the first equation?

110. $y = \sin^{-1} x$ and $y = \sin^{-1} x + 2$

111. $y = \cos^{-1} x$ and $y = \cos^{-1}(x - 1)$

112. $y = \tan^{-1} x$ and $y = -2\tan^{-1} x$

113. $y = \sin^{-1} x$ and $y = \sin^{-1}(x + 2) + 1$

114. Graph $y = \tan^{-1} x$ and its two horizontal asymptotes in a $[-3, 3, 1]$ by $\left[-\pi, \pi, \dfrac{\pi}{2}\right]$ viewing rectangle. Then change the viewing rectangle to $[-50, 50, 5]$ by $\left[-\pi, \pi, \dfrac{\pi}{2}\right]$. What do you observe?

115. Graph $y = \sin^{-1} x + \cos^{-1} x$ in a $[-2, 2, 1]$ by $[0, 3, 1]$ viewing rectangle. What appears to be true about the sum of the inverse sine and inverse cosine for values between -1 and 1, inclusive?

Critical Thinking Exercises

Make Sense? In Exercises 116–119, determine whether each statement makes sense or does not make sense, and explain your reasoning.

116. Because $y = \sin x$ has an inverse function if x is restricted to $\left[-\dfrac{\pi}{2}, \dfrac{\pi}{2}\right]$, they should make restrictions easier to remember by also using $\left[-\dfrac{\pi}{2}, \dfrac{\pi}{2}\right]$ as the restriction for $y = \cos x$.

117. Because $y = \sin x$ has an inverse function if x is restricted to $\left[-\dfrac{\pi}{2}, \dfrac{\pi}{2}\right]$, they should make restrictions easier to remember by also using $\left[-\dfrac{\pi}{2}, \dfrac{\pi}{2}\right]$ as the restriction for $y = \tan x$.

118. Although $\sin^{-1}\left(-\dfrac{1}{2}\right)$ is negative, $\cos^{-1}\left(-\dfrac{1}{2}\right)$ is positive.

119. I used $f^{-1}(f(x)) = x$ and concluded that $\sin^{-1}\left(\sin\dfrac{5\pi}{4}\right) = \dfrac{5\pi}{4}$.

120. Solve $y = 2\sin^{-1}(x - 5)$ for x in terms of y.

121. Solve for x: $2\sin^{-1} x = \dfrac{\pi}{4}$.

122. Prove that if $x > 0$, $\tan^{-1} x + \tan^{-1}\dfrac{1}{x} = \dfrac{\pi}{2}$.

123. Derive the formula for θ, your viewing angle at the movie theater, in Exercise 93. *Hint:* Use the figure shown and represent the acute angle on the left in the smaller right triangle by α. Find expressions for $\tan \alpha$ and $\tan(\alpha + \theta)$.

Retaining the Concepts

124. Given $\tan \theta = -\dfrac{2}{3}$ and $\cos \theta < 0$, find $\sin \theta$ and $\sec \theta$.
(Section 1.3, Example 4)

125. Use a reference angle to find the exact value of $\sin 210°$.
(Section 1.3, Example 7)

126. Determine the amplitude and period of $y = 3\cos 2\pi x$. Then graph the function for $-4 \leq x \leq 4$.
(Section 2.1, Example 5)

Preview Exercises

Exercises 127–129 will help you prepare for the material covered in the next section.

127. Use trigonometric functions to find a and c to two decimal places.

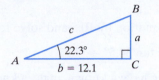

128. Find θ to the nearest tenth of a degree.

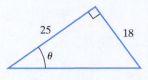

129. Determine the amplitude and period of $y = 10 \cos \dfrac{\pi}{6} x$.

Section 2.4 Applications of Trigonometric Functions

What am I supposed to learn?

After studying this section, you should be able to:

❶ Solve a right triangle.

❷ Solve problems involving bearings.

❸ Model simple harmonic motion.

In the late 1960s, popular musicians were searching for new sounds. Film composers were looking for ways to create unique sounds as well. From these efforts, synthesizers that electronically reproduce musical sounds were born. From providing the backbone of today's most popular music to providing the strange sounds for the most experimental music, synthesizing programs now available on computers are at the forefront of music technology.

If we did not understand the periodic nature of sinusoidal functions, the synthesizing programs used in almost all forms of music would not exist. In this section, we look at applications of trigonometric functions in solving right triangles and in modeling periodic phenomena such as sound.

❶ Solve a right triangle.

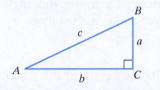

FIGURE 2.42 Labeling right triangles

Solving Right Triangles

Solving a right triangle means finding the missing lengths of its sides and the measurements of its angles. We will label right triangles so that side a is opposite angle A, side b is opposite angle B, and side c, the hypotenuse, is opposite right angle C. **Figure 2.42** illustrates this labeling.

When solving a right triangle, we will use the sine, cosine, and tangent functions, rather than their reciprocals. Example 1 shows how to solve a right triangle when we know the length of a side and the measure of an acute angle.

EXAMPLE 1   Solving a Right Triangle

Solve the right triangle shown in **Figure 2.43**, rounding lengths to two decimal places.

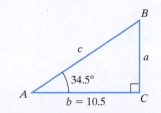

FIGURE 2.43 Find B, a, and c.

SOLUTION

We begin by finding the measure of angle B. We do not need a trigonometric function to do so. Because $C = 90°$ and the sum of a triangle's angles is $180°$, we see that $A + B = 90°$. Thus,

$$B = 90° - A = 90° - 34.5° = 55.5°.$$

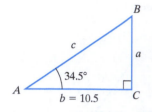

FIGURE 2.43 (repeated)

Now we need to find a. Because we have a known angle, an unknown opposite side, and a known adjacent side, we use the tangent function.

$$\tan 34.5° = \frac{a}{10.5}$$

> Side opposite the 34.5° angle
>
> Side adjacent to the 34.5° angle

Now we multiply both sides of this equation by 10.5 and solve for a.

$$a = 10.5 \tan 34.5° \approx 7.22$$

Finally, we need to find c. Because we have a known angle, a known adjacent side, and an unknown hypotenuse, we use the cosine function.

$$\cos 34.5° = \frac{10.5}{c}$$

> Side adjacent to the 34.5° angle
>
> Hypotenuse

DISCOVERY

There is often more than one correct way to solve a right triangle. In Example 1, find a using angle $B = 55.5°$. Find c using the Pythagorean Theorem.

Now we multiply both sides of $\cos 34.5° = \dfrac{10.5}{c}$ by c and then solve for c.

$$c \cos 34.5° = 10.5 \qquad \text{Multiply both sides by } c.$$

$$c = \frac{10.5}{\cos 34.5°} \approx 12.74 \qquad \begin{array}{l}\text{Divide both sides by } \cos 34.5° \\ \text{and solve for } c.\end{array}$$

In summary, $B = 55.5°$, $a \approx 7.22$, and $c \approx 12.74$.   •••

TECHNOLOGY

When using trigonometric functions to solve for angle measures and lengths of sides, we usually are requried to round the answers. It is good practice to avoid using these rounded values in further calculations. Most calculators will allow you to store the unrounded values and then recall them when they are needed. Consult your manual.

✓ **Check Point 1** In **Figure 2.42** on the previous page, let $A = 62.7°$ and $a = 8.4$. Solve the right triangle, rounding lengths to two decimal places.

Trigonometry was first developed to measure heights and distances that were inconvenient or impossible to measure directly. In solving application problems, begin by making a sketch involving a right triangle that illustrates the problem's conditions. Then put your knowledge of solving right triangles to work and find the required distance or height.

EXAMPLE 2   Finding a Side of a Right Triangle

From a point on level ground 125 feet from the base of a tower, the angle of elevation is 57.2°. Approximate the height of the tower to the nearest foot.

SOLUTION

A sketch is shown in **Figure 2.44**, where a represents the height of the tower. In the right triangle, we have a known angle, an unknown opposite side, and a known adjacent side. Therefore, we use the tangent function.

$$\tan 57.2° = \frac{a}{125}$$

> Side opposite the 57.2° angle
>
> Side adjacent to the 57.2° angle

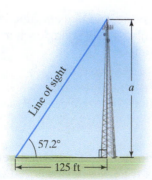

FIGURE 2.44 Determining height without using direct measurement

Now we multiply both sides of this equation by 125 and solve for a.

$$a = 125 \tan 57.2° \approx 194$$

The tower is approximately 194 feet high.   •••

✓ **Check Point 2** From a point on level ground 80 feet from the base of the Eiffel Tower, the angle of elevation is 85.4°. Approximate the height of the Eiffel Tower to the nearest foot.

Example 3 illustrates how to find the measure of an acute angle of a right triangle if the lengths of two sides are known.

EXAMPLE 3   Finding an Angle of a Right Triangle

A kite flies at a height of 30 feet when 65 feet of string is out. If the string is in a straight line, find the angle that it makes with the ground. Round to the nearest tenth of a degree.

SOLUTION

A sketch is shown in **Figure 2.45**, where A represents the angle the string makes with the ground. In the right triangle, we have an unknown angle, a known opposite side, and a known hypotenuse. Therefore, we use the sine function.

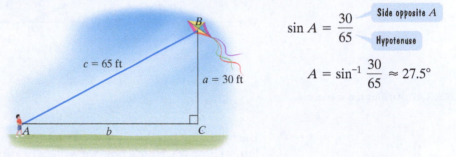

$$\sin A = \frac{30}{65} \quad \begin{array}{l}\text{Side opposite } A\\[4pt]\text{Hypotenuse}\end{array}$$

$$A = \sin^{-1}\frac{30}{65} \approx 27.5°$$

FIGURE 2.45 Flying a kite

The string makes an angle of approximately 27.5° with the ground.   •••

✓ **Check Point 3** A guy wire is 13.8 yards long and is attached from the ground to a pole 6.7 yards above the ground. Find the angle, to the nearest tenth of a degree, that the wire makes with the ground.

EXAMPLE 4   Using Two Right Triangles to Solve a Problem

You are taking your first hot-air balloon ride. Your friend is standing on level ground, 100 feet away from your point of launch, making a video of the terrified look on your rapidly ascending face. How rapidly? At one instant, the angle of elevation from the video camera to your face is 31.7°. One minute later, the angle of elevation is 76.2°. How far did you travel, to the nearest tenth of a foot, during that minute?

SOLUTION

A sketch that illustrates the problem is shown in **Figure 2.46**. We need to determine $b - a$, the distance traveled during the one-minute period. We find a using the small right triangle. Because we have a known angle, an unknown opposite side, and a known adjacent side, we use the tangent function.

$$\tan 31.7° = \frac{a}{100} \quad \begin{array}{l}\text{Side opposite the 31.7° angle}\\[4pt]\text{Side adjacent to the 31.7° angle}\end{array}$$

$$a = 100 \tan 31.7° \approx 61.8$$

We find b using the tangent function in the large right triangle.

$$\tan 76.2° = \frac{b}{100} \quad \begin{array}{l}\text{Side opposite the 76.2° angle}\\[4pt]\text{Side adjacent to the 76.2° angle}\end{array}$$

$$b = 100 \tan 76.2° \approx 407.1$$

The balloon traveled $407.1 - 61.8$, or approximately 345.3 feet, during the minute.   •••

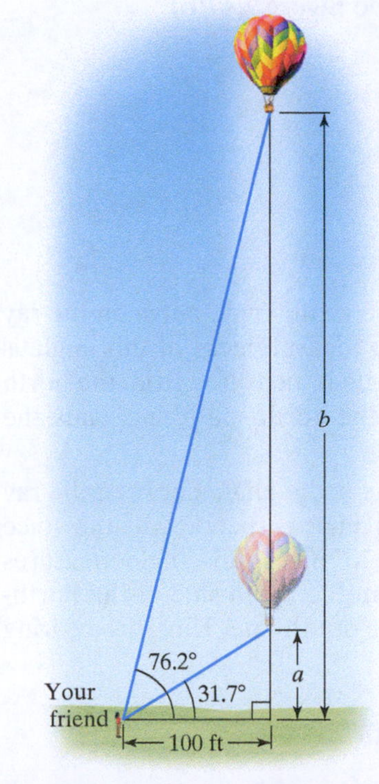

FIGURE 2.46 Ascending in a hot-air balloon

✓ **Check Point 4** You are standing on level ground 800 feet from Mt. Rushmore, looking at the sculpture of Abraham Lincoln's face. The angle of elevation to the bottom of the sculpture is 32° and the angle of elevation to the top is 35°. Find the height of the sculpture of Lincoln's face to the nearest tenth of a foot.

2 Solve problems involving bearings.

Trigonometry and Bearings

In navigation and surveying problems, the term *bearing* is used to specify the location of one point relative to another. The **bearing** from point O to point P is the acute angle, measured in degrees, between ray OP and a north-south line.

The bearing from O to P can also be described using the phrase "the bearing of P from O." **Figure 2.47** illustrates some examples of bearings. The north-south line and the east-west line intersect at right angles.

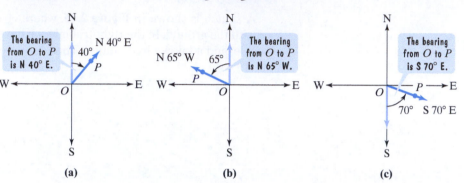

FIGURE 2.47 An illustration of three bearings

(a)   (b)   (c)

Each bearing has three parts: a letter (N or S), the measure of an acute angle, and a letter (E or W). Here's how we write a bearing:

- If the acute angle is measured from the *north side* of the north-south line, then we write N first. [See **Figure 2.47(a)**.] If the acute angle is measured from the *south side* of the north-south line, then we write S first. [See **Figure 2.47(c)**.]
- Second, we write the measure of the acute angle.
- If the acute angle is measured on the *east side* of the north-south line, then we write E last. [See **Figure 2.47(a)**]. If the acute angle is measured on the *west side* of the north-south line, then we write W last. [See **Figure 2.47(b)**.]

EXAMPLE 5   Understanding Bearings

Use **Figure 2.48** to find each of the following:

a. the bearing from O to B

b. the bearing from O to A.

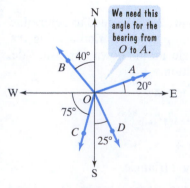

FIGURE 2.48 Finding bearings

SOLUTION

a. To find the bearing from O to B, we need the acute angle between the ray OB and the north-south line through O. The measurement of this angle is given to be 40°. **Figure 2.48** shows that the angle is measured from the north side of the north-south line and lies west of the north-south line. Thus, the bearing from O to B is N 40° W.

b. To find the bearing from O to A, we need the acute angle between the ray OA and the north-south line through O. This angle is specified by the voice balloon in **Figure 2.48**. Because of the given 20° angle, this angle measures 90° − 20°, or 70°. This angle is measured from the north side of the north-south line. This angle is also east of the north-south line. Thus, the bearing from O to A is N 70° E.   • • •

✓ **Check Point 5** Use **Figure 2.48** to find each of the following:

a. the bearing from O to D

b. the bearing from O to C.

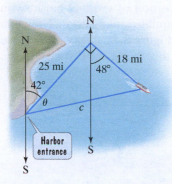

FIGURE 2.49 Finding a boat's bearing from the harbor entrance

GREAT QUESTION!

I can follow Example 6 because Figure 2.49 is given. What should I do if I have to draw the figure?

When making a diagram showing bearings, draw a north-south line through each point at which a change in course occurs. The north side of the line lies above each point. The south side of the line lies below each point.

EXAMPLE 6 Finding the Bearing of a Boat

A boat leaves the entrance to a harbor and travels 25 miles on a bearing of N 42° E. **Figure 2.49** shows that the captain then turns the boat 90° clockwise and travels 18 miles on a bearing of S 48° E. At that time:

a. How far is the boat, to the nearest tenth of a mile, from the harbor entrance?

b. What is the bearing, to the nearest tenth of a degree, of the boat from the harbor entrance?

SOLUTION

a. The boat's distance from the harbor entrance is represented by c in **Figure 2.49**. Because we know the length of two sides of the right triangle, we find c using the Pythagorean Theorem. We have

$$c^2 = a^2 + b^2 = 25^2 + 18^2 = 949$$
$$c = \sqrt{949} \approx 30.8.$$

The boat is approximately 30.8 miles from the harbor entrance.

b. The bearing of the boat from the harbor entrance means the bearing from the harbor entrance to the boat. Look at the north-south line passing through the harbor entrance on the left in **Figure 2.49**. The acute angle from this line to the ray on which the boat lies is $42° + \theta$. Because we are measuring the angle from the north side of the line and the boat is east of the harbor, its bearing from the harbor entrance is N$(42° + \theta)$E. To find θ, we use the right triangle shown in **Figure 2.49** and the tangent function.

$$\tan \theta = \frac{\text{side opposite } \theta}{\text{side adjacent to } \theta} = \frac{18}{25}$$
$$\theta = \tan^{-1} \frac{18}{25}$$

We can use a calculator in degree mode to find the value of θ: $\theta \approx 35.8°$. Thus, $42° + \theta \approx 42° + 35.8° = 77.8°$. The bearing of the boat from the harbor entrance is N 77.8° E.

• • •

✓ **Check Point 6** You leave the entrance to a system of hiking trails and hike 2.3 miles on a bearing of S 31° W. Then the trail turns 90° clockwise and you hike 3.5 miles on a bearing of N 59° W. At that time:

a. How far are you, to the nearest tenth of a mile, from the entrance to the trail system?

b. What is your bearing, to the nearest tenth of a degree, from the entrance to the trail system?

③ Model simple harmonic motion.

Simple Harmonic Motion

Because of their periodic nature, trigonometric functions are used to model phenomena that occur again and again. This includes vibratory or oscillatory motion, such as the motion of a vibrating guitar string, the swinging of a pendulum, or the bobbing of an object attached to a spring. Trigonometric functions are also used to describe radio waves from your favorite FM station, television waves from your not-to-be-missed weekly sitcom, and sound waves from your most-prized CDs.

To see how trigonometric functions are used to model vibratory motion, consider this: A ball is attached to a spring hung from the ceiling. You pull the ball down 4 inches and then release it. If we neglect the effects of friction and air resistance, the ball will continue bobbing up and down on the end of the spring. These up-and-down oscillations are called **simple harmonic motion**.

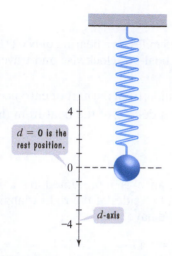

FIGURE 2.50 Using a *d*-axis to describe a ball's distance from its rest position

To better understand this motion, we use a *d*-axis, where *d* represents distance. This axis is shown in **Figure 2.50**. On this axis, the position of the ball before you pull it down is $d = 0$. This rest position is called the **equilibrium position**. Now you pull the ball down 4 inches to $d = -4$ and release it. **Figure 2.51** shows a sequence of "photographs" taken at one-second time intervals illustrating the distance of the ball from its rest position, *d*.

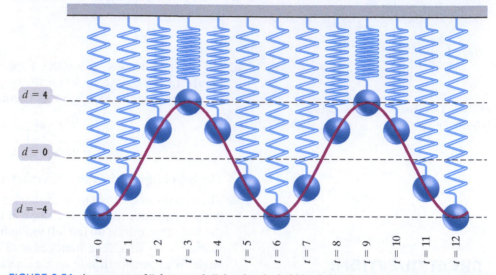

FIGURE 2.51 A sequence of "photographs" showing the bobbing ball's distance from the rest position, taken at one-second intervals

The curve in **Figure 2.51** shows how the ball's distance from its rest position changes over time. The curve is sinusoidal and the motion can be described using a cosine or a sine function.

Simple Harmonic Motion

An object that moves on a coordinate axis is in **simple harmonic motion** if its distance from the origin, *d*, at time *t* is given by either

$$d = a \cos \omega t \quad \text{or} \quad d = a \sin \omega t.$$

The motion has **amplitude** $|a|$, the maximum displacement of the object from its rest position. The **period** of the motion is $\dfrac{2\pi}{\omega}$, where $\omega > 0$. The period gives the time it takes for the motion to go through one complete cycle.

In describing simple harmonic motion, the equation with the cosine function, $d = a \cos \omega t$, is used if the object is at its greatest distance from rest position, the origin, at $t = 0$. By contrast, the equation with the sine function, $d = a \sin \omega t$, is used if the object is at its rest position, the origin, at $t = 0$.

EXAMPLE 7   Finding an Equation for an Object in Simple Harmonic Motion

A ball on a spring is pulled 4 inches below its rest position and then released. The period of the motion is 6 seconds. Write the equation for the ball's simple harmonic motion.

SOLUTION

We need to write an equation that describes *d*, the distance of the ball from its rest position, after *t* seconds. (The motion is illustrated by the "photo" sequence in **Figure 2.51**.) When the object is released ($t = 0$), the ball's distance from its rest position is 4 inches down. Because it is *down* 4 inches, *d* is negative: When

Blitzer Bonus ||

Diminishing Motion with Increasing Time

Due to friction and other resistive forces, the motion of an oscillating object decreases over time. The function

$$d = 3e^{-0.1t} \cos 2t$$

models this type of motion. The graph of the function is shown in a $t = [0, 10, 1]$ by $d = [-3, 3, 1]$ viewing rectangle. Notice how the amplitude is decreasing with time as the moving object loses energy.

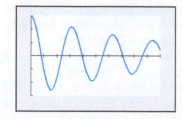

$t = 0$, $d = -4$. Notice that the greatest distance from rest position occurs at $t = 0$. Thus, we will use the equation with the cosine function,

$$d = a \cos \omega t,$$

to model the ball's simple harmonic motion.

Now we determine values for a and ω. Recall that $|a|$ is the maximum displacement. Because the ball is initially below rest position, $a = -4$.

The value of ω in $d = a \cos \omega t$ can be found using the formula for the period.

$$\text{period} = \frac{2\pi}{\omega} = 6 \qquad \text{\color{blue}We are given that the period of the motion is 6 seconds.}$$

$$2\pi = 6\omega \qquad \text{\color{blue}Multiply both sides by } \omega.$$

$$\omega = \frac{2\pi}{6} = \frac{\pi}{3} \qquad \text{\color{blue}Divide both sides by 6 and solve for } \omega.$$

We see that $a = -4$ and $\omega = \frac{\pi}{3}$. Substitute these values into $d = a \cos \omega t$. The equation for the ball's simple harmonic motion is

$$d = -4 \cos \frac{\pi}{3} t. \qquad \bullet\bullet\bullet$$

✅ **Check Point 7** A ball on a spring is pulled 6 inches below its rest position and then released. The period for the motion is 4 seconds. Write the equation for the ball's simple harmonic motion.

The period of the harmonic motion in Example 7 was 6 seconds. It takes 6 seconds for the moving object to complete one cycle. Thus, $\frac{1}{6}$ of a cycle is completed every second. We call $\frac{1}{6}$ the *frequency* of the moving object. **Frequency** describes the number of complete cycles per unit time and is the reciprocal of the period.

> **Frequency of an Object in Simple Harmonic Motion**
>
> An object in simple harmonic motion given by
>
> $$d = a \cos \omega t \quad \text{or} \quad d = a \sin \omega t$$
>
> has **frequency** f given by
>
> $$f = \frac{\omega}{2\pi}, \omega > 0.$$
>
> Equivalently,
>
> $$f = \frac{1}{\text{period}}.$$

EXAMPLE 8   Analyzing Simple Harmonic Motion

Figure 2.52 shows a mass on a smooth surface attached to a spring. The mass moves in simple harmonic motion described by

$$d = 10 \cos \frac{\pi}{6} t,$$

where t is measured in seconds and d in centimeters. Find:

a. the maximum displacement

b. the frequency

c. the time required for one cycle.

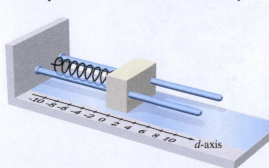

FIGURE 2.52 A mass attached to a spring, moving in simple harmonic motion

Blitzer Bonus ‖

Modeling Music

Sounds are caused by vibrating objects that result in variations in pressure in the surrounding air. Areas of high and low pressure moving through the air are modeled by the harmonic motion formulas. When these vibrations reach our eardrums, the eardrums' vibrations send signals to our brains, which create the sensation of hearing.

Whether a sound is heard as music, speech, noise, or static depends on the various sine waves that combine to make up the sound. French mathematician John Fourier (1768–1830) proved that all musical sounds—instrumental and vocal—could be modeled by sums involving sine functions. Modeling musical sounds with sinusoidal functions is used by synthesizing programs available on computers to electronically produce sounds unobtainable from ordinary musical instruments.

FIGURE 2.52 (repeated)

SOLUTION

We begin by identifying values for a and ω.

$$d = 10 \cos \frac{\pi}{6} t$$

> The form of this equation is $d = a \cos \omega t$ with $a = 10$ and $\omega = \frac{\pi}{6}$.

a. The maximum displacement from the rest position is the amplitude. Because $a = 10$, the maximum displacement is 10 centimeters.

b. The frequency, f, is

$$f = \frac{\omega}{2\pi} = \frac{\frac{\pi}{6}}{2\pi} = \frac{\pi}{6} \cdot \frac{1}{2\pi} = \frac{1}{12}.$$

The frequency is $\frac{1}{12}$ cycle (or oscillation) per second.

c. The time required for one cycle is the period.

$$\text{period} = \frac{2\pi}{\omega} = \frac{2\pi}{\frac{\pi}{6}} = 2\pi \cdot \frac{6}{\pi} = 12$$

The time required for one cycle is 12 seconds. This value can also be obtained by taking the reciprocal of the frequency in part (b). ● ● ●

✓ **Check Point 8** An object moves in simple harmonic motion described by $d = 12 \cos \frac{\pi}{4} t$, where t is measured in seconds and d in centimeters. Find **a.** the maximum displacement, **b.** the frequency, and **c.** the time required for one cycle.

Blitzer Bonus ‖ Resisting Damage of Simple Harmonic Motion

Simple harmonic motion from an earthquake caused this highway in Oakland, California, to collapse. By studying the harmonic motion of the soil under the highway, engineers learn to build structures that can resist damage.

ACHIEVING SUCCESS

Be sure to use the Chapter Test Prep on YouTube for each chapter test. The Chapter Test Prep videos provide step-by-step solutions to every exercise in the test and let you review any exercises you miss.

Are you using any of the other textbook supplements for help and additional study? These include:

- The Student Solutions Manual. This contains fully worked solutions to the odd-numbered section exercises plus all Check Points, Concept and Vocabulary Checks, Review/Preview Exercises, Mid-Chapter Check Points, Chapter Reviews, Chapter Tests, and Cumulative Reviews.
- Objective Videos. These interactive videos highlight important concepts from each objective of the text.
- MyMathLab is a text-specific online course. Math XL is an online homework, tutorial, and assessment system. Ask your instructor whether these are available to you.

CONCEPT AND VOCABULARY CHECK

Fill in each blank so that the resulting statement is true.

1. Solving a right triangle means finding the missing lengths of its _____ and the measurements of its _____.

2. The bearing from point O to point P is the acute angle, measured in degrees, between ray OP and a _____-_____ line.

3. An object that moves on a coordinate axis is in _____ motion if its distance from the origin, d, at time t is given by either

$$d = a \cos \omega t \quad \text{or} \quad d = a \sin \omega t.$$

The motion has amplitude _____, the maximum displacement of the object from its rest position. The period of the motion is _____ and the frequency f is given by $f = $ _____, where $\omega > 0$.

EXERCISE SET 2.4

Practice Exercises

In Exercises 1–12, solve the right triangle shown in the figure. Round lengths to two decimal places and express angles to the nearest tenth of a degree.

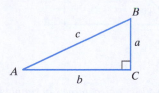

1. $A = 23.5°$, $b = 10$
2. $A = 41.5°$, $b = 20$
3. $A = 52.6°$, $c = 54$
4. $A = 54.8°$, $c = 80$
5. $B = 16.8°$, $b = 30.5$
6. $B = 23.8°$, $b = 40.5$
7. $a = 30.4$, $c = 50.2$
8. $a = 11.2$, $c = 65.8$
9. $a = 10.8$, $b = 24.7$
10. $a = 15.3$, $b = 17.6$
11. $b = 2$, $c = 7$
12. $b = 4$, $c = 9$

Use the figure shown to solve Exercises 13–16.

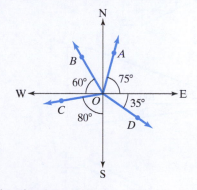

13. Find the bearing from O to A.
14. Find the bearing from O to B.
15. Find the bearing from O to C.
16. Find the bearing from O to D.

In Exercises 17–20, an object is attached to a coiled spring. In Exercises 17–18, the object is pulled down (negative direction from the rest position) and then released. In Exercises 19–20, the object is propelled downward from its rest position at time t = 0. Write an equation for the distance of the object from its rest position after t seconds.

Distance from Rest Position at t = 0	Amplitude	Period
17. 6 centimeters	6 centimeters	4 seconds
18. 8 inches	8 inches	2 seconds
19. 0 inches	3 inches	1.5 seconds
20. 0 centimeters	5 centimeters	2.5 seconds

In Exercises 21–28, an object moves in simple harmonic motion described by the given equation, where t is measured in seconds and d in inches. In each exercise, find the following:

   a. the maximum displacement

   b. the frequency

   c. the time required for one cycle.

21. $d = 5 \cos \dfrac{\pi}{2} t$ **22.** $d = 10 \cos 2\pi t$

23. $d = -6 \cos 2\pi t$ **24.** $d = -8 \cos \dfrac{\pi}{2} t$

25. $d = \frac{1}{2} \sin 2t$ **26.** $d = \frac{1}{3} \sin 2t$

27. $d = -5 \sin \dfrac{2\pi}{3} t$ **28.** $d = -4 \sin \dfrac{3\pi}{2} t$

Practice Plus

In Exercises 29–36, find the length x to the nearest whole unit.

29.

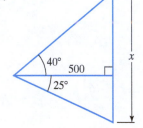

30.

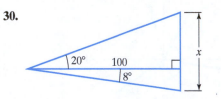

31.

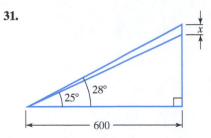

32.

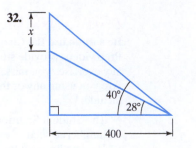

33.

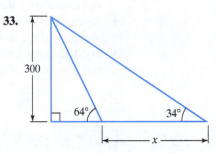

34.

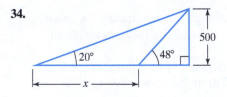

35.

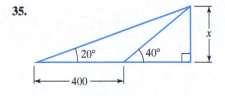

36.

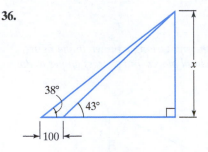

In Exercises 37–40, an object moves in simple harmonic motion described by the given equation, where t is measured in seconds and d in inches. In each exercise, graph one period of the equation. Then find the following:

   a. the maximum displacement

   b. the frequency

   c. the time required for one cycle

   d. the phase shift of the motion.

Describe how (a) through (d) are illustrated by your graph.

37. $d = 4 \cos\left(\pi t - \dfrac{\pi}{2}\right)$ **38.** $d = 3 \cos\left(\pi t + \dfrac{\pi}{2}\right)$

39. $d = -2 \sin\left(\dfrac{\pi t}{4} + \dfrac{\pi}{2}\right)$ **40.** $d = -\dfrac{1}{2}\sin\left(\dfrac{\pi t}{4} - \dfrac{\pi}{2}\right)$

Application Exercises

41. The tallest television transmitting tower in the world is in North Dakota. From a point on level ground 5280 feet (1 mile) from the base of the tower, the angle of elevation is 21.3°. Approximate the height of the tower to the nearest foot.

42. From a point on level ground 30 yards from the base of a building, the angle of elevation is 38.7°. Approximate the height of the building to the nearest foot.

43. The Statue of Liberty is approximately 305 feet tall. If the angle of elevation from a ship to the top of the statue is 23.7°, how far, to the nearest foot, is the ship from the statue's base?

44. A 200-foot cliff drops vertically into the ocean. If the angle of elevation from a ship to the top of the cliff is 22.3°, how far off shore, to the nearest foot, is the ship?

45. A helicopter hovers 1000 feet above a small island. The figure shows that the angle of depression from the helicopter to point *P* on the coast is 36°. How far off the coast, to the nearest foot, is the island?

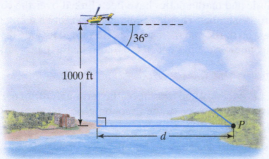

46. A police helicopter is flying at 800 feet. A stolen car is sighted at an angle of depression of 72°. Find the distance of the stolen car, to the nearest foot, from a point directly below the helicopter.

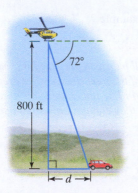

47. A wheelchair ramp is to be built beside the steps to the campus library. Find the angle of elevation of the 23-foot ramp, to the nearest tenth of a degree, if its final height is 6 feet.

48. A building that is 250 feet high casts a shadow 40 feet long. Find the angle of elevation, to the nearest tenth of a degree, of the Sun at this time.

49. A hot-air balloon is rising vertically. From a point on level ground 125 feet from the point directly under the passenger compartment, the angle of elevation to the balloon changes from 19.2° to 31.7°. How far, to the nearest tenth of a foot, does the balloon rise during this period?

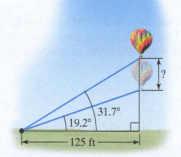

50. A flagpole is situated on top of a building. The angle of elevation from a point on level ground 330 feet from the building to the top of the flagpole is 63°. The angle of elevation from the same point to the bottom of the flagpole is 53°. Find the height of the flagpole to the nearest tenth of a foot.

51. A boat leaves the entrance to a harbor and travels 150 miles on a bearing of N 53° E. How many miles north and how many miles east from the harbor has the boat traveled?

52. A boat leaves the entrance to a harbor and travels 40 miles on a bearing of S 64° E. How many miles south and how many miles east from the harbor has the boat traveled?

53. A forest ranger sights a fire directly to the south. A second ranger, 7 miles east of the first ranger, also sights the fire. The bearing from the second ranger to the fire is S 28° W. How far, to the nearest tenth of a mile, is the first ranger from the fire?

54. A ship sights a lighthouse directly to the south. A second ship, 9 miles east of the first ship, also sights the lighthouse. The bearing from the second ship to the lighthouse is S 34° W. How far, to the nearest tenth of a mile, is the first ship from the lighthouse?

55. You leave your house and run 2 miles due west followed by 1.5 miles due north. At that time, what is your bearing from your house?

56. A ship is 9 miles east and 6 miles south of a harbor. What bearing should be taken to sail directly to the harbor?

57. A jet leaves a runway whose bearing is N 35° E from the control tower. After flying 5 miles, the jet turns 90° and files on a bearing of S 55° E for 7 miles. At that time, what is the bearing of the jet from the control tower?

58. A ship leaves port with a bearing of S 40° W. After traveling 7 miles, the ship turns 90° and travels on a bearing of N 50° W for 11 miles. At that time, what is the bearing of the ship from port?

59. An object in simple harmonic motion has a frequency of $\frac{1}{2}$ oscillation per minute and an amplitude of 6 feet. Write an equation in the form $d = a \sin \omega t$ for the object's simple harmonic motion.

60. An object in simple harmonic motion has a frequency of $\frac{1}{4}$ oscillation per minute and an amplitude of 8 feet. Write an equation in the form $d = a \sin \omega t$ for the object's simple harmonic motion.

61. A piano tuner uses a tuning fork. If middle C has a frequency of 264 vibrations per second, write an equation in the form $d = \sin \omega t$ for the simple harmonic motion.

62. A radio station, 98.1 on the FM dial, has radio waves with a frequency of 98.1 million cycles per second. Write an equation in the form $d = \sin \omega t$ for the simple harmonic motion of the radio waves.

Explaining the Concepts

63. What does it mean to solve a right triangle?

64. Explain how to find one of the acute angles of a right triangle if two sides are known.

65. Describe a situation in which a right triangle and a trigonometric function are used to measure a height or distance that would otherwise be inconvenient or impossible to measure.

66. What is meant by the bearing from point O to point P? Give an example with your description.

67. What is simple harmonic motion? Give an example with your description.

68. Explain the period and the frequency of simple harmonic motion. How are they related?

69. Explain how the photograph of the damaged highway on page 134 illustrates simple harmonic motion.

Technology Exercises

The functions in Exercises 70–71 model motion in which the amplitude decreases with time due to friction or other resistive forces. Graph each function in the given viewing rectangle. How many complete oscillations occur on the time interval $0 \le x \le 10$?

70. $y = 4e^{-0.1x} \cos 2x$; $[0, 10, 1]$ by $[-4, 4, 1]$

71. $y = -6e^{-0.09x} \cos 2\pi x$; $[0, 10, 1]$ by $[-6, 6, 1]$

Critical Thinking Exercises

Make Sense? *In Exercises 72–75, determine whether each statement makes sense or does not make sense, and explain your reasoning.*

72. A wheelchair ramp must be constructed so that the slope is not more than 1 inch of rise for every 1 foot of run, so I used the tangent function to determine the maximum angle that the ramp can make with the ground.

73. The bearing from O to A is N 103° W.

74. The bearing from O to B is E 70° S.

75. I analyzed simple harmonic motion in which the period was 10 seconds and the frequency was 0.2 oscillation per second.

76. The figure shows a satellite circling 112 miles above Earth. When the satellite is directly above point B, angle A measures 76.6°. Find Earth's radius to the nearest mile.

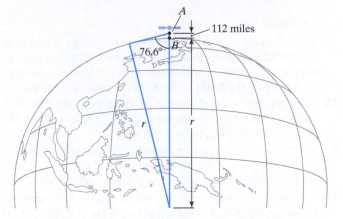

77. The angle of elevation to the top of a building changes from 20° to 40° as an observer advances 75 feet toward the building. Find the height of the building to the nearest foot.

Group Exercise

78. Music and mathematics have been linked over the centuries. Group members should research and present a seminar to the class on music and mathematics. Be sure to include the role of trigonometric functions in the music-mathematics link.

Retaining the Concepts

79. Let $P = (-3, 4)$ be a point on the terminal side of θ. Find the exact value of each of the trigonometric functions of θ.
(Section 1.3, Example 1)

80. Use a reference angle to find the exact value of $\tan \frac{13\pi}{3}$.
(Section 1.3, Example 8)

81. Find the measure of the side of the right triangle whose length is designated by c. Round the answer to the nearest whole number.
(Section 1.2, Example 9)

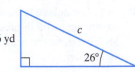

Preview Exercises

Exercises 82–84 will help you prepare for the material covered in the first section of the next chapter. The exercises use identities, introduced in Section 1.2, that enable you to rewrite trigonometric expressions so that they contain only sines and cosines:

$$\csc x = \frac{1}{\sin x} \qquad \sec x = \frac{1}{\cos x}$$

$$\tan x = \frac{\sin x}{\cos x} \qquad \cot x = \frac{\cos x}{\sin x}.$$

In Exercises 82–84, rewrite each expression by changing to sines and cosines. Then simplify the resulting expression.

82. $\sec x \cot x$

83. $\tan x \csc x \cos x$

84. $\sec x + \tan x$

CHAPTER 2	**Summary, Review, and Test**

SUMMARY

DEFINITIONS AND CONCEPTS	**EXAMPLES**

2.1 and 2.2   Graphs of the Trigonometric Functions

a. Graphs of the six trigonometric functions, with a description of the domain, range, and period of each function, are given in Table 2.4 on page 104.

b. The graph of $y = A \sin(Bx - C), B > 0$, can be obtained using amplitude $= |A|$, period $= \dfrac{2\pi}{B}$, and phase shift $= \dfrac{C}{B}$. See the illustration in the box on page 78.	Ex. 1, p. 74; Ex. 2, p. 75; Ex. 3, p. 77; Ex. 4, p. 79
c. The graph of $y = A \cos(Bx - C), B > 0$, can be obtained using amplitude $= |A|$, period $= \dfrac{2\pi}{B}$, and phase shift $= \dfrac{C}{B}$. See the illustration in the box on page 84.	Ex. 5, p. 82; Ex. 6, p. 84
d. The constant D in $y = A \sin(Bx - C) + D$ and $y = A \cos(Bx - C) + D$ causes vertical shifts in the graphs in the preceding items (b) and (c). If $D > 0$, the shift is D units upward and if $D < 0$, the shift is $|D|$ units downward. Oscillation is about the horizontal line $y = D$.	Ex. 7, p. 85
e. The method of adding y-coordinates is used to graph the sum of trigonometric functions.	Ex. 8, p. 87
f. The graph of $y = A \tan(Bx - C), B > 0$, is obtained using the procedure in the box on page 98. Consecutive asymptotes $\left(\text{solve } -\dfrac{\pi}{2} < Bx - C < \dfrac{\pi}{2}; \text{ consecutive asymptotes occur at } Bx - C = -\dfrac{\pi}{2} \text{ and } Bx - C = \dfrac{\pi}{2} \right)$ and an x-intercept midway between them play a key role in the graphing process.	Ex. 1, p. 98; Ex. 2, p. 99
g. The graph of $y = A \cot(Bx - C), B > 0$, is obtained using the procedure in the lower box on page 100. Consecutive asymptotes (solve $0 < Bx - C < \pi$; consecutive asymptotes occur at $Bx - C = 0$ and $Bx - C = \pi$) and an x-intercept midway between them play a key role in the graphing process.	Ex. 3, p. 100
h. To graph a cosecant curve, begin by graphing the corresponding sine curve. Draw vertical asymptotes through x-intercepts, using asymptotes as guides to sketch the graph. To graph a secant curve, first graph the corresponding cosine curve and use the same procedure.	Ex. 4, p. 102; Ex. 5, p. 103

2.3   Inverse Trigonometric Functions

a. On the restricted domain $-\dfrac{\pi}{2} \le x \le \dfrac{\pi}{2}, y = \sin x$ has an inverse function, defined in the box on page 112. Think of $\sin^{-1} x$ as the angle in $\left[-\dfrac{\pi}{2}, \dfrac{\pi}{2} \right]$ whose sine is x. A procedure for finding exact values of $\sin^{-1} x$ is given in the box on page 113.	Ex. 1, p. 113; Ex. 2, p. 114
b. On the restricted domain $0 \le x \le \pi, y = \cos x$ has an inverse function, defined in the upper box on page 115. Think of $\cos^{-1} x$ as the angle in $[0, \pi]$ whose cosine is x. A procedure for finding exact values of $\cos^{-1} x$ is given in the lower box on page 115.	Ex. 3, p. 115
c. On the restricted domain $-\dfrac{\pi}{2} < x < \dfrac{\pi}{2}, y = \tan x$ has an inverse function, defined in the box on page 116. Think of $\tan^{-1} x$ as the angle in $\left(-\dfrac{\pi}{2}, \dfrac{\pi}{2} \right)$ whose tangent is x. A procedure for finding exact values of $\tan^{-1} x$ is given in the box on page 117.	Ex. 4, p. 117

DEFINITIONS AND CONCEPTS	**EXAMPLES**
d. Graphs of the three basic inverse trigonometric functions, with a description of the domain and range of each function, are given in Table 2.8 on page 118.	
e. Inverse properties are given in the box on page 120. Points on terminal sides of angles in standard position and right triangles are used to find exact values of the composition of a function and a different inverse function.	Ex. 6, p. 120; Ex. 7, p. 121; Ex. 8, p. 122; Ex. 9, p. 123

2.4 Applications of Trigonometric Functions

a. Solving a right triangle means finding the missing lengths of its sides and the measurements of its angles. The Pythagorean Theorem, two acute angles whose sum is 90°, and appropriate trigonometric functions are used in this process.	Ex. 1, p. 127; Ex. 2, p. 128; Ex. 3, p. 129; Ex. 4, p. 129
b. The bearing from point O to point P is the acute angle between ray OP and a north-south line, shown in Figure 2.47 on page 130.	Ex. 5, p. 130; Ex. 6, p. 131
c. Simple harmonic motion, described in the box on page 132, is modeled by $d = a \cos \omega t$ or $d = a \sin \omega t$, with amplitude $= |a|$, period $= \dfrac{2\pi}{\omega}$, and frequency $= \dfrac{\omega}{2\pi} = \dfrac{1}{\text{period}}$.	Ex. 7, p. 132; Ex. 8, p. 133

ACHIEVING SUCCESS

Chapters 1 and 2 contained a great deal of new information. Here's where you can find the essential content to achieve success as you continue studying trigonometry.

Much of the essential information in these chapters can be found in three places:

- The Achieving Success feature on page 50, showing special angles and how to obtain exact values of trigonometric functions at these angles
- **Table 2.4** on page 104, showing the graphs of the six trigonometric functions, with their domains, ranges, and periods
- **Table 2.8** on page 118, showing graphs of the three basic inverse trigonometric functions, with their domains and ranges.

Make copies of these pages and mount them on cardstock. Use this reference sheet as you work the review exercises until you have all the information on the reference sheet memorized for the chapter test.

REVIEW EXERCISES

2.1

In Exercises 1–6, determine the amplitude and period of each function. Then graph one period of the function.

1. $y = 3 \sin 4x$

2. $y = -2 \cos 2x$

3. $y = 2 \cos \dfrac{1}{2} x$

4. $y = \dfrac{1}{2} \sin \dfrac{\pi}{3} x$

5. $y = -\sin \pi x$

6. $y = 3 \cos \dfrac{x}{3}$

In Exercises 7–11, determine the amplitude, period, and phase shift of each function. Then graph one period of the function.

7. $y = 2 \sin(x - \pi)$ **8.** $y = -3 \cos(x + \pi)$

9. $y = \dfrac{3}{2} \cos\left(2x + \dfrac{\pi}{4}\right)$

10. $y = \dfrac{5}{2} \sin\left(2x + \dfrac{\pi}{2}\right)$

11. $y = -3 \sin\left(\dfrac{\pi}{3} x - 3\pi\right)$

In Exercises 12–13, use a vertical shift to graph one period of the function.

12. $y = \sin 2x + 1$

13. $y = 2 \cos \frac{1}{3} x - 2$

In Exercises 14–15, use the method of adding y-coordinates to graph each function for $0 \le x \le 2\pi$

14. $y = 3 \sin x + \cos x$

15. $y = \sin x + \cos \dfrac{1}{2} x$

16. The function

$$y = 98.6 + 0.3 \sin\left(\frac{\pi}{12}x - \frac{11\pi}{12}\right)$$

models variation in body temperature, y, in °F, x hours after midnight.
   a. What is body temperature at midnight?
   b. What is the period of the body temperature cycle?
   c. When is body temperature highest? What is the body temperature at this time?
   d. When is body temperature lowest? What is the body temperature at this time?
   e. Graph one period of the body temperature function.

17. Light waves can be modeled by sine functions. The graphs show waves of red and blue light. Write an equation in the form $y = A \sin Bx$ that models each of these light waves.

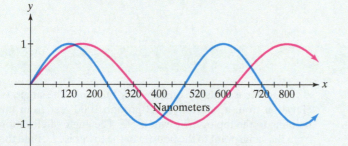

2.2

In Exercises 18–24, graph two full periods of the given tangent or cotangent function.

18. $y = 4 \tan 2x$

19. $y = -2 \tan \frac{\pi}{4}x$

20. $y = \tan(x + \pi)$

21. $y = -\tan\left(x - \frac{\pi}{4}\right)$

22. $y = 2 \cot 3x$

23. $y = -\frac{1}{2}\cot\frac{\pi}{2}x$

24. $y = 2 \cot\left(x + \frac{\pi}{2}\right)$

In Exercises 25–28, graph two full periods of the given cosecant or secant function.

25. $y = 3 \sec 2\pi x$

26. $y = -2 \csc \pi x$

27. $y = 3 \sec(x + \pi)$

28. $y = \frac{5}{2}\csc(x - \pi)$

2.3

In Exercises 29–47, find the exact value of each expression. Do not use a calculator.

29. $\sin^{-1} 1$

30. $\cos^{-1} 1$

31. $\tan^{-1} 1$

32. $\sin^{-1}\left(-\frac{\sqrt{3}}{2}\right)$

33. $\cos^{-1}\left(-\frac{1}{2}\right)$

34. $\tan^{-1}\left(-\frac{\sqrt{3}}{3}\right)$

35. $\cos\left(\sin^{-1}\frac{\sqrt{2}}{2}\right)$

36. $\sin(\cos^{-1} 0)$

37. $\tan\left[\sin^{-1}\left(-\frac{1}{2}\right)\right]$

38. $\tan\left[\cos^{-1}\left(-\frac{\sqrt{3}}{2}\right)\right]$

39. $\csc\left(\tan^{-1}\frac{\sqrt{3}}{3}\right)$

40. $\cos\left(\tan^{-1}\frac{3}{4}\right)$

41. $\sin\left(\cos^{-1}\frac{3}{5}\right)$

42. $\tan\left[\sin^{-1}\left(-\frac{3}{5}\right)\right]$

43. $\tan\left[\cos^{-1}\left(-\frac{4}{5}\right)\right]$

44. $\sin\left[\tan^{-1}\left(-\frac{1}{3}\right)\right]$

45. $\sin^{-1}\left(\sin\frac{\pi}{3}\right)$

46. $\sin^{-1}\left(\sin\frac{2\pi}{3}\right)$

47. $\sin^{-1}\left(\cos\frac{2\pi}{3}\right)$

In Exercises 48–49, use a right triangle to write each expression as an algebraic expression. Assume that x is positive and that the given inverse trigonometric function is defined for the expression in x.

48. $\cos\left(\tan^{-1}\frac{x}{2}\right)$

49. $\sec\left(\sin^{-1}\frac{1}{x}\right)$

2.4

In Exercises 50–53, solve the right triangle shown in the figure. Round lengths to two decimal places and express angles to the nearest tenth of a degree.

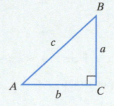

50. $A = 22.3°$, $c = 10$

51. $B = 37.4°$, $b = 6$

52. $a = 2$, $c = 7$

53. $a = 1.4$, $b = 3.6$

54. From a point on level ground 80 feet from the base of a building, the angle of elevation is 25.6°. Approximate the height of the building to the nearest foot.

55. Two buildings with flat roofs are 60 yards apart. The height of the shorter building is 40 yards. From its roof, the angle of elevation to the edge of the roof of the taller building is 40°. Find the height of the taller building to the nearest yard.

56. You want to measure the height of an antenna on the top of a 125-foot building. From a point in front of the building, you measure the angle of elevation to the top of the building to be 68° and the angle of elevation to the top of the antenna to be 71°. How tall is the antenna, to the nearest tenth of a foot?

In Exercises 57–58, use the figures shown to find the bearing from O to A.

57.

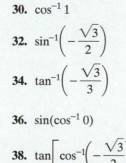

58.

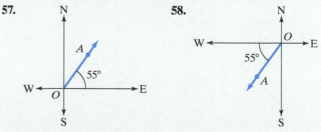

59. A ship is due west of a lighthouse. A second ship is 12 miles south of the first ship. The bearing from the second ship to the lighthouse is N 64° E. How far, to the nearest tenth of a mile, is the first ship from the lighthouse?

60. From city A to city B, a plane flies 850 miles at a bearing of N 58° E. From city B to city C, the plane flies 960 miles at a bearing of S 32° E.
   a. Find, to the nearest tenth of a mile, the distance from city A to city C.
   b. What is the bearing from city A to city C?

In Exercises 61–62, an object moves in simple harmonic motion described by the given equation, where t is measured in seconds and d in centimeters. In each exercise, find:
   a. the maximum displacement
   b. the frequency
   c. the time required for one cycle.

61. $d = 20 \cos \dfrac{\pi}{4} t$

62. $d = \frac{1}{2} \sin 4t$

In Exercises 63–64, an object is attached to a coiled spring. In Exercise 63, the object is pulled down (negative direction from the rest position) and then released. In Exercise 64, the object is propelled downward from its rest position. Write an equation for the distance of the object from its rest position after t seconds.

Distance from Rest Position at $t = 0$		Amplitude	Period
63.	30 inches	30 inches	2 seconds
64.	0 inches	$\frac{1}{4}$ inch	5 seconds

CHAPTER 2 TEST

In Exercises 1–4, graph one period of each function.

1. $y = 3 \sin 2x$

2. $y = -2 \cos\left(x - \dfrac{\pi}{2}\right)$

3. $y = 2 \tan \dfrac{x}{2}$

4. $y = -\frac{1}{2} \csc \pi x$

5. Graph $y = \frac{1}{2} \sin x + 2 \cos x, 0 \le x \le 2\pi$.

6. Find the exact value of $\tan\left[\cos^{-1}\left(-\frac{1}{2}\right)\right]$.

7. Write $\sin\left(\cos^{-1} \dfrac{x}{3}\right)$ as an algebraic expression. Assume that $x > 0$ and $\dfrac{x}{3}$ is in the domain of the inverse cosine function.

8. Solve the right triangle in the figure shown. Round lengths to one decimal place.

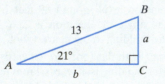

9. A TV tower is situated on top of a building. The angle of elevation from a point on level ground 1000 feet from the building to the top of the tower is 56°. The angle of elevation from the same point to the bottom of the tower is 51°. Find the height of the TV tower to the nearest tenth of a foot.

10. Use the figure to find the bearing from O to P.

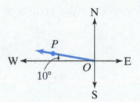

11. An object moves in simple harmonic motion described by $d = -6 \cos \pi t$, where t is measured in seconds and d in inches. Find **a.** the maximum displacement, **b.** the frequency, and **c.** the time required for one oscillation.

Trigonometric Identities and Equations

You enjoy watching your friend participate in the shot put at college track and field events. After a few full turns in a circle, she throws ("puts") an 8-pound, 13-ounce shot from the shoulder. The range of her throwing distance continues to improve. Knowing that you are studying trigonometry, she asks if there is some way that a trigonometric expression might help achieve her best possible distance in the event.

HERE'S WHERE YOU'LL FIND THIS APPLICATION:

This problem appears as Exercise 79 in Exercise Set 3.3. In the solution, you will obtain critical information about athletic performance using a trigonometric identity. In this chapter, we derive important categories of identities involving trigonometric functions. You will learn how to use these identities to better understand your periodic world.

What am I supposed to learn?

After studying this section, you should be able to:

1 Use the fundamental trigonometric identities to verify identities.

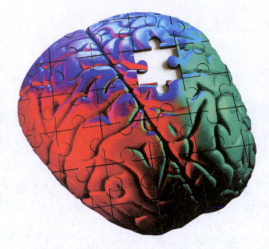

Do you enjoy solving puzzles? The process is a natural way to develop problem-solving skills that are important in every area of our lives. Engaging in problem solving for sheer pleasure releases chemicals in the brain that enhance our feeling of well-being. Perhaps this is why puzzles have fascinated people for over 12,000 years.

Thousands of relationships exist among the six trigonometric functions. Verifying these relationships is like solving a puzzle. Why? There are no rigid rules for the process. Thus, proving a trigonometric relationship requires you to be creative in your approach to problem solving. By learning to establish these relationships, you will become a better, more confident problem solver. Furthermore, you may enjoy the feeling of satisfaction that accompanies solving each "puzzle."

The Fundamental Identities

In Chapter 1, we used right triangles to establish relationships among the trigonometric functions. Although we limited domains to acute angles, the fundamental identities listed in the following box are true for all values of x for which the expressions are defined.

Fundamental Trigonometric Identities

Reciprocal Identities

$$\sin x = \frac{1}{\csc x} \quad \cos x = \frac{1}{\sec x} \quad \tan x = \frac{1}{\cot x}$$

$$\csc x = \frac{1}{\sin x} \quad \sec x = \frac{1}{\cos x} \quad \cot x = \frac{1}{\tan x}$$

Quotient Identities

$$\tan x = \frac{\sin x}{\cos x} \quad \cot x = \frac{\cos x}{\sin x}$$

Pythagorean Identities

$$\sin^2 x + \cos^2 x = 1 \quad 1 + \tan^2 x = \sec^2 x \quad 1 + \cot^2 x = \csc^2 x$$

Even-Odd Identities

$$\sin(-x) = -\sin x \quad \cos(-x) = \cos x \quad \tan(-x) = -\tan x$$

$$\csc(-x) = -\csc x \quad \sec(-x) = \sec x \quad \cot(-x) = -\cot x$$

1 Use the fundamental trigonometric identities to verify identities.

Using Fundamental Identities to Verify Other Identities

The fundamental trigonometric identities are used to establish other relationships among trigonometric functions. To **verify an identity**, we show that one side of the identity can be simplified so that it is identical to the other side. Each side of the equation is manipulated independently of the other side of the equation. Start with the side containing the more complicated expression. If you substitute one or more fundamental identities on the more complicated side, you will often be able to rewrite it in a form identical to that of the other side.

No one method or technique can be used to verify every identity. Some identities can be verified by rewriting the more complicated side so that it contains only sines and cosines.

EXAMPLE 1   Changing to Sines and Cosines to Verify an Identity

Verify the identity: $\sec x \cot x = \csc x$.

SOLUTION

The left side of the equation contains the more complicated expression. Thus, we work with the left side. Let us express this side of the identity in terms of sines and cosines. Perhaps this strategy will enable us to transform the left side into $\csc x$, the expression on the right.

$$\sec x \cot x = \frac{1}{\cos x} \cdot \frac{\cos x}{\sin x} \qquad \text{Apply a reciprocal identity: } \sec x = \frac{1}{\cos x} \text{ and a quotient identity: } \cot x = \frac{\cos x}{\sin x}.$$

$$= \frac{1}{\cancel{\cos x}^{1}} \cdot \frac{\cancel{\cos x}^{1}}{\sin x} \qquad \text{Divide both the numerator and the denominator by } \cos x, \text{ the common factor.}$$

$$= \frac{1}{\sin x} \qquad \text{Multiply the remaining factors in the numerator and denominator.}$$

$$= \csc x \qquad \text{Apply a reciprocal identity: } \csc x = \frac{1}{\sin x}.$$

By working with the left side and simplifying it so that it is identical to the right side, we have verified the given identity.   • • •

TECHNOLOGY

Numeric and Graphic Connections

You can use a graphing utility to provide evidence of an identity. Enter each side of the identity separately under y_1 and y_2. Then use the TABLE feature or the graphs. The table should show that the function values are the same except for those values of x for which y_1, y_2, or both, are undefined. The graphs should appear to be identical.

Let's check the identity in Example 1:

$$\sec x \cot x = \csc x.$$

$y_1 = \sec x \cot x$
Enter $\sec x$ as $\dfrac{1}{\cos x}$
and $\cot x$ as $\dfrac{1}{\tan x}$.

$y_2 = \csc x$
Enter $\csc x$ as $\dfrac{1}{\sin x}$.

Numeric Check

Display a table for y_1 and y_2. We started our table at $-\pi$ and used $\Delta Tbl = \dfrac{\pi}{8}$.

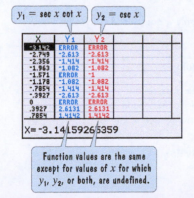

$y_1 = \sec x \cot x$ $y_2 = \csc x$

X	Y1	Y2
-3.142	ERROR	ERROR
-2.749	-2.613	-2.613
-2.356	-1.414	-1.414
-1.963	-1.082	-1.082
-1.571	ERROR	-1
-1.178	-1.082	-1.082
-.7854	-1.414	-1.414
-.3927	-2.613	-2.613
0	ERROR	ERROR
.3927	2.6131	2.6131
.7854	1.4142	1.4142

X=-3.1415926535 9

Function values are the same except for values of x for which y_1, y_2, or both, are undefined.

Graphic Check

Display graphs for y_1 and y_2.

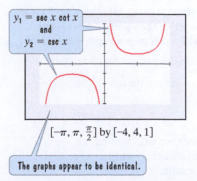

$y_1 = \sec x \cot x$
and
$y_2 = \csc x$

$[-\pi, \pi, \frac{\pi}{2}]$ by $[-4, 4, 1]$

The graphs appear to be identical.

✓ **Check Point 1** Verify the identity: $\csc x \tan x = \sec x$.

In verifying an identity, stay focused on your goal. When manipulating one side of the equation, continue to look at the other side to keep the desired form of the result in mind.

GREAT QUESTION!

What's the difference between solving a conditional equation and verifying that an equation is an identity?

You solve equations by working with both sides at once, adding, subtracting, multiplying, or dividing the sides by the same expression. You verify an identity by manipulating each side independently of the other side. Because you are familiar with solving conditional equations, it may feel strange to verify identities by working separately with the sides of the equation.

Here is an algebraic example that illustrates the difference between solving an equation and verifying an identity:

Solving an Equation	**Verifying an Identity**

Solving an Equation

Solve:

$5(3x + 2) - 4 = -24$.

$15x + 10 - 4 = -24$ Use the distributive property. Continue working with both sides.

$15x + 6 = -24$ Simplify.

$15x = -30$ Subtract 6 from both sides.

$x = -2$ Divide both sides by 15.

The solution set is $\{-2\}$.

Verifying an Identity

Verify:

$5(3x + 2) - 4 = 15x + 6$.

Work with the left side of the equation.

$5(3x + 2) - 4 = 15x + 10 - 4$ Use the distributive property.

$= 15x + 6$ Simplify.

By working with the left side and simplifying it so that it is identical to the right side, we have verified the identity.

Why can't you verify an identity by such methods as adding the same expression to each side and obtaining a true statement? If you do this, you have already assumed that the given statement is true. You do not know that it is true until after you have verified it.

EXAMPLE 2 Changing to Sines and Cosines to Verify an Identity

Verify the identity: $\sin x \tan x + \cos x = \sec x$.

SOLUTION

The left side is more complicated, so we start with it. Notice that the left side contains the sum of two terms, but the right side contains only one term. This means that somewhere during the verification process, the two terms on the left side must be combined to form one term.

Let's begin by expressing the left side of the identity so that it contains only sines and cosines. Thus, we apply a quotient identity and replace $\tan x$ by $\dfrac{\sin x}{\cos x}$. Perhaps this strategy will enable us to transform the left side into $\sec x$, the expression on the right.

GREAT QUESTION!

When proving identities, do I have to write the variable associated with each trigonometric function?

Yes. Do not get lazy and write

$$\sin \tan + \cos$$

for

$$\sin x \tan x + \cos x$$

because sin, tan, and cos are meaningless without specified variables.

$$\sin x \tan x + \cos x = \sin x \left(\frac{\sin x}{\cos x} \right) + \cos x \quad \text{Apply a quotient identity: } \tan x = \frac{\sin x}{\cos x}.$$

$$= \frac{\sin^2 x}{\cos x} + \cos x \quad \text{Multiply.}$$

$$= \frac{\sin^2 x}{\cos x} + \cos x \cdot \frac{\cos x}{\cos x} \quad \text{The least common denominator is } \cos x. \text{ Write the second expression with a denominator of } \cos x.$$

$$= \frac{\sin^2 x}{\cos x} + \frac{\cos^2 x}{\cos x} \quad \text{Multiply.}$$

$$= \frac{\sin^2 x + \cos^2 x}{\cos x}$$ Add numerators and place this sum over the least common denominator.

$$= \frac{1}{\cos x}$$ Apply a Pythagorean identity: $\sin^2 x + \cos^2 x = 1$.

$$= \sec x$$ Apply a reciprocal identity: $\sec x = \frac{1}{\cos x}$.

By working with the left side and arriving at the right side, the identity is verified.

• • •

✓ **Check Point 2** Verify the identity: $\cos x \cot x + \sin x = \csc x$.

Some identities are verified using factoring to simplify a trigonometric expression.

EXAMPLE 3 **Using Factoring to Verify an Identity**

Verify the identity: $\cos x - \cos x \sin^2 x = \cos^3 x$.

SOLUTION

We start with the more complicated side, the left side. Factor out the greatest common factor, $\cos x$, from each of the two terms.

$$\cos x - \cos x \sin^2 x = \cos x(1 - \sin^2 x)$$ Factor cos x from the two terms.

$$= \cos x \cdot \cos^2 x$$ Use a variation of $\sin^2 x + \cos^2 x = 1$. Solving for $\cos^2 x$, we obtain $\cos^2 x = 1 - \sin^2 x$.

$$= \cos^3 x$$ Multiply.

We worked with the left side and arrived at the right side. Thus, the identity is verified.

• • •

✓ **Check Point 3** Verify the identity: $\sin x - \sin x \cos^2 x = \sin^3 x$.

There is often more than one technique that can be used to verify an identity.

EXAMPLE 4 **Using Two Techniques to Verify an Identity**

Verify the identity: $\dfrac{1 + \sin \theta}{\cos \theta} = \sec \theta + \tan \theta$.

SOLUTION

Method 1.   Separating a Single-Term Quotient into Two Terms Let's separate the quotient on the left side into two terms using

$$\frac{a + b}{c} = \frac{a}{c} + \frac{b}{c}.$$

Perhaps this strategy will enable us to transform the left side into $\sec \theta + \tan \theta$, the sum on the right.

$$\frac{1 + \sin \theta}{\cos \theta} = \frac{1}{\cos \theta} + \frac{\sin \theta}{\cos \theta}$$ Divide each term in the numerator by cos θ.

$$= \sec \theta + \tan \theta$$ Apply a reciprocal identity and a quotient identity: $\sec \theta = \dfrac{1}{\cos \theta}$ and $\tan \theta = \dfrac{\sin \theta}{\cos \theta}$.

We worked with the left side and arrived at the right side. Thus, the identity is verified. (Example 4 continues on page 149.)

A Brief Review • Multiplying Polynomials

- A polynomial in x is an algebraic expression of the form

$$a_n x^n + a_{n-1} x^{n-1} + a_{n-2} x^{n-2} + \cdots + a_1 x + a_0,$$

where $a_n, a_{n-1}, a_{n-2}, \ldots, a_1,$ and a_0 are real numbers, $a_n \neq 0$, and n is a nonnegative integer. The polynomial is of degree n, a_n is the leading coefficient, and a_0 is the constant term.

- A simplified polynomial that has two terms is called a binomial. Any two binomials can be quickly multiplied by using the FOIL method, in which F represents the product of the first terms in each binomial, O represents the product of the outside terms, I represents the product of the inside terms, and L represents the product of the last, or second terms in each binomial.

EXAMPLE

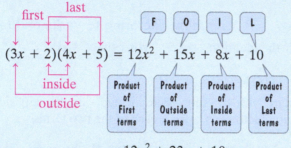

$$= 12x^2 + 23x + 10$$

- The Product of the Sum and Difference of Two Terms

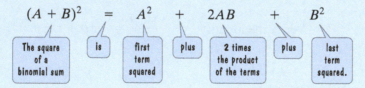

EXAMPLE

$$(4x + 3)(4x - 3) = (4x)^2 - 3^2 = 16x^2 - 9$$

- Squaring Binomials

The Square of a Binomial Sum

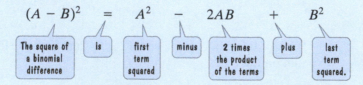

The Square of a Binomial Difference

$$(A - B)^2 = A^2 - 2AB + B^2$$

| The square of a binomial difference | is | first term squared | minus | 2 times the product of the terms | plus | last term squared. |

EXAMPLES

$$(3x + 7)^2 = (3x)^2 + 2(3x)(7) + 7^2 = 9x^2 + 42x + 49$$

$$(5x - 3)^2 = (5x)^2 - 2(5x)(3) + 3^2 = 25x^2 - 30x + 9$$

Method 2. Changing to Sines and Cosines We need to verify the identity $\frac{1 + \sin \theta}{\cos \theta} = \sec \theta + \tan \theta$. Let's work with the right side and express it so that it contains only sines and cosines.

$$\sec \theta + \tan \theta = \frac{1}{\cos \theta} + \frac{\sin \theta}{\cos \theta}$$

Apply a reciprocal identity and a quotient identity: $\sec \theta = \frac{1}{\cos \theta}$ and $\tan \theta = \frac{\sin \theta}{\cos \theta}$.

$$= \frac{1 + \sin \theta}{\cos \theta}$$

Add numerators. Put this sum over the common denominator.

We worked with the right side and arrived at the left side. Thus, the identity is verified. •••

✓ **Check Point 4** Verify the identity: $\dfrac{1 + \cos \theta}{\sin \theta} = \csc \theta + \cot \theta$.

How do we verify identities in which sums or differences of fractions with trigonometric functions appear on one side? Use the least common denominator and combine the fractions. This technique is especially useful when the other side of the identity contains only one term.

EXAMPLE 5   Combining Fractional Expressions to Verify an Identity

Verify the identity: $\dfrac{\cos x}{1 + \sin x} + \dfrac{1 + \sin x}{\cos x} = 2 \sec x$.

SOLUTION

We start with the more complicated side, the left side. The least common denominator of the fractions is $(1 + \sin x)(\cos x)$. We express each fraction in terms of this least common denominator by multiplying the numerator and denominator by the extra factor needed to form $(1 + \sin x)(\cos x)$.

$$\frac{\cos x}{1 + \sin x} + \frac{1 + \sin x}{\cos x}$$

The least common denominator is $(1 + \sin x)(\cos x)$.

$$= \frac{\cos x(\cos x)}{(1 + \sin x)(\cos x)} + \frac{(1 + \sin x)(1 + \sin x)}{(1 + \sin x)(\cos x)}$$

Rewrite each fraction with the least common denominator.

$$= \frac{\cos^2 x}{(1 + \sin x)(\cos x)} + \frac{1 + 2 \sin x + \sin^2 x}{(1 + \sin x)(\cos x)}$$

Use the FOIL method to multiply $(1 + \sin x)(1 + \sin x)$.

$$= \frac{\cos^2 x + 1 + 2 \sin x + \sin^2 x}{(1 + \sin x)(\cos x)}$$

Add numerators. Put this sum over the least common denominator.

$$= \frac{(\sin^2 x + \cos^2 x) + 1 + 2 \sin x}{(1 + \sin x)(\cos x)}$$

Regroup terms to apply a Pythagorean identity.

$$= \frac{1 + 1 + 2 \sin x}{(1 + \sin x)(\cos x)}$$

Apply a Pythagorean identity: $\sin^2 x + \cos^2 x = 1$.

$$= \frac{2 + 2 \sin x}{(1 + \sin x)(\cos x)}$$

Add constant terms in the numerator: $1 + 1 = 2$.

$$= \frac{2(1 + \sin x)}{(1 + \sin x)(\cos x)}$$

Factor and simplify.

$$= \frac{2}{\cos x}$$

$$= 2 \sec x$$

Apply a reciprocal identity: $\sec x = \frac{1}{\cos x}$.

We worked with the left side and arrived at the right side. Thus, the identity is verified. •••

⊘ **Check Point 5** Verify the identity: $\dfrac{\sin x}{1 + \cos x} + \dfrac{1 + \cos x}{\sin x} = 2 \csc x$.

Some identities are verified using a technique that may remind you of rationalizing a denominator.

EXAMPLE 6 Multiplying the Numerator and Denominator by the Same Factor to Verify an Identity

Verify the identity: $\dfrac{\sin x}{1 + \cos x} = \dfrac{1 - \cos x}{\sin x}$.

SOLUTION

The suggestions given in the previous examples do not apply here. Everything is already expressed in terms of sines and cosines. Furthermore, there are no fractions to combine and neither side looks more complicated than the other. Let's solve the puzzle by working with the left side and making it look like the expression on the right. The expression on the right contains $1 - \cos x$ in the numerator. This suggests multiplying the numerator and denominator of the left side by $1 - \cos x$. By doing this, we obtain a factor of $1 - \cos x$ in the numerator, as in the numerator on the right.

DISCOVERY

Verify the identity in Example 6 by making the right side look like the left side. Start with the expression on the right. Multiply the numerator and denominator by $1 + \cos x$.

$$\dfrac{\sin x}{1 + \cos x} = \dfrac{\sin x}{1 + \cos x} \cdot \dfrac{1 - \cos x}{1 - \cos x} \qquad \text{Multiply numerator and denominator by } 1 - \cos x.$$

$$= \dfrac{\sin x (1 - \cos x)}{1 - \cos^2 x} \qquad \text{Multiply. Use } (A + B)(A - B) = A^2 - B^2, \text{ with } A = 1 \text{ and } B = \cos x, \text{ to multiply denominators.}$$

$$= \dfrac{\sin x (1 - \cos x)}{\sin^2 x} \qquad \text{Use a variation of } \sin^2 x + \cos^2 x = 1. \text{ Solving for } \sin^2 x, \text{ we obtain } \sin^2 x = 1 - \cos^2 x.$$

$$= \dfrac{1 - \cos x}{\sin x} \qquad \text{Simplify: } \dfrac{\sin x}{\sin^2 x} = \dfrac{\sin x}{\sin x \cdot \sin x} = \dfrac{1}{\sin x}.$$

We worked with the left side and arrived at the right side. Thus, the identity is verified. • • •

⊘ **Check Point 6** Verify the identity: $\dfrac{\cos x}{1 + \sin x} = \dfrac{1 - \sin x}{\cos x}$.

DISCOVERY

Try simplifying

$$\dfrac{\dfrac{\sin x}{\cos x} + \sin x}{1 + \cos x}$$

by multiplying the two terms in the numerator and the two terms in the denominator by $\cos x$. This method for simplifying the complex fraction involves multiplying the numerator and the denominator by the least common denominator of all fractions in the expression. Do you prefer this simplification procedure over the method used on the next page?

EXAMPLE 7 Changing to Sines and Cosines to Verify an Identity

Verify the identity: $\dfrac{\tan x - \sin(-x)}{1 + \cos x} = \tan x$.

SOLUTION

We begin with the left side. Our goal is to obtain $\tan x$, the expression on the right.

$$\dfrac{\tan x - \sin(-x)}{1 + \cos x} = \dfrac{\tan x - (-\sin x)}{1 + \cos x} \qquad \text{The sine function is odd: } \sin(-x) = -\sin x.$$

$$= \dfrac{\tan x + \sin x}{1 + \cos x} \qquad \text{Simplify.}$$

$$= \dfrac{\dfrac{\sin x}{\cos x} + \sin x}{1 + \cos x} \qquad \text{Apply a quotient identity: } \tan x = \dfrac{\sin x}{\cos x}.$$

$$= \frac{\dfrac{\sin x}{\cos x} + \dfrac{\sin x \cos x}{\cos x}}{1 + \cos x}$$

Express the terms in the numerator with the least common denominator, cos x.

$$= \frac{\dfrac{\sin x + \sin x \cos x}{\cos x}}{1 + \cos x}$$

Add in the numerator.

$$= \frac{\sin x + \sin x \cos x}{\cos x} \div \frac{1 + \cos x}{1}$$

Rewrite the main fraction bar as ÷.

$$= \frac{\sin x + \sin x \cos x}{\cos x} \cdot \frac{1}{1 + \cos x}$$

Invert the divisor and multiply.

$$= \frac{\sin x (1 + \overset{1}{\cancel{\cos x}})}{\cos x} \cdot \frac{1}{\underset{1}{\cancel{1 + \cos x}}}$$

Factor and simplify.

$$= \frac{\sin x}{\cos x}$$

Multiply the remaining factors in the numerator and in the denominator.

$$= \tan x$$

Apply a quotient identity.

The left side simplifies to tan x, the right side of the given equation. Thus, the identity is verified.   ● ● ●

✅ **Check Point 7** Verify the identity: $\dfrac{\sec x + \csc(-x)}{\sec x \csc x} = \sin x - \cos x.$

Is every identity verified by working with only one side? No. You can sometimes work with each side separately and show that both sides are equal to the same trigonometric expression. This is illustrated in Example 8.

EXAMPLE 8   Working with Both Sides Separately to Verify an Identity

Verify the identity: $\dfrac{1}{1 + \cos \theta} + \dfrac{1}{1 - \cos \theta} = 2 + 2 \cot^2 \theta.$

SOLUTION

We begin by working with the left side.

$$\frac{1}{1 + \cos \theta} + \frac{1}{1 - \cos \theta}$$

The least common denominator is $(1 + \cos \theta)(1 - \cos \theta)$.

$$= \frac{1(1 - \cos \theta)}{(1 + \cos \theta)(1 - \cos \theta)} + \frac{1(1 + \cos \theta)}{(1 + \cos \theta)(1 - \cos \theta)}$$

Rewrite each fraction with the least common denominator.

$$= \frac{1 - \cos \theta + 1 + \cos \theta}{(1 + \cos \theta)(1 - \cos \theta)}$$

Add numerators. Put this sum over the least common denominator.

$$= \frac{2}{(1 + \cos \theta)(1 - \cos \theta)}$$

Simplify the numerator: $-\cos \theta + \cos \theta = 0$ and $1 + 1 = 2$.

$$= \frac{2}{1 - \cos^2 \theta}$$

Multiply the factors in the denominator.

Now we work with the right side, $2 + 2\cot^2\theta$. Our goal is to transform this side into the simplified form attained for the left side, $\dfrac{2}{1 - \cos^2\theta}$.

$$2 + 2\cot^2\theta = 2 + 2\left(\frac{\cos^2\theta}{\sin^2\theta}\right) \qquad \text{Use a quotient identity: } \cot\theta = \frac{\cos\theta}{\sin\theta}.$$

$$= \frac{2\sin^2\theta}{\sin^2\theta} + \frac{2\cos^2\theta}{\sin^2\theta} \qquad \text{Rewrite each term with the least common denominator, } \sin^2\theta.$$

$$= \frac{2\sin^2\theta + 2\cos^2\theta}{\sin^2\theta} \qquad \text{Add numerators. Put this sum over the least common denominator.}$$

$$= \frac{2(\sin^2\theta + \cos^2\theta)}{\sin^2\theta} \qquad \text{Factor out the greatest common factor, 2.}$$

$$= \frac{2}{\sin^2\theta} \qquad \text{Apply a Pythagorean identity: } \sin^2\theta + \cos^2\theta = 1.$$

$$= \frac{2}{1 - \cos^2\theta} \qquad \text{Use a variation of } \sin^2\theta + \cos^2\theta = 1 \text{ and solve for } \sin^2\theta: \sin^2\theta = 1 - \cos^2\theta.$$

The identity is verified because both sides are equal to $\dfrac{2}{1 - \cos^2\theta}$. • • •

☑ **Check Point 8** Verify the identity: $\dfrac{1}{1 + \sin\theta} + \dfrac{1}{1 - \sin\theta} = 2 + 2\tan^2\theta$.

Guidelines for Verifying Trigonometric Identities

There is often more than one correct way to solve a puzzle, although one method may be shorter and more efficient than another. The same is true for verifying an identity. For example, how would you verify

$$\frac{\csc^2 x - 1}{\csc^2 x} = \cos^2 x?$$

One approach is to use a Pythagorean identity, $1 + \cot^2 x = \csc^2 x$, on the left side. Then change the resulting expression to sines and cosines.

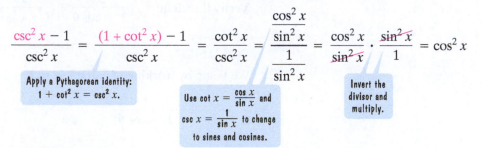

$$\frac{\csc^2 x - 1}{\csc^2 x} = \frac{(1 + \cot^2 x) - 1}{\csc^2 x} = \frac{\cot^2 x}{\csc^2 x} = \frac{\dfrac{\cos^2 x}{\sin^2 x}}{\dfrac{1}{\sin^2 x}} = \frac{\cos^2 x}{\sin^2 x} \cdot \frac{\sin^2 x}{1} = \cos^2 x$$

Apply a Pythagorean identity: $1 + \cot^2 x = \csc^2 x$.

Use $\cot x = \dfrac{\cos x}{\sin x}$ and $\csc x = \dfrac{1}{\sin x}$ to change to sines and cosines.

Invert the divisor and multiply.

A more efficient strategy for verifying this identity may not be apparent at first glance. Work with the left side and divide each term in the numerator by the denominator, $\csc^2 x$.

$$\frac{\csc^2 x - 1}{\csc^2 x} = \frac{\csc^2 x}{\csc^2 x} - \frac{1}{\csc^2 x} = 1 - \sin^2 x = \cos^2 x$$

Apply a reciprocal identity: $\sin x = \dfrac{1}{\csc x}$.

Use $\sin^2 x + \cos^2 x = 1$ and solve for $\cos^2 x$.

With this strategy, we again obtain $\cos^2 x$, the expression on the right side, and it takes fewer steps than the first approach. (This discussion continues on page 155 after the review on factoring polynomials.)

A Brief Review • Factoring Polynomials

- Factoring a polynomial expressed as a sum means finding an equivalent expression that is a product. Polynomials that cannot be factored using integer coefficients are called irreducible over the integers, or prime.
- The first step in any factoring problem is to look for the greatest common factor (GCF), an expression of the highest degree that divides each term of the polynomial. The distributive property, $ab + ac = a(b + c)$, is used to factor out the GCF.

EXAMPLE

Factor: $18x^3 + 27x^2$.

    The GCF of the two terms of the polynomial is $9x^2$.

$$18x^3 + 27x^2 = 9x^2(2x) + 9x^2(3) = 9x^2(2x + 3)$$

- Factoring by Grouping

  Some polynomials have only a greatest common factor of 1. However, by a suitable grouping of the terms, it still may be possible to factor. This process is called factoring by grouping.

EXAMPLE

Factor: $x^3 + 4x^2 + 3x + 12$.

$$\boxed{x^3 + 4x^2} \; + \; \boxed{3x + 12}$$

Common factor is x^2.   Common factor is 3.

$$
\begin{aligned}
&x^3 + 4x^2 + 3x + 12 \\
&= (x^3 + 4x^2) + (3x + 12) \quad &&\text{Group terms with common factors.} \\
&= x^2(x + 4) + 3(x + 4) \quad &&\text{Factor out the greatest common factor} \\
& &&\text{from the grouped terms. The remaining two} \\
& &&\text{terms have } x + 4 \text{ as a common binomial factor.} \\
&= (x + 4)(x^2 + 3) \quad &&\text{Factor out the GCF, } x + 4.
\end{aligned}
$$

- Factoring Trinomials

  To factor a trinomial, a polynomial with three terms, of the form $ax^2 + bx + c$, a little trial and error may be necessary.

A Strategy for Factoring $ax^2 + bx + c$

Assume, for the moment, that there is no greatest common factor.

1. Find two **First** terms whose product is ax^2:

$$(\Box x + \;)(\Box x + \;) = ax^2 + bx + c.$$

2. Find two **Last** terms whose product is c:

$$(\Box x + \Box)(\Box x + \Box) = ax^2 + bx + c.$$

3. By trial and error, perform steps 1 and 2 until the sum of the **Outside** product and **Inside** product is bx:

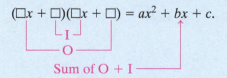

If no such combination exists, the polynomial is prime.

EXAMPLE

Factor: $8x^2 - 10x - 3$.

Step 1   Find two First terms whose product is $8x^2$.

$$8x^2 - 10x - 3 \overset{?}{=} (8x \quad)(x \quad)$$

$$8x^2 - 10x - 3 \overset{?}{=} (4x \quad)(2x \quad)$$

Step 2   Find two Last terms whose product is -3. The possible factorizations are $1(-3)$ and $-1(3)$.

Step 3   Try various combinations of these factors. The correct factorization of $8x^2 - 10x - 3$ is the one in which the sum of the **O**utside and **I**nside products is equal to $-10x$. Here is a list of the possible factorizations:

Possible Factorizations of $8x^2 - 10x - 3$	Sum of Outside and Inside Products (Should Equal $-10x$)
$(8x + 1)(x - 3)$	$-24x + x = -23x$
$(8x - 3)(x + 1)$	$8x - 3x = 5x$
$(8x - 1)(x + 3)$	$24x - x = 23x$
$(8x + 3)(x - 1)$	$-8x + 3x = -5x$
$(4x + 1)(2x - 3)$	$-12x + 2x = -10x$
$(4x - 3)(2x + 1)$	$4x - 6x = -2x$
$(4x - 1)(2x + 3)$	$12x - 2x = 10x$
$(4x + 3)(2x - 1)$	$-4x + 6x = 2x$

These four factorizations use $(8x \quad)(x \quad)$ with $1(-3)$ and $-1(3)$ as factorizations of **-3**.

These four factorizations use $(4x \quad)(2x \quad)$ with $1(-3)$ and $-1(3)$ as factorizations of **-3**.

This is the required middle term.

Thus, $8x^2 - 10x - 3 = (4x + 1)(2x - 3)$ or $(2x - 3)(4x + 1)$.

Use FOIL multiplication to check either of these factorizations.

• Factoring the Difference of Two Squares

$$A^2 - B^2 = (A + B)(A - B)$$

The difference of the squares of two terms factors as the product of a sum and a difference of those terms.

EXAMPLE

Factor: $81x^2 - 49$.

$$81x^2 - 49 = (9x)^2 - 7^2 = (9x + 7)(9x - 7)$$

• Factoring Perfect Square Trinomials

The trinomials that are factored using this technique are called perfect square trinomials.

1. $A^2 + 2AB + B^2 = (A + B)^2$ **2.** $A^2 - 2AB + B^2 = (A - B)^2$

Same sign   Same sign

EXAMPLE

Factor: **a.** $x^2 + 6x + 9$ **b.** $25x^2 - 60x + 36$.

a. $x^2 + 6x + 9 = x^2 + 2 \cdot x \cdot 3 + 3^2 = (x + 3)^2$

$A^2 \ + \ 2AB \ + \ B^2 \ = \ (A \ + \ B)^2$

b. $25x^2 - 60x + 36 = (5x)^2 - 2 \cdot 5x \cdot 6 + 6^2 = (5x - 6)^2$

$A^2 \ - \ 2AB \ + \ B^2 \ = \ (A \ - \ B)^2$

An even longer strategy to verify $\dfrac{\csc^2 x - 1}{\csc^2 x} = \cos^2 x$, but one that works, is to replace each of the two occurrences of $\csc^2 x$ on the left side by $\dfrac{1}{\sin^2 x}$. This may be the approach that you first consider, particularly if you become accustomed to rewriting the more complicated side in terms of sines and cosines. The selection of an appropriate fundamental identity to solve the puzzle most efficiently is learned through lots of practice.

The more identities you prove, the more confident and efficient you will become. Although practice is the only way to learn how to verify identities, there are some guidelines developed throughout the section that should help you get started.

Guidelines for Verifying Trigonometric Identities

- Work with each side of the equation independently of the other side. Start with the more complicated side and transform it in a step-by-step fashion until it looks exactly like the other side.
- Analyze the identity and look for opportunities to apply the fundamental identities.
- Try using one or more of the following techniques:
  1. Rewrite the more complicated side in terms of sines and cosines.
  2. Factor out the greatest common factor.
  3. Separate a single-term quotient into two terms:
     $$\frac{a + b}{c} = \frac{a}{c} + \frac{b}{c} \quad \text{and} \quad \frac{a - b}{c} = \frac{a}{c} - \frac{b}{c}.$$
  4. Combine fractional expressions using the least common denominator.
  5. Multiply the numerator and the denominator by a binomial factor that appears on the other side of the identity.
- Don't be afraid to stop and start over again if you are not getting anywhere. Creative puzzle solvers know that strategies leading to dead ends often provide good problem-solving ideas.

CONCEPT AND VOCABULARY CHECK

Fill in each blank so that the resulting statement is true.

1. To verify an identity, start with the more _____ side and transform it in a step-by-step fashion until it is identical to the _____ side.

2. It is sometimes helpful to verify an identity by rewriting one of the sides in terms of _____ and _____, and then simplifying the result.

3. True or false: To verify the identity
$$\frac{\sin x}{1 + \cos x} = \frac{1 - \cos x}{\sin x},$$
we should begin by multiplying both sides by $\sin x(1 + \cos x)$, the least common denominator. _____

4. $\dfrac{1}{\csc x - 1} - \dfrac{1}{\csc x + 1}$ can be simplified using _____ as the least common denominator.

5. You can use your graphing calculator to provide evidence of an identity. Graph the left side and the right side separately, and see if the two graphs are _____.

EXERCISE SET 3.1

Practice Exercises

In Exercises 1–60, verify each identity.

1. $\sin x \sec x = \tan x$
2. $\cos x \csc x = \cot x$
3. $\tan(-x)\cos x = -\sin x$
4. $\cot(-x) \sin x = -\cos x$
5. $\tan x \csc x \cos x = 1$
6. $\cot x \sec x \sin x = 1$
7. $\sec x - \sec x \sin^2 x = \cos x$
8. $\csc x - \csc x \cos^2 x = \sin x$
9. $\cos^2 x - \sin^2 x = 1 - 2\sin^2 x$
10. $\cos^2 x - \sin^2 x = 2\cos^2 x - 1$
11. $\csc \theta - \sin \theta = \cot \theta \cos \theta$
12. $\tan \theta + \cot \theta = \sec \theta \csc \theta$

13. $\dfrac{\tan \theta \cot \theta}{\csc \theta} = \sin \theta$
14. $\dfrac{\cos \theta \sec \theta}{\cot \theta} = \tan \theta$
15. $\sin^2 \theta(1 + \cot^2 \theta) = 1$
16. $\cos^2 \theta(1 + \tan^2 \theta) = 1$
17. $\sin t \tan t = \dfrac{1 - \cos^2 t}{\cos t}$
18. $\cos t \cot t = \dfrac{1 - \sin^2 t}{\sin t}$
19. $\dfrac{\csc^2 t}{\cot t} = \csc t \sec t$
20. $\dfrac{\sec^2 t}{\tan t} = \sec t \csc t$
21. $\dfrac{\tan^2 t}{\sec t} = \sec t - \cos t$
22. $\dfrac{\cot^2 t}{\csc t} = \csc t - \sin t$

23. $\dfrac{1 - \cos \theta}{\sin \theta} = \csc \theta - \cot \theta$ **24.** $\dfrac{1 - \sin \theta}{\cos \theta} = \sec \theta - \tan \theta$

25. $\dfrac{\sin t}{\csc t} + \dfrac{\cos t}{\sec t} = 1$ **26.** $\dfrac{\sin t}{\tan t} + \dfrac{\cos t}{\cot t} = \sin t + \cos t$

27. $\tan t + \dfrac{\cos t}{1 + \sin t} = \sec t$ **28.** $\cot t + \dfrac{\sin t}{1 + \cos t} = \csc t$

29. $1 - \dfrac{\sin^2 x}{1 + \cos x} = \cos x$ **30.** $1 - \dfrac{\cos^2 x}{1 + \sin x} = \sin x$

31. $\dfrac{\cos x}{1 - \sin x} + \dfrac{1 - \sin x}{\cos x} = 2 \sec x$

32. $\dfrac{\sin x}{\cos x + 1} + \dfrac{\cos x - 1}{\sin x} = 0$

33. $\sec^2 x \csc^2 x = \sec^2 x + \csc^2 x$

34. $\csc^2 x \sec x = \sec x + \csc x \cot x$

35. $\dfrac{\sec x - \csc x}{\sec x + \csc x} = \dfrac{\tan x - 1}{\tan x + 1}$ **36.** $\dfrac{\csc x - \sec x}{\csc x + \sec x} = \dfrac{\cot x - 1}{\cot x + 1}$

37. $\dfrac{\sin^2 x - \cos^2 x}{\sin x + \cos x} = \sin x - \cos x$

38. $\dfrac{\tan^2 x - \cot^2 x}{\tan x + \cot x} = \tan x - \cot x$

39. $\tan^2 2x + \sin^2 2x + \cos^2 2x = \sec^2 2x$

40. $\cot^2 2x + \cos^2 2x + \sin^2 2x = \csc^2 2x$

41. $\dfrac{\tan 2\theta + \cot 2\theta}{\csc 2\theta} = \sec 2\theta$ **42.** $\dfrac{\tan 2\theta + \cot 2\theta}{\sec 2\theta} = \csc 2\theta$

43. $\dfrac{\tan x + \tan y}{1 - \tan x \tan y} = \dfrac{\sin x \cos y + \cos x \sin y}{\cos x \cos y - \sin x \sin y}$

44. $\dfrac{\cot x + \cot y}{1 - \cot x \cot y} = \dfrac{\cos x \sin y + \sin x \cos y}{\sin x \sin y - \cos x \cos y}$

45. $(\sec x - \tan x)^2 = \dfrac{1 - \sin x}{1 + \sin x}$

46. $(\csc x - \cot x)^2 = \dfrac{1 - \cos x}{1 + \cos x}$

47. $\dfrac{\sec t + 1}{\tan t} = \dfrac{\tan t}{\sec t - 1}$

48. $\dfrac{\csc t - 1}{\cot t} = \dfrac{\cot t}{\csc t + 1}$

49. $\dfrac{1 + \cos t}{1 - \cos t} = (\csc t + \cot t)^2$

50. $\dfrac{\cos^2 t + 4 \cos t + 4}{\cos t + 2} = \dfrac{2 \sec t + 1}{\sec t}$

51. $\cos^4 t - \sin^4 t = 1 - 2 \sin^2 t$

52. $\sin^4 t - \cos^4 t = 1 - 2 \cos^2 t$

53. $\dfrac{\sin \theta - \cos \theta}{\sin \theta} + \dfrac{\cos \theta - \sin \theta}{\cos \theta} = 2 - \sec \theta \csc \theta$

54. $\dfrac{\sin \theta}{1 - \cot \theta} - \dfrac{\cos \theta}{\tan \theta - 1} = \sin \theta + \cos \theta$

55. $(\tan^2 \theta + 1)(\cos^2 \theta + 1) = \tan^2 \theta + 2$

56. $(\cot^2 \theta + 1)(\sin^2 \theta + 1) = \cot^2 \theta + 2$

57. $(\cos \theta - \sin \theta)^2 + (\cos \theta + \sin \theta)^2 = 2$

58. $(3 \cos \theta - 4 \sin \theta)^2 + (4 \cos \theta + 3 \sin \theta)^2 = 25$

59. $\dfrac{\cos^2 x - \sin^2 x}{1 - \tan^2 x} = \cos^2 x$

60. $\dfrac{\sin x + \cos x}{\sin x} - \dfrac{\cos x - \sin x}{\cos x} = \sec x \csc x$

Practice Plus

In Exercises 61–66, half of an identity and the graph of this half are given. Use the graph to make a conjecture as to what the right side of the identity should be. Then prove your conjecture.

61. $\dfrac{(\sec x + \tan x)(\sec x - \tan x)}{\sec x} = ?$

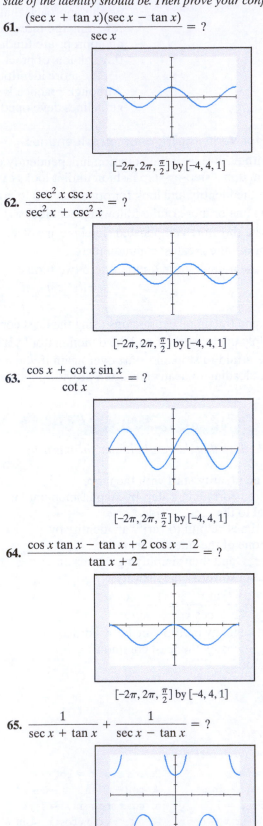

$[-2\pi, 2\pi, \frac{\pi}{2}]$ by $[-4, 4, 1]$

62. $\dfrac{\sec^2 x \csc x}{\sec^2 x + \csc^2 x} = ?$

$[-2\pi, 2\pi, \frac{\pi}{2}]$ by $[-4, 4, 1]$

63. $\dfrac{\cos x + \cot x \sin x}{\cot x} = ?$

$[-2\pi, 2\pi, \frac{\pi}{2}]$ by $[-4, 4, 1]$

64. $\dfrac{\cos x \tan x - \tan x + 2 \cos x - 2}{\tan x + 2} = ?$

$[-2\pi, 2\pi, \frac{\pi}{2}]$ by $[-4, 4, 1]$

65. $\dfrac{1}{\sec x + \tan x} + \dfrac{1}{\sec x - \tan x} = ?$

$[-2\pi, 2\pi, \frac{\pi}{2}]$ by $[-4, 4, 1]$

66. $\dfrac{1 + \cos x}{\sin x} + \dfrac{\sin x}{1 + \cos x} = \,?$

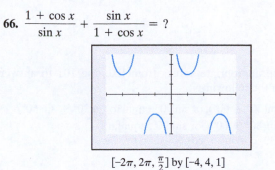

$[-2\pi, 2\pi, \frac{\pi}{2}]$ by $[-4, 4, 1]$

In Exercises 67–74, rewrite each expression in terms of the given function or functions.

67. $\dfrac{\tan x + \cot x}{\csc x}; \cos x$

68. $\dfrac{\sec x + \csc x}{1 + \tan x}; \sin x$

69. $\dfrac{\cos x}{1 + \sin x} + \tan x; \cos x$

70. $\dfrac{1}{\sin x \cos x} - \cot x; \cot x$

71. $\dfrac{1}{1 - \cos x} - \dfrac{\cos x}{1 + \cos x}; \csc x$

72. $(\sec x + \csc x)(\sin x + \cos x) - 2 - \cot x; \tan x$

73. $\dfrac{1}{\csc x - \sin x}; \sec x \text{ and } \tan x$

74. $\dfrac{1 - \sin x}{1 + \sin x} - \dfrac{1 + \sin x}{1 - \sin x}; \sec x \text{ and } \tan x$

Explaining the Concepts

75. Explain how to verify an identity.

76. Describe two strategies that can be used to verify identities.

77. Describe how you feel when you successfully verify a difficult identity. What other activities do you engage in that evoke the same feelings?

78. A 10-point question on a quiz asks students to verify the identity

$$\dfrac{\sin^2 x - \cos^2 x}{\sin x + \cos x} = \sin x - \cos x.$$

One student begins with the left side and obtains the right side as follows:

$$\dfrac{\sin^2 x - \cos^2 x}{\sin x + \cos x} = \dfrac{\sin^2 x}{\sin x} - \dfrac{\cos^2 x}{\cos x} = \sin x - \cos x.$$

How many points (out of 10) would you give this student? Explain your answer.

Technology Exercises

In Exercises 79–87, graph each side of the equation in the same viewing rectangle. If the graphs appear to coincide, verify that the equation is an identity. If the graphs do not appear to coincide, this indicates the equation is not an identity. In these exercises, find a value of x for which both sides are defined but not equal.

79. $\tan x = \sec x (\sin x - \cos x) + 1$

80. $\sin x = -\cos x \tan(-x)$

81. $\sin\left(x + \dfrac{\pi}{4}\right) = \sin x + \sin \dfrac{\pi}{4}$

82. $\cos\left(x + \dfrac{\pi}{4}\right) = \cos x + \cos \dfrac{\pi}{4}$

83. $\cos(x + \pi) = \cos x$ **84.** $\sin(x + \pi) = \sin x$

85. $\dfrac{\sin x}{1 - \cos^2 x} = \csc x$

86. $\sin x - \sin x \cos^2 x = \sin^3 x$

87. $\sqrt{\sin^2 x + \cos^2 x} = \sin x + \cos x$

Critical Thinking Exercises

Make Sense? *In Exercises 88–91, determine whether each statement makes sense or does not make sense, and explain your reasoning.*

88. The word *identity* is used in different ways in additive identity, multiplicative identity, and trigonometric identity.

89. To prove a trigonometric identity, I select one side of the equation and transform it until it is the other side of the equation, or I manipulate both sides to a common trigonometric expression.

90. In order to simplify $\dfrac{\cos x}{1 - \sin x} - \dfrac{\sin x}{\cos x}$, I need to know how to subtract rational expressions with unlike denominators.

91. The most efficient way that I can simplify $\dfrac{(\sec x + 1)(\sec x - 1)}{\sin^2 x}$ is to immediately rewrite the expression in terms of cosines and sines.

In Exercises 92–95, verify each identity.

92. $\dfrac{\sin^3 x - \cos^3 x}{\sin x - \cos x} = 1 + \sin x \cos x$

93. $\dfrac{\sin x - \cos x + 1}{\sin x + \cos x - 1} = \dfrac{\sin x + 1}{\cos x}$

94. $\ln|\sec x| = -\ln|\cos x|$ **95.** $\ln e^{\tan^2 x - \sec^2 x} = -1$

96. Use one of the fundamental identities in the box on page 144 to create an original identity.

Group Exercise

97. Group members are to write a helpful list of items for a pamphlet called "The Underground Guide to Verifying Identities." The pamphlet will be used primarily by students who sit, stare, and freak out every time they are asked to verify an identity. List easy ways to remember the fundamental identities. What helpful guidelines can you offer from the perspective of a student that you probably won't find in math books? If you have your own strategies that work particularly well, include them in the pamphlet.

Retaining the Concepts

98. Use a sketch to find the exact value of $\sec(\sin^{-1} \frac{1}{2})$.
(Section 2.3, Example 7)

99. Determine the amplitude, period, and phase shift of $y = 3 \sin(x + \pi)$. Then graph one period of the function. (Section 2.1, Example 4)

100. A point $P(x, y)$ is shown on the unit circle corresponding to a real number t. Find the values of the trigonometric functions at t.
(Section 1.4, Example 1)

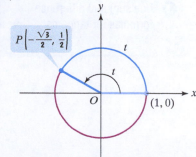

Preview Exercises

Exercises 101–103 will help you prepare for the material covered in the next section.

101. Give exact values for cos 30°, sin 30°, cos 60°, sin 60°, cos 90°, and sin 90°.

102. Use the appropriate values from Exercise 101 to answer each of the following.

   a. Is cos(30° + 60°), or cos 90°, equal to cos 30° + cos 60°?

   b. Is cos(30° + 60°), or cos 90°, equal to cos 30° cos 60° − sin 30° sin 60°?

103. Use the appropriate values from Exercise 101 to answer each of the following.

   a. Is sin(30° + 60°), or sin 90°, equal to sin 30° + sin 60°?

   b. Is sin(30° + 60°), or sin 90°, equal to sin 30° cos 60° + cos 30° sin 60°?

A Brief Review • The Distance Formula

The distance, d, between the points (x_1, y_1) and (x_2, y_2) in the rectangular coordinate system is

$$d = \sqrt{(x_2 - x_1)^2 + (y_2 - y_1)^2}.$$

EXAMPLE

Find the distance between $(-1, 4)$ and $(3, -2)$.

Let $(x_1, y_1) = (-1, 4)$ and $(x_2, y_2) = (3, -2)$. It does not matter which point you call (x_1, y_1) and which you call (x_2, y_2).

$$d = \sqrt{(x_2 - x_1)^2 + (y_2 - y_1)^2} = \sqrt{[3 - (-1)]^2 + (-2 - 4)^2} = \sqrt{4^2 + (-6)^2} = \sqrt{16 + 36} = \sqrt{52}$$
$$= \sqrt{4 \cdot 13} = 2\sqrt{13} \approx 7.21$$

Section 3.2   Sum and Difference Formulas

What am I supposed to learn?

After studying this section, you should be able to:

1. Use the formula for the cosine of the difference of two angles.

2. Use sum and difference formulas for cosines and sines.

3. Use sum and difference formulas for tangents.

Listen to the same note played on a piano and a violin. The notes have a different quality or "tone." Tone depends on the way an instrument vibrates. However, the less than 1% of the population with amusia, or true tone deafness, cannot tell the two sounds apart. Even simple, familiar tunes such as "*Happy Birthday*" and "*Jingle Bells*" are mystifying to amusics.

When a note is played, it vibrates at a specific fundamental frequency and has a particular amplitude. Amusics cannot tell the difference between two sounds from tuning forks modeled by $p = 3 \sin 2t$ and $p = 2 \sin(2t + \pi)$, respectively. However, they can recognize the difference between the two equations. Notice that the second equation contains the sine of the sum of two angles. In this section, we will be developing identities involving the sums or differences of two angles. These formulas are called the **sum and difference formulas**. We begin with $\cos(\alpha - \beta)$, the cosine of the difference of two angles.

The Cosine of the Difference of Two Angles

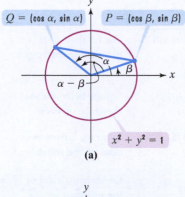

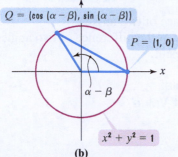

FIGURE 3.1 Using the unit circle and PQ to develop a formula for $\cos(\alpha - \beta)$

> ### The Cosine of the Difference of Two Angles
>
> $$\cos(\alpha - \beta) = \cos \alpha \cos \beta + \sin \alpha \sin \beta$$
>
> The cosine of the difference of two angles equals the cosine of the first angle times the cosine of the second angle plus the sine of the first angle times the sine of the second angle.

We use **Figure 3.1** to prove the identity in the box. The graph in **Figure 3.1(a)** shows a unit circle, $x^2 + y^2 = 1$. The figure uses the definitions of the cosine and sine functions as the x- and y-coordinates of points along the unit circle. For example, point P corresponds to angle β. By definition, the x-coordinate of P is $\cos \beta$ and the y-coordinate is $\sin \beta$. Similarly, point Q corresponds to angle α. By definition, the x-coordinate of Q is $\cos \alpha$ and the y-coordinate is $\sin \alpha$.

Note that if we draw a line segment between points P and Q in **Figure 3.1(a)**, a triangle is formed. Angle $\alpha - \beta$ is one of the angles of this triangle. What happens if we rotate this triangle so that point P falls on the x-axis at $(1, 0)$? This rotation changes the coordinates of points P and Q. However, it has no effect on the length of line segment PQ.

We can use the distance formula, $d = \sqrt{(x_2 - x_1)^2 + (y_2 - y_1)^2}$, to find an expression for PQ in **Figure 3.1(a)** and in **Figure 3.1(b)**. By equating the two expressions for PQ, we will obtain the identity for the cosine of the difference of two angles, $\alpha - \beta$. We first apply the distance formula in **Figure 3.1(a)**.

$$PQ = \sqrt{(\cos \alpha - \cos \beta)^2 + (\sin \alpha - \sin \beta)^2}$$

Apply the distance formula, $d = \sqrt{(x_2 - x_1)^2 + (y_2 - y_1)^2}$, to find the distance between $(\cos \beta, \sin \beta)$ and $(\cos \alpha, \sin \alpha)$.

$$= \sqrt{\cos^2 \alpha - 2 \cos \alpha \cos \beta + \cos^2 \beta + \sin^2 \alpha - 2 \sin \alpha \sin \beta + \sin^2 \beta}$$

Square each expression using $(A - B)^2 = A^2 - 2AB + B^2$.

$$= \sqrt{(\sin^2 \alpha + \cos^2 \alpha) + (\sin^2 \beta + \cos^2 \beta) - 2 \cos \alpha \cos \beta - 2 \sin \alpha \sin \beta}$$

Regroup terms to apply a Pythagorean identity.

$$= \sqrt{1 + 1 - 2 \cos \alpha \cos \beta - 2 \sin \alpha \sin \beta}$$

Because $\sin^2 x + \cos^2 x = 1$, each expression in parentheses equals 1.

$$= \sqrt{2 - 2 \cos \alpha \cos \beta - 2 \sin \alpha \sin \beta}$$

Simplify.

Next, we apply the distance formula in **Figure 3.1(b)** to obtain a second expression for PQ. We let $(x_1, y_1) = (1, 0)$ and $(x_2, y_2) = (\cos(\alpha - \beta), \sin(\alpha - \beta))$.

$$PQ = \sqrt{[\cos(\alpha - \beta) - 1]^2 + [\sin(\alpha - \beta) - 0]^2}$$

Apply the distance formula to find the distance between $(1, 0)$ and $(\cos(\alpha - \beta), \sin(\alpha - \beta))$.

$$= \sqrt{\cos^2(\alpha - \beta) - 2 \cos(\alpha - \beta) + 1 + \sin^2(\alpha - \beta)}$$

Square each expression.

Using a Pythagorean identity, $\sin^2(\alpha - \beta) + \cos^2(\alpha - \beta) = 1$.

$$= \sqrt{1 - 2 \cos(\alpha - \beta) + 1}$$

Use a Pythagorean identity.

$$= \sqrt{2 - 2 \cos(\alpha - \beta)}$$

Simplify.

Now we equate the two expressions for PQ.

$$\sqrt{2 - 2\cos(\alpha - \beta)} = \sqrt{2 - 2\cos\alpha\cos\beta - 2\sin\alpha\sin\beta}$$ The rotation does not change the length of PQ.

$$2 - 2\cos(\alpha - \beta) = 2 - 2\cos\alpha\cos\beta - 2\sin\alpha\sin\beta$$ Square both sides to eliminate radicals.

$$-2\cos(\alpha - \beta) = -2\cos\alpha\cos\beta - 2\sin\alpha\sin\beta$$ Subtract 2 from both sides of the equation.

$$\cos(\alpha - \beta) = \cos\alpha\cos\beta + \sin\alpha\sin\beta$$ Divide both sides of the equation by -2.

① Use the formula for the cosine of the difference of two angles.

This proves the identity for the cosine of the difference of two angles.

Now that we see where the identity for the cosine of the difference of two angles comes from, let's look at some applications of this result.

EXAMPLE 1 Using the Difference Formula for Cosines to Find an Exact Value

Find the exact value of $\cos 15°$.

SOLUTION

We know exact values for trigonometric functions of $60°$ and $45°$. Thus, we write $15°$ as $60° - 45°$ and use the difference formula for cosines.

$$\cos 15° = \cos(60° - 45°)$$

$$= \cos 60° \cos 45° + \sin 60° \sin 45°$$ $\cos(\alpha - \beta) = \cos\alpha\cos\beta + \sin\alpha\sin\beta$

$$= \frac{1}{2} \cdot \frac{\sqrt{2}}{2} + \frac{\sqrt{3}}{2} \cdot \frac{\sqrt{2}}{2}$$ Substitute exact values from memory or use special right triangles.

$$= \frac{\sqrt{2}}{4} + \frac{\sqrt{6}}{4}$$ Multiply.

$$= \frac{\sqrt{2} + \sqrt{6}}{4}$$ Add. • • •

GREAT QUESTION!

Can I use my calculator to verify that

$$\cos 15° = \frac{\sqrt{2} + \sqrt{6}}{4}?$$

Yes. Find approximations for $\cos 15°$ and $\dfrac{\sqrt{2} + \sqrt{6}}{4}$.

```
cos(15)
            .9659258263
(√2+√6)/4
            .9659258263
■
```

Because the approximations are the same, we have checked that

$$\cos 15° = \frac{\sqrt{2} + \sqrt{6}}{4}.$$

✓ **Check Point 1** We know that $\cos 30° = \dfrac{\sqrt{3}}{2}$. Obtain this exact value using $\cos 30° = \cos(90° - 60°)$ and the difference formula for cosines.

Blitzer Bonus ‖

Sound Quality and Amusia

People with true tone deafness cannot hear the difference among tones produced by a tuning fork, a flute, an oboe, and a violin. They cannot dance or tell the difference between harmony and dissonance. People with amusia appear to have been born without the wiring necessary to process music. Intriguingly, they show no overt signs of brain damage and their brain scans appear normal. Thus, they can visually recognize the difference among sound waves that produce varying sound qualities.

Varying Sound Qualities

• Tuning fork: Sound waves are rounded and regular, giving a pure and gentle tone.

• Flute: Sound waves are smooth and give a fluid tone.

• Oboe: Rapid wave changes give a richer tone.

• Violin: Jagged waves give a brighter harsher tone.

EXAMPLE 2 Using the Difference Formula for Cosines to Find an Exact Value

Find the exact value of $\cos 80° \cos 20° + \sin 80° \sin 20°$.

SOLUTION

The given expression is the right side of the formula for $\cos(\alpha - \beta)$ with $\alpha = 80°$ and $\beta = 20°$.

$$\cos(\alpha - \beta) = \cos \alpha \cos \beta + \sin \alpha \sin \beta$$

$$\cos 80° \cos 20° + \sin 80° \sin 20° = \cos(80° - 20°) = \cos 60° = \frac{1}{2}$$ • • •

✓ **Check Point 2** Find the exact value of

$$\cos 70° \cos 40° + \sin 70° \sin 40°.$$

EXAMPLE 3 Verifying an Identity

Verify the identity: $\dfrac{\cos(\alpha - \beta)}{\sin \alpha \cos \beta} = \cot \alpha + \tan \beta$.

SOLUTION

We work with the left side.

$$\frac{\cos(\alpha - \beta)}{\sin \alpha \cos \beta} = \frac{\cos \alpha \cos \beta + \sin \alpha \sin \beta}{\sin \alpha \cos \beta}$$ Use the formula for $\cos(\alpha - \beta)$.

$$= \frac{\cos \alpha \cos \beta}{\sin \alpha \cos \beta} + \frac{\sin \alpha \sin \beta}{\sin \alpha \cos \beta}$$ Divide each term in the numerator by $\sin \alpha \cos \beta$.

$$= \frac{\cos \alpha}{\sin \alpha} \cdot \frac{\cos \beta}{\cos \beta} + \frac{\sin \alpha}{\sin \alpha} \cdot \frac{\sin \beta}{\cos \beta}$$ This step can be done mentally. We wanted you to see the substitutions that follow.

$$= \cot \alpha \cdot 1 + 1 \cdot \tan \beta$$ Use quotient identities.

$$= \cot \alpha + \tan \beta$$ Simplify.

We worked with the left side and arrived at the right side. Thus, the identity is verified. • • •

✓ **Check Point 3** Verify the identity: $\dfrac{\cos(\alpha - \beta)}{\cos \alpha \cos \beta} = 1 + \tan \alpha \tan \beta$.

The difference formula for cosines is used to establish other identities. For example, in our work with right triangles, we noted that cofunctions of complements are equal. Thus, because $\frac{\pi}{2} - \theta$ and θ are complements,

$$\cos\left(\frac{\pi}{2} - \theta\right) = \sin \theta.$$

We can use the formula for $\cos(\alpha - \beta)$ to prove this cofunction identity for all angles.

Apply $\cos(\alpha - \beta)$ with $\alpha = \frac{\pi}{2}$ and $\beta = \theta$.
$\cos(\alpha - \beta) = \cos \alpha \cos \beta + \sin \alpha \sin \beta$

$$\cos\left(\frac{\pi}{2} - \theta\right) = \cos \frac{\pi}{2} \cos \theta + \sin \frac{\pi}{2} \sin \theta$$

$$= 0 \cdot \cos \theta + 1 \cdot \sin \theta$$

$$= \sin \theta$$

TECHNOLOGY

Graphic Connections

The graphs of

$$y = \cos\left(\frac{\pi}{2} - x\right)$$

and

$$y = \sin x$$

are shown in the same viewing rectangle. The graphs are the same. The displayed math on the right below with the voice balloon on top shows the equivalence algebraically.

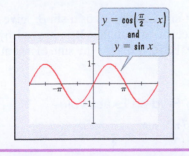

$y = \cos\left(\frac{\pi}{2} - x\right)$
and
$y = \sin x$

② Use sum and difference formulas for cosines and sines.

Sum and Difference Formulas for Cosines and Sines

Our formula for $\cos(\alpha - \beta)$ can be used to verify an identity for a sum involving cosines, as well as identities for a sum and a difference for sines.

> ### Sum and Difference Formulas for Cosines and Sines
> 1. $\cos(\alpha + \beta) = \cos \alpha \cos \beta - \sin \alpha \sin \beta$
> 2. $\cos(\alpha - \beta) = \cos \alpha \cos \beta + \sin \alpha \sin \beta$
> 3. $\sin(\alpha + \beta) = \sin \alpha \cos \beta + \cos \alpha \sin \beta$
> 4. $\sin(\alpha - \beta) = \sin \alpha \cos \beta - \cos \alpha \sin \beta$

Up to now, we have concentrated on the second formula in the preceding box, $\cos(\alpha - \beta) = \cos \alpha \cos \beta + \sin \alpha \sin \beta$. The first identity, $\cos(\alpha + \beta) = \cos \alpha \cos \beta - \sin \alpha \sin \beta$, gives a formula for the cosine of the sum of two angles. It is proved as follows:

$$\cos(\alpha + \beta) = \cos[\alpha - (-\beta)] \qquad \text{Express addition as subtraction of an additive inverse.}$$

$$= \cos \alpha \cos(-\beta) + \sin \alpha \sin(-\beta) \qquad \text{Use the difference formula for cosines.}$$

$$= \cos \alpha \cos \beta + \sin \alpha(-\sin \beta) \qquad \begin{array}{l}\text{Cosine is even: } \cos(-\beta) = \cos \beta. \\ \text{Sine is odd: } \sin(-\beta) = -\sin \beta.\end{array}$$

$$= \cos \alpha \cos \beta - \sin \alpha \sin \beta. \qquad \text{Simplify.}$$

Thus, the cosine of the sum of two angles equals the cosine of the first angle times the cosine of the second angle minus the sine of the first angle times the sine of the second angle.

The third identity in the box, $\sin(\alpha + \beta) = \sin \alpha \cos \beta + \cos \alpha \sin \beta$, gives a formula for $\sin(\alpha + \beta)$, the sine of the sum of two angles. It is proved as follows:

$$\sin(\alpha + \beta) = \cos\left[\frac{\pi}{2} - (\alpha + \beta)\right] \qquad \begin{array}{l}\text{Use a cofunction identity:} \\ \sin \theta = \cos\left(\dfrac{\pi}{2} - \theta\right).\end{array}$$

$$= \cos\left[\left(\frac{\pi}{2} - \alpha\right) - \beta\right] \qquad \text{Regroup.}$$

$$= \cos\left(\frac{\pi}{2} - \alpha\right)\cos \beta + \sin\left(\frac{\pi}{2} - \alpha\right)\sin \beta \qquad \begin{array}{l}\text{Use the difference formula for cosines.}\end{array}$$

$$= \sin \alpha \cos \beta + \cos \alpha \sin \beta. \qquad \text{Use cofunction identities.}$$

Thus, the sine of the sum of two angles equals the sine of the first angle times the cosine of the second angle plus the cosine of the first angle times the sine of the second angle.

The final identity in the box, $\sin(\alpha - \beta) = \sin \alpha \cos \beta - \cos \alpha \sin \beta$, gives a formula for $\sin(\alpha - \beta)$, the sine of the difference of two angles. It is proved by writing $\sin(\alpha - \beta)$ as $\sin[\alpha + (-\beta)]$ and then using the formula for the sine of a sum.

EXAMPLE 4   Using the Sine of a Sum to Find an Exact Value

Find the exact value of $\sin \dfrac{7\pi}{12}$ using the fact that $\dfrac{7\pi}{12} = \dfrac{\pi}{3} + \dfrac{\pi}{4}$.

SOLUTION

We apply the formula for the sine of a sum.

$$\sin\frac{7\pi}{12} = \sin\left(\frac{\pi}{3} + \frac{\pi}{4}\right)$$

$$= \sin\frac{\pi}{3}\cos\frac{\pi}{4} + \cos\frac{\pi}{3}\sin\frac{\pi}{4} \qquad \sin(\alpha + \beta) = \sin\alpha\cos\beta + \cos\alpha\sin\beta$$

$$= \frac{\sqrt{3}}{2}\cdot\frac{\sqrt{2}}{2} + \frac{1}{2}\cdot\frac{\sqrt{2}}{2} \qquad \text{Substitute exact values.}$$

$$= \frac{\sqrt{6} + \sqrt{2}}{4} \qquad \text{Simplify.} \qquad \bullet\bullet\bullet$$

✓ **Check Point 4** Find the exact value of $\sin\dfrac{5\pi}{12}$ using the fact that

$$\frac{5\pi}{12} = \frac{\pi}{6} + \frac{\pi}{4}.$$

EXAMPLE 5 Finding Exact Values

Suppose that $\sin\alpha = \frac{12}{13}$ for a quadrant II angle α and $\sin\beta = \frac{3}{5}$ for a quadrant I angle β. Find the exact value of each of the following:

a. $\cos\alpha$ **b.** $\cos\beta$ **c.** $\cos(\alpha + \beta)$ **d.** $\sin(\alpha + \beta)$.

SOLUTION

a. We find $\cos\alpha$ using a sketch that illustrates

$$\sin\alpha = \frac{12}{13} = \frac{y}{r}.$$

Figure 3.2 shows a quadrant II angle α with $\sin\alpha = \frac{12}{13}$. We find x using $x^2 + y^2 = r^2$. Because α lies in quadrant II, x is negative.

$$x^2 + 12^2 = 13^2 \qquad x^2 + y^2 = r^2$$

$$x^2 + 144 = 169 \qquad \text{Square 12 and 13, respectively.}$$

$$x^2 = 25 \qquad \text{Subtract 144 from both sides.}$$

$$x = -\sqrt{25} = -5 \qquad \text{If } x^2 = 25, \text{ then } x = \pm\sqrt{25} = \pm 5.$$
Choose $x = -\sqrt{25}$ because in quadrant II, x is negative.

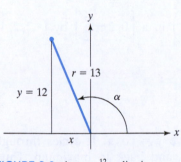

FIGURE 3.2 $\sin\alpha = \frac{12}{13}$: α lies in quadrant II.

Thus,

$$\cos\alpha = \frac{x}{r} = \frac{-5}{13} = -\frac{5}{13}.$$

b. We find $\cos\beta$ using a sketch that illustrates

$$\sin\beta = \frac{3}{5} = \frac{y}{r}.$$

Figure 3.3 shows a quadrant I angle β with $\sin\beta = \frac{3}{5}$. We find x using $x^2 + y^2 = r^2$.

$$x^2 + 3^2 = 5^2 \qquad x^2 + y^2 = r^2$$

$$x^2 + 9 = 25 \qquad \text{Square 3 and 5, respectively.}$$

$$x^2 = 16 \qquad \text{Subtract 9 from both sides.}$$

$$x = \sqrt{16} = 4 \qquad \text{If } x^2 = 16, \text{ then } x = \pm\sqrt{16} = \pm 4.$$
Choose $x = \sqrt{16}$ because in quadrant I, x is positive.

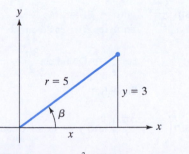

FIGURE 3.3 $\sin\beta = \frac{3}{5}$: β lies in quadrant I.

Thus, $\cos\beta = \dfrac{x}{r} = \dfrac{4}{5}.$

We use the given values and the exact values that we determined to find exact values for $\cos(\alpha + \beta)$ and $\sin(\alpha + \beta)$.

These values are given.

These are the values we found.

$$\sin \alpha = \frac{12}{13}, \ \sin \beta = \frac{3}{5} \qquad \cos \alpha = -\frac{5}{13}, \ \cos \beta = \frac{4}{5}$$

c. We use the formula for the cosine of a sum.

$$\cos(\alpha + \beta) = \cos \alpha \cos \beta - \sin \alpha \sin \beta$$

$$= \left(-\frac{5}{13}\right)\left(\frac{4}{5}\right) - \frac{12}{13}\left(\frac{3}{5}\right) = -\frac{56}{65}$$

d. We use the formula for the sine of a sum.

$$\sin(\alpha + \beta) = \sin \alpha \cos \beta + \cos \alpha \sin \beta$$

$$= \frac{12}{13} \cdot \frac{4}{5} + \left(-\frac{5}{13}\right) \cdot \frac{3}{5} = \frac{33}{65}$$

● ● ●

✓ **Check Point 5** Suppose that $\sin \alpha = \frac{4}{5}$ for a quadrant II angle α and $\sin \beta = \frac{1}{2}$ for a quadrant I angle β. Find the exact value of each of the following:

a. $\cos \alpha$ **b.** $\cos \beta$ **c.** $\cos(\alpha + \beta)$ **d.** $\sin(\alpha + \beta)$.

EXAMPLE 6   Verifying Observations on a Graphing Utility

Figure 3.4 shows the graph of $y = \sin\left(x - \frac{3\pi}{2}\right)$ in a $\left[0, 2\pi, \frac{\pi}{2}\right]$ by $[-2, 2, 1]$ viewing rectangle.

a. Describe the graph using another equation.
b. Verify that the two equations are equivalent.

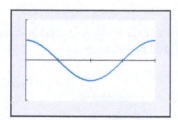

FIGURE 3.4 The graph of $y = \sin\left(x - \frac{3\pi}{2}\right)$ in a $\left[0, 2\pi, \frac{\pi}{2}\right]$ by $[-2, 2, 1]$ viewing rectangle.

SOLUTION

a. The graph appears to be the cosine curve $y = \cos x$. It cycles through maximum, intercept, minimum, intercept, and back to maximum. Thus, $y = \cos x$ also describes the graph.

b. We must show that

$$\sin\left(x - \frac{3\pi}{2}\right) = \cos x.$$

We apply the formula for the sine of a difference on the left side.

$$\sin\left(x - \frac{3\pi}{2}\right) = \sin x \cos \frac{3\pi}{2} - \cos x \sin \frac{3\pi}{2} \qquad \begin{array}{l} \sin(\alpha - \beta) = \\ \sin \alpha \cos \beta - \cos \alpha \sin \beta \end{array}$$

$$= \sin x \cdot 0 - \cos x(-1) \qquad \cos \frac{3\pi}{2} = 0 \text{ and } \sin \frac{3\pi}{2} = -1$$

$$= \cos x \qquad \text{Simplify.}$$

This verifies our observation that $y = \sin\left(x - \frac{3\pi}{2}\right)$ and $y = \cos x$ describe the same graph.

● ● ●

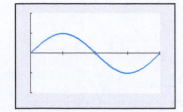

FIGURE 3.5

③ Use sum and difference formulas for tangents.

DISCOVERY

Derive the sum and difference formulas for tangents by working Exercises 55 and 56 in Exercise Set 3.2.

✓ Check Point **6** **Figure 3.5** shows the graph of $y = \cos\left(x + \dfrac{3\pi}{2}\right)$ in a $\left[0, 2\pi, \dfrac{\pi}{2}\right]$ by $[-2, 2, 1]$ viewing rectangle.

    a. Describe the graph using another equation.
    b. Verify that the two equations are equivalent.

Sum and Difference Formulas for Tangents

By writing $\tan(\alpha + \beta)$ as the quotient of $\sin(\alpha + \beta)$ and $\cos(\alpha + \beta)$, we can develop a formula for the tangent of a sum. Writing subtraction as addition of an inverse leads to a formula for the tangent of a difference.

Sum and Difference Formulas for Tangents

$$\tan(\alpha + \beta) = \frac{\tan \alpha + \tan \beta}{1 - \tan \alpha \tan \beta}$$

The tangent of the sum of two angles equals the tangent of the first angle plus the tangent of the second angle divided by 1 minus their product.

$$\tan(\alpha - \beta) = \frac{\tan \alpha - \tan \beta}{1 + \tan \alpha \tan \beta}$$

The tangent of the difference of two angles equals the tangent of the first angle minus the tangent of the second angle divided by 1 plus their product.

EXAMPLE 7 Verifying an Identity

Verify the identity: $\tan\left(x - \dfrac{\pi}{4}\right) = \dfrac{\tan x - 1}{\tan x + 1}$.

SOLUTION

We work with the left side.

$$\tan\left(x - \frac{\pi}{4}\right) = \frac{\tan x - \tan\dfrac{\pi}{4}}{1 + \tan x \tan\dfrac{\pi}{4}} \qquad \tan(\alpha - \beta) = \frac{\tan \alpha - \tan \beta}{1 + \tan \alpha \tan \beta}$$

$$= \frac{\tan x - 1}{1 + \tan x \cdot 1} \qquad \tan\frac{\pi}{4} = 1$$

$$= \frac{\tan x - 1}{\tan x + 1}$$

• • •

✓ Check Point **7** Verify the identity: $\tan(x + \pi) = \tan x$.

ACHIEVING SUCCESS

Analyze the errors you make on quizzes and tests.

For each error, write out the correct solution along with a description of the concept needed to solve the problem correctly. Do your mistakes indicate gaps in understanding concepts or do you at times believe that you are just not a good test taker? Are you repeatedly making the same kinds of mistakes on tests? Keeping track of errors should increase your understanding of the material, resulting in improved test scores.

CONCEPT AND VOCABULARY CHECK

Fill in each blank so that the resulting statement is true.

1. $\cos(x + y) = $ _____

2. $\cos(x - y) = $ _____

3. $\sin(C + D) = $ _____

4. $\sin(C - D) = $ _____

5. $\tan(\theta + \phi) = $ _____

6. $\tan(\theta - \phi) = $ _____

7. True or false: The cosine of the sum of two angles equals the sum of the cosines of those angles. _____

8. True or false: $\tan 75° = \tan 30° + \tan 45°$ _____

EXERCISE SET 3.2

Practice Exercises

Use the formula for the cosine of the difference of two angles to solve Exercises 1–12.

In Exercises 1–4, find the exact value of each expression.

1. $\cos(45° - 30°)$

2. $\cos(120° - 45°)$

3. $\cos\left(\dfrac{3\pi}{4} - \dfrac{\pi}{6}\right)$

4. $\cos\left(\dfrac{2\pi}{3} - \dfrac{\pi}{6}\right)$

In Exercises 5–8, each expression is the right side of the formula for $\cos(\alpha - \beta)$ with particular values for α and β.

   a. Identify α and β in each expression.

   b. Write the expression as the cosine of an angle.

   c. Find the exact value of the expression.

5. $\cos 50° \cos 20° + \sin 50° \sin 20°$

6. $\cos 50° \cos 5° + \sin 50° \sin 5°$

7. $\cos\dfrac{5\pi}{12}\cos\dfrac{\pi}{12} + \sin\dfrac{5\pi}{12}\sin\dfrac{\pi}{12}$

8. $\cos\dfrac{5\pi}{18}\cos\dfrac{\pi}{9} + \sin\dfrac{5\pi}{18}\sin\dfrac{\pi}{9}$

In Exercises 9–12, verify each identity.

9. $\dfrac{\cos(\alpha - \beta)}{\cos \alpha \sin \beta} = \tan \alpha + \cot \beta$

10. $\dfrac{\cos(\alpha - \beta)}{\sin \alpha \sin \beta} = \cot \alpha \cot \beta + 1$

11. $\cos\left(x - \dfrac{\pi}{4}\right) = \dfrac{\sqrt{2}}{2}(\cos x + \sin x)$

12. $\cos\left(x - \dfrac{5\pi}{4}\right) = -\dfrac{\sqrt{2}}{2}(\cos x + \sin x)$

Use one or more of the six sum and difference identities to solve Exercises 13–54.

In Exercises 13–24, find the exact value of each expression.

13. $\sin(45° - 30°)$

14. $\sin(60° - 45°)$

15. $\sin 105°$

16. $\sin 75°$

17. $\cos(135° + 30°)$

18. $\cos(240° + 45°)$

19. $\cos 75°$

20. $\cos 105°$

21. $\tan\left(\dfrac{\pi}{6} + \dfrac{\pi}{4}\right)$

22. $\tan\left(\dfrac{\pi}{3} + \dfrac{\pi}{4}\right)$

23. $\tan\left(\dfrac{4\pi}{3} - \dfrac{\pi}{4}\right)$

24. $\tan\left(\dfrac{5\pi}{3} - \dfrac{\pi}{4}\right)$

In Exercises 25–32, write each expression as the sine, cosine, or tangent of an angle. Then find the exact value of the expression.

25. $\sin 25° \cos 5° + \cos 25° \sin 5°$

26. $\sin 40° \cos 20° + \cos 40° \sin 20°$

27. $\dfrac{\tan 10° + \tan 35°}{1 - \tan 10° \tan 35°}$

28. $\dfrac{\tan 50° - \tan 20°}{1 + \tan 50° \tan 20°}$

29. $\sin\dfrac{5\pi}{12}\cos\dfrac{\pi}{4} - \cos\dfrac{5\pi}{12}\sin\dfrac{\pi}{4}$

30. $\sin\dfrac{7\pi}{12}\cos\dfrac{\pi}{12} - \cos\dfrac{7\pi}{12}\sin\dfrac{\pi}{12}$

31. $\dfrac{\tan\dfrac{\pi}{5} - \tan\dfrac{\pi}{30}}{1 + \tan\dfrac{\pi}{5}\tan\dfrac{\pi}{30}}$

32. $\dfrac{\tan\dfrac{\pi}{5} + \tan\dfrac{4\pi}{5}}{1 - \tan\dfrac{\pi}{5}\tan\dfrac{4\pi}{5}}$

In Exercises 33–54, verify each identity.

33. $\sin\left(x + \dfrac{\pi}{2}\right) = \cos x$

34. $\sin\left(x + \dfrac{3\pi}{2}\right) = -\cos x$

35. $\cos\left(x - \dfrac{\pi}{2}\right) = \sin x$

36. $\cos(\pi - x) = -\cos x$

37. $\tan(2\pi - x) = -\tan x$ **38.** $\tan(\pi - x) = -\tan x$

39. $\sin(\alpha + \beta) + \sin(\alpha - \beta) = 2 \sin \alpha \cos \beta$

40. $\cos(\alpha + \beta) + \cos(\alpha - \beta) = 2 \cos \alpha \cos \beta$

41. $\dfrac{\sin(\alpha - \beta)}{\cos \alpha \cos \beta} = \tan \alpha - \tan \beta$

42. $\dfrac{\sin(\alpha + \beta)}{\cos \alpha \cos \beta} = \tan \alpha + \tan \beta$

43. $\tan\left(\theta + \dfrac{\pi}{4}\right) = \dfrac{\cos \theta + \sin \theta}{\cos \theta - \sin \theta}$

44. $\tan\left(\dfrac{\pi}{4} - \theta\right) = \dfrac{\cos \theta - \sin \theta}{\cos \theta + \sin \theta}$

45. $\cos(\alpha + \beta) \cos(\alpha - \beta) = \cos^2 \beta - \sin^2 \alpha$

46. $\sin(\alpha + \beta) \sin(\alpha - \beta) = \cos^2 \beta - \cos^2 \alpha$

47. $\dfrac{\sin(\alpha + \beta)}{\sin(\alpha - \beta)} = \dfrac{\tan \alpha + \tan \beta}{\tan \alpha - \tan \beta}$

48. $\dfrac{\cos(\alpha + \beta)}{\cos(\alpha - \beta)} = \dfrac{1 - \tan \alpha \tan \beta}{1 + \tan \alpha \tan \beta}$

49. $\dfrac{\cos(x + h) - \cos x}{h} = \cos x \dfrac{\cos h - 1}{h} - \sin x \dfrac{\sin h}{h}$

50. $\dfrac{\sin(x + h) - \sin x}{h} = \cos x \dfrac{\sin h}{h} + \sin x \dfrac{\cos h - 1}{h}$

51. $\sin 2\alpha = 2 \sin \alpha \cos \alpha$

Hint: Write $\sin 2\alpha$ as $\sin(\alpha + \alpha)$.

52. $\cos 2\alpha = \cos^2 \alpha - \sin^2 \alpha$

Hint: Write $\cos 2\alpha$ as $\cos(\alpha + \alpha)$.

53. $\tan 2\alpha = \dfrac{2 \tan \alpha}{1 - \tan^2 \alpha}$

Hint: Write $\tan 2\alpha$ as $\tan(\alpha + \alpha)$.

54. $\tan\left(\dfrac{\pi}{4} + \alpha\right) - \tan\left(\dfrac{\pi}{4} - \alpha\right) = 2 \tan 2\alpha$

Hint: Use the result in Exercise 53.

55. Derive the identity for $\tan(\alpha + \beta)$ using

$$\tan(\alpha + \beta) = \dfrac{\sin(\alpha + \beta)}{\cos(\alpha + \beta)}.$$

After applying the formulas for sums of sines and cosines, divide the numerator and denominator by $\cos \alpha \cos \beta$.

56. Derive the identity for $\tan(\alpha - \beta)$ using

$$\tan(\alpha - \beta) = \tan[\alpha + (-\beta)].$$

After applying the formula for the tangent of the sum of two angles, use the fact that the tangent is an odd function.

In Exercises 57–64, find the exact value of the following under the given conditions:

a. $\cos(\alpha + \beta)$ **b.** $\sin(\alpha + \beta)$ **c.** $\tan(\alpha + \beta)$.

57. $\sin \alpha = \frac{3}{5}, \alpha$ lies in quadrant I, and $\sin \beta = \frac{5}{13}, \beta$ lies in quadrant II.

58. $\sin \alpha = \frac{4}{5}, \alpha$ lies in quadrant I, and $\sin \beta = \frac{7}{25}, \beta$ lies in quadrant II.

59. $\tan \alpha = -\frac{3}{4}, \alpha$ lies in quadrant II, and $\cos \beta = \frac{1}{3}, \beta$ lies in quadrant I.

60. $\tan \alpha = -\frac{4}{3}, \alpha$ lies in quadrant II, and $\cos \beta = \frac{2}{3}, \beta$ lies in quadrant I.

61. $\cos \alpha = \frac{8}{17}, \alpha$ lies in quadrant IV, and $\sin \beta = -\frac{1}{2}, \beta$ lies in quadrant III.

62. $\cos \alpha = \frac{1}{2}, \alpha$ lies in quadrant IV, and $\sin \beta = -\frac{1}{3}, \beta$ lies in quadrant III.

63. $\tan \alpha = \frac{3}{4}, \pi < \alpha < \frac{3\pi}{2}$, and $\cos \beta = \frac{1}{4}, \frac{3\pi}{2} < \beta < 2\pi$.

64. $\sin \alpha = \frac{5}{6}, \frac{\pi}{2} < \alpha < \pi$, and $\tan \beta = \frac{3}{7}, \pi < \beta < \frac{3\pi}{2}$.

In Exercises 65–68, the graph with the given equation is shown in a $\left[0, 2\pi, \dfrac{\pi}{2}\right]$ by $[-2, 2, 1]$ viewing rectangle.

   a. Describe the graph using another equation.

   b. Verify that the two equations are equivalent.

65. $y = \sin(\pi - x)$

66. $y = \cos(x - 2\pi)$

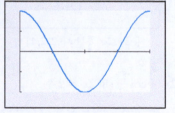

67. $y = \sin\left(x + \dfrac{\pi}{2}\right) + \sin\left(\dfrac{\pi}{2} - x\right)$

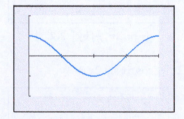

68. $y = \cos\left(x - \dfrac{\pi}{2}\right) - \cos\left(x + \dfrac{\pi}{2}\right)$

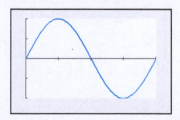

Practice Plus

In Exercises 69–74, rewrite each expression as a simplified expression containing one term.

69. $\cos(\alpha + \beta) \cos \beta + \sin(\alpha + \beta) \sin \beta$

70. $\sin(\alpha - \beta) \cos \beta + \cos(\alpha - \beta) \sin \beta$

71. $\dfrac{\sin(\alpha + \beta) - \sin(\alpha - \beta)}{\cos(\alpha + \beta) + \cos(\alpha - \beta)}$

72. $\dfrac{\cos(\alpha - \beta) + \cos(\alpha + \beta)}{-\sin(\alpha - \beta) + \sin(\alpha + \beta)}$

73. $\cos\left(\dfrac{\pi}{6} + \alpha\right) \cos\left(\dfrac{\pi}{6} - \alpha\right) - \sin\left(\dfrac{\pi}{6} + \alpha\right) \sin\left(\dfrac{\pi}{6} - \alpha\right)$

(Do not use four different identities to solve this exercise.)

74. $\sin\left(\dfrac{\pi}{3} - \alpha\right) \cos\left(\dfrac{\pi}{3} + \alpha\right) + \cos\left(\dfrac{\pi}{3} - \alpha\right) \sin\left(\dfrac{\pi}{3} + \alpha\right)$

(Do not use four different identities to solve this exercise.)

In Exercises 75–78, half of an identity and the graph of this half are given. Use the graph to make a conjecture as to what the right side of the identity should be. Then prove your conjecture.

75. $\cos 2x \cos 5x + \sin 2x \sin 5x = ?$

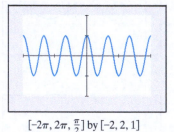

$[-2\pi, 2\pi, \frac{\pi}{2}]$ by $[-2, 2, 1]$

76. $\sin 5x \cos 2x - \cos 5x \sin 2x = ?$

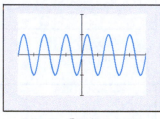

$[-2\pi, 2\pi, \frac{\pi}{2}]$ by $[-2, 2, 1]$

77. $\sin\dfrac{5x}{2} \cos 2x - \cos\dfrac{5x}{2} \sin 2x = ?$

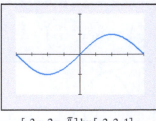

$[-2\pi, 2\pi, \frac{\pi}{2}]$ by $[-2, 2, 1]$

78. $\cos\dfrac{5x}{2} \cos 2x + \sin\dfrac{5x}{2} \sin 2x = ?$

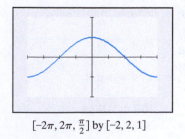

$[-2\pi, 2\pi, \frac{\pi}{2}]$ by $[-2, 2, 1]$

Application Exercises

79. A ball attached to a spring is raised 2 feet and released with an initial vertical velocity of 3 feet per second. The distance of the ball from its rest position after t seconds is given by $d = 2 \cos t + 3 \sin t$. Show that

$$2 \cos t + 3 \sin t = \sqrt{13} \cos(t - \theta),$$

where θ lies in quadrant I and $\tan \theta = \frac{3}{2}$. Use the identity to find the amplitude and the period of the ball's motion.

80. A tuning fork is held a certain distance from your ears and struck. Your eardrums' vibrations after t seconds are given by $p = 3 \sin 2t$. When a second tuning fork is struck, the formula $p = 2 \sin(2t + \pi)$ describes the effects of the sound on the eardrums' vibrations. The total vibrations are given by $p = 3 \sin 2t + 2 \sin(2t + \pi)$.

   a. Simplify p to a single term containing the sine.

   b. If the amplitude of p is zero, no sound is heard. Based on your equation in part (a), does this occur with the two tuning forks in this exercise? Explain your answer.

Explaining the Concepts

In Exercises 81–86, use words to describe the formula for each of the following:

81. the cosine of the difference of two angles.

82. the cosine of the sum of two angles.

83. the sine of the sum of two angles.

84. the sine of the difference of two angles.

85. the tangent of the difference of two angles.

86. the tangent of the sum of two angles.

87. The distance formula and the definitions for cosine and sine are used to prove the formula for the cosine of the difference of two angles. This formula logically leads the way to the other sum and difference identities. Using this development of ideas and formulas, describe a characteristic of mathematical logic.

Technology Exercises

In Exercises 88–93, graph each side of the equation in the same viewing rectangle. If the graphs appear to coincide, verify that the equation is an identity. If the graphs do not appear to coincide, this indicates that the equation is not an identity. In these exercises, find a value of x for which both sides are defined but not equal.

88. $\cos\left(\dfrac{3\pi}{2} - x\right) = -\sin x$

89. $\tan(\pi - x) = -\tan x$

90. $\sin\left(x + \dfrac{\pi}{2}\right) = \sin x + \sin\dfrac{\pi}{2}$

91. $\cos\left(x + \dfrac{\pi}{2}\right) = \cos x + \cos\dfrac{\pi}{2}$

92. $\cos 1.2x \cos 0.8x - \sin 1.2x \sin 0.8x = \cos 2x$

93. $\sin 1.2x \cos 0.8x + \cos 1.2x \sin 0.8x = \sin 2x$

Critical Thinking Exercises

Make Sense? *In Exercises 94–97, determine whether each statement makes sense or does not make sense, and explain your reasoning.*

94. I've noticed that for sine, cosine, and tangent, the trig function for the sum of two angles is not equal to that trig function of the first angle plus that trig function of the second angle.

95. After using an identity to determine the exact value of sin 105°, I verified the result with a calculator.

96. Using sum and difference formulas, I can find exact values for sine, cosine, and tangent at any angle.

97. After the difference formula for cosines is verified, I noticed that the other sum and difference formulas are verified relatively quickly.

98. Verify the identity:

$$\frac{\sin(x-y)}{\cos x \cos y} + \frac{\sin(y-z)}{\cos y \cos z} + \frac{\sin(z-x)}{\cos z \cos x} = 0.$$

In Exercises 99–102, find the exact value of each expression. Do not use a calculator.

99. $\sin\left(\cos^{-1}\dfrac{1}{2} + \sin^{-1}\dfrac{3}{5}\right)$

100. $\sin\left[\sin^{-1}\dfrac{3}{5} - \cos^{-1}\left(-\dfrac{4}{5}\right)\right]$

101. $\cos\left(\tan^{-1}\dfrac{4}{3} + \cos^{-1}\dfrac{5}{13}\right)$

102. $\cos\left[\cos^{-1}\left(-\dfrac{\sqrt{3}}{2}\right) - \sin^{-1}\left(-\dfrac{1}{2}\right)\right]$

In Exercises 103–105, write each trigonometric expression as an algebraic expression (that is, without any trigonometric functions). Assume that x and y are positive and in the domain of the given inverse trigonometric function.

103. $\cos(\sin^{-1}x - \cos^{-1}y)$

104. $\sin(\tan^{-1}x - \sin^{-1}y)$

105. $\tan(\sin^{-1}x + \cos^{-1}y)$

Group Exercise

106. Remembering the six sum and difference identities can be difficult. Did you have problems with some exercises because the identity you were using in your head turned out to be an incorrect formula? Are there easy ways to remember the six new identities presented in this section? Group members should address this question, considering one identity at a time. For each formula, list ways to make it easier to remember.

Retaining the Concepts

107. A hot-air balloon is rising vertically. From a point on level ground 120 feet from the point directly under the passenger compartment, the angle of elevation to the balloon changes from 37.1° to 62.4°. How far, to the nearest tenth of a foot, does the balloon rise during this period? (Section 2.4, Example 4)

108. Find an equation of the form $y = A \cos Bx$ that produces the graph shown in the figure. (Section 2.1, Example 9)

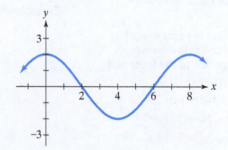

109. Use a sketch to find the exact value of $\cos\left(\tan^{-1}\dfrac{3}{4}\right)$. (Section 2.3, Example 7)

Preview Exercises

Exercises 110–112 will help you prepare for the material covered in the next section.

110. Give exact values for sin 30°, cos 30°, sin 60°, and cos 60°.

111. Use the appropriate values from Exercise 110 to answer each of the following.

    a. Is sin(2 · 30°), or sin 60°, equal to 2 sin 30°?

    b. Is sin(2 · 30°), or sin 60°, equal to 2 sin 30° cos 30°?

112. Use the appropriate values from Exercise 110 to answer each of the following.

    a. Is cos(2 · 30°), or cos 60°, equal to 2 cos 30°?

    b. Is cos(2 · 30°), or cos 60°, equal to $\cos^2 30° - \sin^2 30°$?

Section 3.3

Double-Angle, Power-Reducing, and Half-Angle Formulas

What am I supposed to learn?

After studying this section, you should be able to:

1. Use the double-angle formulas.

2. Use the power-reducing formulas.

3. Use the half-angle formulas.

We have a long history of throwing things. Prior to 400 B.C., the Greeks competed in games that included discus throwing. In the seventeenth century, English soldiers organized cannonball-throwing competitions. In 1827, a Yale University student, disappointed over failing an exam, took out his frustrations at the passing of a collection plate in chapel. Seizing the monetary tray, he flung it in the direction of a large open space on campus. Yale students see this act of frustration as the origin of the Frisbee.

In this section, we develop other important classes of identities, called the double-angle, power-reducing, and half-angle formulas. We will see how one of these formulas can be used by athletes to increase throwing distance.

1. **Use the double-angle formulas.**

Double-Angle Formulas

A number of basic identities follow from the sum formulas for sine, cosine, and tangent. The first category of identities involves **double-angle formulas**.

GREAT QUESTION!

Isn't it easier to write $\sin 2\theta = 2 \sin \theta$ and not bother memorizing the double-angle formula?

No. The 2 that appears in each of the double-angle expressions cannot be pulled to the front and written as a coefficient.

Incorrect!

~~$\sin 2\theta = 2 \sin \theta$~~
~~$\cos 2\theta = 2 \cos \theta$~~
~~$\tan 2\theta = 2 \tan \theta$~~

The figure shows that the graphs of

$$y = \sin 2x$$

and

$$y = 2 \sin x$$

do not coincide: $\sin 2x \neq 2 \sin x$.

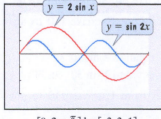

$[0, 2\pi, \frac{\pi}{2}]$ by $[-3, 3, 1]$

> **Double-Angle Formulas**
>
> $$\sin 2\theta = 2 \sin \theta \cos \theta$$
>
> $$\cos 2\theta = \cos^2 \theta - \sin^2 \theta$$
>
> $$\tan 2\theta = \frac{2 \tan \theta}{1 - \tan^2 \theta}$$

To prove each of these formulas, we replace α and β by θ in the sum formulas for $\sin(\alpha + \beta)$, $\cos(\alpha + \beta)$, and $\tan(\alpha + \beta)$.

- $\sin 2\theta = \sin(\theta + \theta) = \sin \theta \cos \theta + \cos \theta \sin \theta = 2 \sin \theta \cos \theta$

  We use $\sin(\alpha + \beta) = \sin \alpha \cos \beta + \cos \alpha \sin \beta$.

- $\cos 2\theta = \cos(\theta + \theta) = \cos \theta \cos \theta - \sin \theta \sin \theta = \cos^2 \theta - \sin^2 \theta$

  We use $\cos(\alpha + \beta) = \cos \alpha \cos \beta - \sin \alpha \sin \beta$.

- $\tan 2\theta = \tan(\theta + \theta) = \dfrac{\tan \theta + \tan \theta}{1 - \tan \theta \tan \theta} = \dfrac{2 \tan \theta}{1 - \tan^2 \theta}$

  We use $\tan(\alpha + \beta) = \dfrac{\tan \alpha + \tan \beta}{1 - \tan \alpha \tan \beta}$.

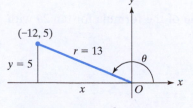

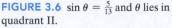

FIGURE 3.6 $\sin \theta = \frac{5}{13}$ and θ lies in quadrant II.

DISCOVERY

Use a quotient identity and the results from parts (a) and (b) of Example 1 to find $\tan 2\theta$. Do you get the result in part (c)?

EXAMPLE 1   Using Double-Angle Formulas to Find Exact Values

If $\sin \theta = \frac{5}{13}$ and θ lies in quadrant II, find the exact value of each of the following:

   a. $\sin 2\theta$ **b.** $\cos 2\theta$ **c.** $\tan 2\theta$.

SOLUTION

We begin with a sketch that illustrates

$$\sin \theta = \frac{5}{13} = \frac{y}{r}.$$

Figure 3.6 shows a quadrant II angle θ for which $\sin \theta = \frac{5}{13}$. We find x using $x^2 + y^2 = r^2$. Because θ lies in quadrant II, x is negative.

$$x^2 + 5^2 = 13^2 \qquad \textcolor{teal}{x^2 + y^2 = r^2}$$

$$x^2 + 25 = 169 \qquad \textcolor{teal}{\text{Square 5 and 13, respectively.}}$$

$$x^2 = 144 \qquad \textcolor{teal}{\text{Subtract 25 from both sides.}}$$

$$x = -\sqrt{144} = -12 \qquad \textcolor{teal}{\begin{array}{l}\text{If } x^2 = 144, \text{ then } x = \pm\sqrt{144} = \pm12. \\ \text{Choose } x = -\sqrt{144} \text{ because in} \\ \text{quadrant II, } x \text{ is negative.}\end{array}}$$

Now we can use values for x, y, and r to find the required values. We will use $\cos \theta = \dfrac{x}{r} = -\dfrac{12}{13}$ and $\tan \theta = \dfrac{y}{x} = -\dfrac{5}{12}$. We were given $\sin \theta = \dfrac{5}{13}$.

   a. $\sin 2\theta = 2 \sin \theta \cos \theta = 2\left(\dfrac{5}{13}\right)\left(-\dfrac{12}{13}\right) = -\dfrac{120}{169}$

   b. $\cos 2\theta = \cos^2 \theta - \sin^2 \theta = \left(-\dfrac{12}{13}\right)^2 - \left(\dfrac{5}{13}\right)^2 = \dfrac{144}{169} - \dfrac{25}{169} = \dfrac{119}{169}$

   c. $\tan 2\theta = \dfrac{2 \tan \theta}{1 - \tan^2 \theta} = \dfrac{2\left(-\dfrac{5}{12}\right)}{1 - \left(-\dfrac{5}{12}\right)^2} = \dfrac{-\dfrac{5}{6}}{1 - \dfrac{25}{144}} = \dfrac{-\dfrac{5}{6}}{\dfrac{119}{144}}$

$$= \left(-\dfrac{5}{6}\right)\left(\dfrac{144}{119}\right) = -\dfrac{120}{119} \qquad \bullet\bullet\bullet$$

✅ **Check Point 1** If $\sin \theta = \frac{4}{5}$ and θ lies in quadrant II, find the exact value of each of the following:

   a. $\sin 2\theta$ **b.** $\cos 2\theta$ **c.** $\tan 2\theta$.

EXAMPLE 2   Using the Double-Angle Formula for
                Tangent to Find an Exact Value

Find the exact value of $\dfrac{2 \tan 15°}{1 - \tan^2 15°}$.

SOLUTION

The given expression, $\dfrac{2 \tan 15°}{1 - \tan^2 15°}$, is the right side of the formula for $\tan 2\theta$ with $\theta = 15°$.

$$\tan 2\theta = \dfrac{2 \tan \theta}{1 - \tan^2 \theta}$$

$$\dfrac{2 \tan 15°}{1 - \tan^2 15°} = \tan(2 \cdot 15°) = \tan 30° = \dfrac{\sqrt{3}}{3}$$

$\bullet\bullet\bullet$

✓ **Check Point 2** Find the exact value of $\cos^2 15° - \sin^2 15°$.

There are three forms of the double-angle formula for $\cos 2\theta$. The form we have seen involves both the cosine and the sine:

$$\cos 2\theta = \cos^2 \theta - \sin^2 \theta.$$

There are situations where it is more efficient to express $\cos 2\theta$ in terms of just one trigonometric function. Using the Pythagorean identity $\sin^2 \theta + \cos^2 \theta = 1$, we can write $\cos 2\theta = \cos^2 \theta - \sin^2 \theta$ in terms of the cosine only. We substitute $1 - \cos^2 \theta$ for $\sin^2 \theta$.

$$\cos 2\theta = \cos^2 \theta - \sin^2 \theta = \cos^2 \theta - (1 - \cos^2 \theta)$$

$$= \cos^2 \theta - 1 + \cos^2 \theta = 2 \cos^2 \theta - 1$$

We can also use a Pythagorean identity to write $\cos 2\theta$ in terms of sine only. We substitute $1 - \sin^2 \theta$ for $\cos^2 \theta$.

$$\cos 2\theta = \cos^2 \theta - \sin^2 \theta = 1 - \sin^2 \theta - \sin^2 \theta = 1 - 2 \sin^2 \theta$$

Three Forms of the Double-Angle Formula for $\cos 2\theta$

$$\cos 2\theta = \cos^2 \theta - \sin^2 \theta$$
$$\cos 2\theta = 2 \cos^2 \theta - 1$$
$$\cos 2\theta = 1 - 2 \sin^2 \theta$$

EXAMPLE 3 Verifying an Identity

Verify the identity: $\cos 3\theta = 4 \cos^3 \theta - 3 \cos \theta$.

SOLUTION

We begin by working with the left side. In order to obtain an expression for $\cos 3\theta$, we use the sum formula and write 3θ as $2\theta + \theta$.

$\cos 3\theta = \cos(2\theta + \theta)$ Write 3θ as $2\theta + \theta$.

$\qquad = \cos 2\theta \cos \theta - \sin 2\theta \sin \theta$ $\cos(\alpha + \beta)$
$\qquad\qquad\qquad\qquad\qquad\qquad\qquad\qquad = \cos \alpha \cos \beta - \sin \alpha \sin \beta$

[$2 \cos^2 \theta - 1$]   [$2 \sin \theta \cos \theta$]

$\qquad = (2 \cos^2 \theta - 1) \cos \theta - 2 \sin \theta \cos \theta \sin \theta$ Substitute double-angle formulas. Because the right side of the given equation involves cosines only, use this form for $\cos 2\theta$.

$\qquad = 2 \cos^3 \theta - \cos \theta - 2 \sin^2 \theta \cos \theta$ Multiply.

[$1 - \cos^2 \theta$]

$$= 2\cos^3\theta - \cos\theta - 2(1 - \cos^2\theta)\cos\theta$$

To get cosines only, use $\sin^2\theta + \cos^2\theta = 1$ and substitute $1 - \cos^2\theta$ for $\sin^2\theta$.

$$= 2\cos^3\theta - \cos\theta - 2\cos\theta + 2\cos^3\theta$$

Multiply.

$$= 4\cos^3\theta - 3\cos\theta$$

Simplify:
$2\cos^3\theta + 2\cos^3\theta = 4\cos^3\theta$
and
$-\cos\theta - 2\cos\theta = -3\cos\theta$.

We were required to verify $\cos 3\theta = 4\cos^3\theta - 3\cos\theta$. By working with the left side, $\cos 3\theta$, and expressing it in a form identical to the right side, we have verified the identity.   • • •

⊘ **Check Point 3** Verify the identity: $\sin 3\theta = 3\sin\theta - 4\sin^3\theta$.

② Use the power-reducing formulas.

Power-Reducing Formulas

The double-angle formulas are used to derive the **power-reducing formulas**:

Power-Reducing Formulas

$$\sin^2\theta = \frac{1 - \cos 2\theta}{2} \qquad \cos^2\theta = \frac{1 + \cos 2\theta}{2} \qquad \tan^2\theta = \frac{1 - \cos 2\theta}{1 + \cos 2\theta}$$

We can prove the first two formulas in the box by working with two forms of the double-angle formula for $\cos 2\theta$.

This is the form with sine only.

This is the form with cosine only.

$$\cos 2\theta = 1 - 2\sin^2\theta \qquad\qquad \cos 2\theta = 2\cos^2\theta - 1$$

Solve the formula on the left for $\sin^2\theta$. Solve the formula on the right for $\cos^2\theta$.

$$2\sin^2\theta = 1 - \cos 2\theta \qquad\qquad 2\cos^2\theta = 1 + \cos 2\theta$$
$$\sin^2\theta = \frac{1 - \cos 2\theta}{2} \qquad\qquad \cos^2\theta = \frac{1 + \cos 2\theta}{2}$$

Divide both sides of each equation by 2.

These are the first two formulas in the box. The third formula in the box is proved by writing the tangent as the quotient of the sine and the cosine.

$$\tan^2\theta = \frac{\sin^2\theta}{\cos^2\theta} = \frac{\dfrac{1 - \cos 2\theta}{2}}{\dfrac{1 + \cos 2\theta}{2}} = \frac{1 - \cos 2\theta}{2} \cdot \frac{\overset{1}{\cancel{2}}}{\underset{1}{\cancel{2}}} \cdot \frac{1}{1 + \cos 2\theta} = \frac{1 - \cos 2\theta}{1 + \cos 2\theta}$$

Power-reducing formulas are quite useful in calculus. By reducing the power of trigonometric functions, calculus can better explore the relationship between a function and how it is changing at every single instant in time.

EXAMPLE 4   Reducing the Power of a Trigonometric Function

Write an equivalent expression for $\cos^4 x$ that does not contain powers of trigonometric functions greater than 1.

SOLUTION

Our goal is to rewrite $\cos^4 x$ without powers of trigonometric functions greater than 1. To achieve this goal, we will apply the formula for $\cos^2 \theta$ twice.

$$\cos^4 x = (\cos^2 x)^2$$

$$= \left(\frac{1 + \cos 2x}{2}\right)^2 \qquad \text{Use } \cos^2 \theta = \frac{1 + \cos 2\theta}{2} \text{ with } \theta = x.$$

$$= \frac{1 + 2\cos 2x + \cos^2 2x}{4} \qquad \begin{array}{l}\text{Square the numerator:}\\ (A + B)^2 = A^2 + 2AB + B^2.\\ \text{Square the denominator.}\end{array}$$

$$= \frac{1}{4} + \frac{1}{2}\cos 2x + \frac{1}{4}\cos^2 2x \qquad \text{Divide each term in the numerator by 4.}$$

We can reduce the power of $\cos^2 2x$ using $\cos^2 \theta = \frac{1 + \cos 2\theta}{2}$ with $\theta = 2x$.

$$= \frac{1}{4} + \frac{1}{2}\cos 2x + \frac{1}{4}\left[\frac{1 + \cos 2(2x)}{2}\right] \qquad \begin{array}{l}\text{Use the power-reducing formula for}\\ \cos^2 \theta \text{ with } \theta = 2x.\end{array}$$

$$= \frac{1}{4} + \frac{1}{2}\cos 2x + \frac{1}{8}(1 + \cos 4x) \qquad \text{Multiply.}$$

$$= \frac{1}{4} + \frac{1}{2}\cos 2x + \frac{1}{8} + \frac{1}{8}\cos 4x \qquad \text{Distribute } \tfrac{1}{8} \text{ throughout parentheses.}$$

$$= \frac{3}{8} + \frac{1}{2}\cos 2x + \frac{1}{8}\cos 4x \qquad \text{Simplify: } \tfrac{1}{4} + \tfrac{1}{8} = \tfrac{2}{8} + \tfrac{1}{8} = \tfrac{3}{8}.$$

Thus, $\cos^4 x = \frac{3}{8} + \frac{1}{2}\cos 2x + \frac{1}{8}\cos 4x$. The expression for $\cos^4 x$ does not contain powers of trigonometric functions greater than 1. •••

✅ **Check Point 4** Write an equivalent expression for $\sin^4 x$ that does not contain powers of trigonometric functions greater than 1.

Half-Angle Formulas

Useful equivalent forms of the power-reducing formulas can be obtained by replacing θ with $\frac{\alpha}{2}$. Then solve for the trigonometric function on the left sides of the equations. The resulting identities are called the **half-angle formulas**:

> **Half-Angle Formulas**
>
> $$\sin \frac{\alpha}{2} = \pm\sqrt{\frac{1 - \cos \alpha}{2}}$$
>
> $$\cos \frac{\alpha}{2} = \pm\sqrt{\frac{1 + \cos \alpha}{2}}$$
>
> $$\tan \frac{\alpha}{2} = \pm\sqrt{\frac{1 - \cos \alpha}{1 + \cos \alpha}}$$

The $\pm$ symbol in each formula does not mean that there are two possible values for each function. Instead, the $\pm$ indicates that you must determine the sign of the trigonometric function, $+$ or $-$, based on the quadrant in which the half-angle $\frac{\alpha}{2}$ lies.

GREAT QUESTION!

Isn't it easier to write $\sin \frac{\theta}{2} = \frac{1}{2}\sin \theta$ and not bother memorizing the half-angle formula?

No. The $\frac{1}{2}$ that appears in each of the half-angle formulas cannot be pulled to the front and written as a coefficient.

Incorrect!

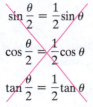

$$\sin \frac{\theta}{2} = \frac{1}{2}\sin \theta$$

$$\cos \frac{\theta}{2} = \frac{1}{2}\cos \theta$$

$$\tan \frac{\theta}{2} = \frac{1}{2}\tan \theta$$

The figure shows that the graphs of $y = \sin \frac{x}{2}$ and $y = \frac{1}{2}\sin x$ do not coincide: $\sin \frac{x}{2} \neq \frac{1}{2}\sin x$.

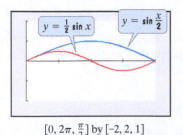

$[0, 2\pi, \frac{\pi}{2}]$ by $[-2, 2, 1]$

3 Use the half-angle formulas.

If we know the exact value for the cosine of an angle, we can use the half-angle formulas to find exact values of sine, cosine, and tangent for half of that angle. For example, we know that $\cos 225° = -\dfrac{\sqrt{2}}{2}$. In the next example, we find the exact value of the cosine of half of 225°, or $\cos 112.5°$.

EXAMPLE 5   Using a Half-Angle Formula to Find an Exact Value

Find the exact value of $\cos 112.5°$.

SOLUTION

Because $112.5° = \dfrac{225°}{2}$, we use the half-angle formula for $\cos\dfrac{\alpha}{2}$ with $\alpha = 225°$. What sign should we use when we apply the formula? Because 112.5° lies in quadrant II, where only the sine and cosecant are positive, $\cos 112.5° < 0$. Thus, we use the $-$ sign in the half-angle formula.

$$\cos 112.5° = \cos\frac{225°}{2}$$

$$= -\sqrt{\frac{1 + \cos 225°}{2}} \qquad \text{Use } \cos\frac{\alpha}{2} = -\sqrt{\frac{1 + \cos\alpha}{2}} \text{ with } \alpha = 225°.$$

$$= -\sqrt{\frac{1 + \left(-\dfrac{\sqrt{2}}{2}\right)}{2}} \qquad \cos 225° = -\frac{\sqrt{2}}{2}$$

$$= -\sqrt{\frac{2 - \sqrt{2}}{4}} \qquad \text{Multiply the radicand by } \tfrac{2}{2}:$$
$$\frac{1 + \left(-\dfrac{\sqrt{2}}{2}\right)}{2} \cdot \frac{2}{2} = \frac{2 - \sqrt{2}}{4}.$$

$$= -\frac{\sqrt{2 - \sqrt{2}}}{2} \qquad \text{Simplify: } \sqrt{4} = 2. \qquad \bullet\bullet\bullet$$

GREAT QUESTION!

What's the relationship between α and the signs I need to work with in the half-angle formulas?

The sign *outside* the radical is determined by the half angle $\dfrac{\alpha}{2}$. By contrast, the sign of $\cos\alpha$, which appears *under* the radical, is determined by the full angle α.

$$\sin\frac{\alpha}{2} = \pm\sqrt{\frac{1 - \cos\alpha}{2}}$$

The sign of $\cos\alpha$ is determined by the quadrant of α.

The sign is determined by the quadrant of $\dfrac{\alpha}{2}$.

⊘ **Check Point 5** Use $\cos 210° = -\dfrac{\sqrt{3}}{2}$ to find the exact value of $\cos 105°$.

There are alternate formulas for $\tan\dfrac{\alpha}{2}$ that do not require us to determine which sign to use when applying the formula. These formulas are logically connected to the identities in Example 6 and Check Point 6.

EXAMPLE 6   Verifying an Identity

Verify the identity: $\tan\theta = \dfrac{1 - \cos 2\theta}{\sin 2\theta}$.

SOLUTION

We work with the right side of $\tan \theta = \dfrac{1 - \cos 2\theta}{\sin 2\theta}$.

$$\dfrac{1 - \cos 2\theta}{\sin 2\theta} = \dfrac{1 - (1 - 2\sin^2 \theta)}{2\sin \theta \cos \theta}$$

The form $\cos 2\theta = 1 - 2\sin^2 \theta$ is used because it produces only one term in the numerator. Use the double-angle formula for sine in the denominator.

$$= \dfrac{2\sin^2 \theta}{2\sin \theta \cos \theta}$$

Simplify the numerator.

$$= \dfrac{\sin \theta}{\cos \theta}$$

Divide the numerator and denominator by $2\sin \theta$.

$$= \tan \theta$$

Use a quotient identity: $\tan \theta = \dfrac{\sin \theta}{\cos \theta}$.

The right side simplifies to $\tan \theta$, the expression on the left side. Thus, the identity is verified. • • •

✅ **Check Point 6** Verify the identity: $\tan \theta = \dfrac{\sin 2\theta}{1 + \cos 2\theta}$.

Half-angle formulas for $\tan \dfrac{\alpha}{2}$ can be obtained using the identities in Example 6 and Check Point 6:

$$\tan \theta = \dfrac{1 - \cos 2\theta}{\sin 2\theta} \quad \text{and} \quad \tan \theta = \dfrac{\sin 2\theta}{1 + \cos 2\theta}.$$

Do you see how to do this? Replace each occurrence of θ with $\dfrac{\alpha}{2}$. This results in the following identities:

> **Half-Angle Formulas for Tangent**
>
> $$\tan \dfrac{\alpha}{2} = \dfrac{1 - \cos \alpha}{\sin \alpha}$$
>
> $$\tan \dfrac{\alpha}{2} = \dfrac{\sin \alpha}{1 + \cos \alpha}$$

GREAT QUESTION!

I'm suffering from identity overload! Where can I find a complete list of all the trigonometric identities I should know?

We've provided a box at the end of this section that contains all identities presented so far.

EXAMPLE 7 Verifying an Identity

Verify the identity: $\tan \dfrac{\alpha}{2} = \csc \alpha - \cot \alpha$.

SOLUTION

We begin with the right side.

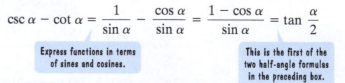

$$\csc \alpha - \cot \alpha = \dfrac{1}{\sin \alpha} - \dfrac{\cos \alpha}{\sin \alpha} = \dfrac{1 - \cos \alpha}{\sin \alpha} = \tan \dfrac{\alpha}{2}$$

Express functions in terms of sines and cosines.

This is the first of the two half-angle formulas in the preceding box.

We worked with the right side and arrived at the left side. Thus, the identity is verified. • • •

✅ **Check Point 7** Verify the identity: $\tan \dfrac{\alpha}{2} = \dfrac{\sec \alpha}{\sec \alpha \csc \alpha + \csc \alpha}$.

We conclude with a summary of the principal trigonometric identities developed in this section and the previous section. The fundamental identities can be found in the box on page 144.

GREAT QUESTION!

Any hint to help remember the correct sign in the numerator in the first two power-reducing formulas and the first two half-angle formulas?

Remember *sinus-minus*–the sine is minus.

Principal Trigonometric Identities

Sum and Difference Formulas

$$\sin(\alpha + \beta) = \sin \alpha \cos \beta + \cos \alpha \sin \beta \qquad \sin(\alpha - \beta) = \sin \alpha \cos \beta - \cos \alpha \sin \beta$$

$$\cos(\alpha + \beta) = \cos \alpha \cos \beta - \sin \alpha \sin \beta \qquad \cos(\alpha - \beta) = \cos \alpha \cos \beta + \sin \alpha \sin \beta$$

$$\tan(\alpha + \beta) = \frac{\tan \alpha + \tan \beta}{1 - \tan \alpha \tan \beta} \qquad \tan(\alpha - \beta) = \frac{\tan \alpha - \tan \beta}{1 + \tan \alpha \tan \beta}$$

Double-Angle Formulas

$$\sin 2\theta = 2 \sin \theta \cos \theta$$

$$\cos 2\theta = \cos^2 \theta - \sin^2 \theta = 2 \cos^2 \theta - 1 = 1 - 2 \sin^2 \theta$$

$$\tan 2\theta = \frac{2 \tan \theta}{1 - \tan^2 \theta}$$

Power-Reducing Formulas

$$\sin^2 \theta = \frac{1 - \cos 2\theta}{2} \qquad \cos^2 \theta = \frac{1 + \cos 2\theta}{2} \qquad \tan^2 \theta = \frac{1 - \cos 2\theta}{1 + \cos 2\theta}$$

Half-Angle Formulas

$$\sin \frac{\alpha}{2} = \pm \sqrt{\frac{1 - \cos \alpha}{2}} \qquad \cos \frac{\alpha}{2} = \pm \sqrt{\frac{1 + \cos \alpha}{2}}$$

$$\tan \frac{\alpha}{2} = \pm \sqrt{\frac{1 - \cos \alpha}{1 + \cos \alpha}} = \frac{1 - \cos \alpha}{\sin \alpha} = \frac{\sin \alpha}{1 + \cos \alpha}$$

ACHIEVING SUCCESS

Read your lecture notes before starting your homework.

Often homework problems, and later the test problems, are variations of the ones done by your professor in class.

CONCEPT AND VOCABULARY CHECK

Fill in each blank so that the resulting statement is true.

1. $\sin 2x = $ _____

2. $\cos 2A = \cos^2 A - $ _____ $= $ _____ $- 1 = 1 - $ _____

3. $\tan 2B = $ _____

4. $\sin^2 \alpha = \dfrac{\rule{2cm}{0.4pt}}{2}$

5. $\cos^2 \alpha = \dfrac{\rule{2cm}{0.4pt}}{2}$

6. $\tan^2 y = \dfrac{\rule{2cm}{0.4pt}}{1 + \cos 2y}$

7. $\sin \dfrac{x}{2} = \pm \sqrt{\dfrac{\rule{2cm}{0.4pt}}{2}}$

8. $\cos \dfrac{y}{2} = \pm \sqrt{\dfrac{\rule{2cm}{0.4pt}}{2}}$

9. $\tan \dfrac{\alpha}{2} = \pm \sqrt{\dfrac{\rule{2cm}{0.4pt}}{1 + \cos \alpha}} = \dfrac{\rule{2cm}{0.4pt}}{\sin \alpha} = \dfrac{\sin \alpha}{\rule{1.5cm}{0.4pt}}$

10. True or false: The sine of twice an angle equals twice the sine of the angle. _____

11. True or false: $\dfrac{\cos 2A}{2} = \cos A$ _____

12. True or false: The tangent of half an angle equals half the tangent of the angle. _____

In Exercises 13–15, determine whether the positive or negative sign results in a true statement.

13. $\sin 100° = \pm \sqrt{\dfrac{1 - \cos 200°}{2}}$ _____

14. $\cos 100° = \pm \sqrt{\dfrac{1 + \cos 200°}{2}}$ _____

15. $\tan 200° = \pm \sqrt{\dfrac{1 - \cos 400°}{1 + \cos 400°}}$ _____

EXERCISE SET 3.3

Practice Exercises

In Exercises 1–6, use the figures to find the exact value of each trigonometric function.

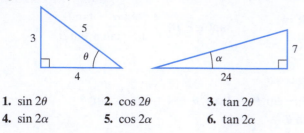

1. $\sin 2\theta$ **2.** $\cos 2\theta$ **3.** $\tan 2\theta$

4. $\sin 2\alpha$ **5.** $\cos 2\alpha$ **6.** $\tan 2\alpha$

In Exercises 7–14, use the given information to find the exact value of each of the following:

    a. $\sin 2\theta$ **b.** $\cos 2\theta$ **c.** $\tan 2\theta$.

7. $\sin \theta = \frac{15}{17}$, θ lies in quadrant II.

8. $\sin \theta = \frac{12}{13}$, θ lies in quadrant II.

9. $\cos \theta = \frac{24}{25}$, θ lies in quadrant IV.

10. $\cos \theta = \frac{40}{41}$, θ lies in quadrant IV.

11. $\cot \theta = 2$, θ lies in quadrant III.

12. $\cot \theta = 3$, θ lies in quadrant III.

13. $\sin \theta = -\frac{9}{41}$, θ lies in quadrant III.

14. $\sin \theta = -\frac{2}{3}$, θ lies in quadrant III.

In Exercises 15–22, write each expression as the sine, cosine, or tangent of a double angle. Then find the exact value of the expression.

15. $2 \sin 15° \cos 15°$

16. $2 \sin 22.5° \cos 22.5°$

17. $\cos^2 75° - \sin^2 75°$

18. $\cos^2 105° - \sin^2 105°$

19. $2 \cos^2 \frac{\pi}{8} - 1$

20. $1 - 2 \sin^2 \frac{\pi}{12}$

21. $\dfrac{2 \tan \frac{\pi}{12}}{1 - \tan^2 \frac{\pi}{12}}$

22. $\dfrac{2 \tan \frac{\pi}{8}}{1 - \tan^2 \frac{\pi}{8}}$

In Exercises 23–34, verify each identity.

23. $\sin 2\theta = \dfrac{2 \tan \theta}{1 + \tan^2 \theta}$

24. $\sin 2\theta = \dfrac{2 \cot \theta}{1 + \cot^2 \theta}$

25. $(\sin \theta + \cos \theta)^2 = 1 + \sin 2\theta$

26. $(\sin \theta - \cos \theta)^2 = 1 - \sin 2\theta$

27. $\sin^2 x + \cos 2x = \cos^2 x$

28. $1 - \tan^2 x = \dfrac{\cos 2x}{\cos^2 x}$

29. $\cot x = \dfrac{\sin 2x}{1 - \cos 2x}$

30. $\cot x = \dfrac{1 + \cos 2x}{\sin 2x}$

31. $\sin 2t - \tan t = \tan t \cos 2t$

32. $\sin 2t - \cot t = -\cot t \cos 2t$

33. $\sin 4t = 4 \sin t \cos^3 t - 4 \sin^3 t \cos t$

34. $\cos 4t = 8 \cos^4 t - 8 \cos^2 t + 1$

In Exercises 35–38, use the power-reducing formulas to rewrite each expression as an equivalent expression that does not contain powers of trigonometric functions greater than 1.

35. $6 \sin^4 x$

36. $10 \cos^4 x$

37. $\sin^2 x \cos^2 x$

38. $8 \sin^2 x \cos^2 x$

In Exercises 39–46, use a half-angle formula to find the exact value of each expression.

39. $\sin 15°$ **40.** $\cos 22.5°$

41. $\cos 157.5°$ **42.** $\sin 105°$

43. $\tan 75°$ **44.** $\tan 112.5°$

45. $\tan \dfrac{7\pi}{8}$ **46.** $\tan \dfrac{3\pi}{8}$

In Exercises 47–54, use the figures to find the exact value of each trigonometric function.

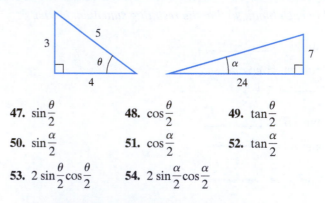

47. $\sin \dfrac{\theta}{2}$ **48.** $\cos \dfrac{\theta}{2}$ **49.** $\tan \dfrac{\theta}{2}$

50. $\sin \dfrac{\alpha}{2}$ **51.** $\cos \dfrac{\alpha}{2}$ **52.** $\tan \dfrac{\alpha}{2}$

53. $2 \sin \dfrac{\theta}{2} \cos \dfrac{\theta}{2}$ **54.** $2 \sin \dfrac{\alpha}{2} \cos \dfrac{\alpha}{2}$

In Exercises 55–58, use the given information to find the exact value of each of the following:

    a. $\sin \dfrac{\alpha}{2}$ **b.** $\cos \dfrac{\alpha}{2}$ **c.** $\tan \dfrac{\alpha}{2}$.

55. $\tan \alpha = \frac{4}{3}$, $180° < \alpha < 270°$

56. $\tan \alpha = \frac{8}{15}$, $180° < \alpha < 270°$

57. $\sec \alpha = -\frac{13}{5}$, $\frac{\pi}{2} < \alpha < \pi$

58. $\sec \alpha = -3$, $\frac{\pi}{2} < \alpha < \pi$

In Exercises 59–68, verify each identity.

59. $\sin^2\dfrac{\theta}{2} = \dfrac{\sec\theta - 1}{2\sec\theta}$

60. $\sin^2\dfrac{\theta}{2} = \dfrac{\csc\theta - \cot\theta}{2\csc\theta}$

61. $\cos^2\dfrac{\theta}{2} = \dfrac{\sin\theta + \tan\theta}{2\tan\theta}$

62. $\cos^2\dfrac{\theta}{2} = \dfrac{\sec\theta + 1}{2\sec\theta}$

63. $\tan\dfrac{\alpha}{2} = \dfrac{\tan\alpha}{\sec\alpha + 1}$

64. $2\tan\dfrac{\alpha}{2} = \dfrac{\sin^2\alpha + 1 - \cos^2\alpha}{\sin\alpha(1 + \cos\alpha)}$

65. $\cot\dfrac{x}{2} = \dfrac{\sin x}{1 - \cos x}$

66. $\cot\dfrac{x}{2} = \dfrac{1 + \cos x}{\sin x}$

67. $\tan\dfrac{x}{2} + \cot\dfrac{x}{2} = 2\csc x$

68. $\tan\dfrac{x}{2} - \cot\dfrac{x}{2} = -2\cot x$

Practice Plus

In Exercises 69–78, half of an identity and the graph of this half are given. Use the graph to make a conjecture as to what the right side of the identity should be. Then prove your conjecture.

69. $\dfrac{\cot x - \tan x}{\cot x + \tan x} = ?$

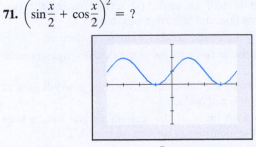

$[-2\pi, 2\pi, \frac{\pi}{2}]$ by $[-3, 3, 1]$

70. $\dfrac{2(\tan x - \cot x)}{\tan^2 x - \cot^2 x} = ?$

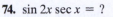

$[-2\pi, 2\pi, \frac{\pi}{2}]$ by $[-3, 3, 1]$

71. $\left(\sin\dfrac{x}{2} + \cos\dfrac{x}{2}\right)^2 = ?$

$[-2\pi, 2\pi, \frac{\pi}{2}]$ by $[-3, 3, 1]$

72. $\sin^2\dfrac{x}{2} - \cos^2\dfrac{x}{2} = ?$

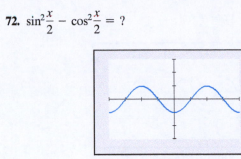

$[-2\pi, 2\pi, \frac{\pi}{2}]$ by $[-3, 3, 1]$

73. $\dfrac{\sin 2x}{\sin x} - \dfrac{\cos 2x}{\cos x} = ?$

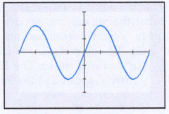

$[-2\pi, 2\pi, \frac{\pi}{2}]$ by $[-3, 3, 1]$

74. $\sin 2x \sec x = ?$

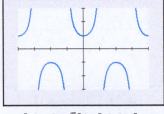

$[-2\pi, 2\pi, \frac{\pi}{2}]$ by $[-3, 3, 1]$

75. $\dfrac{\csc^2 x}{\cot x} = ?$

$[-2\pi, 2\pi, \frac{\pi}{2}]$ by $[-3, 3, 1]$

76. $\tan x + \cot x = ?$

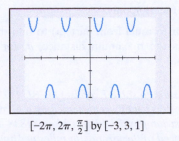

$[-2\pi, 2\pi, \frac{\pi}{2}]$ by $[-3, 3, 1]$

77. $\sin x(4\cos^2 x - 1) = ?$

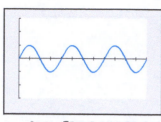

$[0, 2\pi, \frac{\pi}{6}]$ by $[-3, 3, 1]$

78. $1 - 8\sin^2 x \cos^2 x = ?$

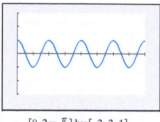

$[0, 2\pi, \frac{\pi}{8}]$ by $[-3, 3, 1]$

Application Exercises

79. Throwing events in track and field include the shot put, the discus throw, the hammer throw, and the javelin throw. The distance that the athlete can achieve depends on the initial speed of the object thrown and the angle above the horizontal at which the object leaves the hand. This angle is represented by θ in the figure shown. The distance, d, in feet, that the athlete throws is modeled by the formula

$$d = \frac{v_0^2}{16}\sin\theta\cos\theta,$$

in which v_0 is the initial speed of the object thrown, in feet per second, and θ is the angle, in degrees, at which the object leaves the hand.

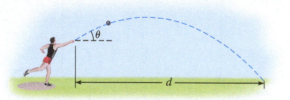

a. Use an identity to express the formula so that it contains the sine function only.

b. Use your formula from part (a) to find the angle, θ, that produces the maximum distance, d, for a given initial speed, v_0.

Use this information to solve Exercises 80–81: The speed of a supersonic aircraft is usually represented by a Mach number, named after Austrian physicist Ernst Mach (1838–1916). A Mach number is the speed of the aircraft, in miles per hour, divided by the speed of sound, approximately 740 miles per hour. Thus, a plane flying at twice the speed of sound has a speed, M, of Mach 2.

Concorde
Mach 2.03

SR-71 Blackbird
Mach 3.3

If an aircraft has a speed greater than Mach 1, a sonic boom is heard, created by sound waves that form a cone with a vertex angle θ, shown in the figure.

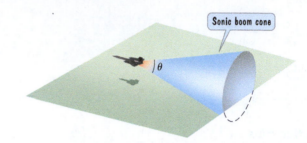

Sonic boom cone

The relationship between the cone's vertex angle, θ, and the Mach speed, M, of an aircraft that is flying faster than the speed of sound is given by

$$\sin\frac{\theta}{2} = \frac{1}{M}.$$

80. If $\theta = \frac{\pi}{6}$, determine the Mach speed, M, of the aircraft. Express the speed as an exact value and as a decimal to the nearest tenth.

81. If $\theta = \frac{\pi}{4}$, determine the Mach speed, M, of the aircraft. Express the speed as an exact value and as a decimal to the nearest tenth.

Explaining the Concepts

In Exercises 82–89, use words to describe the formula for:

82. the sine of double an angle.

83. the cosine of double an angle. (Describe one of the three formulas.)

84. the tangent of double an angle.

85. the power-reducing formula for the sine squared of an angle.

86. the power-reducing formula for the cosine squared of an angle.

87. the sine of half an angle.

88. the cosine of half an angle.

89. the tangent of half an angle. (Describe one of the two formulas that does not involve a square root.)

90. Explain how the double-angle formulas are derived.

91. How can there be three forms of the double-angle formula for $\cos 2\theta$?

92. Without showing algebraic details, describe in words how to reduce the power of $\cos^4 x$.

93. Describe one or more of the techniques you use to help remember the identities in the box on page 177.

94. Your friend is about to compete as a shot-putter in a college field event. Using Exercise 79(b), write a short description to your friend on how to achieve the best distance possible in the throwing event.

Technology Exercises

In Exercises 95–98, graph each side of the equation in the same viewing rectangle. If the graphs appear to coincide, verify that the equation is an identity. If the graphs do not appear to coincide, find a value of x for which both sides are defined but not equal.

95. $3 - 6\sin^2 x = 3\cos 2x$

96. $4\cos^2\dfrac{x}{2} = 2 + 2\cos x$

97. $\sin\dfrac{x}{2} = \dfrac{1}{2}\sin x$

98. $\cos\dfrac{x}{2} = \dfrac{1}{2}\cos x$

In Exercises 99–101, graph each equation in a $\left[-2\pi, 2\pi, \dfrac{\pi}{2}\right]$ *by*

$[-3, 3, 1]$ *viewing rectangle. Then* **a.** *Describe the graph using another equation, and* **b.** *Verify that the two equations are equivalent.*

99. $y = \dfrac{1 - 2\cos 2x}{2\sin x - 1}$

100. $y = \dfrac{2\tan\dfrac{x}{2}}{1 + \tan^2\dfrac{x}{2}}$

101. $y = \csc x - \cot x$

Critical Thinking Exercises

Make Sense? *In Exercises 102–105, determine whether each statement makes sense or does not make sense, and explain your reasoning.*

102. The double-angle identities are derived from the sum identities by adding an angle to itself.

103. I simplified a double-angle trigonometric expression by pulling 2 to the front and treating it as a coefficient.

104. When using the half-angle formulas for trigonometric functions of $\dfrac{\alpha}{2}$, I determine the sign based on the quadrant in which α lies.

105. I used a half-angle formula to find the exact value of $\cos 100°$.

106. Verify the identity:

$$\sin^3 x + \cos^3 x = (\sin x + \cos x)\left(1 - \dfrac{\sin 2x}{2}\right).$$

In Exercises 107–110, find the exact value of each expression. Do not use a calculator.

107. $\sin\left(2\sin^{-1}\dfrac{\sqrt{3}}{2}\right)$

108. $\cos\left[2\tan^{-1}\left(-\dfrac{4}{3}\right)\right]$

109. $\cos^2\left(\dfrac{1}{2}\sin^{-1}\dfrac{3}{5}\right)$

110. $\sin^2\left(\dfrac{1}{2}\cos^{-1}\dfrac{3}{5}\right)$

111. Use a right triangle to write $\sin(2\sin^{-1}x)$ as an algebraic expression. Assume that x is positive and in the domain of the given inverse trigonometric function.

112. Use the power-reducing formulas to rewrite $\sin^6 x$ as an equivalent expression that does not contain powers of trigonometric functions greater than 1.

Retaining the Concepts

113. Determine the amplitude and period of $y = 3\sin\dfrac{1}{2}x$.

Then graph the function for $0 \le x \le 4\pi$.

(Section 2.1, Example 3)

114. From a point on level ground 120 feet from the base of a tower, the angle of elevation is 48.3°. Approximate the height of the tower to the nearest foot. (Section 2.4, Example 2)

115. Use a reference angle to find the exact value of $\sin\dfrac{23\pi}{4}$.

(Section 1.3, Example 8)

Preview Exercises

Exercises 116–118 will help you prepare for the material covered in the next section. In each exercise, use exact values of trigonometric functions to show that the statement is true. Notice that each statement expresses the product of sines and/or cosines as a sum or a difference.

116. $\sin 60° \sin 30° = \tfrac{1}{2}[\cos(60° - 30°) - \cos(60° + 30°)]$

117. $\cos\dfrac{\pi}{2}\cos\dfrac{\pi}{3} = \dfrac{1}{2}\left[\cos\left(\dfrac{\pi}{2} - \dfrac{\pi}{3}\right) + \cos\left(\dfrac{\pi}{2} + \dfrac{\pi}{3}\right)\right]$

118. $\sin \pi \cos\dfrac{\pi}{2} = \dfrac{1}{2}\left[\sin\left(\pi + \dfrac{\pi}{2}\right) + \sin\left(\pi - \dfrac{\pi}{2}\right)\right]$

CHAPTER 3 | Mid-Chapter Check Point

WHAT YOU KNOW: Verifying an identity means showing that the expressions on each side are identical. Like solving puzzles, the process can be intriguing because there are sometimes several "best" ways to proceed. We presented some guidelines to help you get started (see page 155). We used fundamental trigonometric identities (see page 144), as well as sum and difference formulas, double-angle formulas, power-reducing formulas, and half-angle formulas (see page 177) to verify identities. We also used these formulas to find exact values of trigonometric functions.

ACHIEVING SUCCESS

Here's a way to organize what you should know for solving the exercises in this Mid-Chapter Check Point.

Make copies of the boxes on pages 144 and 177 that contain the essential trigonometric identities. Mount these boxes on cardstock and add this reference sheet to the one you prepared for Chapter 2. (If you didn't prepare a reference sheet for Chapter 2, it's not too late: See the Achieving Success feature on page 140.)

In Exercises 1–18, verify each identity.

1. $\cos x(\tan x + \cot x) = \csc x$

2. $\dfrac{\sin(x + \pi)}{\cos\left(x + \dfrac{3\pi}{2}\right)} = \tan^2 x - \sec^2 x$

3. $(\sin\theta + \cos\theta)^2 + (\sin\theta - \cos\theta)^2 = 2$

4. $\dfrac{\sin t - 1}{\cos t} = \dfrac{\cos t - \cot t}{\cos t \cot t}$

5. $\dfrac{1 - \cos 2x}{\sin 2x} = \tan x$

6. $\sin\theta \cos\theta + \cos^2\theta = \dfrac{\cos\theta(1 + \cot\theta)}{\csc\theta}$

7. $\dfrac{\sin x}{\tan x} + \dfrac{\cos x}{\cot x} = \sin x + \cos x$

8. $\sin^2\dfrac{t}{2} = \dfrac{\tan t - \sin t}{2\tan t}$

9. $\sin\alpha \cos\beta = \dfrac{1}{2}[\sin(\alpha + \beta) + \sin(\alpha - \beta)]$

10. $\dfrac{1 + \csc x}{\sec x} - \cot x = \cos x$

11. $\dfrac{\cot x - 1}{\cot x + 1} = \dfrac{1 - \tan x}{1 + \tan x}$

12. $2\sin^3\theta \cos\theta + 2\sin\theta \cos^3\theta = \sin 2\theta$

13. $\dfrac{\sin t + \cos t}{\sec t + \csc t} = \dfrac{\sin t}{\sec t}$

14. $\sec 2x = \dfrac{\sec^2 x}{2 - \sec^2 x}$

15. $\tan(\alpha + \beta)\tan(\alpha - \beta) = \dfrac{\tan^2\alpha - \tan^2\beta}{1 - \tan^2\alpha \tan^2\beta}$

16. $\csc\theta + \cot\theta = \dfrac{\sin\theta}{1 - \cos\theta}$

17. $\dfrac{1}{\csc 2x} = \dfrac{2\tan x}{1 + \tan^2 x}$

18. $\dfrac{\sec t - 1}{t\sec t} = \dfrac{1 - \cos t}{t}$

Use the following conditions to solve Exercises 19–22:

$$\sin\alpha = \frac{3}{5}, \qquad \frac{\pi}{2} < \alpha < \pi$$

$$\cos\beta = -\frac{12}{13}, \qquad \pi < \beta < \frac{3\pi}{2}.$$

Find the exact value of each of the following.

19. $\cos(\alpha - \beta)$

20. $\tan(\alpha + \beta)$

21. $\sin 2\alpha$

22. $\cos\dfrac{\beta}{2}$

In Exercises 23–26, find the exact value of each expression. Do not use a calculator.

23. $\sin\left(\dfrac{3\pi}{4} + \dfrac{5\pi}{6}\right)$

24. $\cos^2 15° - \sin^2 15°$

25. $\cos\dfrac{5\pi}{12}\cos\dfrac{\pi}{12} + \sin\dfrac{5\pi}{12}\sin\dfrac{\pi}{12}$

26. $\tan 22.5°$

Section 3.4

Product-to-Sum and Sum-to-Product Formulas

What am I supposed to learn?

After studying this section, you should be able to:

① Use the product-to-sum formulas.

② Use the sum-to-product formulas.

James K. Polk
Born November 2, 1795

Warren G. Harding
Born November 2, 1865

Of all the U.S. presidents, two share a birthday (same month and day). The probability of two or more people in a group sharing a birthday rises sharply as the group's size increases. Above 50 people, the probability approaches certainty. So, come November 2, we salute Presidents Polk and Harding with

112, 163-, 112, 196-, 110, 8521-, 008, 121-.

Were you aware that each button on a touch-tone phone produces a unique sound? If we treat the commas as pauses and the hyphens as held notes, this sequence of numbers is "Happy Birthday" on a touch-tone phone.

Although "Happy Birthday" isn't Mozart or Sondheim, it is sinusoidal. Each of its touch-tone musical sounds can be described by the sum of two sine functions or the product of sines and cosines. In this section, we develop identities that enable us to use both descriptions. They are called the product-to-sum and sum-to-product formulas.

① Use the product-to-sum formulas.

The Product-to-Sum Formulas

How do we write the products of sines and/or cosines as sums or differences? We use the following identities, which are called **product-to-sum formulas**:

GREAT QUESTION!

Do I have to memorize the formulas in this section?

Not necessarily. When you need these formulas, you can either refer to one of the two boxes in the section or perhaps even derive them using the methods shown.

Product-to-Sum Formulas

$$\sin \alpha \sin \beta = \tfrac{1}{2}[\cos(\alpha - \beta) - \cos(\alpha + \beta)]$$
$$\cos \alpha \cos \beta = \tfrac{1}{2}[\cos(\alpha - \beta) + \cos(\alpha + \beta)]$$
$$\sin \alpha \cos \beta = \tfrac{1}{2}[\sin(\alpha + \beta) + \sin(\alpha - \beta)]$$
$$\cos \alpha \sin \beta = \tfrac{1}{2}[\sin(\alpha + \beta) - \sin(\alpha - \beta)]$$

Although these formulas are difficult to remember, they are fairly easy to derive. For example, let's derive the first identity in the box,

$$\sin \alpha \sin \beta = \tfrac{1}{2}[\cos(\alpha - \beta) - \cos(\alpha + \beta)].$$

We begin with the difference and sum formulas for the cosine, and subtract the second identity from the first:

$$\cos(\alpha - \beta) = \cos \alpha \cos \beta + \sin \alpha \sin \beta$$
$$-[\cos(\alpha + \beta) = \cos \alpha \cos \beta - \sin \alpha \sin \beta]$$
$$\cos(\alpha - \beta) - \cos(\alpha + \beta) = \quad 0 \quad + 2 \sin \alpha \sin \beta.$$

> **Subtract terms on the left side.** | **Subtract terms on the right side:** $\cos \alpha \cos \beta - \cos \alpha \cos \beta = 0.$ | **Subtract terms on the right side:** $\sin \alpha \sin \beta - (-\sin \alpha \sin \beta) = 2 \sin \alpha \sin \beta.$

Now we use this result to derive the product-to-sum formula for $\sin \alpha \sin \beta$.

$2 \sin \alpha \sin \beta = \cos(\alpha - \beta) - \cos(\alpha + \beta)$ *Reverse the sides in the preceding equation.*

$\sin \alpha \sin \beta = \frac{1}{2}[\cos(\alpha - \beta) - \cos(\alpha + \beta)]$ *Multiply each side by $\frac{1}{2}$.*

This last equation is the desired formula. Likewise, we can derive the product-to-sum formula for cosine, $\cos \alpha \cos \beta = \frac{1}{2}[\cos(\alpha - \beta) + \cos(\alpha + \beta)]$. As we did for the previous derivation, begin with the difference and sum formulas for cosine. However, we *add* the formulas rather than subtracting them. Reversing both sides of this result and multiplying each side by $\frac{1}{2}$ produces the formula for $\cos \alpha \cos \beta$. The last two product-to-sum formulas, $\sin \alpha \cos \beta = \frac{1}{2}[\sin (\alpha + \beta) + \sin (\alpha - \beta)]$ and $\cos \alpha \sin \beta = \frac{1}{2}[\sin (\alpha + \beta) - \sin (\alpha - \beta)]$, are derived using the sum and difference formulas for sine in a similar manner.

EXAMPLE 1   Using the Product-to-Sum Formulas

Express each of the following products as a sum or difference:

a. $\sin 8x \sin 3x$ **b.** $\sin 4x \cos x$.

SOLUTION

The product-to-sum formula that we are using is shown in each of the voice balloons.

a.

> $\sin \alpha \sin \beta = \frac{1}{2}[\cos(\alpha - \beta) - \cos(\alpha + \beta)]$

$$\sin 8x \sin 3x = \frac{1}{2}[\cos(8x - 3x) - \cos(8x + 3x)] = \frac{1}{2}(\cos 5x - \cos 11x)$$

b.

> $\sin \alpha \cos \beta = \frac{1}{2}[\sin(\alpha + \beta) + \sin(\alpha - \beta)]$

$$\sin 4x \cos x = \frac{1}{2}[\sin(4x + x) + \sin(4x - x)] = \frac{1}{2}(\sin 5x + \sin 3x)$$

• • •

✓ **Check Point 1** Express each of the following products as a sum or difference:

a. $\sin 5x \sin 2x$ **b.** $\cos 7x \cos x$.

TECHNOLOGY

Graphic Connections

The graphs of

$$y = \sin 8x \sin 3x$$

and

$$y = \frac{1}{2}(\cos 5x - \cos 11x)$$

are shown in a $\left[-2\pi, 2\pi, \dfrac{\pi}{2}\right]$ by $[-1, 1, 1]$ viewing rectangle. The graphs coincide. This supports our algebraic work in Example 1(a).

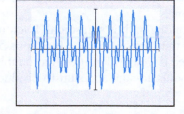

② Use the sum-to-product formulas.

The Sum-to-Product Formulas

How do we write the sum or difference of sines and/or cosines as products? We use the following identities, which are called the **sum-to-product formulas**:

Sum-to-Product Formulas

$$\sin \alpha + \sin \beta = 2 \sin \frac{\alpha + \beta}{2} \cos \frac{\alpha - \beta}{2}$$

$$\sin \alpha - \sin \beta = 2 \sin \frac{\alpha - \beta}{2} \cos \frac{\alpha + \beta}{2}$$

$$\cos \alpha + \cos \beta = 2 \cos \frac{\alpha + \beta}{2} \cos \frac{\alpha - \beta}{2}$$

$$\cos \alpha - \cos \beta = -2 \sin \frac{\alpha + \beta}{2} \sin \frac{\alpha - \beta}{2}$$

We verify these formulas using the product-to-sum formulas. Let's verify the first sum-to-product formula

$$\sin \alpha + \sin \beta = 2 \sin \frac{\alpha + \beta}{2} \cos \frac{\alpha - \beta}{2}.$$

We start with the right side of the formula, the side with the product. We can apply the product-to-sum formula for $\sin A \cos B$ to this expression. By doing so, we obtain the left side of the formula, $\sin \alpha + \sin \beta$. Here's how:

$$\sin A \cos B = \tfrac{1}{2}[\sin(A + B) + \sin(A - B)]$$

$$2 \sin \frac{\alpha + \beta}{2} \cos \frac{\alpha - \beta}{2} = 2 \cdot \frac{1}{2}\left[\sin\left(\frac{\alpha + \beta}{2} + \frac{\alpha - \beta}{2}\right) + \sin\left(\frac{\alpha + \beta}{2} - \frac{\alpha - \beta}{2}\right)\right]$$

$$= \sin\left(\frac{\alpha + \beta + \alpha - \beta}{2}\right) + \sin\left(\frac{\alpha + \beta - \alpha + \beta}{2}\right)$$

$$= \sin \frac{2\alpha}{2} + \sin \frac{2\beta}{2} = \sin \alpha + \sin \beta.$$

The three other sum-to-product formulas in the preceding box are verified in a similar manner. Start with the right side and obtain the left side using an appropriate product-to-sum formula.

EXAMPLE 2   Using the Sum-to-Product Formulas

Express each sum or difference as a product:

   a. $\sin 9x + \sin 5x$ **b.** $\cos 4x - \cos 3x$.

SOLUTION

The sum-to-product formula that we are using is shown in each of the voice balloons.

a.

$$\sin \alpha + \sin \beta = 2 \sin \frac{\alpha + \beta}{2} \cos \frac{\alpha - \beta}{2}$$

$$\sin 9x + \sin 5x = 2 \sin \frac{9x + 5x}{2} \cos \frac{9x - 5x}{2}$$

$$= 2 \sin \frac{14x}{2} \cos \frac{4x}{2} = 2 \sin 7x \cos 2x$$

b.

$$\cos \alpha - \cos \beta = -2 \sin \frac{\alpha + \beta}{2} \sin \frac{\alpha - \beta}{2}$$

$$\cos 4x - \cos 3x = -2 \sin \frac{4x + 3x}{2} \sin \frac{4x - 3x}{2}$$

$$= -2 \sin \frac{7x}{2} \sin \frac{x}{2}$$

• • •

✓ **Check Point 2** Express each sum as a product:

   a. $\sin 7x + \sin 3x$ **b.** $\cos 3x + \cos 2x$.

Some identities contain a fraction on one side with sums and differences of sines and/or cosines. Applying the sum-to-product formulas in the numerator and the denominator is often helpful in verifying these identities.

EXAMPLE 3 Using Sum-to-Product Formulas to Verify an Identity

Verify the identity: $\dfrac{\cos 3x - \cos 5x}{\sin 3x + \sin 5x} = \tan x$.

SOLUTION

Because the left side is more complicated, we will work with it. We use sum-to-product formulas for the numerator and the denominator of the fraction on this side.

$$\frac{\cos 3x - \cos 5x}{\sin 3x + \sin 5x}$$

$$\cos \alpha - \cos \beta = -2 \sin \frac{\alpha + \beta}{2} \sin \frac{\alpha - \beta}{2}$$

$$= \frac{-2 \sin \dfrac{3x + 5x}{2} \sin \dfrac{3x - 5x}{2}}{\sin 3x + \sin 5x}$$

$$\sin \alpha + \sin \beta = 2 \sin \frac{\alpha + \beta}{2} \cos \frac{\alpha - \beta}{2}$$

$$= \frac{-2 \sin \dfrac{3x + 5x}{2} \sin \dfrac{3x - 5x}{2}}{2 \sin \dfrac{3x + 5x}{2} \cos \dfrac{3x - 5x}{2}}$$

$$= \frac{-2 \sin \dfrac{8x}{2} \sin\left(\dfrac{-2x}{2}\right)}{2 \sin \dfrac{8x}{2} \cos\left(\dfrac{-2x}{2}\right)} \qquad \text{Perform the indicated additions and subtractions.}$$

$$= \frac{-2 \sin 4x \sin(-x)}{2 \sin 4x \cos(-x)} \qquad \text{Simplify.}$$

$$= \frac{-(-\sin x)}{\cos x} \qquad \text{The sine function is odd: } \sin(-x) = -\sin x. \text{ The cosine function is even: } \cos(-x) = \cos x.$$

$$= \frac{\sin x}{\cos x} \qquad \text{Simplify.}$$

$$= \tan x \qquad \text{Apply a quotient identity: } \tan x = \frac{\sin x}{\cos x}.$$

We were required to verify $\dfrac{\cos 3x - \cos 5x}{\sin 3x + \sin 5x} = \tan x$. We worked with the left side and arrived at the right side, tan x. Thus, the identity is verified. • • •

✓ **Check Point 3** Verify the identity: $\dfrac{\cos 3x - \cos x}{\sin 3x + \sin x} = -\tan x$.

Blitzer Bonus || Sinusoidal Sounds

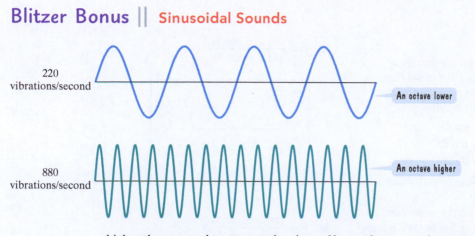

220 vibrations/second

An octave lower

880 vibrations/second

An octave higher

Music is all around us. A mere snippet of a song from the past can trigger vivid memories, inducing emotions ranging from unabashed joy to deep sorrow. Trigonometric functions can explain how sound travels from its source and describe its pitch, loudness, and quality. Still unexplained is the remarkable influence music has on the brain, including the deepest question of all: Why do we appreciate music?

When a note is played, it disturbs nearby air molecules, creating regions of higher-than-normal pressure and regions of lower-than-normal pressure. If we graph pressure, y, versus time, t, we get a sine wave that represents the note. The frequency of the sine wave is the number of high-low disturbances, or vibrations, per second. The greater the frequency, the higher the pitch; the lesser the frequency, the lower the pitch.

The amplitude of a note's sine wave is related to its loudness. The amplitude for the two sine waves shown above is the same. Thus, the notes have the same loudness, although they differ in pitch. The greater the amplitude, the louder the sound; the lesser the amplitude, the softer the sound. The amplitude and frequency are characteristic of every note—and thus of its graph—until the note dissipates.

CONCEPT AND VOCABULARY CHECK

Because you may not be required to memorize the identities in this section, it's often tempting to pay no attention to them at all! Exercises 1–4 are provided to familiarize you with what these identities do. Fill in each blank using the word sum, difference, product, *or* quotient.

1. The formula

$$\sin \alpha \sin \beta = \frac{1}{2}[\cos(\alpha - \beta) - \cos(\alpha + \beta)]$$

can be used to change a _____ of two sines into the _____ of two cosine expressions.

2. The formula

$$\cos \alpha \cos \beta = \frac{1}{2}[\cos(\alpha - \beta) + \cos(\alpha + \beta)]$$

can be used to change a _____ of two cosines into the _____ of two cosine expressions.

3. The formula

$$\sin \alpha \cos \beta = \frac{1}{2}[\sin(\alpha + \beta) + \sin(\alpha - \beta)]$$

can be used to change a _____ of a sine and a cosine into the _____ of two sine expressions.

4. The formula

$$\cos \alpha \sin \beta = \frac{1}{2}[\sin(\alpha + \beta) - \sin(\alpha - \beta)]$$

can be used to change a _____ of a cosine and a sine into the _____ of two sine expressions.

Exercises 5–8 are provided to familiarize you with the second set of identities presented in this section. Fill in each blank using the word sum, difference, product, *or* quotient.

5. The formula

$$\sin \alpha + \sin \beta = 2 \sin \frac{\alpha + \beta}{2} \cos \frac{\alpha - \beta}{2}$$

can be used to change a _____ of two sines into the _____ of a sine and a cosine expression.

6. The formula

$$\sin \alpha - \sin \beta = 2 \sin \frac{\alpha - \beta}{2} \cos \frac{\alpha + \beta}{2}$$

can be used to change a _____ of two sines into the _____ of a sine and a cosine expression.

7. The formula

$$\cos \alpha + \cos \beta = 2 \cos \frac{\alpha + \beta}{2} \cos \frac{\alpha - \beta}{2}$$

can be used to change a _____ of two cosines into the _____ of two cosine expressions.

8. The formula

$$\cos \alpha - \cos \beta = -2 \sin \frac{\alpha + \beta}{2} \sin \frac{\alpha - \beta}{2}$$

can be used to change a _____ of two cosines into the _____ of two sine expressions.

EXERCISE SET 3.4

Practice Exercises

Be sure that you've familiarized yourself with the first set of formulas presented in this section by working 1–4 in the Concept and Vocabulary Check. In Exercises 1–8, use the appropriate formula to express each product as a sum or difference.

1. $\sin 6x \sin 2x$ **2.** $\sin 8x \sin 4x$

3. $\cos 7x \cos 3x$ **4.** $\cos 9x \cos 2x$

5. $\sin x \cos 2x$ **6.** $\sin 2x \cos 3x$

7. $\cos \dfrac{3x}{2} \sin \dfrac{x}{2}$ **8.** $\cos \dfrac{5x}{2} \sin \dfrac{x}{2}$

Be sure that you've familiarized yourself with the second set of formulas presented in this section by working 5–8 in the Concept and Vocabulary Check. In Exercises 9–22, express each sum or difference as a product. If possible, find this product's exact value.

9. $\sin 6x + \sin 2x$ **10.** $\sin 8x + \sin 2x$

11. $\sin 7x - \sin 3x$ **12.** $\sin 11x - \sin 5x$

13. $\cos 4x + \cos 2x$ **14.** $\cos 9x - \cos 7x$

15. $\sin x + \sin 2x$ **16.** $\sin x - \sin 2x$

17. $\cos \dfrac{3x}{2} + \cos \dfrac{x}{2}$ **18.** $\sin \dfrac{3x}{2} + \sin \dfrac{x}{2}$

19. $\sin 75° + \sin 15°$ **20.** $\cos 75° - \cos 15°$

21. $\sin \dfrac{\pi}{12} - \sin \dfrac{5\pi}{12}$ **22.** $\cos \dfrac{\pi}{12} - \cos \dfrac{5\pi}{12}$

In Exercises 23–30, verify each identity.

23. $\dfrac{\sin 3x - \sin x}{\cos 3x - \cos x} = -\cot 2x$ **24.** $\dfrac{\sin x + \sin 3x}{\cos x + \cos 3x} = \tan 2x$

25. $\dfrac{\sin 2x + \sin 4x}{\cos 2x + \cos 4x} = \tan 3x$ **26.** $\dfrac{\cos 4x - \cos 2x}{\sin 2x - \sin 4x} = \tan 3x$

27. $\dfrac{\sin x - \sin y}{\sin x + \sin y} = \tan \dfrac{x - y}{2} \cot \dfrac{x + y}{2}$

28. $\dfrac{\sin x + \sin y}{\sin x - \sin y} = \tan \dfrac{x + y}{2} \cot \dfrac{x - y}{2}$

29. $\dfrac{\sin x + \sin y}{\cos x + \cos y} = \tan \dfrac{x + y}{2}$

30. $\dfrac{\sin x - \sin y}{\cos x - \cos y} = -\cot \dfrac{x + y}{2}$

Practice Plus

In Exercises 31–36, the graph with the given equation is shown in a $\left[0, 2\pi, \dfrac{\pi}{2} \right]$ by $[-2, 2, 1]$ viewing rectangle.

    a. *Describe the graph using another equation.*
    b. *Verify that the two equations are equivalent.*

31. $y = \dfrac{\sin x + \sin 3x}{2 \sin 2x}$

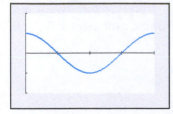

32. $y = \dfrac{\cos x - \cos 3x}{\sin x + \sin 3x}$

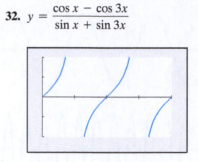

33. $y = \dfrac{\cos x - \cos 5x}{\sin x + \sin 5x}$

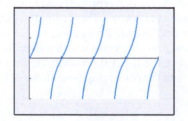

34. $y = \dfrac{\cos 5x - \cos 3x}{\sin 5x + \sin 3x}$

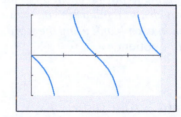

35. $y = \dfrac{\sin x - \sin 3x}{\cos x - \cos 3x}$

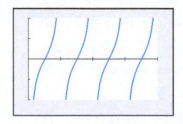

36. $y = \dfrac{\sin 2x + \sin 6x}{\cos 6x - \cos 2x}$

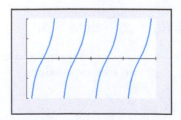

Application Exercises

Use this information to solve Exercises 37–38. The sound produced by touching each button on a touch-tone phone is described by

$$y = \sin 2\pi l t + \sin 2\pi h t,$$

where l and h are the low and high frequencies in the figure shown. For example, what sound is produced by touching 5? The low frequency is l = 770 cycles per second and the high frequency is h = 1336 cycles per second. The sound produced by touching 5 is described by

$$y = \sin 2\pi(770)t + \sin 2\pi(1336)t.$$

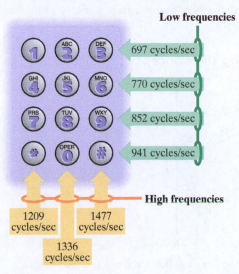

Low frequencies

697 cycles/sec
770 cycles/sec
852 cycles/sec
941 cycles/sec

High frequencies

1209 cycles/sec   1477 cycles/sec
1336 cycles/sec

37. The touch-tone phone sequence for that most naive of melodies is given as follows:

Mary Had a Little Lamb

3212333,222,399,3212333322321.

a. Many numbers do not appear in this sequence, including 7. If you accidently touch 7 for one of the notes, describe this sound as the sum of sines.

b. Describe this accidental sound as a product of sines and cosines.

38. The touch-tone phone sequence for "Jingle Bells" is given as follows:

Jingle Bells

333,333,39123,666-663333322329,333,333,39123,666-6633,399621.

a. The first six notes of the song are produced by repeatedly touching 3. Describe this repeated sound as the sum of sines.

b. Describe the repeated sound as a product of sines and cosines.

Explaining the Concepts

In Exercises 39–42, use words to describe the given formula.

39. $\sin \alpha \sin \beta = \frac{1}{2}[\cos(\alpha - \beta) - \cos(\alpha + \beta)]$

40. $\cos \alpha \cos \beta = \frac{1}{2}[\cos(\alpha - \beta) + \cos(\alpha + \beta)]$

41. $\sin \alpha + \sin \beta = 2 \sin\dfrac{\alpha + \beta}{2}\cos\dfrac{\alpha - \beta}{2}$

42. $\cos \alpha + \cos \beta = 2 \cos\dfrac{\alpha + \beta}{2}\cos\dfrac{\alpha - \beta}{2}$

43. Describe identities that can be verified using the sum-to-product formulas.

44. Why do the sounds produced by touching each button on a touch-tone phone have the same loudness? Answer the question using the equation described for Exercises 37 and 38, $y = \sin 2\pi l t + \sin 2\pi h t$, and determine the maximum value of *y* for each sound.

Technology Exercises

In Exercises 45–48, graph each side of the equation in the same viewing rectangle. If the graphs appear to coincide, verify that the equation is an identity. If the graphs do not appear to coincide, find a value of x for which both sides are defined but not equal.

45. $\sin x + \sin 2x = \sin 3x$

46. $\cos x + \cos 2x = \cos 3x$

47. $\sin x + \sin 3x = 2 \sin 2x \cos x$

48. $\cos x + \cos 3x = 2 \cos 2x \cos x$

49. In Exercise 37(a), you wrote an equation for the sound produced by touching 7 on a touch-tone phone. Graph the equation in a [0, 0.01, 0.001] by [−2, 2, 1] viewing rectangle.

50. In Exercise 38(a), you wrote an equation for the sound produced by touching 3 on a touch-tone phone. Graph the equation in a [0, 0.01, 0.001] by [−2, 2, 1] viewing rectangle.

51. In this section, we saw how sums could be expressed as products. Sums of trigonometric functions can also be used to describe functions that are not trigonometric. French mathematician Jean Fourier (1768–1830) showed that *any function* can be described by a series of trigonometric functions. For example, the basic linear function $f(x) = x$ can also be represented by

$$f(x) = 2\left(\frac{\sin x}{1} - \frac{\sin 2x}{2} + \frac{\sin 3x}{3} - \frac{\sin 4x}{4} + \cdots\right).$$

a. Graph

$$y = 2\left(\frac{\sin x}{1}\right),$$

$$y = 2\left(\frac{\sin x}{1} - \frac{\sin 2x}{2}\right),$$

$$y = 2\left(\frac{\sin x}{1} - \frac{\sin 2x}{2} + \frac{\sin 3x}{3}\right)$$

and

$$y = 2\left(\frac{\sin x}{1} - \frac{\sin 2x}{2} + \frac{\sin 3x}{3} - \frac{\sin 4x}{4}\right)$$

in a $\left[-\pi, \pi, \dfrac{\pi}{2}\right]$ by [−3, 3, 1] viewing rectangle. What patterns do you observe?

b. Graph

$$y = 2\left(\frac{\sin x}{1} - \frac{\sin 2x}{2} + \frac{\sin 3x}{3} - \frac{\sin 4x}{4} + \frac{\sin 5x}{5} - \frac{\sin 6x}{6}\right.$$
$$\left. + \frac{\sin 7x}{7} - \frac{\sin 8x}{8} + \frac{\sin 9x}{9} - \frac{\sin 10x}{10}\right)$$

in a $\left[-\pi, \pi, \frac{\pi}{2}\right]$ by $[-3, 3, 1]$ viewing rectangle. Is a portion of the graph beginning to look like the graph of $f(x) = x$? Obtain a better approximation for the line by graphing functions that contain more and more terms involving sines of multiple angles.

c. Use

$$x = 2\left(\frac{\sin x}{1} - \frac{\sin 2x}{2} + \frac{\sin 3x}{3} - \frac{\sin 4x}{4} + \cdots\right)$$

and substitute $\frac{\pi}{2}$ for x to obtain a formula for $\frac{\pi}{2}$. Show at least four nonzero terms. Then multiply both sides of your formula by 2 to write a nonending series of subtractions and additions that approaches π. Use this series to obtain an approximation for π that is more accurate than the one given by your graphing utility.

Critical Thinking Exercises

Make Sense? *In Exercises 52–55, determine whether each statement makes sense or does not make sense, and explain your reasoning.*

52. The product-to-sum formulas are difficult to remember because they are all so similar to one another.

53. I can use the sum and difference formulas for cosines and sines to derive the product-to-sum formulas.

54. I expressed $\sin 13° \cos 48°$ as $\frac{1}{2}(\sin 61° - \sin 35°)$.

55. I expressed $\cos 47° + \cos 59°$ as $2 \cos 53° \cos 6°$.

Use the identities for $\sin(\alpha + \beta)$ and $\sin(\alpha - \beta)$ to solve Exercises 56–57.

56. Add the left and right sides of the identities and derive the product-to-sum formula for $\sin \alpha \cos \beta$.

57. Subtract the left and right sides of the identities and derive the product-to-sum formula for $\cos \alpha \sin \beta$.

In Exercises 58–59, verify the given sum-to-product formula. Start with the right side and obtain the expression on the left side by using an appropriate product-to-sum formula.

58. $\sin \alpha - \sin \beta = 2 \sin\dfrac{\alpha - \beta}{2}\cos\dfrac{\alpha + \beta}{2}$

59. $\cos \alpha + \cos \beta = 2 \cos\dfrac{\alpha + \beta}{2}\cos\dfrac{\alpha - \beta}{2}$

In Exercises 60–61, verify each identity.

60. $\dfrac{\sin 2x + (\sin 3x + \sin x)}{\cos 2x + (\cos 3x + \cos x)} = \tan 2x$

61. $4 \cos x \cos 2x \sin 3x = \sin 2x + \sin 4x + \sin 6x$

Group Exercise

62. This activity should result in an unusual group display entitled "'Frère Jacques', A New Perspective." Here is the touch-tone phone sequence:

Frère Jacques

4564,4564,69#,69#,#*#964,#*#964,414,414.

Group members should write every sound in the sequence as both the sum-of-sines and the product of sines and cosines. Use the sum-of-sines form and a graphing utility with a $[0, 0.01, 0.001]$ by $[-2, 2, 1]$ viewing rectangle to obtain a graph for every sound. Download these graphs. Use the graphs and equations to create your display in such a way that adults find the trigonometry of this naive melody interesting.

Retaining the Concepts

63.

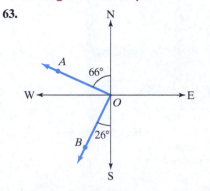

Use the figure to find each of the following:

a. the bearing from O to A.

b. the bearing from O to B.

(Section 2.4, Example 5)

64. Use a sketch to find the exact value of $\tan\left(\sin^{-1}\frac{5}{13}\right)$. (Section 2.3, Example 7)

65. A circle has a radius of 12 inches. Find the length of the arc intercepted by a central angle of 225°. Express arc length in terms of π. Then round your answer to two decimal places. (Section 1.1, Example 8)

Preview Exercises

Exercises 66–68 will help you prepare for the material covered in the next section.

66. Solve: $2(1 - u^2) + 3u = 0$.

67. Solve: $u^3 - 3u = 0$.

68. Solve: $u^2 - u - 1 = 0$.

A Brief Review • Quadratic Equations

- A quadratic equation in x can be expressed in the form $ax^2 + bx + c = 0$, $a \neq 0$.

EXAMPLE

$$3x^2 + 5x - 2 = 0$$

$a = 3$ $b = 5$ $c = -2$

- Some quadratic equations can be solved by factoring and using the zero-product principle: If $AB = 0$, then $A = 0$ or $B = 0$.

EXAMPLE

Solve:

$$3x^2 + 5x - 2 = 0.$$

$(3x - 1)(x + 2) = 0$ Factor.

$3x - 1 = 0$ or $x + 2 = 0$ Use the zero-product principle and set each factor equal to zero.

$3x = 1 \qquad x = -2$ Solve the resulting equations.

$$x = \frac{1}{3}$$

The solutions are $\frac{1}{3}$ and -2.

- Quadratic equations in the form $ax^2 + c = 0$ can be solved by isolating x^2 and applying the square root property: If $u^2 = d$, then $u = \sqrt{d}$ or $u = -\sqrt{d}$.

EXAMPLE

Solve:

$$3x^2 - 15 = 0.$$ We begin by isolating x^2.

$$3x^2 = 15$$ Add 15 to both sides.

$$x^2 = 5$$ Divide both sides by 3.

$$x = \sqrt{5} \text{ or } x = -\sqrt{5}$$ Apply the square root property. Equivalently, $x = \pm\sqrt{5}$.

The solutions are $-\sqrt{5}$ and $\sqrt{5}$.

- All quadratic equations can be solved by the quadratic formula

$$x = \frac{-b \pm \sqrt{b^2 - 4ac}}{2a}.$$

EXAMPLE

Solve:

$$x^2 - 2x - 6 = 0$$

$a = 1$ $b = -2$ $c = -6$

$$x = \frac{-b \pm \sqrt{b^2 - 4ac}}{2a} = \frac{-(-2) \pm \sqrt{(-2)^2 - 4(1)(-6)}}{2(1)}$$

$$= \frac{2 \pm \sqrt{4 - (-24)}}{2} = \frac{2 \pm \sqrt{4 + 24}}{2} = \frac{2 \pm \sqrt{28}}{2}$$

$$= \frac{2 \pm \sqrt{4}\sqrt{7}}{2} = \frac{2 \pm 2\sqrt{7}}{2} = \frac{2(1 \pm \sqrt{7})}{2} = 1 \pm \sqrt{7}$$

The solutions are $1 + \sqrt{7}$ and $1 - \sqrt{7}$.

Section 3.5 · Trigonometric Equations

What am I supposed to learn?

After studying this section, you should be able to:

1. Find all solutions of a trigonometric equation.

2. Solve equations with multiple angles.

3. Solve trigonometric equations quadratic in form.

4. Use factoring to separate different functions in trigonometric equations.

5. Use identities to solve trigonometric equations.

6. Use a calculator to solve trigonometric equations.

Exponential functions display the manic energies of uncontrolled growth. By contrast, trigonometric functions repeat their behavior. Do they embody, in their regularity, some basic rhythm of the universe? The cycles of periodic phenomena provide events that we can comfortably count on. When will the moon look just as it does at this moment? When can I count on 13.5 hours of daylight? When will my breathing be exactly as it is right now? Models with trigonometric functions embrace the periodic rhythms of our world. Equations containing trigonometric functions are used to answer questions about these models.

Trigonometric Equations and Their Solutions

A **trigonometric equation** is an equation that contains a trigonometric expression with a variable, such as $\sin x$. We have seen that some trigonometric equations are identities, such as $\sin^2 x + \cos^2 x = 1$. These equations are true for every value of the variable for which the expressions are defined. In this section, we consider trigonometric equations that are true for only some values of the variable. The values that satisfy such an equation are its **solutions**. (There are trigonometric equations that have no solution.)

An example of a trigonometric equation is

$$\sin x = \tfrac{1}{2}.$$

A solution of this equation is $\tfrac{\pi}{6}$ because $\sin \tfrac{\pi}{6} = \tfrac{1}{2}$. By contrast, π is not a solution because $\sin \pi = 0 \neq \tfrac{1}{2}$.

Is $\tfrac{\pi}{6}$ the only solution of $\sin x = \tfrac{1}{2}$? The answer is no. Because of the periodic nature of the sine function, there are infinitely many values of x for which $\sin x = \tfrac{1}{2}$. **Figure 3.7** shows five of the solutions, including $\tfrac{\pi}{6}$, for $-\tfrac{3\pi}{2} \leq x \leq \tfrac{7\pi}{2}$. Notice that the x-coordinates of the points where the graph of $y = \sin x$ intersects the line $y = \tfrac{1}{2}$ are the solutions of the equation $\sin x = \tfrac{1}{2}$.

1 Find all solutions of a trigonometric equation.

FIGURE 3.7 The equation $\sin x = \tfrac{1}{2}$ has five solutions when x is restricted to the interval $\left[-\tfrac{3\pi}{2}, \tfrac{7\pi}{2}\right]$.

How do we represent all solutions of $\sin x = \tfrac{1}{2}$? Because the period of the sine function is 2π, first find all solutions in $[0, 2\pi)$. The solutions are

$$x = \frac{\pi}{6} \quad \text{and} \quad x = \pi - \frac{\pi}{6} = \frac{5\pi}{6}.$$

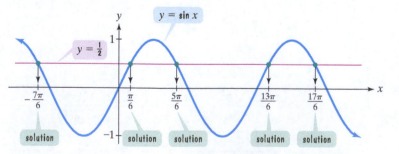

The sine is positive in quadrants I and II.

Any multiple of 2π can be added to these values and the sine is still $\frac{1}{2}$. Thus, all solutions of $\sin x = \frac{1}{2}$ are given by

$$x = \frac{\pi}{6} + 2n\pi \quad \text{or} \quad x = \frac{5\pi}{6} + 2n\pi,$$

where n is any integer. By choosing any two integers, such as $n = 0$ and $n = 1$, we can find some solutions of $\sin x = \frac{1}{2}$. Thus, four of the solutions are determined as follows:

Let $n = 0$.   Let $n = 1$.

$$x = \frac{\pi}{6} + 2 \cdot 0\pi \qquad x = \frac{5\pi}{6} + 2 \cdot 0\pi \qquad x = \frac{\pi}{6} + 2 \cdot 1\pi \qquad x = \frac{5\pi}{6} + 2 \cdot 1\pi$$

$$= \frac{\pi}{6} \qquad\qquad = \frac{5\pi}{6} \qquad\qquad = \frac{\pi}{6} + 2\pi \qquad\qquad = \frac{5\pi}{6} + 2\pi$$

$$\qquad\qquad\qquad\qquad\qquad\qquad = \frac{\pi}{6} + \frac{12\pi}{6} = \frac{13\pi}{6} \qquad = \frac{5\pi}{6} + \frac{12\pi}{6} = \frac{17\pi}{6}.$$

These four solutions are shown among the five solutions in **Figure 3.7**.

A Brief Review • Solving Linear Equations in One Variable

- A linear equation in one variable x is an equation that can be written in the form $ax + b = 0$, where a and b are real numbers, $a \neq 0$.

EXAMPLE

$4x + 12 = 0$ is a linear equation in one variable.
- Solving an equation in x involves determining all values of x that result in a true statement when substituted into the equation. Such values are solutions, or roots, of the equation.
- Two or more equations that have the same solution are called equivalent equations.

EXAMPLE

$4x + 12 = 0$, $4x = -12$, and $x = -3$ are equivalent equations. The solution for each equation is -3.

- An equation can be transformed into an equivalent equation by one or more of the following operations:

Operation	Example
1. Simply an expression by removing grouping symbols and combining like terms.	$3(x - 6) = 6x - x$ $3x - 18 = 5x$
2. Add (or subtract) the same real number or variable expression on *both* sides of the equation.	$3x - 18 = 5x$ $3x - 18 - 3x = 5x - 3x$ Subtract $3x$ from both sides of the equation. $-18 = 2x$
3. Multiply (or divide) by the same *nonzero* quantity on *both* sides of the equation.	$-18 = 2x$ $\dfrac{-18}{2} = \dfrac{2x}{2}$ Divide both sides of the equation by 2. $-9 = x$
4. Interchange the two sides of the equation.	$-9 = x$ $x = -9$

If you look closely at the equations in the box, you will notice that we have solved the equation $3(x - 6) = 6x - x$. The final equation, $x = -9$, with x isolated on the left side, shows that -9 is the solution. The idea in solving a linear equation is to get the variable by itself on one side of the equal sign and a number by itself on the other side.

Equations Involving a Single Trigonometric Function

To solve an equation containing a single trigonometric function:

- Isolate the function on one side of the equation.
- Solve for the variable.

EXAMPLE 1   Finding All Solutions of a Trigonometric Equation

Solve the equation: $3 \sin x - 2 = 5 \sin x - 1$.

SOLUTION

The equation contains a single trigonometric function, $\sin x$.

Step 1   Isolate the function on one side of the equation. We can solve for $\sin x$ by collecting terms with $\sin x$ on the left side and constant terms on the right side.

$$3 \sin x - 2 = 5 \sin x - 1 \qquad \text{This is the given equation.}$$
$$3 \sin x - 5 \sin x - 2 = 5 \sin x - 5 \sin x - 1 \qquad \text{Subtract 5 sin x from both sides.}$$
$$-2 \sin x - 2 = -1 \qquad \text{Simplify.}$$
$$-2 \sin x = 1 \qquad \text{Add 2 to both sides.}$$
$$\sin x = -\tfrac{1}{2} \qquad \text{Divide both sides by } -2 \text{ and solve for sin x.}$$

Step 2   Solve for the variable. We must solve for x in $\sin x = -\dfrac{1}{2}$. Because $\sin \dfrac{\pi}{6} = \dfrac{1}{2}$, the solutions of $\sin x = -\dfrac{1}{2}$ in $[0, 2\pi)$ are

$$x = \pi + \frac{\pi}{6} = \frac{6\pi}{6} + \frac{\pi}{6} = \frac{7\pi}{6} \qquad x = 2\pi - \frac{\pi}{6} = \frac{12\pi}{6} - \frac{\pi}{6} = \frac{11\pi}{6}.$$

The sine is negative in quadrant III.

The sine is negative in quadrant IV.

Because the period of the sine function is 2π, the solutions of the equation are given by

$$x = \frac{7\pi}{6} + 2n\pi \quad \text{and} \quad x = \frac{11\pi}{6} + 2n\pi,$$

where n is any integer.   • • •

GREAT QUESTION!

Why did you add $\dfrac{\pi}{6}$ to π but subtract $\dfrac{\pi}{6}$ from 2π?

We are using an acute reference angle, $\dfrac{\pi}{6}$, to find angles in different quadrants. We are interested in solutions where the sine is negative, namely in quadrants III and IV. Adding $\dfrac{\pi}{6}$ to π puts us in quadrant III. Subtracting $\dfrac{\pi}{6}$ from 2π puts us in quadrant IV.

✓ **Check Point 1** Solve the equation: $5 \sin x = 3 \sin x + \sqrt{3}$.

Now we will concentrate on finding solutions of trigonometric equations for $0 \le x < 2\pi$. You can use a graphing utility to check the solutions of these equations. Graph the left side and graph the right side. The solutions are the x-coordinates of the points where the graphs intersect.

② Solve equations with multiple angles.

Equations Involving Multiple Angles

Here are examples of two equations that include multiple angles:

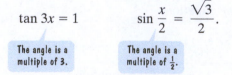

$$\tan 3x = 1 \qquad \sin \frac{x}{2} = \frac{\sqrt{3}}{2}.$$

The angle is a multiple of 3.

The angle is a multiple of $\frac{1}{2}$.

We will solve each equation for $0 \le x < 2\pi$. The period of the function plays an important role in ensuring that we do not leave out any solutions.

EXAMPLE 2   Solving an Equation with a Multiple Angle

Solve the equation: $\tan 3x = 1$, $0 \le x < 2\pi$.

SOLUTION

The period of the tangent function is π. In the interval $[0, \pi)$, the only value for which the tangent function is 1 is $\frac{\pi}{4}$. This means that $3x = \frac{\pi}{4}$. Because the period is π, all the solutions to $\tan 3x = 1$ are given by

$$3x = \frac{\pi}{4} + n\pi. \quad \text{\textit{n} is any integer.}$$

$$x = \frac{\pi}{12} + \frac{n\pi}{3} \quad \text{\color{blue}Divide both sides by 3 and solve for } x.$$

In the interval $[0, 2\pi)$, we obtain the solutions of $\tan 3x = 1$ as follows:

Let $n = 0$.

$$x = \frac{\pi}{12} + \frac{0\pi}{3}$$
$$= \frac{\pi}{12}$$

Let $n = 1$.

$$x = \frac{\pi}{12} + \frac{1\pi}{3}$$
$$= \frac{\pi}{12} + \frac{4\pi}{12} = \frac{5\pi}{12}$$

Let $n = 2$.

$$x = \frac{\pi}{12} + \frac{2\pi}{3}$$
$$= \frac{\pi}{12} + \frac{8\pi}{12} = \frac{9\pi}{12} = \frac{3\pi}{4}$$

Let $n = 3$.

$$x = \frac{\pi}{12} + \frac{3\pi}{3}$$
$$= \frac{\pi}{12} + \frac{12\pi}{12} = \frac{13\pi}{12}$$

Let $n = 4$.

$$x = \frac{\pi}{12} + \frac{4\pi}{3}$$
$$= \frac{\pi}{12} + \frac{16\pi}{12} = \frac{17\pi}{12}$$

Let $n = 5$.

$$x = \frac{\pi}{12} + \frac{5\pi}{3}$$
$$= \frac{\pi}{12} + \frac{20\pi}{12} = \frac{21\pi}{12} = \frac{7\pi}{4}.$$

If you let $n = 6$, you will obtain $x = \frac{25\pi}{12}$. This value exceeds 2π. In the interval $[0, 2\pi)$, the solutions of $\tan 3x = 1$ are $\frac{\pi}{12}, \frac{5\pi}{12}, \frac{3\pi}{4}, \frac{13\pi}{12}, \frac{17\pi}{12}$, and $\frac{7\pi}{4}$. These solutions are illustrated by the six intersection points in the technology box. •••

✅ **Check Point 2** Solve the equation: $\tan 2x = \sqrt{3}, 0 \le x < 2\pi$.

EXAMPLE 3 Solving an Equation with a Multiple Angle

Solve the equation: $\sin \frac{x}{2} = \frac{\sqrt{3}}{2}, 0 \le x < 2\pi$.

SOLUTION

We are interested in solving $\sin \frac{x}{2} = \frac{\sqrt{3}}{2}, 0 \le x < 2\pi$. The period of the sine function is 2π. In the interval $[0, 2\pi)$, there are two values at which the sine function is $\frac{\sqrt{3}}{2}$. One of these values is $\frac{\pi}{3}$. The sine is positive in quadrant II; thus, the other value is $\pi - \frac{\pi}{3}$, or $\frac{2\pi}{3}$. This means that $\frac{x}{2} = \frac{\pi}{3}$ or $\frac{x}{2} = \frac{2\pi}{3}$. Because the period is 2π, all the solutions of $\sin \frac{x}{2} = \frac{\sqrt{3}}{2}$ are given by

$$\frac{x}{2} = \frac{\pi}{3} + 2n\pi \quad \text{or} \quad \frac{x}{2} = \frac{2\pi}{3} + 2n\pi \quad \text{\textit{n} is any integer.}$$

$$x = \frac{2\pi}{3} + 4n\pi \qquad x = \frac{4\pi}{3} + 4n\pi. \quad \text{\color{blue}Multiply both sides by 2 and solve for } x.$$

We see that $x = \frac{2\pi}{3} + 4n\pi$ or $x = \frac{4\pi}{3} + 4n\pi$. If $n = 0$, we obtain $x = \frac{2\pi}{3}$ from the first equation and $x = \frac{4\pi}{3}$ from the second equation. If we let $n = 1$, we are adding $4 \cdot 1 \cdot \pi$, or 4π, to $\frac{2\pi}{3}$ and $\frac{4\pi}{3}$. These values of x exceed 2π. Thus, in the interval $[0, 2\pi)$, the only solutions of $\sin \frac{x}{2} = \frac{\sqrt{3}}{2}$ are $\frac{2\pi}{3}$ and $\frac{4\pi}{3}$. •••

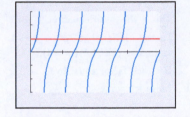

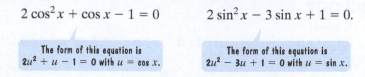 **Check Point 3** Solve the equation: $\sin\frac{x}{3} = \frac{1}{2}, 0 \leq x < 2\pi$.

③ Solve trigonometric equations quadratic in form.

Trigonometric Equations Quadratic in Form

Some trigonometric equations are in the form of a quadratic equation $au^2 + bu + c = 0$, where u is a trigonometric function and $a \neq 0$. Here are two examples of trigonometric equations that are quadratic in form:

$$2\cos^2 x + \cos x - 1 = 0 \qquad 2\sin^2 x - 3\sin x + 1 = 0.$$

> The form of this equation is $2u^2 + u - 1 = 0$ with $u = \cos x$.

> The form of this equation is $2u^2 - 3u + 1 = 0$ with $u = \sin x$.

To solve this kind of equation, try using factoring. If the trigonometric expression does not factor, use another method, such as the quadratic formula or the square root property.

TECHNOLOGY

Graphic Connections

The graph of

$$y = 2\cos^2 x + \cos x - 1$$

is shown in a $\left[0, 2\pi, \frac{\pi}{2}\right]$ by $[-3, 3, 1]$ viewing rectangle. The x-intercepts, $\frac{\pi}{3}$, π, and $\frac{5\pi}{3}$, verify the three solutions of $2\cos^2 x + \cos x - 1 = 0$ in $[0, 2\pi)$.

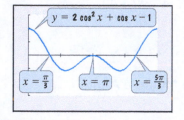

EXAMPLE 4 Solving a Trigonometric Equation Quadratic in Form

Solve the equation: $2\cos^2 x + \cos x - 1 = 0, \quad 0 \leq x < 2\pi$.

SOLUTION

The given equation is in quadratic form $2u^2 + u - 1 = 0$ with $u = \cos x$. Let us attempt to solve the equation by factoring.

$$2\cos^2 x + \cos x - 1 = 0 \qquad \text{This is the given equation.}$$
$$(2\cos x - 1)(\cos x + 1) = 0 \qquad \text{Factor: Notice that } 2u^2 + u - 1 \text{ factors as } (2u - 1)(u + 1).$$

$$2\cos x - 1 = 0 \quad \text{or} \quad \cos x + 1 = 0 \qquad \text{Set each factor equal to 0.}$$
$$2\cos x = 1 \qquad\qquad \cos x = -1 \qquad \text{Solve for } \cos x.$$
$$\cos x = \tfrac{1}{2}$$

$$x = \frac{\pi}{3} \qquad x = 2\pi - \frac{\pi}{3} = \frac{5\pi}{3} \qquad x = \pi \qquad \text{Solve each equation for } x, \; 0 \leq x < 2\pi.$$

> The cosine is positive in quadrants I and IV.

The solutions in the interval $[0, 2\pi)$ are $\frac{\pi}{3}$, π, and $\frac{5\pi}{3}$.   •••

 Check Point 4 Solve the equation: $2\sin^2 x - 3\sin x + 1 = 0, \quad 0 \leq x < 2\pi$.

EXAMPLE 5 Solving a Trigonometric Equation Quadratic in Form

Solve the equation: $4\sin^2 x - 1 = 0, \quad 0 \leq x < 2\pi$.

SOLUTION

The given equation is in quadratic form $4u^2 - 1 = 0$ with $u = \sin x$. We can solve this equation by the square root property: If $u^2 = c$, then $u = \pm\sqrt{c}$.

$$4\sin^2 x - 1 = 0 \qquad \text{This is the given equation.}$$
$$4\sin^2 x = 1 \qquad \text{Add 1 to both sides.}$$
$$\sin^2 x = \frac{1}{4} \qquad \text{Divide both sides by 4 and solve for } \sin^2 x.$$

$$\sin x = \sqrt{\frac{1}{4}} = \frac{1}{2} \quad \text{or} \quad \sin x = -\sqrt{\frac{1}{4}} = -\frac{1}{2} \qquad \text{Apply the square root property: If } u^2 = c, \text{ then } u = \sqrt{c} \text{ or } u = -\sqrt{c}.$$

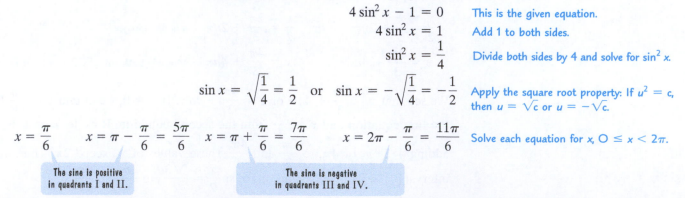

$$x = \frac{\pi}{6} \qquad x = \pi - \frac{\pi}{6} = \frac{5\pi}{6} \qquad x = \pi + \frac{\pi}{6} = \frac{7\pi}{6} \qquad x = 2\pi - \frac{\pi}{6} = \frac{11\pi}{6} \qquad \text{Solve each equation for } x, \; 0 \leq x < 2\pi.$$

> The sine is positive in quadrants I and II.

> The sine is negative in quadrants III and IV.

The solutions in the interval $[0, 2\pi)$ are $\dfrac{\pi}{6}, \dfrac{5\pi}{6}, \dfrac{7\pi}{6}$, and $\dfrac{11\pi}{6}$. • • •

TECHNOLOGY

Numeric Connections

You can use a graphing utility's TABLE feature to verify that the solutions of $4\sin^2 x - 1 = 0$ in $[0, 2\pi)$ are $\dfrac{\pi}{6}, \dfrac{5\pi}{6}, \dfrac{7\pi}{6}$, and $\dfrac{11\pi}{6}$. The table for $y = 4\sin^2 x - 1$, shown on the right, verifies that $\dfrac{\pi}{6}, \dfrac{5\pi}{6}$, and $\dfrac{7\pi}{6}$ are solutions. Scroll through the table to verify the other solution.

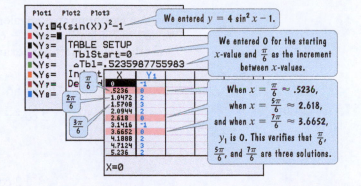

✓ **Check Point 5** Solve the equation: $4\cos^2 x - 3 = 0$, $0 \le x < 2\pi$.

4 Use factoring to separate different functions in trigonometric equations.

Using Factoring to Separate Two Different Trigonometric Functions in an Equation

We have seen that factoring is used to solve some trigonometric equations that are quadratic in form. Factoring can also be used to solve some trigonometric equations that contain two different functions such as

$$\tan x \sin^2 x = 3 \tan x.$$

In such a case, move all terms to one side and obtain zero on the other side. Then try to use factoring to separate the different functions. Example 6 shows how this is done.

GREAT QUESTION!

Can I begin solving

$$\tan x \sin^2 x = 3 \tan x$$

by dividing both sides by tan x?

No. Division by zero is undefined. If you divide by $\tan x$, you lose the two solutions for which $\tan x = 0$, namely 0 and π.

EXAMPLE 6   Using Factoring to Separate Different Functions

Solve the equation: $\tan x \sin^2 x = 3 \tan x$, $0 \le x < 2\pi$.

SOLUTION

Move all terms to one side and obtain zero on the other side.

$$\tan x \sin^2 x = 3 \tan x \qquad \text{This is the given equation.}$$
$$\tan x \sin^2 x - 3 \tan x = 0 \qquad \text{Subtract 3 tan x from both sides.}$$

We now have $\tan x \sin^2 x - 3 \tan x = 0$, which contains both tangent and sine functions. Use factoring to separate the two functions.

$$\tan x(\sin^2 x - 3) = 0 \qquad \text{Factor out tan x from the two terms on the left side.}$$

$$\tan x = 0 \quad \text{or} \quad \sin^2 x - 3 = 0 \qquad \text{Set each factor equal to 0.}$$
$$x = 0 \quad x = \pi \qquad \qquad \sin^2 x = 3 \qquad \text{Solve for x.}$$
$$\sin x = \pm\sqrt{3}$$

This equation has no solution because $\sin x$ cannot be greater than 1 or less than −1.

The solutions in the interval $[0, 2\pi)$ are 0 and π. • • •

✓ **Check Point 6** Solve the equation: $\sin x \tan x = \sin x$, $0 \le x < 2\pi$.

5 Use identities to solve trigonometric equations.

Using Identities to Solve Trigonometric Equations

Some trigonometric equations contain more than one function on the same side and these functions cannot be separated by factoring. For example, consider the equation

$$2 \cos^2 x + 3 \sin x = 0.$$

How can we obtain an equivalent equation that has only one trigonometric function? We use the identity $\sin^2 x + \cos^2 x = 1$ and substitute $1 - \sin^2 x$ for $\cos^2 x$. This forms the basis of our next example.

EXAMPLE 7   Using an Identity to Solve a Trigonometric Equation

Solve the equation: $2 \cos^2 x + 3 \sin x = 0,\quad 0 \le x < 2\pi$.

SOLUTION

$2 \cos^2 x + 3 \sin x = 0$	This is the given equation.
$2(1 - \sin^2 x) + 3 \sin x = 0$	$\cos^2 x = 1 - \sin^2 x$
$2 - 2 \sin^2 x + 3 \sin x = 0$	Use the distributive property.
$-2 \sin^2 x + 3 \sin x + 2 = 0$	Write the equation in descending powers of $\sin x$.
$2 \sin^2 x - 3 \sin x - 2 = 0$	Multiply both sides by -1. The equation is in quadratic form $2u^2 - 3u - 2 = 0$ with $u = \sin x$.
$(2 \sin x + 1)(\sin x - 2) = 0$	Factor. Notice that $2u^2 - 3u - 2$ factors as $(2u + 1)(u - 2)$.
$2 \sin x + 1 = 0\quad$ or $\quad \sin x - 2 = 0$	Set each factor equal to 0.
$2 \sin x = -1 \qquad\qquad\qquad \sin x = 2$	Solve $2 \sin x + 1 = 0$ and $\sin x - 2 = 0$ for $\sin x$.
$\sin x = -\dfrac{1}{2}$	

It's easier to factor with a positive leading coefficient.

This equation has no solution because $\sin x$ cannot be greater than 1.

$$x = \pi + \frac{\pi}{6} = \frac{7\pi}{6} \qquad x = 2\pi - \frac{\pi}{6} = \frac{11\pi}{6} \qquad \text{Solve for } x.$$

$\sin \frac{\pi}{6} = \frac{1}{2}$. The sine is negative in quadrants III and IV.

The solutions of $2 \cos^2 x + 3 \sin x = 0$ in the interval $[0, 2\pi)$ are $\dfrac{7\pi}{6}$ and $\dfrac{11\pi}{6}$.   • • •

✓ **Check Point 7** Solve the equation: $2 \sin^2 x - 3 \cos x = 0,\quad 0 \le x < 2\pi$.

EXAMPLE 8   Using an Identity to Solve a Trigonometric Equation

Solve the equation: $\cos 2x + 3 \sin x - 2 = 0,\quad 0 \le x < 2\pi$.

SOLUTION

The given equation contains a cosine function and a sine function. The cosine is a function of $2x$ and the sine is a function of x. We want one trigonometric function of the same angle. This can be accomplished by using the double-angle identity $\cos 2x = 1 - 2 \sin^2 x$ to obtain an equivalent equation involving $\sin x$ only.

$\cos 2x + 3 \sin x - 2 = 0$	This is the given equation.
$1 - 2 \sin^2 x + 3 \sin x - 2 = 0$	$\cos 2x = 1 - 2 \sin^2 x$
$-2 \sin^2 x + 3 \sin x - 1 = 0$	Combine like terms.

$$2 \sin^2 x - 3 \sin x + 1 = 0$$ Multiply both sides by −1. The equation is in quadratic form $2u^2 - 3u + 1 = 0$ with $u = \sin x$.

$$(2 \sin x - 1)(\sin x - 1) = 0$$ Factor. Notice that $2u^2 - 3u + 1$ factors as $(2u - 1)(u - 1)$.

$$2 \sin x - 1 = 0 \quad \text{or} \quad \sin x - 1 = 0$$ Set each factor equal to O.

$$\sin x = \tfrac{1}{2} \qquad\qquad \sin x = 1$$ Solve for sin x.

$$x = \frac{\pi}{6} \qquad x = \pi - \frac{\pi}{6} = \frac{5\pi}{6} \qquad x = \frac{\pi}{2}$$ Solve each equation for x, $0 \le x < 2\pi$.

> The sine is positive in quadrants I and II.

The solutions in the interval $[0, 2\pi)$ are $\dfrac{\pi}{6}, \dfrac{\pi}{2}$, and $\dfrac{5\pi}{6}$. • • •

✓ **Check Point 8** Solve the equation: $\cos 2x + \sin x = 0, \quad 0 \le x < 2\pi$.

Sometimes it is necessary to do something to both sides of a trigonometric equation before using an identity. For example, consider the equation

$$\sin x \cos x = \tfrac{1}{2}.$$

This equation contains both a sine and a cosine function. How can we obtain a single function? Multiply both sides by 2. In this way, we can use the double-angle identity $\sin 2x = 2 \sin x \cos x$ and obtain $\sin 2x$, a single function, on the left side.

TECHNOLOGY

Graphic Connections

Shown below are the graphs of
$$y = \sin x \cos x$$
and
$$y = \tfrac{1}{2}$$
in a $\left[0, 2\pi, \dfrac{\pi}{2}\right]$ by $[-1, 1, 1]$
viewing rectangle. The solutions of
$$\sin x \cos x = \tfrac{1}{2}$$
are shown by the x-coordinates of the two intersection points.

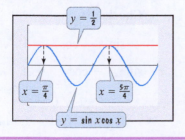

EXAMPLE 9 Using an Identity to Solve a Trigonometric Equation

Solve the equation: $\sin x \cos x = \tfrac{1}{2}, \quad 0 \le x < 2\pi$.

SOLUTION

$$\sin x \cos x = \tfrac{1}{2}$$ This is the given equation.

$$2 \sin x \cos x = 1$$ Multiply both sides by 2 in anticipation of using $\sin 2x = 2 \sin x \cos x$.

$$\sin 2x = 1$$ Use a double-angle identity.

Notice that we have an equation, $\sin 2x = 1$, with $2x$, a multiple angle. The period of the sine function is 2π. In the interval $[0, 2\pi)$, the only value for which the sine function is 1 is $\dfrac{\pi}{2}$. This means that $2x = \dfrac{\pi}{2}$. Because the period is 2π, all the solutions of $\sin 2x = 1$ are given by

$$2x = \frac{\pi}{2} + 2n\pi$$ n is any integer.

$$x = \frac{\pi}{4} + n\pi.$$ Divide both sides by 2 and solve for x.

The solutions of $\sin x \cos x = \dfrac{1}{2}$ in the interval $[0, 2\pi)$ are obtained by letting $n = 0$ and $n = 1$. The solutions are $\dfrac{\pi}{4}$ and $\dfrac{5\pi}{4}$. • • •

✓ **Check Point 9** Solve the equation: $\sin x \cos x = -\tfrac{1}{2}, \quad 0 \le x < 2\pi$.

A Brief Review • Extraneous Solutions

- Extra solutions may be introduced if you square both sides of an equation. Such solutions, which are not solutions of the given equation, are called extraneous solutions or extraneous roots.

EXAMPLE

Consider the equation $x = 4$. If we square both sides, we obtain $x^2 = 16$. Solving this equation, we obtain $x = \pm\sqrt{16} = \pm 4$. The new equation, $x^2 = 16$, has two solutions, -4 and 4. By contrast, only 4 is a solution of the original equation $x = 4$. Thus, -4 is an extraneous solution.

- When squaring both sides of an equation, always check proposed solutions in the original equation.

Let's look at another equation that contains two different functions, $\sin x - \cos x = 1$. Can you think of an identity that can be used to produce only one function? Perhaps $\sin^2 x + \cos^2 x = 1$ might be helpful. The next example shows how we can use this identity after squaring both sides of the given equation. Remember that if we raise both sides of an equation to an even power, we have the possibility of introducing extraneous solutions. Thus, we must check each proposed solution in the given equation. Alternatively, we can use a graphing utility to verify actual solutions.

TECHNOLOGY

Graphic Connections

A graphing utility can be used instead of an algebraic check. Shown are the graphs of

$$y = \sin x - \cos x$$

and

$$y = 1$$

in a $\left[0, 2\pi, \dfrac{\pi}{2}\right]$ by $[-2, 2, 1]$ viewing rectangle. The actual solutions of

$$\sin x - \cos x = 1$$

are shown by the x-coordinates of the two intersection points, $\dfrac{\pi}{2}$ and π.

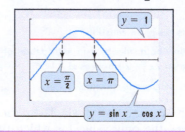

EXAMPLE 10   Using an Identity to Solve a Trigonometric Equation

Solve the equation: $\sin x - \cos x = 1$, $0 \le x < 2\pi$.

SOLUTION

We square both sides of the equation in anticipation of using $\sin^2 x + \cos^2 x = 1$.

$\sin x - \cos x = 1$	This is the given equation.
$(\sin x - \cos x)^2 = 1^2$	Square both sides.
$\sin^2 x - 2\sin x \cos x + \cos^2 x = 1$	Square the left side using $(A - B)^2 = A^2 - 2AB + B^2$.
$\sin^2 x + \cos^2 x - 2\sin x \cos x = 1$	Rearrange terms.
$1 - 2\sin x \cos x = 1$	Apply a Pythagorean identity: $\sin^2 x + \cos^2 x = 1$.
$-2\sin x \cos x = 0$	Subtract 1 from both sides of the equation.
$\sin x \cos x = 0$	Divide both sides of the equation by -2.
$\sin x = 0$ or $\cos x = 0$	Set each factor equal to 0.
$x = 0$ $x = \pi$ $x = \dfrac{\pi}{2}$ $x = \dfrac{3\pi}{2}$	Solve for x in $[0, 2\pi)$.

We check these proposed solutions to see if any are extraneous.

Check 0:

$\sin x - \cos x = 1$

$\sin 0 - \cos 0 \overset{?}{=} 1$

$0 - 1 \overset{?}{=} 1$

$-1 = 1$, false

O is extraneous.

Check $\dfrac{\pi}{2}$:

$\sin x - \cos x = 1$

$\sin \dfrac{\pi}{2} - \cos \dfrac{\pi}{2} \overset{?}{=} 1$

$1 - 0 \overset{?}{=} 1$

$1 = 1$, true

Check π:

$\sin x - \cos x = 1$

$\sin \pi - \cos \pi \overset{?}{=} 1$

$0 - (-1) \overset{?}{=} 1$

$1 = 1$, true

Check $\dfrac{3\pi}{2}$:

$\sin x - \cos x = 1$

$\sin \dfrac{3\pi}{2} - \cos \dfrac{3\pi}{2} \overset{?}{=} 1$

$-1 - 0 \overset{?}{=} 1$

$-1 = 1$, false

$\dfrac{3\pi}{2}$ is extraneous.

The actual solutions of $\sin x - \cos x = 1$ in the interval $[0, 2\pi)$ are $\dfrac{\pi}{2}$ and π.   • • •

✓ **Check Point 10** Solve the equation: $\cos x - \sin x = -1, \quad 0 \leq x < 2\pi$.

Using a Calculator to Solve Trigonometric Equations

In all our previous examples, the equations had solutions that were found by knowing the exact values of trigonometric functions of special angles, such as $\frac{\pi}{6}, \frac{\pi}{4}$, and $\frac{\pi}{3}$. However, not all trigonometric equations involve these special angles. For those that do not, we will use the secondary keys marked $\boxed{\text{SIN}^{-1}}$, $\boxed{\text{COS}^{-1}}$, and $\boxed{\text{TAN}^{-1}}$ on a calculator. Recall that on most calculators, the inverse trigonometric function keys are the secondary functions for the buttons labeled $\boxed{\text{SIN}}$, $\boxed{\text{COS}}$, and $\boxed{\text{TAN}}$, respectively.

6 Use a calculator to solve trigonometric equations.

EXAMPLE 11   Solving Trigonometric Equations with a Calculator

Solve each equation, correct to four decimal places, for $0 \leq x < 2\pi$:

a. $\tan x = 12.8044$ **b.** $\cos x = -0.4317$.

SOLUTION

We begin by using a calculator to find $\theta, 0 \leq \theta < \frac{\pi}{2}$ satisfying the following equations:

$$\tan \theta = 12.8044 \qquad \cos \theta = 0.4317.$$

> These numbers are the absolute values of the given range values.

Once θ is determined, we use our knowledge of the signs of the trigonometric functions to find x in $[0, 2\pi)$ satisfying $\tan x = 12.8044$ and $\cos x = -0.4317$.

GREAT QUESTION!

To find solutions in $[0, 2\pi)$, does my calculator have to be in radian mode?

Yes. Be careful. Most scientific calculators revert to degree mode every time they are cleared.

a. $\tan x = 12.8044$ — This is the given equation.

$\tan \theta = 12.8044$ — Use a calculator to solve this equation for θ, $0 \leq \theta < \frac{\pi}{2}$.

$\theta = \tan^{-1}(12.8044) \approx 1.4929$ 12.8044 $\boxed{\text{2nd}}$ $\boxed{\text{TAN}}$ or $\boxed{\text{2nd}}$ $\boxed{\text{TAN}}$ 12.8044 $\boxed{\text{ENTER}}$

$\tan x = 12.8044$ — Return to the given equation. Because the tangent is positive, x lies in quadrant I or III.

$x \approx 1.4929$ $x \approx \pi + 1.4929 \approx 4.6345$ Solve for x, $0 \leq x < 2\pi$.

> The tangent is positive in quadrant I.

> The tangent is positive in quadrant III.

Correct to four decimal places, the solutions of $\tan x = 12.8044$ in the interval $[0, 2\pi)$ are 1.4929 and 4.6345. (Note: Slight differences in approximate solutions can occur due to rounding. If you don't round $\tan^{-1}(12.8044)$ first, then $x = \pi + \tan^{-1}(12.8044) \approx 4.6344$.)

b. $\cos x = -0.4317$ — This is the given equation.

$\cos \theta = 0.4317$ — Use a calculator to solve this equation for θ, $0 \leq \theta < \frac{\pi}{2}$.

$\theta = \cos^{-1}(0.4317) \approx 1.1244$ .4317 $\boxed{\text{2nd}}$ $\boxed{\text{COS}}$ or $\boxed{\text{2nd}}$ $\boxed{\text{COS}}$ .4317 $\boxed{\text{ENTER}}$

$\cos x = -0.4317$ — Return to the given equation. Because the cosine is negative, x lies in quadrant II or III.

$x \approx \pi - 1.1244 \approx 2.0172$ $x \approx \pi + 1.1244 \approx 4.2660$ Solve for x, $0 \leq x < 2\pi$.

> The cosine is negative in quadrant II.

> The cosine is negative in quadrant III.

Correct to four decimal places, the solutions of $\cos x = -0.4317$ in the interval $[0, 2\pi)$ are 2.0172 and 4.2660.

●●●

☑ **Check Point 11** Solve each equation, correct to four decimal places, for $0 \le x < 2\pi$:

a. $\tan x = 3.1044$

b. $\sin x = -0.2315$.

EXAMPLE 12 Solving a Trigonometric Equation Using the Quadratic Formula and a Calculator

Solve the equation, correct to four decimal places, for $0 \le x < 2\pi$:

$$\sin^2 x - \sin x - 1 = 0.$$

SOLUTION

The given equation is in quadratic form $u^2 - u - 1 = 0$ with $u = \sin x$. We use the quadratic formula to solve for $\sin x$ because $u^2 - u - 1$ cannot be factored. Begin by identifying the values for a, b, and c.

$$\sin^2 x - \sin x - 1 = 0$$

$a = 1 \qquad b = -1 \qquad c = -1$

Substituting these values into the quadratic formula and simplifying gives the values for $\sin x$. Once we obtain these values, we will solve for x.

$$\sin x = \frac{-b \pm \sqrt{b^2 - 4ac}}{2a} = \frac{-(-1) \pm \sqrt{(-1)^2 - 4(1)(-1)}}{2(1)} = \frac{1 \pm \sqrt{1 - (-4)}}{2} = \frac{1 \pm \sqrt{5}}{2}$$

$$\sin x = \frac{1 + \sqrt{5}}{2} \approx 1.6180 \qquad \text{or} \qquad \sin x = \frac{1 - \sqrt{5}}{2} \approx -0.6180$$

This equation has no solution because $\sin x$ cannot be greater than 1.

The sine is negative in quadrants III and IV. Use a calculator to solve $\sin \theta = 0.6180$, $0 \le \theta < \frac{\pi}{2}$.

Using a calculator to solve $\sin \theta = 0.6180$, we have

$$\theta = \sin^{-1}(0.6180) \approx 0.6662.$$

We use 0.6662 to solve $\sin x = -0.6180$, $0 \le x < 2\pi$.

$$x \approx \pi + 0.6662 \approx 3.8078 \qquad\qquad x \approx 2\pi - 0.6662 \approx 5.6170$$

The sine is negative in quadrant III. The sine is negative in quadrant IV.

Correct to four decimal places, the solutions of $\sin^2 x - \sin x - 1 = 0$ in the interval $[0, 2\pi)$ are 3.8078 and 5.6170. ● ● ●

☑ **Check Point 12** Solve the equation, correct to four decimal places, for $0 \le x < 2\pi$:

$$\cos^2 x + 5 \cos x + 3 = 0.$$

ACHIEVING SUCCESS

Assuming that you have done very well preparing for an exam, **there are certain things you can do that will make you a better test taker**.

- Just before the exam, briefly review the relevant material in the chapter summary.
- Bring everything you need to the exam, including two pencils, an eraser, scratch paper (if permitted), a calculator (if you're allowed to use one), water, and a watch.
- Survey the entire exam quickly to get an idea of its length.
- Read the directions to each problem carefully. Make sure that you have answered the specific question asked.
- Work the easy problems first. Then return to the hard problems you are not sure of. Doing the easy problems first will build your confidence. If you get bogged down on any one problem, you may not be able to complete the exam and receive credit for the questions you can easily answer.
- Attempt every problem. There may be partial credit even if you do not obtain the correct answer.
- Work carefully. Show your step-by-step solutions neatly. Check your work and answers.
- Watch the time. Pace yourself and be aware of when half the time is up. Determine how much of the exam you have completed. This will indicate if you're moving at a good pace or need to speed up. Prepare to spend more time on problems worth more points.
- Never turn in a test early. Use every available minute you are given for the test. If you have extra time, double check your arithmetic and look over your solutions.

CONCEPT AND VOCABULARY CHECK

Fill in each blank so that the resulting statement is true.

1. The solutions of $\sin x = \frac{\sqrt{2}}{2}$ in $[0, 2\pi)$ are $x = \frac{\pi}{4}$ and $x = $ _____. If n is any integer, all solutions of $\sin x = \frac{\sqrt{2}}{2}$ are given by $x = $ _____ and $x = $ _____.

2. The solution of $\tan x = -\sqrt{3}$ in $[0, \pi)$ is $x = \pi - \frac{\pi}{3}$, or $x = $ _____. If n is any integer, all solutions of $\tan x = -\sqrt{3}$ are given by _____.

3. True or false: If $3x = \frac{\pi}{4} + n\pi$ for any integer n, then $x = \frac{\pi}{12} + n\pi$. _____

4. True or false: If $\cos \frac{x}{3} = \frac{1}{2}$, then $\frac{x}{3} = \frac{\pi}{3} + 2n\pi$ or $\frac{x}{3} = \frac{5\pi}{3} + 2n\pi$ for any integer n. _____

5. True or false: If $\tan 3x = 1$, then $x = \frac{\pi}{4} + n\pi$ for any integer n. _____

6. If $2\cos^2 x - 9\cos x - 5 = 0$, then _____ $= 0$ or _____ $= 0$. Of these two equations, the equation that has no solution is _____.

7. If $2\sin x \cos x + \sqrt{2}\cos x = 0$, then _____ $= 0$ or _____ $= 0$.

8. The first step in solving the equation $4\cos^2 x + 4\sin x - 5 = 0, 0 \le x < 2\pi$, is to replace _____ with _____.

9. If $\sin 0.9695 \approx 0.8246$, then the solutions of $\sin x = -0.8246, 0 \le x < 2\pi$, are given by $x \approx $ ___ $+ 0.9695$ and $x \approx $ ___ $- 0.9695$.

EXERCISE SET 3.5

Practice Exercises

In Exercises 1–10, use substitution to determine whether the given x-value is a solution of the equation.

1. $\cos x = \frac{\sqrt{2}}{2}, \quad x = \frac{\pi}{4}$

2. $\tan x = \sqrt{3}, \quad x = \frac{\pi}{3}$

3. $\sin x = \frac{\sqrt{3}}{2}, \quad x = \frac{\pi}{6}$

4. $\sin x = \frac{\sqrt{2}}{2}, \quad x = \frac{\pi}{3}$

5. $\cos x = -\frac{1}{2}, \quad x = \frac{2\pi}{3}$

6. $\cos x = -\frac{1}{2}, \quad x = \frac{4\pi}{3}$

7. $\tan 2x = -\frac{\sqrt{3}}{3}, \quad x = \frac{5\pi}{12}$

8. $\cos \frac{2x}{3} = -\frac{1}{2}, \quad x = \pi$

9. $\cos x = \sin 2x, \quad x = \frac{\pi}{3}$

10. $\cos x + 2 = \sqrt{3}\sin x, \quad x = \frac{\pi}{6}$

In Exercises 11–24, find all solutions of each equation.

11. $\sin x = \frac{\sqrt{3}}{2}$

12. $\cos x = \frac{\sqrt{3}}{2}$

13. $\tan x = 1$

14. $\tan x = \sqrt{3}$

15. $\cos x = -\frac{1}{2}$

16. $\sin x = -\frac{\sqrt{2}}{2}$

17. $\tan x = 0$

18. $\sin x = 0$

19. $2 \cos x + \sqrt{3} = 0$

20. $2 \sin x + \sqrt{3} = 0$

21. $4 \sin \theta - 1 = 2 \sin \theta$

22. $5 \sin \theta + 1 = 3 \sin \theta$

23. $3 \sin \theta + 5 = -2 \sin \theta$

24. $7 \cos \theta + 9 = -2 \cos \theta$

Exercises 25–38 involve equations with multiple angles. Solve each equation on the interval $[0, 2\pi)$.

25. $\sin 2x = \dfrac{\sqrt{3}}{2}$

26. $\cos 2x = \dfrac{\sqrt{2}}{2}$

27. $\cos 4x = -\dfrac{\sqrt{3}}{2}$

28. $\sin 4x = -\dfrac{\sqrt{2}}{2}$

29. $\tan 3x = \dfrac{\sqrt{3}}{3}$

30. $\tan 3x = \sqrt{3}$

31. $\tan \dfrac{x}{2} = \sqrt{3}$

32. $\tan \dfrac{x}{2} = \dfrac{\sqrt{3}}{3}$

33. $\sin \dfrac{2\theta}{3} = -1$

34. $\cos \dfrac{2\theta}{3} = -1$

35. $\sec \dfrac{3\theta}{2} = -2$

36. $\cot \dfrac{3\theta}{2} = -\sqrt{3}$

37. $\sin \left(2x + \dfrac{\pi}{6}\right) = \dfrac{1}{2}$

38. $\sin \left(2x - \dfrac{\pi}{4}\right) = \dfrac{\sqrt{2}}{2}$

Exercises 39–52 involve trigonometric equations quadratic in form. Solve each equation on the interval $[0, 2\pi)$.

39. $2 \sin^2 x - \sin x - 1 = 0$

40. $2 \sin^2 x + \sin x - 1 = 0$

41. $2 \cos^2 x + 3 \cos x + 1 = 0$ **42.** $\cos^2 x + 2 \cos x - 3 = 0$

43. $2 \sin^2 x = \sin x + 3$ **44.** $2 \sin^2 x = 4 \sin x + 6$

45. $\sin^2 \theta - 1 = 0$ **46.** $\cos^2 \theta - 1 = 0$

47. $4 \cos^2 x - 1 = 0$ **48.** $4 \sin^2 x - 3 = 0$

49. $9 \tan^2 x - 3 = 0$ **50.** $3 \tan^2 x - 9 = 0$

51. $\sec^2 x - 2 = 0$ **52.** $4 \sec^2 x - 2 = 0$

In Exercises 53–62, solve each equation on the interval $[0, 2\pi)$.

53. $(\tan x - 1)(\cos x + 1) = 0$

54. $(\tan x + 1)(\sin x - 1) = 0$

55. $\left(2 \cos x + \sqrt{3}\right)(2 \sin x + 1) = 0$

56. $\left(2 \cos x - \sqrt{3}\right)(2 \sin x - 1) = 0$

57. $\cot x(\tan x - 1) = 0$ **58.** $\cot x(\tan x + 1) = 0$

59. $\sin x + 2 \sin x \cos x = 0$ **60.** $\cos x - 2 \sin x \cos x = 0$

61. $\tan^2 x \cos x = \tan^2 x$ **62.** $\cot^2 x \sin x = \cot^2 x$

In Exercises 63–84, use an identity to solve each equation on the interval $[0, 2\pi)$.

63. $2 \cos^2 x + \sin x - 1 = 0$ **64.** $2 \cos^2 x - \sin x - 1 = 0$

65. $\sin^2 x - 2 \cos x - 2 = 0$

66. $4 \sin^2 x + 4 \cos x - 5 = 0$

67. $4 \cos^2 x = 5 - 4 \sin x$ **68.** $3 \cos^2 x = \sin^2 x$

69. $\sin 2x = \cos x$ **70.** $\sin 2x = \sin x$

71. $\cos 2x = \cos x$ **72.** $\cos 2x = \sin x$

73. $\cos 2x + 5 \cos x + 3 = 0$ **74.** $\cos 2x + \cos x + 1 = 0$

75. $\sin x \cos x = \dfrac{\sqrt{2}}{4}$

76. $\sin x \cos x = \dfrac{\sqrt{3}}{4}$

77. $\sin x + \cos x = 1$

78. $\sin x + \cos x = -1$

79. $\sin \left(x + \dfrac{\pi}{4}\right) + \sin \left(x - \dfrac{\pi}{4}\right) = 1$

80. $\sin \left(x + \dfrac{\pi}{3}\right) + \sin \left(x - \dfrac{\pi}{3}\right) = 1$

81. $\sin 2x \cos x + \cos 2x \sin x = \dfrac{\sqrt{2}}{2}$

82. $\sin 3x \cos 2x + \cos 3x \sin 2x = 1$

83. $\tan x + \sec x = 1$

84. $\tan x - \sec x = 1$

In Exercises 85–96, use a calculator to solve each equation, correct to four decimal places, on the interval $[0, 2\pi)$.

85. $\sin x = 0.8246$

86. $\sin x = 0.7392$

87. $\cos x = -\dfrac{2}{5}$

88. $\cos x = -\dfrac{4}{7}$

89. $\tan x = -3$

90. $\tan x = -5$

91. $\cos^2 x - \cos x - 1 = 0$

92. $3 \cos^2 x - 8 \cos x - 3 = 0$

93. $4 \tan^2 x - 8 \tan x + 3 = 0$

94. $\tan^2 x - 3 \tan x + 1 = 0$

95. $7 \sin^2 x - 1 = 0$

96. $5 \sin^2 x - 1 = 0$

In Exercises 97–116, use the most appropriate method to solve each equation on the interval $[0, 2\pi)$. Use exact values where possible or give approximate solutions correct to four decimal places.

97. $2 \cos 2x + 1 = 0$ **98.** $2 \sin 3x + \sqrt{3} = 0$

99. $\sin 2x + \sin x = 0$ **100.** $\sin 2x + \cos x = 0$

101. $3 \cos x - 6\sqrt{3} = \cos x - 5\sqrt{3}$

102. $\cos x - 5 = 3 \cos x + 6$

103. $\tan x = -4.7143$

104. $\tan x = -6.2154$

105. $2 \sin^2 x = 3 - \sin x$

106. $2 \sin^2 x = 2 - 3 \sin x$

107. $\cos x \csc x = 2 \cos x$

108. $\tan x \sec x = 2 \tan x$

109. $5 \cot^2 x - 15 = 0$

110. $5 \sec^2 x - 10 = 0$

111. $\cos^2 x + 2 \cos x - 2 = 0$

112. $\cos^2 x + 5 \cos x - 1 = 0$

113. $5 \sin x = 2 \cos^2 x - 4$

114. $7 \cos x = 4 - 2 \sin^2 x$

115. $2 \tan^2 x + 5 \tan x + 3 = 0$

116. $3 \tan^2 x - \tan x - 2 = 0$

Practice Plus

In Exercises 117–120, graph f and g in the same rectangular coordinate system for $0 \le x \le 2\pi$. Then solve a trigonometric equation to determine points of intersection and identify these points on your graphs.

117. $f(x) = 3\cos x, g(x) = \cos x - 1$

118. $f(x) = 3\sin x, g(x) = \sin x - 1$

119. $f(x) = \cos 2x, g(x) = -2\sin x$

120. $f(x) = \cos 2x, g(x) = 1 - \sin x$

In Exercises 121–126, solve each equation on the interval $[0, 2\pi)$.

121. $|\cos x| = \dfrac{\sqrt{3}}{2}$

122. $|\sin x| = \dfrac{1}{2}$

123. $10\cos^2 x + 3\sin x - 9 = 0$

124. $3\cos^2 x - \sin x = \cos^2 x$

125. $2\cos^3 x + \cos^2 x - 2\cos x - 1 = 0$ (*Hint:* Use factoring by grouping.)

126. $2\sin^3 x - \sin^2 x - 2\sin x + 1 = 0$ (*Hint:* Use factoring by grouping.)

In Exercises 127–128, find the x-intercepts, correct to four decimal places, of the graph of each function. Then use the x-intercepts to match the function with its graph. The graphs are labeled (a) and (b).

127. $f(x) = \tan^2 x - 3\tan x + 1$

128. $g(x) = 4\tan^2 x - 8\tan x + 3$

a.

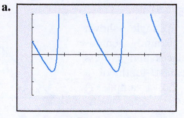

$[0, 2\pi, \frac{\pi}{4}]$ by $[-3, 3, 1]$

b.

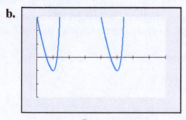

$[0, 2\pi, \frac{\pi}{4}]$ by $[-3, 3, 1]$

Application Exercises

Use this information to solve Exercises 129–130. Our cycle of normal breathing takes place every 5 seconds. Velocity of air flow, y, measured in liters per second, after x seconds is modeled by

$$y = 0.6\sin\frac{2\pi}{5}x.$$

Velocity of air flow is positive when we inhale and negative when we exhale.

129. Within each breathing cycle, when are we inhaling at a rate of 0.3 liter per second? Round to the nearest tenth of a second.

130. Within each breathing cycle, when are we exhaling at a rate of 0.3 liter per second? Round to the nearest tenth of a second.

Use this information to solve Exercises 131–132. The number of hours of daylight in Boston is given by

$$y = 3\sin\left[\frac{2\pi}{365}(x - 79)\right] + 12,$$

where x is the number of days after January 1.

131. Within a year, when does Boston have 10.5 hours of daylight? Give your answer in days after January 1 and round to the nearest day.

132. Within a year, when does Boston have 13.5 hours of daylight? Give your answer in days after January 1 and round to the nearest day.

Use this information to solve Exercises 133–134. A ball on a spring is pulled 4 inches below its rest position and then released. After t seconds, the ball's distance, d, in inches from its rest position is given by

$$d = -4\cos\frac{\pi}{3}t.$$

133. Find all values of t for which the ball is 2 inches above its rest position.

134. Find all values of t for which the ball is 2 inches below its rest position.

Use this information to solve Exercises 135–136. When throwing an object, the distance achieved depends on its initial velocity, v_0, and the angle above the horizontal at which the object is thrown, θ. The distance, d, in feet, that describes the range covered is given by

$$d = \frac{v_0^2}{16}\sin\theta\cos\theta,$$

where v_0 is measured in feet per second.

135. You and your friend are throwing a baseball back and forth. If you throw the ball with an initial velocity of $v_0 = 90$ feet per second, at what angle of elevation, θ, to the nearest degree, should you direct your throw so that it can be easily caught by your friend located 170 feet away?

136. In Exercise 135, you increase the distance between you and your friend to 200 feet. With this increase, at what angle of elevation, θ, to the nearest degree, should you direct your throw?

Explaining the Concepts

137. What are the solutions of a trigonometric equation?

138. Describe the difference between verifying a trigonometric identity and solving a trigonometric equation.

139. Without actually solving the equation, describe how to solve

$$3 \tan x - 2 = 5 \tan x - 1.$$

140. In the interval $[0, 2\pi)$, the solutions of $\sin x = \cos 2x$ are $\dfrac{\pi}{6}, \dfrac{5\pi}{6}$, and $\dfrac{3\pi}{2}$. Explain how to use graphs generated by a graphing utility to check these solutions.

141. Suppose you are solving equations in the interval $[0, 2\pi)$. Without actually solving equations, what is the difference between the number of solutions of $\sin x = \frac{1}{2}$ and $\sin 2x = \frac{1}{2}$? How do you account for this difference?

In Exercises 142–143, describe a general strategy for solving each equation. Do not solve the equation.

142. $2 \sin^2 x + 5 \sin x + 3 = 0$

143. $\sin 2x = \sin x$

144. Describe a natural periodic phenomenon. Give an example of a question that can be answered by a trigonometric equation in the study of this phenomenon.

145. A city's tall buildings and narrow streets reduce the amount of sunlight. If h is the average height of the buildings and w is the width of the street, the angle of elevation from the street to the top of the buildings is given by the trigonometric equation

$$\tan \theta = \frac{h}{w}.$$

A value of $\theta = 63°$ can result in an 85% loss of illumination. Some people experience depression with loss of sunlight. Determine whether such a person should live on a city street that is 80 feet wide with buildings whose heights average 400 feet. Explain your answer and include θ, to the nearest degree, in your argument.

Technology Exercises

146. Use a graphing utility to verify the solutions of any five equations that you solved in Exercises 63–84.

In Exercises 147–151, use a graphing utility to approximate the solutions of each equation in the interval $[0, 2\pi)$. Round to the nearest hundredth of a radian.

147. $15 \cos^2 x + 7 \cos x - 2 = 0$

148. $\cos x = x$

149. $2 \sin^2 x = 1 - 2 \sin x$

150. $\sin 2x = 2 - x^2$

151. $\sin x + \sin 2x + \sin 3x = 0$

Critical Thinking Exercises

Make Sense? *In Exercises 152–155, determine whether each statement makes sense or does not make sense, and explain your reasoning.*

152. I solved $4 \cos^2 x = 5 - 4 \sin x$ by working independently with the left side, applying a Pythagorean identity, and transforming the left side into $5 - 4 \sin x$.

153. There are similarities and differences between solving $4x + 1 = 3$ and $4 \sin \theta + 1 = 3$: In the first equation, I need to isolate x to get the solution. In the trigonometric equation, I need to first isolate $\sin \theta$, but then I must continue to solve for θ.

154. I solved $\cos(x - \frac{\pi}{3}) = -1$ by first applying the formula for the cosine of the difference of two angles.

155. Using the equation for simple harmonic motion described in Exercises 133–134, I need to solve a trigonometric equation to determine the ball's distance from its rest position after 2 seconds.

In Exercises 156–159, determine whether each statement is true or false. If the statement is false, make the necessary change(s) to produce a true statement.

156. The equation $(\sin x - 3)(\cos x + 2) = 0$ has no solution.

157. The equation $\tan x = \dfrac{\pi}{2}$ has no solution.

158. A trigonometric equation with an infinite number of solutions is an identity.

159. The equations $\sin 2x = 1$ and $\sin 2x = \frac{1}{2}$ have the same number of solutions on the interval $[0, 2\pi)$.

In Exercises 160–162, solve each equation on the interval $[0, 2\pi)$. Do not use a calculator.

160. $2 \cos x - 1 + 3 \sec x = 0$

161. $\sin 3x + \sin x + \cos x = 0$

162. $\sin x + 2 \sin \dfrac{x}{2} = \cos \dfrac{x}{2} + 1$

Retaining the Concepts

163. Graph $y = 2 \sin \frac{1}{2}x$. Then use the graph to obtain the graph of $y = 2 \csc \frac{1}{2}x$. (Section 2.2, Example 4)

164. Use a right triangle to write $\sin(\cos^{-1} x)$ as an algebraic expression. Assume that x is positive and that the given inverse trigonometric function is defined for the expression in x. (Section 2.3, Example 9)

165. Solve the right triangle shown in the figure. Round lengths to two decimal places and express angles to the nearest tenth of a degree. (Section 2.4, Example 1)

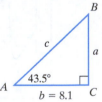

Preview Exercises

Exercises 166–168 will help you prepare for the material covered in the first section of the next chapter. Solve each equation by using the cross-products principle to clear fractions from the proportion:

$$\text{If } \frac{a}{b} = \frac{c}{d}, \text{ then } ad = bc. \ (b \neq 0 \text{ and } d \neq 0)$$

Round to the nearest tenth.

166. Solve for a: $\dfrac{a}{\sin 46°} = \dfrac{56}{\sin 63°}$.

167. Solve for B, $0 < B < 180°$: $\dfrac{81}{\sin 43°} = \dfrac{62}{\sin B}$.

168. Solve for B: $\dfrac{51}{\sin 75°} = \dfrac{71}{\sin B}$.

CHAPTER 3	Summary, Review, and Test

SUMMARY

DEFINITIONS AND CONCEPTS	EXAMPLES
3.1 Verifying Trigonometric Identities	
a. Identities are trigonometric equations that are true for all values of the variable for which the expressions are defined.	
b. Fundamental trigonometric identities are given in the box on page 144.	
c. Guidelines for verifying trigonometric identities are given in the box on page 155.	Ex. 1, p. 145; Ex. 2, p. 146; Ex. 3, p. 147; Ex. 4, p. 147; Ex. 5, p. 149; Ex. 6, p. 150; Ex. 7, p. 150; Ex. 8, p. 151
3.2 Sum and Difference Formulas	
a. Sum and difference formulas are given in the box on page 162 and the box on page 165.	
b. Sum and difference formulas can be used to find exact values of trigonometric functions.	Ex. 1, p. 160; Ex. 2, p. 161; Ex. 4, p. 162; Ex. 5, p. 163
c. Sum and difference formulas can be used to verify trigonometric identities.	Ex. 3, p. 161; Ex. 6, p. 164; Ex. 7, p. 165
3.3 Double-Angle, Power-Reducing, and Half-Angle Formulas	
a. Double-angle, power-reducing, and half-angle formulas are given in the box on page 177.	
b. Double-angle and half-angle formulas can be used to find exact values of trigonometric functions.	Ex. 1, p. 171; Ex. 2, p. 171; Ex. 5, p. 175
c. Double-angle and half-angle formulas can be used to verify trigonometric identities.	Ex. 3, p. 172; Ex. 6, p. 175; Ex. 7, p. 176
d. Power-reducing formulas can be used to reduce the powers of trigonometric functions.	Ex. 4, p. 173
3.4 Product-to-Sum and Sum-to-Product Formulas	
a. The product-to-sum formulas are given in the box on page 183.	Ex. 1, p. 184
b. The sum-to-product formulas are given in the box on page 184. These formulas are useful to verify identities with fractions that contain sums and differences of sines and/or cosines.	Ex. 2, p. 185; Ex. 3, p. 186
3.5 Trigonometric Equations	
a. The values that satisfy a trigonometric equation are its solutions.	
b. To solve an equation containing a single trigonometric function, isolate the function on one side and solve for the variable.	Ex. 1, p. 194
c. When solving equations involving multiple angles, the period plays an important role in ensuring that we do not leave out any solutions.	Ex. 2, p. 194; Ex. 3, p. 195

DEFINITIONS AND CONCEPTS	**EXAMPLES**
d. Trigonometric equations quadratic in form can be expressed as $au^2 + bu + c = 0$, where u is a trigonometric function and $a \neq 0$. Such equations can be solved by factoring, the square root property, or the quadratic formula.	Ex. 4, p. 196; Ex. 5, p. 196; Ex. 12, p. 202
e. Factoring can be used to separate two different trigonometric functions in an equation.	Ex. 6, p. 197
f. Identities are used to solve some trigonometric equations.	Ex. 7, p. 198; Ex. 8, p. 198; Ex. 9, p. 199; Ex. 10, p. 200
g. Some trigonometric equations have solutions that cannot be determined by knowing the exact values of trigonometric functions of special angles. Such equations are solved using a calculator's inverse trigonometric function feature.	Ex. 11, p. 201; Ex. 12, p. 202

REVIEW EXERCISES

3.1

In Exercises 1–13, verify each identity.

1. $\sec x - \cos x = \tan x \sin x$

2. $\cos x + \sin x \tan x = \sec x$

3. $\sin^2 \theta (1 + \cot^2 \theta) = 1$

4. $(\sec \theta - 1)(\sec \theta + 1) = \tan^2 \theta$

5. $\dfrac{1 - \tan x}{\sin x} = \csc x - \sec x$

6. $\dfrac{1}{\sin t - 1} + \dfrac{1}{\sin t + 1} = -2 \tan t \sec t$

7. $\dfrac{1 + \sin t}{\cos^2 t} = \tan^2 t + 1 + \tan t \sec t$

8. $\dfrac{\cos x}{1 - \sin x} = \dfrac{1 + \sin x}{\cos x}$

9. $1 - \dfrac{\sin^2 x}{1 + \cos x} = \cos x$

10. $(\tan \theta + \cot \theta)^2 = \sec^2 \theta + \csc^2 \theta$

11. $\dfrac{1}{\sin \theta + \cos \theta} + \dfrac{1}{\sin \theta - \cos \theta} = \dfrac{2 \sin \theta}{\sin^4 \theta - \cos^4 \theta}$

12. $\dfrac{\cos t}{\cot t - 5 \cos t} = \dfrac{1}{\csc t - 5}$

13. $\dfrac{1 - \cos t}{1 + \cos t} = (\csc t - \cot t)^2$

3.2 and 3.3

In Exercises 14–19, use a sum or difference formula to find the exact value of each expression.

14. $\cos(45° + 30°)$

15. $\sin 195°$

16. $\tan\left(\dfrac{4\pi}{3} - \dfrac{\pi}{4}\right)$

17. $\tan \dfrac{5\pi}{12}$

18. $\cos 65° \cos 5° + \sin 65° \sin 5°$

19. $\sin 80° \cos 50° - \cos 80° \sin 50°$

In Exercises 20–31, verify each identity.

20. $\sin\left(x + \dfrac{\pi}{6}\right) - \cos\left(x + \dfrac{\pi}{3}\right) = \sqrt{3} \sin x$

21. $\tan\left(x + \dfrac{3\pi}{4}\right) = \dfrac{\tan x - 1}{1 + \tan x}$

22. $\sec(\alpha + \beta) = \dfrac{\sec \alpha \sec \beta}{1 - \tan \alpha \tan \beta}$

23. $\dfrac{\cos(\alpha - \beta)}{\cos \alpha \cos \beta} = 1 + \tan \alpha \tan \beta$

24. $\cos^4 t - \sin^4 t = \cos 2t$

25. $\sin t - \cos 2t = (2 \sin t - 1)(\sin t + 1)$

26. $\dfrac{\sin 2\theta - \sin \theta}{\cos 2\theta + \cos \theta} = \dfrac{1 - \cos \theta}{\sin \theta}$

27. $\dfrac{\sin 2\theta}{1 - \sin^2 \theta} = 2 \tan \theta$

28. $\tan 2t = 2 \sin t \cos t \sec 2t$

29. $\cos 4t = 1 - 8 \sin^2 t \cos^2 t$

30. $\tan \dfrac{x}{2}(1 + \cos x) = \sin x$ **31.** $\tan \dfrac{x}{2} = \dfrac{\sec x - 1}{\tan x}$

In Exercises 32–34, the graph with the given equation is shown in a $\left[0, 2\pi, \dfrac{\pi}{2}\right]$ by $[-2, 2, 1]$ viewing rectangle.

   a. Describe the graph using another equation.

   b. Verify that the two equations are equivalent.

32. $y = \sin\left(x - \dfrac{3\pi}{2}\right)$

33. $y = \cos\left(x + \dfrac{\pi}{2}\right)$

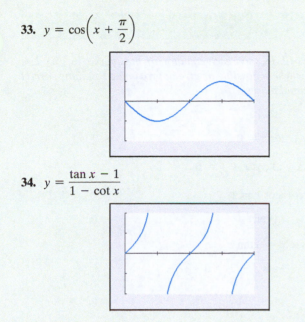

34. $y = \dfrac{\tan x - 1}{1 - \cot x}$

In Exercises 35–38, find the exact value of the following under the given conditions:

    a. $\sin(\alpha + \beta)$

    b. $\cos(\alpha - \beta)$

    c. $\tan(\alpha + \beta)$

    d. $\sin 2\alpha$

    e. $\cos\dfrac{\beta}{2}$

35. $\sin \alpha = \frac{3}{5}, 0 < \alpha < \frac{\pi}{2}$, and $\sin \beta = \frac{12}{13}, \frac{\pi}{2} < \beta < \pi$.

36. $\tan \alpha = \frac{4}{3}, \pi < \alpha < \frac{3\pi}{2}$, and $\tan \beta = \frac{5}{12}, 0 < \beta < \frac{\pi}{2}$.

37. $\tan \alpha = -3, \frac{\pi}{2} < \alpha < \pi$, and $\cot \beta = -3, \frac{3\pi}{2} < \beta < 2\pi$.

38. $\sin \alpha = -\frac{1}{3}, \pi < \alpha < \frac{3\pi}{2}$, and $\cos \beta = -\frac{1}{3}, \pi < \beta < \frac{3\pi}{2}$.

In Exercises 39–42, use double- and half-angle formulas to find the exact value of each expression.

39. $\cos^2 15° - \sin^2 15°$

40. $\dfrac{2 \tan \dfrac{5\pi}{12}}{1 - \tan^2 \dfrac{5\pi}{12}}$

41. $\sin 22.5°$

42. $\tan\dfrac{\pi}{12}$

3.4

In Exercises 43–44, express each product as a sum or difference.

43. $\sin 6x \sin 4x$

44. $\sin 7x \cos 3x$

In Exercises 45–46, express each sum or difference as a product. If possible, find this product's exact value.

45. $\sin 2x - \sin 4x$

46. $\cos 75° + \cos 15°$

In Exercises 47–48, verify each identity.

47. $\dfrac{\cos 3x + \cos 5x}{\cos 3x - \cos 5x} = \cot x \cot 4x$

48. $\dfrac{\sin 2x + \sin 6x}{\sin 2x - \sin 6x} = -\tan 4x \cot 2x$

49. The graph with the given equation is shown in a $\left[0, 2\pi, \dfrac{\pi}{2}\right]$ by $[-2, 2, 1]$ viewing rectangle.

$$y = \dfrac{\cos 3x + \cos x}{\sin 3x - \sin x}$$

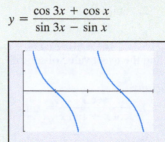

    a. Describe the graph using another equation.

    b. Verify that the two equations are equivalent.

3.5

In Exercises 50–53, find all solutions of each equation.

50. $\cos x = -\dfrac{1}{2}$ **51.** $\sin x = \dfrac{\sqrt{2}}{2}$

52. $2 \sin x + 1 = 0$ **53.** $\sqrt{3} \tan x - 1 = 0$

In Exercises 54–67, solve each equation on the interval $[0, 2\pi)$. Use exact values where possible or give approximate solutions correct to four decimal places.

54. $\cos 2x = -1$ **55.** $\sin 3x = 1$

56. $\tan\dfrac{x}{2} = -1$ **57.** $\tan x = 2 \cos x \tan x$

58. $\cos^2 x - 2 \cos x = 3$ **59.** $2 \cos^2 x - \sin x = 1$

60. $4 \sin^2 x = 1$ **61.** $\cos 2x - \sin x = 1$

62. $\sin 2x = \sqrt{3} \sin x$ **63.** $\sin x = \tan x$

64. $\sin x = -0.6031$ **65.** $5 \cos^2 x - 3 = 0$

66. $\sec^2 x = 4 \tan x - 2$ **67.** $2 \sin^2 x + \sin x - 2 = 0$

68. A ball on a spring is pulled 6 inches below its rest position and then released. After t seconds, the ball's distance, d, in inches from its rest position is given by

$$d = -6 \cos\dfrac{\pi}{2}t.$$

Find all values of t for which the ball is 3 inches below its rest position.

69. You are playing catch with a friend located 100 feet away. If you throw the ball with an initial velocity of $v_0 = 90$ feet per second, at what angle of elevation, θ, to the nearest degree should you direct your throw so that it can be caught easily? Use the formula

$$d = \dfrac{v_0^2}{16} \sin \theta \cos \theta.$$

CHAPTER 3 TEST

Use the following conditions to solve Exercises 1–4:

$$\sin \alpha = \tfrac{4}{5}, \tfrac{\pi}{2} < \alpha < \pi$$
$$\cos \beta = \tfrac{5}{13}, 0 < \beta < \tfrac{\pi}{2}$$

Find the exact value of each of the following.

1. $\cos(\alpha + \beta)$ 2. $\tan(\alpha - \beta)$

3. $\sin 2\alpha$ 4. $\cos \dfrac{\beta}{2}$

5. Use $105° = 135° - 30°$ to find the exact value of $\sin 105°$.

In Exercises 6–11, verify each identity.

6. $\cos x \csc x = \cot x$ 7. $\dfrac{\sec x}{\cot x + \tan x} = \sin x$

8. $1 - \dfrac{\cos^2 x}{1 + \sin x} = \sin x$ 9. $\cos\left(\theta + \dfrac{\pi}{2}\right) = -\sin \theta$

10. $\dfrac{\sin(\alpha - \beta)}{\sin \alpha \cos \beta} = 1 - \cot \alpha \tan \beta$

11. $\sin t \cos t(\tan t + \cot t) = 1$

In Exercises 12–18, solve each equation on the interval $[0, 2\pi)$. Use exact values where possible or give approximate solutions correct to four decimal places.

12. $\sin 3x = -\tfrac{1}{2}$

13. $\sin 2x + \cos x = 0$

14. $2 \cos^2 x - 3 \cos x + 1 = 0$

15. $2 \sin^2 x + \cos x = 1$

16. $\cos x = -0.8092$

17. $\tan x \sec x = 3 \tan x$

18. $\tan^2 x - 3 \tan x - 2 = 0$

CUMULATIVE REVIEW EXERCISES (CHAPTERS 1–3)

In Exercises 1–3, solve equation on the interval $[0, 2\pi)$.

1. $\cos 2x + 3 = 5 \cos x, \quad 0 \le x < 2\pi$

2. $\cos^2 x + \sin x + 1 = 0$

3. $\tan x + \sec^2 x = 3, \quad 0 \le x < 2\pi$

In Exercises 4–7, graph each equation.

4. $y = 3 \cos 2x, \quad -2\pi \le x \le 2\pi$

5. $y = \sin\left(2x + \dfrac{\pi}{2}\right), 0 \le x \le 2\pi$

6. $y = \dfrac{1}{2} \sec 2\pi x, 0 \le x \le 2$

7. $y = 2 \tan 3x$; Graph two complete cycles.

In Exercises 8–10, verify each identity.

8. $\dfrac{5 \cos x + 2}{\sin x} = 2 \csc x + 5 \cot x$

9. $\sin x \tan x + \cos x = \sec x$

10. $\cos\left(x + \dfrac{3\pi}{2}\right) = \sin x$

11. Convert $320°$ to radians.

In Exercises 12–14, find the exact value of each expression. Do not use a calculator.

12. $\sin 225°$ 13. $\tan \dfrac{\pi}{3} \csc \dfrac{\pi}{4}$ 14. $\cos[\tan^{-1}(-\tfrac{3}{4})]$

15. If C is a right angle in triangle ABC with $A = 23°$ and $a = 12$, solve the triangle.

16. From a point on the ground 12 feet from the base of a flagpole, the angle of elevation to the top of the pole is $53°$. Approximate the height of the flagpole to the nearest tenth of a foot.

In Exercises 17–18, let $\sin \theta = a$ and $\cos \theta = b$. Represent each expression as a single fraction in terms of a and b.

17. $\cot \theta - \csc \theta$

18. $\dfrac{2 \sin(-\theta)}{\sin 2\theta - \cos(-\theta)}$

19. Write $\sin(\cos^{-1}x)$ as an algebraic expression. Assume that $x > 0$ and x is in the domain of the inverse cosine

20. Find the area of the sector of a circle of radius 60 feet formed by a $15°$ central angle. Express area in terms of π. Then round your answer to two decimal places.

Laws of Sines and Cosines; Vectors

In 1995, this raging inferno came dangerously close to destroying your author's Northern California writer's cabin. Through the heroic efforts of local firefighters, the cabin was saved and Bob's writing career forged onward. How is this fiery tale related to trigonometry?

HERE'S WHERE YOU'LL FIND THIS APPLICATION:
The role that trigonometry plays in extinguishing the flames of wilderness fires before they explode into raging infernos is discussed in Section 4.1 and developed in Example 7 on page 219.

Section 4.1

The Law of Sines

What am I supposed to learn?

After studying this section, you should be able to:

1. Use the Law of Sines to solve oblique triangles.
2. Use the Law of Sines to solve, if possible, the triangle or triangles in the ambiguous case.
3. Find the area of an oblique triangle using the sine function.
4. Solve applied problems using the Law of Sines.

Point Reyes National Seashore, 40 miles north of San Francisco, consists of 75,000 acres with miles of pristine surf-pummeled beaches, forested ridges, and bays flanked by white cliffs. A few people, inspired by nature in the raw, live on private property adjoining the National Seashore. In 1995, a fire in the park burned 12,350 acres and destroyed 45 homes.

Fire is a necessary part of the life cycle in many wilderness areas. It is also an ongoing threat to those who choose to live surrounded by nature's unspoiled beauty. In this section, we see how trigonometry can be used to locate small wilderness fires before they become raging infernos. To do this, we begin by considering triangles other than right triangles.

GREAT QUESTION!

Does what I know about right triangles also apply to oblique triangles?

No. Up until now, our work with triangles has involved right triangles. **Do not apply relationships that are valid for right triangles to oblique triangles.** Avoid the error of using the Pythagorean Theorem, $a^2 + b^2 = c^2$, to find a missing side of an oblique triangle. This relationship among the three sides applies only to right triangles.

The Law of Sines and Its Derivation

An **oblique triangle** is a triangle that does not contain a right angle. **Figure 4.1** shows that an oblique triangle has either three acute angles or two acute angles and one obtuse angle. Notice that the angles are labeled A, B, and C. The sides opposite each angle are labeled as a, b, and c, respectively.

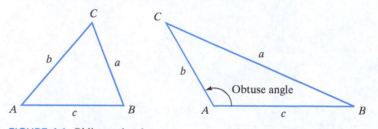

FIGURE 4.1 Oblique triangles

The relationships among the sides and angles of right triangles defined by the trigonometric functions are not valid for oblique triangles. Thus, we must observe and develop new relationships in order to work with oblique triangles.

Many relationships exist among the sides and angles in oblique triangles. One such relationship is called the **Law of Sines**.

GREAT QUESTION!

Do I have to write the Law of Sines with the sines in the denominator?

No. The Law of Sines can be expressed with the sines in the numerator:

$$\frac{\sin A}{a} = \frac{\sin B}{b} = \frac{\sin C}{c}.$$

The Law of Sines

If A, B, and C are the measures of the angles of a triangle, and a, b, and c are the lengths of the sides opposite these angles, then

$$\frac{a}{\sin A} = \frac{b}{\sin B} = \frac{c}{\sin C}.$$

The ratio of the length of the side of any triangle to the sine of the angle opposite that side is the same for all three sides of the triangle.

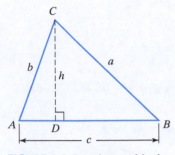

FIGURE 4.2 Drawing an altitude to prove the Law of Sines

To prove the Law of Sines, we draw an altitude of length h from one of the vertices of the triangle. In **Figure 4.2**, the altitude is drawn from vertex C. Two smaller triangles are formed, triangles ACD and BCD. Note that both are right triangles. Thus, we can use the definition of the sine of an angle of a right triangle.

$$\sin B = \frac{h}{a} \qquad \sin A = \frac{h}{b} \qquad \sin \theta = \frac{\text{opposite}}{\text{hypotenuse}}$$

$$h = a \sin B \qquad h = b \sin A \qquad \text{Solve each equation for } h.$$

Because we have found two expressions for h, we can set these expressions equal to each other.

$$a \sin B = b \sin A \qquad \text{Equate the expressions for } h.$$

$$\frac{a \sin B}{\sin A \sin B} = \frac{b \sin A}{\sin A \sin B} \qquad \text{Divide both sides by } \sin A \sin B.$$

$$\frac{a}{\sin A} = \frac{b}{\sin B} \qquad \text{Simplify.}$$

This proves part of the Law of Sines. If we use the same process and draw an altitude of length h from vertex A, we obtain the following result:

$$\frac{b}{\sin B} = \frac{c}{\sin C}.$$

When this equation is combined with the previous equation, we obtain the Law of Sines. Because the sine of an angle is equal to the sine of 180° minus that angle, the Law of Sines is derived in a similar manner if the oblique triangle contains an obtuse angle.

❶ Use the Law of Sines to solve oblique triangles.

Solving Oblique Triangles

Solving an oblique triangle means finding the lengths of its sides and the measurements of its angles. The Law of Sines can be used to solve a triangle in which one side and two angles are known. The three known measurements can be abbreviated using SAA (a side and two angles are known) or ASA (two angles and the side between them are known).

EXAMPLE 1   Solving an SAA Triangle Using the Law of Sines

Solve the triangle shown in **Figure 4.3** with $A = 46°$, $C = 63°$, and $c = 56$ inches. Round lengths of sides to the nearest tenth.

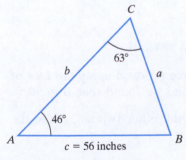

FIGURE 4.3 Solving an oblique SAA triangle

SOLUTION

We begin by finding B, the third angle of the triangle. We do not need the Law of Sines to do this. Instead, we use the fact that the sum of the measures of the interior angles of a triangle is 180°.

$$A + B + C = 180°$$
$$46° + B + 63° = 180° \qquad \text{Substitute the given values:}$$
$$\qquad\qquad\qquad\qquad\qquad A = 46° \text{ and } C = 63°.$$
$$109° + B = 180° \qquad \text{Add.}$$
$$B = 71° \qquad \text{Subtract 109° from both sides.}$$

When we use the Law of Sines, we must be given one of the three ratios. In this example, we are given c and C: $c = 56$ and $C = 63°$. Thus, we use the ratio $\dfrac{c}{\sin C}$, or $\dfrac{56}{\sin 63°}$, to find the other two sides. Use the Law of Sines to find a.

$$\frac{a}{\sin A} = \frac{c}{\sin C} \qquad \text{The ratio of any side to the sine of its opposite angle equals the ratio of any other side to the sine of its opposite angle.}$$

$$\frac{a}{\sin 46°} = \frac{56}{\sin 63°} \qquad A = 46°, c = 56, \text{ and } C = 63°.$$

$$a = \frac{56 \sin 46°}{\sin 63°} \qquad \text{Multiply both sides by } \sin 46° \text{ and solve for } a.$$

$$a \approx 45.2 \text{ inches} \qquad \text{Use a calculator.}$$

GREAT QUESTION!

Do I have to set up the Law of Sines with the unknown side in the upper left position?

No. However, many students find it easier to solve for the unknown sides when they are placed in the upper left position.

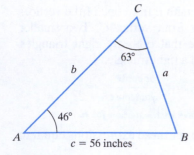

FIGURE 4.3 (repeated)

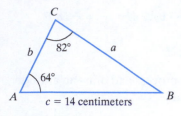

FIGURE 4.4

Use the Law of Sines again, this time to find b.

$$\frac{b}{\sin B} = \frac{c}{\sin C} \qquad \text{We use the given ratio, } \frac{c}{\sin C}, \text{ to find } b.$$

$$\frac{b}{\sin 71°} = \frac{56}{\sin 63°} \qquad \text{We found that } B = 71°. \text{ We are given } c = 56 \text{ and } C = 63°.$$

$$b = \frac{56 \sin 71°}{\sin 63°} \qquad \text{Multiply both sides by } \sin 71° \text{ and solve for } b.$$

$$b \approx 59.4 \text{ inches} \qquad \text{Use a calculator.}$$

The solution is $B = 71°$, $a \approx 45.2$ inches, and $b \approx 59.4$ inches. • • •

✓ **Check Point 1** Solve the triangle shown in **Figure 4.4** with $A = 64°$, $C = 82°$, and $c = 14$ centimeters. Round as in Example 1.

EXAMPLE 2 Solving an ASA Triangle Using the Law of Sines

Solve triangle ABC if $A = 50°$, $C = 33.5°$, and $b = 76$. Round measures to the nearest tenth.

SOLUTION

We begin by drawing a picture of triangle ABC and labeling it with the given information. **Figure 4.5** shows the triangle that we must solve. We begin by finding B.

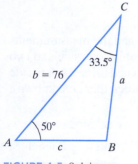

FIGURE 4.5 Solving an ASA triangle

$$A + B + C = 180° \qquad \text{The sum of the measures of a triangle's interior angles is 180°.}$$

$$50° + B + 33.5° = 180° \qquad A = 50° \text{ and } C = 33.5°.$$

$$83.5° + B = 180° \qquad \text{Add.}$$

$$B = 96.5° \qquad \text{Subtract 83.5° from both sides.}$$

Keep in mind that we must be given one of the three ratios to apply the Law of Sines. In this example, we are given that $b = 76$ and we found that $B = 96.5°$. Thus, we use the ratio $\frac{b}{\sin B}$, or $\frac{76}{\sin 96.5°}$, to find the other two sides. Use the Law of Sines to find a and c.

Find a:

$$\frac{a}{\sin A} = \frac{b}{\sin B} \qquad \text{This is the known ratio.}$$

$$\frac{a}{\sin 50°} = \frac{76}{\sin 96.5°}$$

$$a = \frac{76 \sin 50°}{\sin 96.5°} \approx 58.6$$

Find c:

$$\frac{c}{\sin C} = \frac{b}{\sin B}$$

$$\frac{c}{\sin 33.5°} = \frac{76}{\sin 96.5°}$$

$$c = \frac{76 \sin 33.5°}{\sin 96.5°} \approx 42.2$$

The solution is $B = 96.5°$, $a \approx 58.6$, and $c \approx 42.2$. • • •

✓ **Check Point 2** Solve triangle ABC if $A = 40°$, $C = 22.5°$, and $b = 12$. Round as in Example 2.

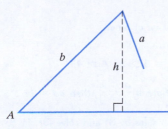

② Use the Law of Sines to solve, if possible, the triangle or triangles in the ambiguous case.

FIGURE 4.6 Given SSA, no triangle may result.

The Ambiguous Case (SSA)

If we are given two sides and an angle opposite one of them (SSA), does this determine a unique triangle? Can we solve this case using the Law of Sines? Such a case is called the **ambiguous case** because the given information may result in one triangle, two triangles, or no triangle at all. For example, in **Figure 4.6**, we are given a, b, and A. Because a is shorter than h, it is not long enough to form a triangle. The number of possible triangles, if any, that can be formed in the SSA case depends on h, the length of the altitude, where $h = b \sin A$.

The Ambiguous Case (SSA)

Consider a triangle in which a, b, and A are given. This information may result in

One Triangle

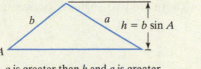

a is greater than h and a is greater than b. One triangle is formed.

One Right Triangle

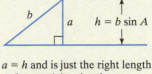

$a = h$ and is just the right length to form a right triangle.

No Triangle

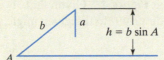

a is less than h and is not long enough to form a triangle.

Two Triangles

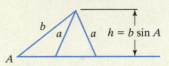

a is greater than h and a is less than b. Two distinct triangles are formed.

In an SSA situation, it is not necessary to draw an accurate sketch like those shown in the box. The Law of Sines determines the number of triangles, if any, and gives the solution for each triangle.

A Brief Review • The Cross-Products Principle for Proportions

- When solving for a variable in a proportion's denominator, it is convenient to apply the cross-products principle:

$$\text{If } \frac{a}{b} = \frac{c}{d}, \text{ then } ad = bc \ (b \neq 0 \text{ and } d \neq 0).$$

EXAMPLE Solve for x: $\dfrac{52}{75} = \dfrac{71}{x}$.

$$52x = (75)(71) \qquad \text{Cross multiply using the cross-products principle.}$$

$$x = \frac{(75)(71)}{52} \approx 102.4$$

EXAMPLE 3 Solving an SSA Triangle Using the Law of Sines (One Solution)

Solve triangle ABC if $A = 43°$, $a = 81$, and $b = 62$. Round lengths of sides to the nearest tenth and angle measures to the nearest degree.

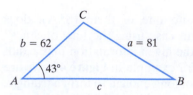

FIGURE 4.7 Solving an SSA triangle; the ambiguous case

SOLUTION

Using $A = 43°$, $a = 81$, and $b = 62$, we begin with the sketch in **Figure 4.7**. The known ratio is $\dfrac{a}{\sin A}$, or $\dfrac{81}{\sin 43°}$. Because side b is given, we use the Law of Sines to find angle B.

$$\frac{a}{\sin A} = \frac{b}{\sin B} \qquad \text{Apply the Law of Sines.}$$

$$\frac{81}{\sin 43°} = \frac{62}{\sin B} \qquad a = 81, b = 62, \text{ and } A = 43°.$$

$$81 \sin B = 62 \sin 43° \qquad \text{Cross multiply: If } \frac{a}{b} = \frac{c}{d}, \text{ then } ad = bc.$$

$$\sin B = \frac{62 \sin 43°}{81} \qquad \text{Divide both sides by 81 and solve for } \sin B.$$

$$\sin B \approx 0.5220 \qquad \text{Use a calculator.}$$

There are two angles B between $0°$ and $180°$ for which $\sin B \approx 0.5220$.

$$B_1 \approx 31° \qquad\qquad B_2 \approx 180° - 31° = 149°$$

> Obtain the acute angle with your calculator in degree mode: $\sin^{-1} 0.5220$.

> The sine is positive in quadrant II.

Look at **Figure 4.7**. Given that $A = 43°$, can you see that $B_2 \approx 149°$ is impossible? By adding $149°$ to the given angle, $43°$, we exceed a $180°$ sum:

$$43° + 149° = 192°.$$

Thus, the only possibility is that $B_1 \approx 31°$. We find C using this approximation for B_1 and the measure that was given for A: $A = 43°$.

$$C = 180° - B_1 - A \approx 180° - 31° - 43° = 106°$$

Side c that lies opposite this $106°$ angle can now be found using the Law of Sines.

$$\frac{c}{\sin C} = \frac{a}{\sin A} \qquad \text{Apply the Law of Sines.}$$

$$\frac{c}{\sin 106°} = \frac{81}{\sin 43°} \qquad a = 81, C \approx 106°, \text{ and } A = 43°.$$

$$c = \frac{81 \sin 106°}{\sin 43°} \approx 114.2 \qquad \begin{array}{l}\text{Multiply both sides by } \sin 106° \\ \text{and solve for } c.\end{array}$$

There is one triangle and the solution is B_1 (or B) $\approx 31°$, $C \approx 106°$, and $c \approx 114.2$.

● ● ●

✓ **Check Point 3** Solve triangle ABC if $A = 57°$, $a = 33$, and $b = 26$. Round as in Example 3.

EXAMPLE 4 Solving an SSA Triangle Using the Law of Sines (No Solution)

Solve triangle ABC if $A = 75°$, $a = 51$, and $b = 71$.

SOLUTION

The known ratio is $\dfrac{a}{\sin A}$, or $\dfrac{51}{\sin 75°}$. Because side b is given, we use the Law of Sines to find angle B.

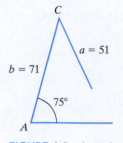

FIGURE 4.8 a is not long enough to form a triangle.

$$\frac{a}{\sin A} = \frac{b}{\sin B}$$ Use the Law of Sines.

$$\frac{51}{\sin 75°} = \frac{71}{\sin B}$$ Substitute the given values.

$$51 \sin B = 71 \sin 75°$$ Cross multiply: If $\frac{a}{b} = \frac{c}{d}$, then $ad = bc$.

$$\sin B = \frac{71 \sin 75°}{51} \approx 1.34$$ Divide by 51 and solve for $\sin B$.

Because the sine can never exceed 1, there is no angle B for which $\sin B \approx 1.34$. There is no triangle with the given measurements, as illustrated in **Figure 4.8**. • • •

☑ **Check Point 4** Solve triangle ABC if $A = 50°$, $a = 10$, and $b = 20$.

EXAMPLE 5   Solving an SSA Triangle Using the Law of Sines (Two Solutions)

Solve triangle ABC if $A = 40°$, $a = 54$, and $b = 62$. Round lengths of sides to the nearest tenth and angle measures to the nearest degree.

SOLUTION

The known ratio is $\frac{a}{\sin A}$, or $\frac{54}{\sin 40°}$. We use the Law of Sines to find angle B.

$$\frac{a}{\sin A} = \frac{b}{\sin B}$$ Use the Law of Sines.

$$\frac{54}{\sin 40°} = \frac{62}{\sin B}$$ Substitute the given values.

$$54 \sin B = 62 \sin 40°$$ Cross multiply: If $\frac{a}{b} = \frac{c}{d}$, then $ad = bc$.

$$\sin B = \frac{62 \sin 40°}{54} \approx 0.7380$$ Divide by 54 and solve for $\sin B$.

There are two angles B between $0°$ and $180°$ for which $\sin B \approx 0.7380$.

$$B_1 \approx 48° \qquad B_2 \approx 180° - 48° = 132°$$

Find $\sin^{-1} 0.7380$ with your calculator.   The sine is positive in quadrant II.

GREAT QUESTION!

Do I have to draw the two triangles in Figure 4.9 to solve Example 5?

The two triangles shown in **Figure 4.9** are helpful in organizing the solutions. However, if you keep track of the two triangles, one with the given information and $B_1 = 48°$, and the other with the given information and $B_2 = 132°$, you do not have to draw the figure to solve the triangles.

If you add either angle to the given angle, $40°$, the sum does not exceed $180°$. Thus, there are two triangles with the given conditions, $A = 40°$, $a = 54$, and $b = 62$, shown in **Figure 4.9(a)**. The triangles, AB_1C_1 and AB_2C_2, are shown separately in **Figure 4.9(b)** and **Figure 4.9(c)**.

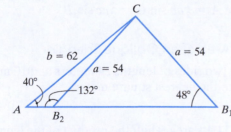

(a) Two triangles are possible with $A = 40°$, $a = 54$, and $b = 62$.

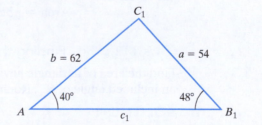

(b) In one possible triangle, $B_1 = 48°$.

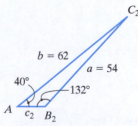

(c) In the second possible triangle, $B_2 = 132°$.

FIGURE 4.9

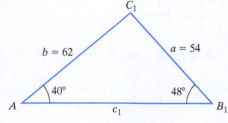

(b) In one possible triangle, $B_1 = 48°$.

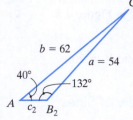

(c) In the second possible triangle, $B_2 = 132°$.

FIGURE 4.9 (b) and (c) (repeated)

We find angles C_1 and C_2 using a 180° angle sum in each of the two triangles.

$$C_1 = 180° - A - B_1 \qquad C_2 = 180° - A - B_2$$
$$\approx 180° - 40° - 48° \qquad \approx 180° - 40° - 132°$$
$$= 92° \qquad\qquad = 8°$$

We use the Law of Sines to find c_1 and c_2.

$$\frac{c_1}{\sin C_1} = \frac{a}{\sin A} \qquad\qquad \frac{c_2}{\sin C_2} = \frac{a}{\sin A}$$

$$\frac{c_1}{\sin 92°} = \frac{54}{\sin 40°} \qquad\qquad \frac{c_2}{\sin 8°} = \frac{54}{\sin 40°}$$

$$c_1 = \frac{54 \sin 92°}{\sin 40°} \approx 84.0 \qquad c_2 = \frac{54 \sin 8°}{\sin 40°} \approx 11.7$$

There are two triangles. In one triangle, the solution is $B_1 \approx 48°, C_1 \approx 92°$, and $c_1 \approx 84.0$. In the other triangle, $B_2 \approx 132°, C_2 \approx 8°$, and $c_2 \approx 11.7$.   •••

⊘ **Check Point 5** Solve triangle ABC if $A = 35°, a = 12$, and $b = 16$. Round as in Example 5.

The Area of an Oblique Triangle

A formula for the area of an oblique triangle can be obtained using the procedure for proving the Law of Sines. We draw an altitude of length h from one of the vertices of the triangle, as shown in **Figure 4.10**. We apply the definition of the sine of angle A, $\dfrac{\text{opposite}}{\text{hypotenuse}}$, in right triangle ACD:

$$\sin A = \frac{h}{b}, \quad \text{so} \quad h = b \sin A.$$

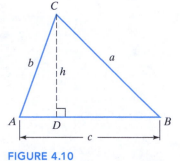

FIGURE 4.10

The area of a triangle is $\frac{1}{2}$ the product of any side and the altitude drawn to that side. Using the altitude h in **Figure 4.10**, we have

$$\text{Area} = \frac{1}{2} ch = \frac{1}{2} cb \sin A.$$

> Use the result from above:
> $h = b \sin A$.

This result, Area $= \frac{1}{2} cb \sin A$, or $\frac{1}{2} bc \sin A$, indicates that the area of the triangle is one-half the product of b and c times the sine of their included angle. If we draw altitudes from the other two vertices, we see that we can use any two sides to compute the area.

❸ Find the area of an oblique triangle using the sine function.

Area of an Oblique Triangle

The area of a triangle equals one-half the product of the lengths of two sides times the sine of their included angle. In **Figure 4.10**, this wording can be expressed by the formulas

$$\text{Area} = \tfrac{1}{2} bc \sin A = \tfrac{1}{2} ab \sin C = \tfrac{1}{2} ac \sin B.$$

EXAMPLE 6   Finding the Area of an Oblique Triangle

Find the area of a triangle having two sides of lengths 24 meters and 10 meters and an included angle of 62°. Round to the nearest square meter.

SOLUTION

The triangle is shown in **Figure 4.11**. Its area is half the product of the lengths of the two sides times the sine of the included angle.

$$\text{Area} = \tfrac{1}{2}(24)(10)(\sin 62°) \approx 106$$

The area of the triangle is approximately 106 square meters.   •••

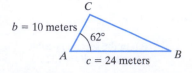

FIGURE 4.11 Finding the area of an SAS triangle

✓ **Check Point 6** Find the area of a triangle having two sides of lengths 8 meters and 12 meters and an included angle of 135°. Round to the nearest square meter.

④ Solve applied problems using the Law of Sines.

Applications of the Law of Sines

We have seen how the trigonometry of right triangles can be used to solve many different kinds of applied problems. The Law of Sines enables us to work with triangles that are not right triangles. As a result, this law can be used to solve problems involving surveying, engineering, astronomy, navigation, and the environment. Example 7 illustrates the use of the Law of Sines in detecting potentially devastating fires.

EXAMPLE 7   An Application of the Law of Sines

Two fire-lookout stations are 20 miles apart, with station B directly east of station A. Both stations spot a fire on a mountain to the north. The bearing from station A to the fire is N50°E (50° east of north). The bearing from station B to the fire is N36°W (36° west of north). How far, to the nearest tenth of a mile, is the fire from station A?

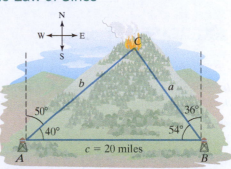

FIGURE 4.12

SOLUTION

Figure 4.12 shows the information given in the problem. The distance from station A to the fire is represented by b. Notice that the angles describing the bearing from each station to the fire, 50° and 36°, are not interior angles of triangle ABC. Using a north–south line, the interior angles are found as follows:

$$A = 90° - 50° = 40° \qquad B = 90° - 36° = 54°.$$

To find b using the Law of Sines, we need a known side and an angle opposite that side. Because $c = 20$ miles, we find angle C using a 180° angle sum in the triangle. Thus,

$$C = 180° - A - B = 180° - 40° - 54° = 86°.$$

The ratio $\dfrac{c}{\sin C}$, or $\dfrac{20}{\sin 86°}$, is now known. We use this ratio and the Law of Sines to find b.

$$\frac{b}{\sin B} = \frac{c}{\sin C} \qquad \text{Use the Law of Sines.}$$

$$\frac{b}{\sin 54°} = \frac{20}{\sin 86°} \qquad c = 20,\ B = 54°,\ \text{and}\ C = 86°.$$

$$b = \frac{20 \sin 54°}{\sin 86°} \approx 16.2 \qquad \text{Multiply both sides by } \sin 54° \text{ and solve for } b.$$

The fire is approximately 16.2 miles from station A.   • • •

✓ **Check Point 7** Two fire-lookout stations are 13 miles apart, with station B directly east of station A. Both stations spot a fire. The bearing of the fire from station A is N35°E and the bearing of the fire from station B is N49°W. How far, to the nearest tenth of a mile, is the fire from station B?

ACHIEVING SUCCESS

Avoid coursus interruptus.

Now that you're well into trigonometry, don't interrupt the sequence until you have completed all your required math classes. You'll have better results if you take your math courses without a break. If you start, stop, and start again, it's easy to forget what you've learned and lose your momentum.

CONCEPT AND VOCABULARY CHECK

Fill in each blank so that the resulting statement is true.

1. A triangle that does not contain a right angle is called a/an _____ triangle. Solving such a triangle means finding the lengths of its _____ and the measurements of its _____.

2. If A, B, and C are the measures of the angles of a triangle, and a, b, and c are the lengths of the sides opposite these angles, then the Law of Sines states that

   _____.

3. We can always use the Law of Sines to find missing parts of triangles in which one _____ and two _____ are known.

4. True or false: A triangle in which two sides and an angle opposite one of them are given (SSA) always results in at least one triangle. _____

5. If two sides a and b and the included angle C are known in a triangle, then the area of the triangle is found using the formula Area = _____.

EXERCISE SET 4.1

Practice Exercises

In Exercises 1–8, solve each triangle. Round lengths of sides to the nearest tenth and angle measures to the nearest degree.

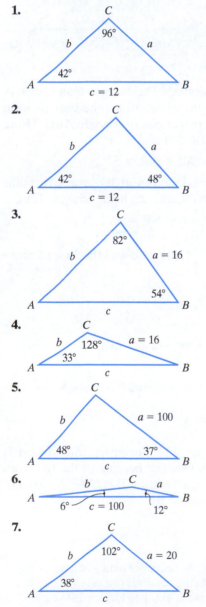

1.
2.
3.
4.
5.
6.
7.

8.

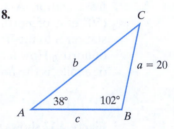

In Exercises 9–16, solve each triangle. Round lengths to the nearest tenth and angle measures to the nearest degree.

9. $A = 44°$, $B = 25°$, $a = 12$
10. $A = 56°$, $C = 24°$, $a = 22$
11. $B = 85°$, $C = 15°$, $b = 40$
12. $A = 85°$, $B = 35°$, $c = 30$
13. $A = 115°$, $C = 35°$, $c = 200$
14. $B = 5°$, $C = 125°$, $b = 200$
15. $A = 65°$, $B = 65°$, $c = 6$
16. $B = 80°$, $C = 10°$, $a = 8$

In Exercises 17–32, two sides and an angle (SSA) of a triangle are given. Determine whether the given measurements produce one triangle, two triangles, or no triangle at all. Solve each triangle that results. Round to the nearest tenth and the nearest degree for sides and angles, respectively.

17. $a = 20$, $b = 15$, $A = 40°$
18. $a = 30$, $b = 20$, $A = 50°$
19. $a = 10$, $c = 8.9$, $A = 63°$
20. $a = 57.5$, $c = 49.8$, $A = 136°$
21. $a = 42.1$, $c = 37$, $A = 112°$
22. $a = 6.1$, $b = 4$, $A = 162°$
23. $a = 10$, $b = 40$, $A = 30°$
24. $a = 10$, $b = 30$, $A = 150°$
25. $a = 16$, $b = 18$, $A = 60°$
26. $a = 30$, $b = 40$, $A = 20°$
27. $a = 12$, $b = 16.1$, $A = 37°$
28. $a = 7$, $b = 28$, $A = 12°$
29. $a = 22$, $c = 24.1$, $A = 58°$
30. $a = 95$, $c = 125$, $A = 49°$
31. $a = 9.3$, $b = 41$, $A = 18°$
32. $a = 1.4$, $b = 2.9$, $A = 142°$

In Exercises 33–38, find the area of the triangle having the given measurements. Round to the nearest square unit.

33. $A = 48°, b = 20$ feet, $c = 40$ feet
34. $A = 22°, b = 20$ feet, $c = 50$ feet
35. $B = 36°, a = 3$ yards, $c = 6$ yards
36. $B = 125°, a = 8$ yards, $c = 5$ yards
37. $C = 124°, a = 4$ meters, $b = 6$ meters
38. $C = 102°, a = 16$ meters, $b = 20$ meters

Practice Plus

In Exercises 39–40, find h to the nearest tenth.

39.

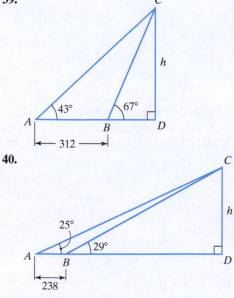

40.

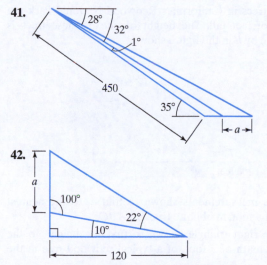

In Exercises 41–42, find a to the nearest tenth.

41.

42.

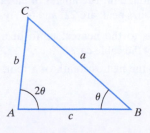

In Exercises 43–44, use the given measurements to solve the following triangle. Round lengths of sides to the nearest tenth and angle measures to the nearest degree.

43. $a = 300, b = 200$
44. $a = 400, b = 300$

In Exercises 45–46, find the area of the triangle with the given vertices. Round to the nearest square unit.

45. $(-3, -2), (2, -2), (1, 2)$ **46.** $(-2, -3), (-2, 2), (2, 1)$

Application Exercises

47. Two fire-lookout stations are 10 miles apart, with station B directly east of station A. Both stations spot a fire. The bearing of the fire from station A is N25°E and the bearing of the fire from station B is N56°W. How far, to the nearest tenth of a mile, is the fire from each lookout station?

48. The Federal Communications Commission is attempting to locate an illegal radio station. It sets up two monitoring stations, A and B, with station B 40 miles east of station A. Station A measures the illegal signal from the radio station as coming from a direction of 48° east of north. Station B measures the signal as coming from a point 34° west of north. How far is the illegal radio station from monitoring stations A and B? Round to the nearest tenth of a mile.

49. The figure shows a 1200-yard-long sand beach and an oil platform in the ocean. The angle made with the platform from one end of the beach is 85° and from the other end is 76°. Find the distance of the oil platform, to the nearest tenth of a yard, from each end of the beach.

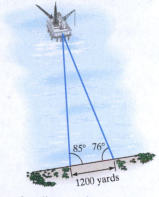

50. A surveyor needs to determine the distance between two points that lie on opposite banks of a river. The figure shows that 300 yards are measured along one bank. The angles from each end of this line segment to a point on the opposite bank are 62° and 53°. Find the distance between A and B to the nearest tenth of a yard.

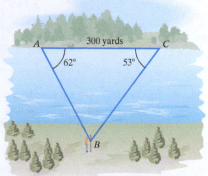

51. The Leaning Tower of Pisa in Italy leans at an angle of about 84.7°. The figure shows that 171 feet from the base of the tower, the angle of elevation to the top is 50°. Find the distance, to the nearest tenth of a foot, from the base to the top of the tower.

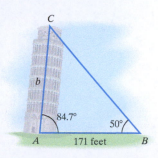

52. A pine tree growing on a hillside makes a 75° angle with the hill. From a point 80 feet up the hill, the angle of elevation to the top of the tree is 62° and the angle of depression to the bottom is 23°. Find, to the nearest tenth of a foot, the height of the tree.

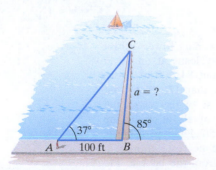

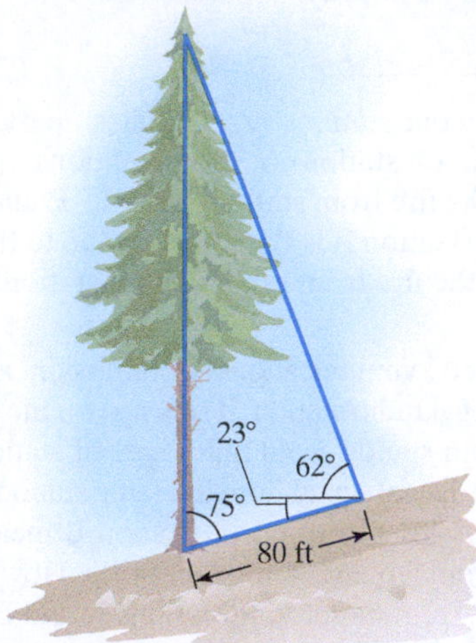

53. The figure shows a shot-put ring. The shot is tossed from *A* and lands at *B*. Using modern electronic equipment, the distance of the toss can be measured without the use of measuring tapes. When the shot lands at *B*, an electronic transmitter placed at *B* sends a signal to a device in the official's booth above the track. The device determines the angles at *B* and *C*. At a track meet, the distance from the official's booth to the shot-put ring is 562 feet. If $B = 85.3°$ and $C = 5.7°$, determine the length of the toss to the nearest tenth of a foot.

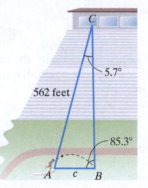

54. A pier forms an 85° angle with a straight shore. At a distance of 100 feet from the pier, the line of sight to the tip forms a 37° angle. Find the length of the pier to the nearest tenth of a foot.

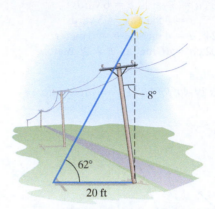

55. When the angle of elevation of the Sun is 62°, a telephone pole that is tilted at an angle of 8° directly away from the Sun casts a shadow 20 feet long. Determine the length of the pole to the nearest tenth of a foot.

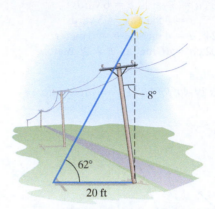

56. A leaning wall is inclined 6° from the vertical. At a distance of 40 feet from the wall, the angle of elevation to the top is 22°. Find the height of the wall to the nearest tenth of a foot.

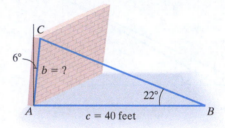

57. Redwood trees in California's Redwood National Park are hundreds of feet tall. The height of one of these trees is represented by *h* in the figure shown.

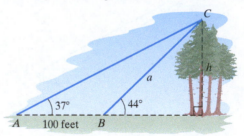

a. Use the measurements shown to find *a*, to the nearest tenth of a foot, in oblique triangle *ABC*.

b. Use the right triangle shown to find the height, to the nearest tenth of a foot, of a typical redwood tree in the park.

58. The figure at the top of the next page shows a cable car that carries passengers from *A* to *C*. Point *A* is 1.6 miles from the base of the mountain. The angles of elevation from *A* and *B* to the mountain's peak are 22° and 66°, respectively.

a. Determine, to the nearest tenth of a foot, the distance covered by the cable car.

b. Find *a*, to the nearest tenth of a foot, in oblique triangle *ABC*.

c. Use the right triangle to find the height of the mountain to the nearest tenth of a foot.

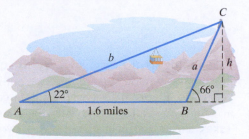

59. Lighthouse B is 7 miles west of lighthouse A. A boat leaves A and sails 5 miles. At this time, it is sighted from B. If the bearing of the boat from B is N62°E, how far from B is the boat? Round to the nearest tenth of a mile.

60. After a wind storm, you notice that your 16-foot flagpole may be leaning, but you are not sure. From a point on the ground 15 feet from the base of the flagpole, you find that the angle of elevation to the top is 48°. Is the flagpole leaning? If so, find the acute angle, to the nearest degree, that the flagpole makes with the ground.

Explaining the Concepts

61. What is an oblique triangle?

62. Without using symbols, state the Law of Sines in your own words.

63. Briefly describe how the Law of Sines is proved.

64. What does it mean to solve an oblique triangle?

65. What do the abbreviations SAA and ASA mean?

66. Why is SSA called the ambiguous case?

67. How is the sine function used to find the area of an oblique triangle?

68. Write an original problem that can be solved using the Law of Sines. Then solve the problem.

69. Use Exercise 53 to describe how the Law of Sines is used for throwing events at track and field meets. Why aren't tape measures used to determine tossing distance?

70. You are cruising in your boat parallel to the coast, looking at a lighthouse. Explain how you can use your boat's speed and a device for measuring angles to determine the distance at any instant from your boat to the lighthouse.

Critical Thinking Exercises

Make Sense? *In Exercises 71–74, determine whether each statement makes sense or does not make sense, and explain your reasoning.*

71. I began using the Law of Sines to solve an oblique triangle in which the measures of two sides and the angle between them were known.

72. If I know the measures of the sides and angles of an oblique triangle, I have three ways of determining the triangle's area.

73. When solving an SSA triangle using the Law of Sines, my calculator gave me both the acute and obtuse angles B for which sin B = 0.5833.

74. Under certain conditions, a fire can be located by superimposing a triangle onto the situation and applying the Law of Sines.

75. If you are given two sides of a triangle and their included angle, you can find the triangle's area. Can the Law of Sines be used to solve the triangle with this given information? Explain your answer.

76. Two buildings of equal height are 800 feet apart. An observer on the street between the buildings measures the angles of elevation to the tops of the buildings as 27° and 41°. How high, to the nearest foot, are the buildings?

77. The figure shows the design for the top of the wing of a jet fighter. The fuselage is 5 feet wide. Find the wing span *CC'*.

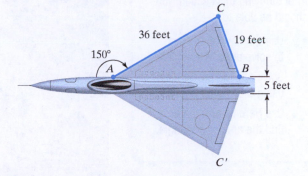

Retaining the Concepts

78. Use $\cos\dfrac{5\pi}{12} = \cos\left(\dfrac{\pi}{6} + \dfrac{\pi}{4}\right)$ and the formula for the cosine of the sum of two angles to find the exact value of $\cos\dfrac{5\pi}{12}$.

(Section 3.2, Example 4)

79. Determine the amplitude, period, and phase shift of $y = -2\cos(2x - \frac{\pi}{2})$. Then graph one period of the function. (Section 2.1, Example 6)

80. An object moves in simple harmonic motion described by $d = 6\cos\frac{3\pi}{2}t$, where *t* is measured in seconds and *d* in inches. Find:

a. the maximum displacement

b. the frequency

c. the time required for one cycle.

(Section 2.4, Example 8)

Preview Exercises

Exercises 81–83 will help you prepare for the material covered in the next section.

81. Find the obtuse angle *B*, rounded to the nearest degree, satisfying

$$\cos B = \frac{6^2 + 4^2 - 9^2}{2 \cdot 6 \cdot 4}.$$

82. Simplify and round to the nearest whole number:

$$\sqrt{26(26 - 12)(26 - 16)(26 - 24)}.$$

83. Two airplanes leave an airport at the same time on different runways. The first plane, flying on a bearing of N66°W, travels 650 miles after two hours. The second plane, flying on a bearing of S26°W, travels 600 miles after two hours. Illustrate the situation with an oblique triangle that shows how far apart the airplanes will be after two hours.

Section 4.2

The Law of Cosines

What am I supposed to learn?

After studying this section, you should be able to:

1. Use the Law of Cosines to solve oblique triangles.
2. Solve applied problems using the Law of Cosines.
3. Use Heron's formula to find the area of a triangle.

Paleontologists use trigonometry to study the movements made by dinosaurs millions of years ago. **Figure 4.13**, based on data collected at Dinosaur Valley State Park in Glen Rose, Texas, shows footprints made by a two-footed carnivorous (meat-eating) dinosaur and the hindfeet of a herbivorous (plant-eating) dinosaur.

For each dinosaur, the figure indicates the *pace* and the *stride*. The pace is the distance from the left footprint to the right footprint, and vice versa. The stride is the distance from the left footprint to the next left footprint or from the right footprint to the next right footprint. Also shown in **Figure 4.13** is the pace angle, designated by θ. Notice that neither dinosaur moves with a pace angle of 180°, meaning that the footprints are directly in line. The footprints show a "zig-zig" pattern that is numerically described by the pace angle. A dinosaur that is an efficient walker has a pace angle close to 180°, minimizing zig-zag motion and maximizing forward motion.

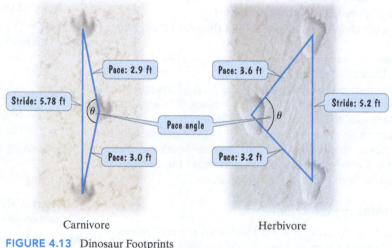

Carnivore · Herbivore

FIGURE 4.13 Dinosaur Footprints
Source: Glen J. Kuban, *An Overview of Dinosaur Tracking*

How can we determine the pace angles for the carnivore and the herbivore in **Figure 4.13**? Problems such as this, in which we know the measures of three sides of a triangle and we need to find the measurement of a missing angle, cannot be solved by the Law of Sines. To numerically describe which dinosaur in **Figure 4.13** made more forward progress with each step, we turn to the Law of Cosines.

The Law of Cosines and Its Derivation

We now look at another relationship that exists among the sides and angles in an oblique triangle. The **Law of Cosines** is used to solve triangles in which two sides and the included angle (SAS) are known, or those in which three sides (SSS) are known.

DISCOVERY

What happens to the Law of Cosines

$$c^2 = a^2 + b^2 - 2ab \cos C$$

if $C = 90°$? What familiar theorem do you obtain?

The Law of Cosines

If A, B, and C are the measures of the angles of a triangle, and a, b, and c are the lengths of the sides opposite these angles, then

$$a^2 = b^2 + c^2 - 2bc \cos A$$
$$b^2 = a^2 + c^2 - 2ac \cos B$$
$$c^2 = a^2 + b^2 - 2ab \cos C.$$

The square of a side of a triangle equals the sum of the squares of the other two sides minus twice their product times the cosine of their included angle.

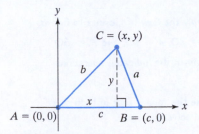

FIGURE 4.14

To prove the Law of Cosines, we place triangle ABC in a rectangular coordinate system. **Figure 4.14** shows a triangle with three acute angles. The vertex A is at the origin and side c lies along the positive x-axis. The coordinates of C are (x, y). Using the right triangle that contains angle A, we apply the definitions of the cosine and the sine.

$$\cos A = \frac{x}{b} \qquad \sin A = \frac{y}{b}$$

$$x = b \cos A \qquad y = b \sin A \qquad \text{Multiply both sides of each equation by } b \text{ and solve for } x \text{ and } y, \text{ respectively.}$$

Thus, the coordinates of C are $(x, y) = (b \cos A, b \sin A)$. Although triangle ABC in **Figure 4.14** shows angle A as an acute angle, if A were obtuse, the coordinates of C would still be $(b \cos A, b \sin A)$. This means that our proof applies to both kinds of oblique triangles.

We now apply the distance formula to the side of the triangle with length a. Notice that a is the distance from (x, y) to $(c, 0)$.

$$a = \sqrt{(x - c)^2 + (y - 0)^2} \qquad \text{Use the distance formula.}$$
$$a^2 = (x - c)^2 + y^2 \qquad \text{Square both sides of the equation.}$$
$$a^2 = (b \cos A - c)^2 + (b \sin A)^2 \qquad x = b \cos A \text{ and } y = b \sin A$$
$$a^2 = b^2 \cos^2 A - 2bc \cos A + c^2 + b^2 \sin^2 A \qquad \text{Square the two expressions.}$$
$$a^2 = b^2 \sin^2 A + b^2 \cos^2 A + c^2 - 2bc \cos A \qquad \text{Rearrange terms.}$$
$$a^2 = b^2(\sin^2 A + \cos^2 A) + c^2 - 2bc \cos A \qquad \text{Factor } b^2 \text{ from the first two terms.}$$
$$a^2 = b^2 + c^2 - 2bc \cos A \qquad \sin^2 A + \cos^2 A = 1$$

The resulting equation is one of the three formulas for the Law of Cosines. The other two formulas are derived in a similar manner.

 Use the Law of Cosines to solve oblique triangles.

Solving Oblique Triangles

If you are given two sides and an included angle (SAS) of an oblique triangle, none of the three ratios in the Law of Sines is known. This means that we do not begin solving the triangle using the Law of Sines. Instead, we apply the Law of Cosines and the following procedure:

Solving an SAS Triangle

1. Use the Law of Cosines to find the side opposite the given angle.
2. Use the Law of Sines to find the angle opposite the shorter of the two given sides. This angle is always acute.
3. Find the third angle by subtracting the measure of the given angle and the angle found in step 2 from 180°.

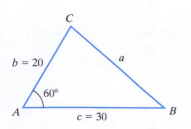

FIGURE 4.15 Solving an SAS triangle

EXAMPLE 1 Solving an SAS Triangle

Solve the triangle in **Figure 4.15** with $A = 60°$, $b = 20$, and $c = 30$. Round lengths of sides to the nearest tenth and angle measures to the nearest degree.

SOLUTION

We are given two sides and an included angle. Therefore, we apply the three-step procedure for solving an SAS triangle.

Step 1 Use the Law of Cosines to find the side opposite the given angle. Thus, we will find a.

> In this example, we know the exact value of $\cos 60°$: $\cos 60° = 0.5$. If the exact value of the cosine is not available, you can calculate $b^2 + c^2 - 2bc \cos A$ in one step with a calculator.

$$a^2 = b^2 + c^2 - 2bc \cos A \qquad \text{Apply the Law of Cosines to find } a.$$
$$a^2 = 20^2 + 30^2 - 2(20)(30) \cos 60° \qquad b = 20, c = 30, \text{ and } A = 60°.$$
$$= 400 + 900 - 1200(0.5) \qquad \text{Perform the indicated operations.}$$
$$= 700$$
$$a = \sqrt{700} \approx 26.5 \qquad \text{Take the square root of both sides and solve for } a.$$

Step 2 Use the Law of Sines to find the angle opposite the shorter of the two given sides. This angle is always acute. The shorter of the two given sides is $b = 20$. Thus, we will find acute angle B.

$$\frac{b}{\sin B} = \frac{a}{\sin A} \qquad \text{Apply the Law of Sines.}$$
$$\frac{20}{\sin B} = \frac{\sqrt{700}}{\sin 60°} \qquad \text{We are given } b = 20 \text{ and } A = 60°. \text{ Use the value of } a, \sqrt{700}, \text{ from step 1.}$$
$$\sqrt{700} \sin B = 20 \sin 60° \qquad \text{Cross multiply: If } \frac{a}{b} = \frac{c}{d}, \text{ then } ad = bc.$$
$$\sin B = \frac{20 \sin 60°}{\sqrt{700}} \approx 0.6547 \qquad \text{Divide by } \sqrt{700} \text{ and solve for } \sin B.$$
$$B \approx 41° \qquad \text{Find } \sin^{-1} 0.6547 \text{ using a calculator.}$$

Step 3 Find the third angle. Subtract the measure of the given angle and the angle found in step 2 from 180°.

$$C = 180° - A - B \approx 180° - 60° - 41° = 79°$$

The solution is $a \approx 26.5$, $B \approx 41°$, and $C \approx 79°$. • • •

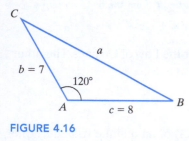

FIGURE 4.16

✓ **Check Point 1** Solve the triangle shown in **Figure 4.16** with $A = 120°$, $b = 7$, and $c = 8$. Round as in Example 1.

If you are given three sides of a triangle (SSS), solving the triangle involves finding the three angles. We use the following procedure:

Solving an SSS Triangle

1. Use the Law of Cosines to find the angle opposite the longest side.
2. Use the Law of Sines to find either of the two remaining acute angles.
3. Find the third angle by subtracting the measures of the angles found in steps 1 and 2 from 180°.

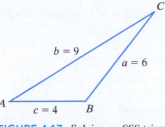

FIGURE 4.17 Solving an SSS triangle

EXAMPLE 2 Solving an SSS Triangle

Solve triangle ABC if $a = 6, b = 9,$ and $c = 4$. Round angle measures to the nearest degree.

SOLUTION

We are given three sides. Therefore, we apply the three-step procedure for solving an SSS triangle. The triangle is shown in **Figure 4.17**.

Step 1 Use the Law of Cosines to find the angle opposite the longest side. The longest side is $b = 9$. Thus, we will find angle B.

$$b^2 = a^2 + c^2 - 2ac \cos B \quad \text{Apply the Law of Cosines to find } B.$$
$$2ac \cos B = a^2 + c^2 - b^2 \quad \text{Solve for } \cos B.$$
$$\cos B = \frac{a^2 + c^2 - b^2}{2ac}$$
$$\cos B = \frac{6^2 + 4^2 - 9^2}{2 \cdot 6 \cdot 4} = -\frac{29}{48} \quad a = 6, b = 9, \text{ and } c = 4.$$

Using a calculator, $\cos^{-1}\left(\frac{29}{48}\right) \approx 53°$. Because $\cos B$ is negative, B is an obtuse angle. Thus,

$$B \approx 180° - 53° = 127°.$$

> Because the domain of $y = \cos^{-1} x$ is $[-1, 1]$, you can use a calculator to find $\cos^{-1}\left(-\frac{29}{48}\right) \approx 127°.$

Step 2 Use the Law of Sines to find either of the two remaining acute angles. We will find angle A.

$$\frac{a}{\sin A} = \frac{b}{\sin B} \quad \text{Apply the Law of Sines.}$$
$$\frac{6}{\sin A} = \frac{9}{\sin 127°} \quad \text{We are given } a = 6 \text{ and } b = 9. \text{ We found that } B \approx 127°.$$
$$9 \sin A = 6 \sin 127° \quad \text{Cross multiply.}$$
$$\sin A = \frac{6 \sin 127°}{9} \approx 0.5324 \quad \text{Divide by 9 and solve for } \sin A.$$
$$A \approx 32° \quad \text{Find } \sin^{-1} 0.5324 \text{ using a calculator.}$$

Step 3 Find the third angle. Subtract the measures of the angles found in steps 1 and 2 from 180°.

$$C = 180° - B - A \approx 180° - 127° - 32° = 21°$$

The solution is $B \approx 127°, A \approx 32°,$ and $C \approx 21°$. • • •

GREAT QUESTION!

In Step 2, do I have to use the Law of Sines to find either of the remaining angles?

No. You can also use the Law of Cosines to find either angle. However, it is simpler to use the Law of Sines. Because the largest angle has been found, the remaining angles must be acute. Thus, there is no need to be concerned about two possible triangles or an ambiguous case.

2 Solve applied problems using the Law of Cosines.

✓ **Check Point 2** Solve triangle ABC if $a = 8, b = 10,$ and $c = 5$. Round angle measures to the nearest degree.

Applications of the Law of Cosines

Applied problems involving SAS and SSS triangles can be solved using the Law of Cosines.

EXAMPLE 3 An Application of the Law of Cosines

Two airplanes leave an airport at the same time on different runways. One flies on a bearing of N66°W at 325 miles per hour. The other airplane flies on a bearing of S26°W at 300 miles per hour. How far apart will the airplanes be after two hours?

SOLUTION

After two hours, the plane flying at 325 miles per hour travels $325 \cdot 2$ miles, or 650 miles. Similarly, the plane flying at 300 miles per hour travels 600 miles. The situation is illustrated in **Figure 4.18**.

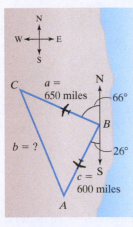

FIGURE 4.18

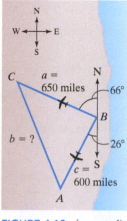

FIGURE 4.18 (repeated)

Let b = the distance between the planes after two hours. We can use a north-south line to find angle B in triangle ABC. Thus,

$$B = 180° - 66° - 26° = 88°.$$

We now have $a = 650$, $c = 600$, and $B = 88°$. We use the Law of Cosines to find b in this SAS situation.

$$b^2 = a^2 + c^2 - 2ac \cos B \qquad \text{Apply the Law of Cosines.}$$
$$b^2 = 650^2 + 600^2 - 2(650)(600) \cos 88° \qquad \text{Substitute: } a = 650, c = 600, \text{ and } B = 88°.$$
$$\approx 755{,}278 \qquad \text{Use a calculator.}$$
$$b \approx \sqrt{755{,}278} \approx 869 \qquad \text{Take the square root and solve for } b.$$

After two hours, the planes are approximately 869 miles apart. • • •

✓ **Check Point 3** Two airplanes leave an airport at the same time on different runways. One flies directly north at 400 miles per hour. The other airplane flies on a bearing of N75°E at 350 miles per hour. How far apart will the airplanes be after two hours?

③ Use Heron's formula to find the area of a triangle.

Heron's Formula

Approximately 2000 years ago, the Greek mathematician Heron of Alexandria derived a formula for the area of a triangle in terms of the lengths of its sides. A more modern derivation uses the Law of Cosines and can be found in the appendix.

> **Heron's Formula for the Area of a Triangle**
>
> The area of a triangle with sides a, b, and c is
>
> $$\text{Area} = \sqrt{s(s - a)(s - b)(s - c)},$$
>
> where s is one-half its perimeter: $s = \frac{1}{2}(a + b + c)$.

EXAMPLE 4 Using Heron's Formula

Find the area of the triangle with $a = 12$ yards, $b = 16$ yards, and $c = 24$ yards. Round to the nearest square yard.

SOLUTION

Begin by calculating one-half the perimeter:

$$s = \tfrac{1}{2}(a + b + c) = \tfrac{1}{2}(12 + 16 + 24) = 26.$$

Use Heron's formula to find the area:

$$\text{Area} = \sqrt{s(s - a)(s - b)(s - c)}$$
$$= \sqrt{26(26 - 12)(26 - 16)(26 - 24)}$$
$$= \sqrt{7280} \approx 85.$$

The area of the triangle is approximately 85 square yards. • • •

✓ **Check Point 4** Find the area of the triangle with $a = 6$ meters, $b = 16$ meters, and $c = 18$ meters. Round to the nearest square meter.

ACHIEVING SUCCESS

Many trigonometry courses cover only selected sections from this chapter and the book's remaining chapter. Regardless of the content requirements for your course, **it's never too early to start thinking about a final exam**. Here are some strategies to help you prepare for your final:

- Review your back exams. Be sure you understand any errors that you made. Seek help with any concepts that are still unclear.
- Ask your professor if there are additional materials to help students review for the final. This includes review sheets and final exams from previous semesters.
- Attend any review sessions conducted by your professor or by the math department.
- Use the strategy introduced earlier in the book: Imagine that your professor will permit two 3 by 5 index cards of notes on the final. Organize and create such a two-card summary for the most vital information in the course, including all important formulas. Refer to the chapter summaries in the textbook to prepare your personalized summary.
- For further review, work the relevant exercises in the Cumulative Review exercises at the end of all chapters covered in your course.
- Write your own final exam with detailed solutions for each item. You can use test questions from back exams in mixed order, exercises in the Cumulative Reviews, exercises in the Chapter Tests, and problems from course handouts. Use your test as a practice final exam.

CONCEPT AND VOCABULARY CHECK

Fill in each blank so that the resulting statement is true.

1. If A, B, and C are the measures of the angles of a triangle, and a, b, and c are the lengths of the sides opposite these angles, then the Law of Cosines states that $a^2 = $ _____.

2. To solve an oblique triangle given two sides and an included angle (SAS), the first step is to find the missing _____ using the Law of _____. Then we use the Law of _____ to find the angle opposite the shorter of the two given sides. This angle is always _____. The third angle is found by subtracting the measure of the given angle and the angle found in the second step from _____.

3. To solve an oblique triangle given three sides (SSS), the first step is to find the angle opposite the longest side using the Law of _____. Then we find either of the two remaining acute angles using the Law of _____.

4. Heron's formula for the area of a triangle with sides a, b, and c is Area $= $ _____, where $s = $ _____.

EXERCISE SET 4.2

Practice Exercises

In Exercises 1–8, solve each triangle. Round lengths of sides to the nearest tenth and angle measures to the nearest degree.

1.
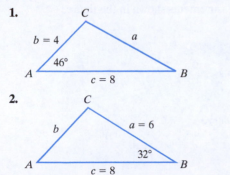

2.

3.
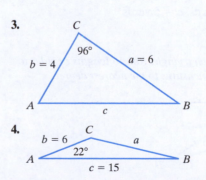

4.

5.

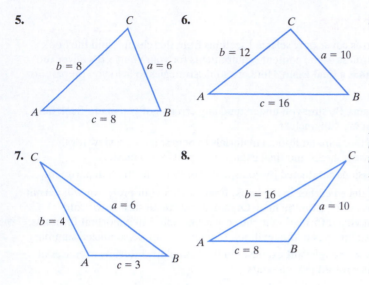

6.

7.

8.

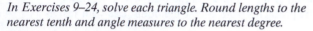

32.

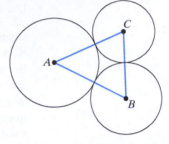

In Exercises 33–34, the three circles are arranged so that they touch each other, as shown in the figure. Use the given radii for the circles with centers A, B, and C, respectively, to solve triangle ABC. Round angle measures to the nearest degree.

33. 5.0, 4.0, 3.5

34. 7.5, 4.3, 3.0

In Exercises 35–36, the three given points are the vertices of a triangle. Solve each triangle, rounding lengths of sides to the nearest tenth and angle measures to the nearest degree.

35. $A(0, 0)$, $B(-3, 4)$, $C(3, -1)$

36. $A(0, 0)$, $B(4, -3)$, $C(1, -5)$

Application Exercises

37. Use **Figure 4.13** on page 224 to find the pace angle, to the nearest degree, for the carnivore. Does the angle indicate that this dinosaur was an efficient walker? Describe your answer.

38. Use **Figure 4.13** on page 224 to find the pace angle, to the nearest degree, for the herbivore. Does the angle indicate that this dinosaur was an efficient walker? Describe your answer.

39. Two ships leave a harbor at the same time. One ship travels on a bearing of S12°W at 14 miles per hour. The other ship travels on a bearing of N75°E at 10 miles per hour. How far apart will the ships be after three hours? Round to the nearest tenth of a mile.

40. A plane leaves airport A and travels 580 miles to airport B on a bearing of N34°E. The plane later leaves airport B and travels to airport C 400 miles away on a bearing of S74°E. Find the distance from airport A to airport C to the nearest tenth of a mile.

41. Find the distance across the lake from A to C, to the nearest yard, using the measurements shown in the figure.

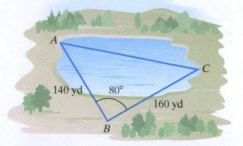

In Exercises 9–24, solve each triangle. Round lengths to the nearest tenth and angle measures to the nearest degree.

9. $a = 5, b = 7, C = 42°$

10. $a = 10, b = 3, C = 15°$

11. $b = 5, c = 3, A = 102°$

12. $b = 4, c = 1, A = 100°$

13. $a = 6, c = 5, B = 50°$

14. $a = 4, c = 7, B = 55°$

15. $a = 5, c = 2, B = 90°$

16. $a = 7, c = 3, B = 90°$

17. $a = 5, b = 7, c = 10$

18. $a = 4, b = 6, c = 9$

19. $a = 3, b = 9, c = 8$

20. $a = 4, b = 7, c = 6$

21. $a = 3, b = 3, c = 3$

22. $a = 5, b = 5, c = 5$

23. $a = 63, b = 22, c = 50$

24. $a = 66, b = 25, c = 45$

In Exercises 25–30, use Heron's formula to find the area of each triangle. Round to the nearest square unit.

25. $a = 4$ feet, $b = 4$ feet, $c = 2$ feet

26. $a = 5$ feet, $b = 5$ feet, $c = 4$ feet

27. $a = 14$ meters, $b = 12$ meters, $c = 4$ meters

28. $a = 16$ meters, $b = 10$ meters, $c = 8$ meters

29. $a = 11$ yards, $b = 9$ yards, $c = 7$ yards

30. $a = 13$ yards, $b = 9$ yards, $c = 5$ yards

Practice Plus

In Exercises 31–32, solve each triangle. Round lengths of sides to the nearest tenth and angle measures to the nearest degree.

31.

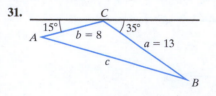

42. To find the distance across a protected cove at a lake, a surveyor makes the measurements shown in the figure. Use these measurements to find the distance from *A* to *B* to the nearest yard.

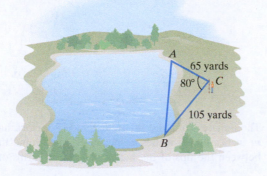

The diagram shows three islands in Florida Bay. You rent a boat and plan to visit each of these remote islands. Use the diagram to solve Exercises 43–44.

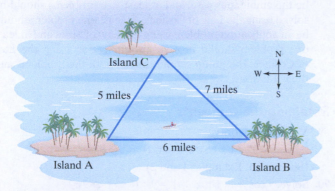

43. If you are on island A, on what bearing should you navigate to go to island C?

44. If you are on island B, on what bearing should you navigate to go to island C?

45. You are on a fishing boat that leaves its pier and heads east. After traveling for 25 miles, there is a report warning of rough seas directly south. The captain turns the boat and follows a bearing of S40°W for 13.5 miles.

    a. At this time, how far are you from the boat's pier? Round to the nearest tenth of a mile.

    b. What bearing could the boat have originally taken to arrive at this spot?

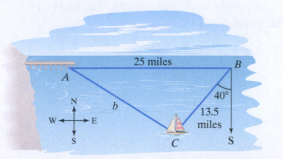

46. You are on a fishing boat that leaves its pier and heads east. After traveling for 30 miles, there is a report warning of rough seas directly south. The captain turns the boat and follows a bearing of S45°W for 12 miles.

    a. At this time, how far are you from the boat's pier? Round to the nearest tenth of a mile.

    b. What bearing could the boat have originally taken to arrive at this spot?

47. The figure shows a 400-foot tower on the side of a hill that forms a 7° angle with the horizontal. Find the length of each of the two guy wires that are anchored 80 feet uphill and downhill from the tower's base and extend to the top of the tower. Round to the nearest tenth of a foot.

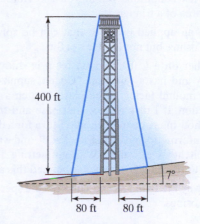

48. The figure shows a 200-foot tower on the side of a hill that forms a 5° angle with the horizontal. Find the length of each of the two guy wires that are anchored 150 feet uphill and downhill from the tower's base and extend to the top of the tower. Round to the nearest tenth of a foot.

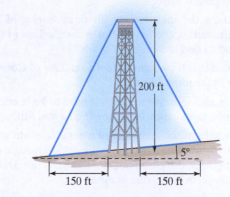

49. A Major League baseball diamond has four bases forming a square whose sides measure 90 feet each. The pitcher's mound is 60.5 feet from home plate on a line joining home plate and second base. Find the distance from the pitcher's mound to first base. Round to the nearest tenth of a foot.

50. A Little League baseball diamond has four bases forming a square whose sides measure 60 feet each. The pitcher's mound is 46 feet from home plate on a line joining home plate and second base. Find the distance from the pitcher's mound to third base. Round to the nearest tenth of a foot.

51. A piece of commercial real estate is priced at $3.50 per square foot. Find the cost, to the nearest dollar, of a triangular lot measuring 240 feet by 300 feet by 420 feet.

52. A piece of commercial real estate is priced at $4.50 per square foot. Find the cost, to the nearest dollar, of a triangular lot measuring 320 feet by 510 feet by 410 feet.

Explaining the Concepts

53. Without using symbols, state the Law of Cosines in your own words.

54. Why can't the Law of Sines be used in the first step to solve an SAS triangle?

55. Describe a strategy for solving an SAS triangle.

56. Describe a strategy for solving an SSS triangle.

57. Under what conditions would you use Heron's formula to find the area of a triangle?

58. Describe an applied problem that can be solved using the Law of Cosines but not the Law of Sines.

59. The pitcher on a Little League team is studying angles in geometry and has a question. "Coach, suppose I'm on the pitcher's mound facing home plate. I catch a fly ball hit in my direction. If I turn to face first base and throw the ball, through how many degrees should I turn for a direct throw?" Use the information given in Exercise 50 and write an answer to the pitcher's question. Without getting too technical, describe to the pitcher how you obtained this angle.

60. Explain why the Pythagorean Theorem is a special case of the Law of Cosines.

Critical Thinking Exercises

Make Sense?   *In Exercises 61–64, determine whether each statement makes sense or does not make sense, and explain your reasoning.*

61. The Law of Cosines is similar to the Law of Sines, with all the sines replaced with cosines.

62. If I know the measures of all three angles of an oblique triangle, neither the Law of Sines nor the Law of Cosines can be used to find the length of a side.

63. I noticed that for a right triangle, the Law of Cosines reduces to the Pythagorean Theorem.

64. Solving an SSS triangle, I do not have to be concerned about the ambiguous case when using the Law of Sines.

65. The lengths of the diagonals of a parallelogram are 20 inches and 30 inches. The diagonals intersect at an angle of 35°. Find the lengths of the parallelogram's sides. (*Hint:* Diagonals of a parallelogram bisect one another.)

66. Use the figure to solve triangle *ABC*. Round lengths of sides to the nearest tenth and angle measures to the nearest degree.

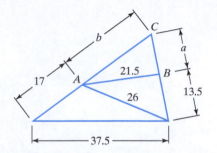

67. The minute hand and the hour hand of a clock have lengths *m* inches and *h* inches, respectively. Determine the distance between the tips of the hands at 10:00 in terms of *m* and *h*.

Group Exercise

68. The group should design five original problems that can be solved using the Laws of Sines and Cosines. At least two problems should be solved using the Law of Sines, one should be the ambiguous case, and at least two problems should be solved using the Law of Cosines. At least one problem should be an application problem using the Law of Sines and at least one problem should involve an application using the Law of Cosines. The group should turn in both the problems and their solutions.

Retaining the Concepts

69. Verify the identity:
$$\csc x \cos^2 x + \sin x = \csc x.$$
(Section 3.1, Examples 1 and 2)

70. Solve: $\cos^2 x + \sin x + 1 = 0, \ 0 \le x \le 2\pi$.
(Section 3.5, Example 7)

71. Graph $y = 3 \tan \dfrac{x}{2}$ for $-\pi < x < 3\pi$.
(Section 2.2, Example 1)

Preview Exercises

Exercises 72–74 will help you prepare for the material covered in the next section.

72. Use the distance formula to determine if the line segment with endpoints $(-3, -3)$ and $(0, 3)$ has the same length as the line segment with endpoints $(0, 0)$ and $(3, 6)$.

73. Use slope to determine if the line through $(-3, -3)$ and $(0, 3)$ is parallel to the line through $(0, 0)$ and $(3, 6)$.

74. Simplify: $4(5x + 4y) - 2(6x - 9y)$.

CHAPTER 4   Mid-Chapter Check Point

WHAT YOU KNOW: We learned to solve oblique triangles using the Law of Sines $\left(\dfrac{a}{\sin A} = \dfrac{b}{\sin B} = \dfrac{c}{\sin C}\right)$ and the Law of Cosines $(a^2 = b^2 + c^2 - 2bc \cos A)$. We applied the Law of Sines to SAA, ASA, and SSA (the ambiguous case) triangles. We applied the Law of Cosines to SAS and SSS triangles. We found areas of SAS triangles $\left(\text{area} = \frac{1}{2}bc \sin A\right)$ and SSS triangles (Heron's formula: area $= \sqrt{s(s-a)(s-b)(s-c)}$, s is $\frac{1}{2}$ the perimeter).

In Exercises 1–6, solve each triangle. Round lengths to the nearest tenth and angle measures to the nearest degree. If no triangle exists, state "no triangle." If two triangles exist, solve each triangle.

1. $A = 32°, B = 41°, a = 20$
2. $A = 42°, a = 63, b = 57$
3. $A = 65°, a = 6, b = 7$
4. $B = 110°, a = 10, c = 16$
5. $C = 42°, a = 16, c = 13$
6. $a = 5.0, b = 7.2, c = 10.1$

In Exercises 7–8, find the area of the triangle having the given measurements. Round to the nearest square unit.

7. $C = 36°, a = 5$ feet, $b = 7$ feet
8. $a = 7$ meters, $b = 9$ meters, $c = 12$ meters
9. Two trains leave a station on different tracks that make an angle of 110° with the station as vertex. The first train travels at an average rate of 50 miles per hour and the second train travels at an average rate of 40 miles per hour. How far apart, to the nearest tenth of a mile, are the trains after 2 hours?
10. Two fire-lookout stations are 16 miles apart, with station B directly east of station A. Both stations spot a fire on a mountain to the south. The bearing from station A to the fire is S56°E. The bearing from station B to the fire is S23°W. How far, to the nearest tenth of a mile, is the fire from station A?
11. A tree that is perpendicular to the ground sits on a straight line between two people located 420 feet apart. The angles of elevation from each person to the top of the tree measure 50° and 66°, respectively. How tall, to the nearest tenth of a foot, is the tree?

A Brief Review • Slope

- The slope, m, of the line through the distinct points (x_1, y_1) and (x_2, y_2) is

$$m = \frac{\text{Change in } y}{\text{Change in } x} = \frac{y_2 - y_1}{x_2 - x_1}, x_2 - x_1 \neq 0.$$

When computing slope, it makes no difference which point you call (x_1, y_1) and which point you call (x_2, y_2).

EXAMPLE

The slope of the line passing through $(-3, -1)$ and $(-2, 4)$ is

$$m = \frac{\text{Change in } y}{\text{Change in } x} = \frac{y_2 - y_1}{x_2 - x_1} = \frac{4 - (-1)}{-2 - (-3)} = \frac{4 + 1}{-2 + 3} = \frac{5}{1} = 5.$$

- A line with a positive slope rises from left to right. A line with a negative slope falls from left to right. A horizontal line has a slope of zero. The slope of a vertical line is undefined.
- If two distinct nonvertical lines have the same slope, then the lines are parallel.
- If the product of the slopes of two lines is −1, then the lines are perpendicular. Equivalently, if the slopes are negative reciprocals, then the lines are perpendicular.

Section 4.3   Vectors

What am I supposed to learn?

After studying this section, you should be able to:

1. Use magnitude and direction to show vectors are equal.

2. Visualize scalar multiplication, vector addition, and vector subtraction as geometric vectors.

3. Represent vectors in the rectangular coordinate system.

4. Perform operations with vectors in terms of **i** and **j**.

5. Find the unit vector in the direction of **v**.

6. Write a vector in terms of its magnitude and direction.

7. Solve applied problems involving vectors.

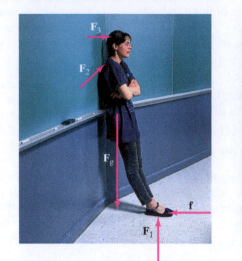

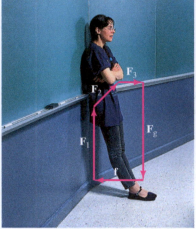

It's been a dynamic lecture, but now that it's over it's obvious that my professor is exhausted. She's slouching motionless against the board and—what's that? The forces acting against her body, including the pull of gravity, are appearing as arrows. I know that mathematics reveals the hidden patterns of the universe, but this is ridiculous. Does the arrangement of the arrows on the right have anything to do with the fact that my wiped-out professor is not sliding down the wall?

Ours is a world of pushes and pulls. For example, suppose you are pulling a cart up a 30° incline, requiring an effort of 100 pounds. This quantity is described by giving its magnitude (a number indicating size, including a unit of measure) and also its direction. The magnitude is 100 pounds and the direction is 30° from the horizontal. Quantities that involve both a magnitude and a direction are called **vector quantities**, or **vectors** for short. Here is another example of a vector:

> You are driving due north at 50 miles per hour. The magnitude is the speed, 50 miles per hour. The direction of motion is due north.

Some quantities can be completely described by giving only their magnitudes. For example, the temperature of the lecture room that you just left is 75°. This temperature has magnitude, 75°, but no direction. Quantities that involve magnitude, but no direction, are called **scalar quantities**, or **scalars** for short. Thus, a scalar has only a numerical value. Another example of a scalar is your professor's height, which you estimate to be 5.5 feet.

In this section and the next, we introduce the world of vectors, which literally surrounds your every move. Because vectors have nonnegative magnitude as well as direction, we begin our discussion with directed line segments.

This sign shows a distance and direction for each city. Thus, the sign defines a vector for each destination.

Directed Line Segments and Geometric Vectors

A line segment to which a direction has been assigned is called a **directed line segment**. **Figure 4.19** shows a directed line segment from P to Q. We call P the **initial point** and Q the **terminal point**. We denote this directed line segment by

$$\overrightarrow{PQ}.$$

FIGURE 4.19 A directed line segment from P to Q

The **magnitude** of the directed line segment $\overrightarrow{PQ}$ is its length. We denote this by $\|\overrightarrow{PQ}\|$. Thus, $\|\overrightarrow{PQ}\|$ is the distance from point P to point Q. Because distance is nonnegative, vectors do not have negative magnitudes.

Geometrically, a **vector** is a directed line segment. Vectors are often denoted by boldface letters, such as **v**. If a vector **v** has the same magnitude and the same direction as the directed line segment $\overrightarrow{PQ}$, we write

$$\mathbf{v} = \overrightarrow{PQ}.$$

GREAT QUESTION!

Because it's impossible for me to write boldface on paper, how should I denote a vector?

Use an arrow over a single letter.

Representing Vectors in Print	Representing Vectors on Paper		Representing Vectors in Print	Representing Vectors on Paper
Vector v **v**	$\vec{v}$		Magnitude of v $\|\mathbf{v}\|$	$\|\vec{v}\|$
Vector w **w**	$\vec{w}$		Magnitude of w $\|\mathbf{w}\|$	$\|\vec{w}\|$

GREAT QUESTION!

What's the difference between a ray and a vector?

A ray is a directed line that has only an initial point and extends forever in one direction. A vector is a directed line segment that has both an initial point and a terminal point.

 Use magnitude and direction to show vectors are equal.

Figure 4.20 shows four possible relationships between vectors **v** and **w**. In **Figure 4.20(a)**, the vectors have the same magnitude and the same direction, and are said to be *equal*. In general, vectors **v** and **w** are **equal** if they have the *same magnitude* and the *same direction*. We write this as $\mathbf{v} = \mathbf{w}$.

(a) $\mathbf{v} = \mathbf{w}$ because the vectors have the same magnitude and same direction.

(b) Vectors **v** and **w** have the same magnitude, but different directions.

(c) Vectors **v** and **w** have the same magnitude, but opposite directions.

(d) Vectors **v** and **w** have the same direction, but different magnitudes.

FIGURE 4.20 Relationships between vectors

EXAMPLE 1   Showing That Two Vectors Are Equal

Use **Figure 4.21** to show that $\mathbf{u} = \mathbf{v}$.

SOLUTION

Equal vectors have the same magnitude and the same direction. Use the distance formula to show that **u** and **v** have the same magnitude.

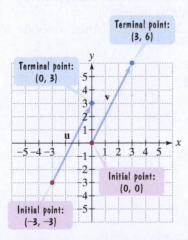

Terminal point: (3, 6)

Terminal point: (0, 3)

Initial point: (0, 0)

Initial point: (−3, −3)

FIGURE 4.21

Magnitude of u
$$\|\mathbf{u}\| = \sqrt{(x_2 - x_1)^2 + (y_2 - y_1)^2} = \sqrt{[0 - (-3)]^2 + [3 - (-3)]^2}$$
$$= \sqrt{3^2 + 6^2} = \sqrt{9 + 36} = \sqrt{45} \quad (\text{or } 3\sqrt{5})$$

Magnitude of v
$$\|\mathbf{v}\| = \sqrt{(x_2 - x_1)^2 + (y_2 - y_1)^2} = \sqrt{(3 - 0)^2 + (6 - 0)^2}$$
$$= \sqrt{3^2 + 6^2} = \sqrt{9 + 36} = \sqrt{45} \quad (\text{or } 3\sqrt{5})$$

Thus, **u** and **v** have the same magnitude: $\|\mathbf{u}\| = \|\mathbf{v}\|$.

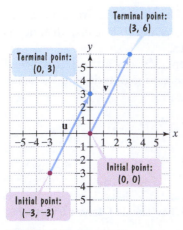

FIGURE 4.21 (repeated)

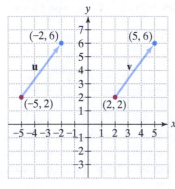

FIGURE 4.22

One way to show that **u** and **v** have the same direction is to find the slopes of the lines on which they lie. We continue to use **Figure 4.21**.

Line on which **u** lies	$m = \dfrac{y_2 - y_1}{x_2 - x_1} = \dfrac{3 - (-3)}{0 - (-3)} = \dfrac{6}{3} = 2$	**u** lies on a line passing through $(-3, -3)$ and $(0, 3)$.
Line on which **v** lies	$m = \dfrac{y_2 - y_1}{x_2 - x_1} = \dfrac{6 - 0}{3 - 0} = \dfrac{6}{3} = 2$	**v** lies on a line passing through $(0, 0)$ and $(3, 6)$.

Because **u** and **v** are both directed toward the upper right on lines having the same slope, 2, they have the same direction.

Thus, **u** and **v** have the same magnitude and direction, and **u** = **v**. • • •

✅ **Check Point 1** Use **Figure 4.22** to show that **u** = **v**.

A vector can be multiplied by a real number. **Figure 4.23** shows three such multiplications: $2\mathbf{v}$, $\frac{1}{2}\mathbf{v}$, and $-\frac{3}{2}\mathbf{v}$. **Multiplying a vector by any positive real number (except for 1) changes the magnitude of the vector but not its direction.** This can be seen by the blue and green vectors in **Figure 4.23**. Compare the black and blue vectors. Can you see that $2\mathbf{v}$ has the same direction as **v** but is twice the magnitude of **v**? Now, compare the black and green vectors: $\frac{1}{2}\mathbf{v}$ has the same direction as **v** but is half the magnitude of **v**.

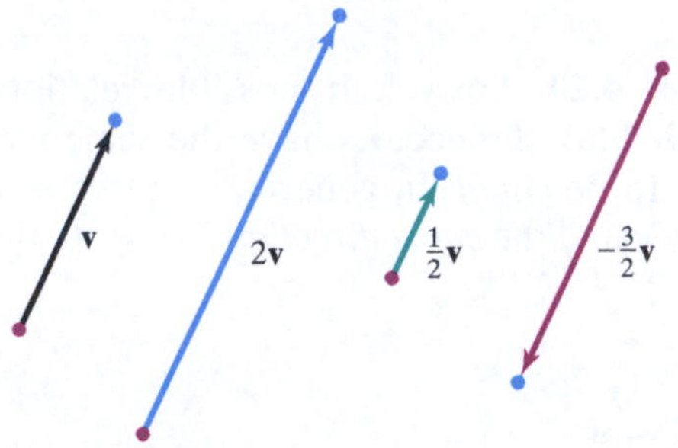

FIGURE 4.23 Multiplying vector **v** by real numbers

Now compare the black and red vectors in **Figure 4.23**. **Multiplying a vector by a negative number reverses the direction of the vector.** Notice that $-\frac{3}{2}\mathbf{v}$ has the opposite direction as **v** and is $\frac{3}{2}$ the magnitude of **v**.

The multiplication of a real number k and a vector **v** is called **scalar multiplication**. We write this product as $k\mathbf{v}$.

② Visualize scalar multiplication, vector addition, and vector subtraction as geometric vectors.

Scalar Multiplication

If k is a real number and **v** a vector, the vector $k\mathbf{v}$ is called a **scalar multiple** of the vector **v**. The magnitude and direction of $k\mathbf{v}$ are given as follows:

The vector $k\mathbf{v}$ has a *magnitude* of $|k|\,\|\mathbf{v}\|$. We describe this as the absolute value of k times the magnitude of vector **v**.

The vector $k\mathbf{v}$ has a *direction* that is

- the same as the direction of **v** if $k > 0$, and
- opposite the direction of **v** if $k < 0$.

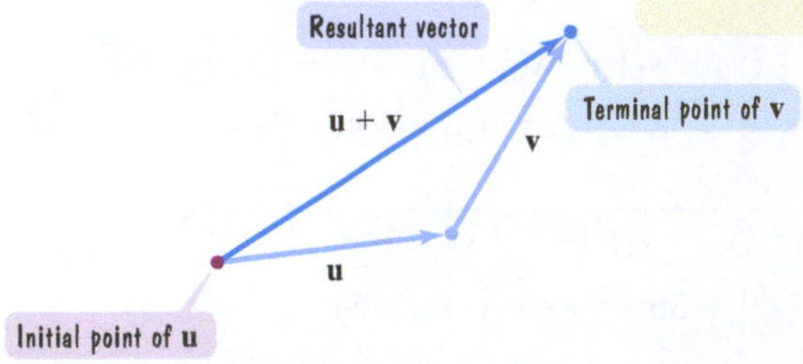

FIGURE 4.24 Vector addition **u** + **v**; the terminal point of **u** coincides with the initial point of **v**.

A geometric method for adding two vectors is shown in **Figure 4.24**. The sum of **u** and **v**, denoted by **u** + **v** is called the **resultant vector**. Here is how we find this vector:

1. Position **u** and **v**, so that the terminal point of **u** coincides with the initial point of **v**.

2. The resultant vector, **u** + **v**, extends from the initial point of **u** to the terminal point of **v**.

The **difference of two vectors**, $\mathbf{v} - \mathbf{u}$, is defined as $\mathbf{v} - \mathbf{u} = \mathbf{v} + (-\mathbf{u})$, where $-\mathbf{u}$ is the scalar multiplication of $\mathbf{u}$ and -1: $-1\mathbf{u}$. The difference $\mathbf{v} - \mathbf{u}$ is shown geometrically in **Figure 4.25**.

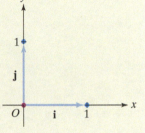

FIGURE 4.25 Vector subtraction $\mathbf{v} - \mathbf{u}$; the terminal point of $\mathbf{v}$ coincides with the initial point of $-\mathbf{u}$.

Vectors in the Rectangular Coordinate System

As you saw in Example 1, vectors can be shown in the rectangular coordinate system. Now let's see how we can use the rectangular coordinate system to represent vectors. We begin with two vectors that both have a magnitude of 1. Such vectors are called **unit vectors**.

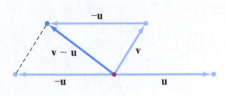

The **i** and **j** Unit Vectors

Vector **i** is the unit vector whose direction is along the positive x-axis. Vector **j** is the unit vector whose direction is along the positive y-axis.

Why are the unit vectors **i** and **j** important? Vectors in the rectangular coordinate system can be represented in terms of **i** and **j**. For example, consider vector $\mathbf{v}$ with initial point at the origin, $(0, 0)$, and terminal point at $P = (a, b)$. The vector $\mathbf{v}$ is shown in **Figure 4.26**. We can represent $\mathbf{v}$ using **i** and **j** as $\mathbf{v} = a\mathbf{i} + b\mathbf{j}$.

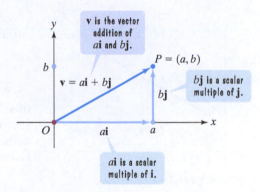

FIGURE 4.26 Using vector addition, vector $\mathbf{v}$ is represented as $\mathbf{v} = a\mathbf{i} + b\mathbf{j}$.

③ Represent vectors in the rectangular coordinate system.

Representing Vectors in Rectangular Coordinates

Vector $\mathbf{v}$, from $(0, 0)$ to (a, b), is represented as

$$\mathbf{v} = a\mathbf{i} + b\mathbf{j}.$$

The real numbers a and b are called the **scalar components of v**. Note that

- a is the **horizontal component** of $\mathbf{v}$, and
- b is the **vertical component** of $\mathbf{v}$.

The vector sum $a\mathbf{i} + b\mathbf{j}$ is called a **linear combination** of the vectors **i** and **j**. The magnitude of $\mathbf{v} = a\mathbf{i} + b\mathbf{j}$ is given by

$$\|\mathbf{v}\| = \sqrt{a^2 + b^2}.$$

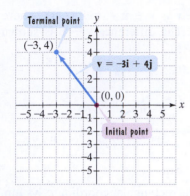

FIGURE 4.27 Sketching $\mathbf{v} = -3\mathbf{i} + 4\mathbf{j}$ in the rectangular coordinate system

EXAMPLE 2   Representing a Vector in Rectangular Coordinates and Finding Its Magnitude

Sketch the vector $\mathbf{v} = -3\mathbf{i} + 4\mathbf{j}$ and find its magnitude.

SOLUTION

For the given vector $\mathbf{v} = -3\mathbf{i} + 4\mathbf{j}$, $a = -3$ and $b = 4$. The vector can be represented with its initial point at the origin, $(0, 0)$, as shown in **Figure 4.27**. The vector's terminal point is then $(a, b) = (-3, 4)$. We sketch the vector by drawing an arrow from $(0, 0)$ to $(-3, 4)$. We determine the magnitude of the vector by using the distance formula. Thus, the magnitude is

$$\|\mathbf{v}\| = \sqrt{a^2 + b^2} = \sqrt{(-3)^2 + 4^2} = \sqrt{9 + 16} = \sqrt{25} = 5.$$ • • •

GREAT QUESTION!

In Example 2, since $\mathbf{v} = -3\mathbf{i} + 4\mathbf{j}$, is it ok if I write the magnitude of v as $\|\mathbf{v}\| = \sqrt{(-3\mathbf{i})^2 + (4\mathbf{j})^2}$?

No. The vectors $\mathbf{i}$ and $\mathbf{j}$ are not included when determining the magnitude of $\mathbf{v} = a\mathbf{i} + b\mathbf{j}$.

Correct	**Incorrect**
$\mathbf{v} = a\mathbf{i} + b\mathbf{j}$	~~$\mathbf{v} = a\mathbf{i} + b\mathbf{j}$~~
$\|\mathbf{v}\| = \sqrt{a^2 + b^2}$	~~$\|\mathbf{v}\| = \sqrt{(a\mathbf{i})^2 + (b\mathbf{j})^2}$~~

✅ **Check Point 2**   Sketch the vector $\mathbf{v} = 3\mathbf{i} - 3\mathbf{j}$ and find its magnitude.

The vector in Example 2 was represented with its initial point at the origin. A vector whose initial point is at the origin is called a **position vector**. Any vector in rectangular coordinates whose initial point is not at the origin can be shown to be equal to a position vector. As shown in the following box, this gives us a way to represent vectors between any two points.

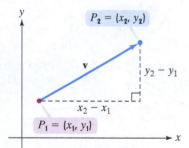

FIGURE 4.28(a)

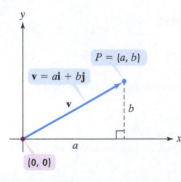

FIGURE 4.28(b)

Representing Vectors in Rectangular Coordinates

Vector $\mathbf{v}$ with initial point $P_1 = (x_1, y_1)$ and terminal point $P_2 = (x_2, y_2)$ is equal to the position vector

$$\mathbf{v} = (x_2 - x_1)\mathbf{i} + (y_2 - y_1)\mathbf{j}.$$

We can use congruent triangles, triangles with the same size and shape, to derive this formula. Begin with the right triangle in **Figure 4.28(a)**. This triangle shows vector $\mathbf{v}$ from $P_1 = (x_1, y_1)$ to $P_2 = (x_2, y_2)$. In **Figure 4.28(b)**, we move vector $\mathbf{v}$, without changing its magnitude or its direction, so that its initial point is at the origin. Using this position vector in **Figure 4.28(b)**, we see that

$$\mathbf{v} = a\mathbf{i} + b\mathbf{j},$$

where a and b are the components of $\mathbf{v}$. The equal vectors and the right angles in the right triangles in **Figure 4.28(a)** and **(b)** result in congruent triangles. The corresponding sides of these congruent triangles are equal, so that $a = x_2 - x_1$ and $b = y_2 - y_1$. This means that $\mathbf{v}$ may be expressed as

$$\mathbf{v} = a\mathbf{i} + b\mathbf{j} = (x_2 - x_1)\mathbf{i} + (y_2 - y_1)\mathbf{j}.$$

Horizontal component: x-coordinate of terminal point minus x-coordinate of initial point

Vertical component: y-coordinate of terminal point minus y-coordinate of initial point

Thus, any vector between two points in rectangular coordinates can be expressed in terms of $\mathbf{i}$ and $\mathbf{j}$. In rectangular coordinates, the term *vector* refers to the position vector expressed in terms of $\mathbf{i}$ and $\mathbf{j}$ that is equal to it.

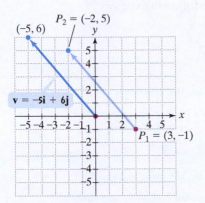

FIGURE 4.29 Representing the vector from $(3, -1)$ to $(-2, 5)$ as a position vector

EXAMPLE 3   Representing a Vector in Rectangular Coordinates

Let **v** be the vector from initial point $P_1 = (3, -1)$ to terminal point $P_2 = (-2, 5)$. Write **v** in terms of **i** and **j**.

SOLUTION

We identify the values for the variables in the formula.

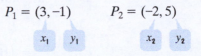

$$P_1 = (3, -1) \qquad P_2 = (-2, 5)$$

Using these values, we write **v** in terms of **i** and **j** as follows:

$$\mathbf{v} = (x_2 - x_1)\mathbf{i} + (y_2 - y_1)\mathbf{j} = (-2 - 3)\mathbf{i} + [5 - (-1)]\mathbf{j} = -5\mathbf{i} + 6\mathbf{j}.$$

Figure 4.29 shows the vector from $P_1 = (3, -1)$ to $P_2 = (-2, 5)$ represented in terms of **i** and **j** and as a position vector.   • • •

GREAT QUESTION!

When representing a vector from an initial point to a terminal point, does the order in which I perform the subtractions make a difference?

Yes. When writing the vector from $P_1 = (x_1, y_1)$ to $P_2 = (x_2, y_2)$, P_2 must be the terminal point and the order in the subtractions is important:

$$\mathbf{v} = (x_2 - x_1)\mathbf{i} + (y_2 - y_1)\mathbf{j}.$$

(x_2, y_2), the terminal point, is used first in each subtraction.

Notice how this differs from finding the distance from $P_1 = (x_1, y_1)$ to $P_2 = (x_2, y_2)$, where the order in which the subtractions are performed makes no difference:

$$d = \sqrt{(x_2 - x_1)^2 + (y_2 - y_1)^2} \quad \text{or} \quad d = \sqrt{(x_1 - x_2)^2 + (y_1 - y_2)^2}.$$

✓ **Check Point 3** Let **v** be the vector from initial point $P_1 = (-1, 3)$ to terminal point $P_2 = (2, 7)$. Write **v** in terms of **i** and **j**.

④ Perform operations with vectors in terms of **i** and **j**.

Operations with Vectors in Terms of i and j

If vectors are expressed in terms of **i** and **j**, we can easily carry out operations such as vector addition, vector subtraction, and scalar multiplication. Recall the geometric definitions of these operations given earlier. Based on these ideas, we can add and subtract vectors using the following procedure:

Adding and Subtracting Vectors in Terms of i and j

If $\mathbf{v} = a_1\mathbf{i} + b_1\mathbf{j}$ and $\mathbf{w} = a_2\mathbf{i} + b_2\mathbf{j}$, then

$$\mathbf{v} + \mathbf{w} = (a_1 + a_2)\mathbf{i} + (b_1 + b_2)\mathbf{j}$$

$$\mathbf{v} - \mathbf{w} = (a_1 - a_2)\mathbf{i} + (b_1 - b_2)\mathbf{j}.$$

EXAMPLE 4 Adding and Subtracting Vectors

If $\mathbf{v} = 5\mathbf{i} + 4\mathbf{j}$ and $\mathbf{w} = 6\mathbf{i} - 9\mathbf{j}$, find each of the following vectors:

    a. $\mathbf{v} + \mathbf{w}$ **b.** $\mathbf{v} - \mathbf{w}$.

SOLUTION

  a. $\mathbf{v} + \mathbf{w} = (5\mathbf{i} + 4\mathbf{j}) + (6\mathbf{i} - 9\mathbf{j})$ These are the given vectors.

               $= (5 + 6)\mathbf{i} + [4 + (-9)]\mathbf{j}$ Add the horizontal components.
                                         Add the vertical components.

               $= 11\mathbf{i} - 5\mathbf{j}$ Simplify.

  b. $\mathbf{v} - \mathbf{w} = (5\mathbf{i} + 4\mathbf{j}) - (6\mathbf{i} - 9\mathbf{j})$ These are the given vectors.

               $= (5 - 6)\mathbf{i} + [4 - (-9)]\mathbf{j}$ Subtract the horizontal components.
                                         Subtract the vertical components.

               $= -\mathbf{i} + 13\mathbf{j}$ Simplify.   •••

✅ **Check Point 4** If $\mathbf{v} = 7\mathbf{i} + 3\mathbf{j}$ and $\mathbf{w} = 4\mathbf{i} - 5\mathbf{j}$, find each of the following vectors:

    a. $\mathbf{v} + \mathbf{w}$ **b.** $\mathbf{v} - \mathbf{w}$.

How do we perform scalar multiplication if vectors are expressed in terms of $\mathbf{i}$ and $\mathbf{j}$? We use the following procedure to multiply the vector $\mathbf{v}$ by the scalar k:

Scalar Multiplication with a Vector in Terms of $\mathbf{i}$ and $\mathbf{j}$

If $\mathbf{v} = a\mathbf{i} + b\mathbf{j}$ and k is a real number, then the scalar multiplication of the vector $\mathbf{v}$ and the scalar k is

$$k\mathbf{v} = (ka)\mathbf{i} + (kb)\mathbf{j}.$$

EXAMPLE 5 Scalar Multiplication

If $\mathbf{v} = 5\mathbf{i} + 4\mathbf{j}$, find each of the following vectors:

    a. $6\mathbf{v}$ **b.** $-3\mathbf{v}$.

SOLUTION

  a. $6\mathbf{v} = 6(5\mathbf{i} + 4\mathbf{j})$ The scalar multiplication is expressed with the given vector.

          $= (6 \cdot 5)\mathbf{i} + (6 \cdot 4)\mathbf{j}$ Multiply each component by 6.
          $= 30\mathbf{i} + 24\mathbf{j}$ Simplify.

  b. $-3\mathbf{v} = -3(5\mathbf{i} + 4\mathbf{j})$ The scalar multiplication is expressed with the given vector.

           $= (-3 \cdot 5)\mathbf{i} + (-3 \cdot 4)\mathbf{j}$ Multiply each component by -3.
           $= -15\mathbf{i} - 12\mathbf{j}$ Simplify.   •••

✅ **Check Point 5** If $\mathbf{v} = 7\mathbf{i} + 10\mathbf{j}$, find each of the following vectors:

    a. $8\mathbf{v}$ **b.** $-5\mathbf{v}$.

EXAMPLE 6 Vector Operations

If $\mathbf{v} = 5\mathbf{i} + 4\mathbf{j}$ and $\mathbf{w} = 6\mathbf{i} - 9\mathbf{j}$, find $4\mathbf{v} - 2\mathbf{w}$.

SOLUTION

$$4\mathbf{v} - 2\mathbf{w} = 4(5\mathbf{i} + 4\mathbf{j}) - 2(6\mathbf{i} - 9\mathbf{j}) \qquad \text{Operations are expressed with the given vectors.}$$

$$= 20\mathbf{i} + 16\mathbf{j} - 12\mathbf{i} + 18\mathbf{j} \qquad \text{Perform each scalar multiplication.}$$

$$= (20 - 12)\mathbf{i} + (16 + 18)\mathbf{j} \qquad \text{Add horizontal and vertical components to perform the vector addition.}$$

$$= 8\mathbf{i} + 34\mathbf{j} \qquad \text{Simplify.}$$

• • •

Check Point **6** If $\mathbf{v} = 7\mathbf{i} + 3\mathbf{j}$ and $\mathbf{w} = 4\mathbf{i} - 5\mathbf{j}$, find $6\mathbf{v} - 3\mathbf{w}$.

Properties involving vector operations resemble familiar properties of real numbers. For example, the order in which vectors are added makes no difference:

$$\mathbf{u} + \mathbf{v} = \mathbf{v} + \mathbf{u}.$$

Does this remind you of the commutative property $a + b = b + a$?

Just as 0 plays an important role in the properties of real numbers, the **zero vector 0** plays exactly the same role in the properties of vectors.

The Zero Vector

The vector whose magnitude is 0 is called the **zero vector, 0**. The zero vector is assigned no direction. It can be expressed in terms of $\mathbf{i}$ and $\mathbf{j}$ using

$$\mathbf{0} = 0\mathbf{i} + 0\mathbf{j}.$$

Properties of vector addition and scalar multiplication are given as follows:

Properties of Vector Addition and Scalar Multiplication

If $\mathbf{u}$, $\mathbf{v}$, and $\mathbf{w}$ are vectors, and c and d are scalars, then the following properties are true.

Vector Addition Properties

1. $\mathbf{u} + \mathbf{v} = \mathbf{v} + \mathbf{u}$ Commutative property
2. $(\mathbf{u} + \mathbf{v}) + \mathbf{w} = \mathbf{u} + (\mathbf{v} + \mathbf{w})$ Associative property
3. $\mathbf{u} + \mathbf{0} = \mathbf{0} + \mathbf{u} = \mathbf{u}$ Additive identity
4. $\mathbf{u} + (-\mathbf{u}) = (-\mathbf{u}) + \mathbf{u} = \mathbf{0}$ Additive inverse

Scalar Multiplication Properties

1. $(cd)\mathbf{u} = c(d\mathbf{u})$ Associative property
2. $c(\mathbf{u} + \mathbf{v}) = c\mathbf{u} + c\mathbf{v}$ Distributive property
3. $(c + d)\mathbf{u} = c\mathbf{u} + d\mathbf{u}$ Distributive property
4. $1\mathbf{u} = \mathbf{u}$ Multiplicative identity
5. $0\mathbf{u} = \mathbf{0}$ Multiplication property of zero
6. $\|c\mathbf{v}\| = |c|\,\|\mathbf{v}\|$ Magnitude property

---

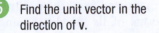
5 Find the unit vector in the direction of **v**.

Unit Vectors

A **unit vector** is defined to be a vector whose magnitude is one. In many applications of vectors, it is helpful to find the unit vector that has the same direction as a given vector.

DISCOVERY

To find out why the procedure in the box produces a unit vector, work Exercise 112 in Exercise Set 4.3.

Finding the Unit Vector That Has the Same Direction as a Given Nonzero Vector v

For any nonzero vector $\mathbf{v}$, the vector

$$\frac{\mathbf{v}}{\|\mathbf{v}\|}$$

is the unit vector that has the same direction as $\mathbf{v}$. To find this vector, divide $\mathbf{v}$ by its magnitude.

EXAMPLE 7 Finding a Unit Vector

Find the unit vector in the same direction as $\mathbf{v} = 5\mathbf{i} - 12\mathbf{j}$. Then verify that the vector has magnitude 1.

SOLUTION

We find the unit vector in the same direction as $\mathbf{v}$ by dividing $\mathbf{v}$ by its magnitude. We first find the magnitude of $\mathbf{v}$.

$$\|\mathbf{v}\| = \sqrt{a^2 + b^2} = \sqrt{5^2 + (-12)^2} = \sqrt{25 + 144} = \sqrt{169} = 13$$

The unit vector in the same direction as $\mathbf{v}$ is

$$\frac{\mathbf{v}}{\|\mathbf{v}\|} = \frac{5\mathbf{i} - 12\mathbf{j}}{13} = \frac{5}{13}\mathbf{i} - \frac{12}{13}\mathbf{j}. \quad \text{This is the scalar multiplication of } v \text{ and } \tfrac{1}{13}.$$

Now we must verify that the magnitude of this vector is 1. Recall that the magnitude of $a\mathbf{i} + b\mathbf{j}$ is $\sqrt{a^2 + b^2}$. Thus, the magnitude of $\frac{5}{13}\mathbf{i} - \frac{12}{13}\mathbf{j}$ is

$$\sqrt{\left(\frac{5}{13}\right)^2 + \left(-\frac{12}{13}\right)^2} = \sqrt{\frac{25}{169} + \frac{144}{169}} = \sqrt{\frac{169}{169}} = \sqrt{1} = 1. \quad \bullet\bullet\bullet$$

✓ **Check Point 7** Find the unit vector in the same direction as $\mathbf{v} = 4\mathbf{i} - 3\mathbf{j}$. Then verify that the vector has magnitude 1.

6 Write a vector in terms of its magnitude and direction.

Writing a Vector in Terms of Its Magnitude and Direction

Consider the vector $\mathbf{v} = a\mathbf{i} + b\mathbf{j}$. The components a and b can be expressed in terms of the magnitude of $\mathbf{v}$ and the angle θ that $\mathbf{v}$ makes with the positive x-axis. This angle is called the **direction angle** of $\mathbf{v}$ and is shown in **Figure 4.30**. By the definitions of sine and cosine, we have

$$\cos \theta = \frac{a}{\|\mathbf{v}\|} \qquad \text{and} \qquad \sin \theta = \frac{b}{\|\mathbf{v}\|}$$

$$a = \|\mathbf{v}\| \cos \theta \qquad\qquad b = \|\mathbf{v}\| \sin \theta.$$

Thus,

$$\mathbf{v} = a\mathbf{i} + b\mathbf{j} = \|\mathbf{v}\| \cos \theta \mathbf{i} + \|\mathbf{v}\| \sin \theta \mathbf{j}.$$

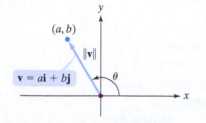

FIGURE 4.30 Expressing a vector in terms of its magnitude, $\|\mathbf{v}\|$, and its direction angle, θ

Writing a Vector in Terms of Its Magnitude and Direction

Let $\mathbf{v}$ be a nonzero vector. If θ is the direction angle measured from the positive x-axis to $\mathbf{v}$, then the vector can be expressed in terms of its magnitude and direction angle as

$$\mathbf{v} = \|\mathbf{v}\| \cos \theta \mathbf{i} + \|\mathbf{v}\| \sin \theta \mathbf{j}.$$

A vector that represents the direction and speed of an object in motion is called a **velocity vector**. In Example 8, we express a wind's velocity vector in terms of the wind's magnitude and direction.

Writing a Vector Whose Magnitude and Direction Are Given

The wind is blowing at 20 miles per hour in the direction N30°W. Express its velocity as a vector **v** in terms of **i** and **j**.

SOLUTION

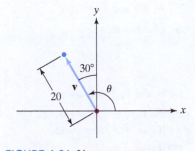

The vector **v** is shown in **Figure 4.31**. The vector's direction angle, from the positive x-axis to **v**, is

$$\theta = 90° + 30° = 120°.$$

Because the wind is blowing at 20 miles per hour, the magnitude of **v** is 20 miles per hour: $\|\mathbf{v}\| = 20$. Thus,

$$\mathbf{v} = \|\mathbf{v}\| \cos \theta \mathbf{i} + \|\mathbf{v}\| \sin \theta \mathbf{j} \qquad \text{Use the formula for a vector in terms of magnitude and direction.}$$

$$= 20 \cos 120°\mathbf{i} + 20 \sin 120°\mathbf{j} \qquad \|\mathbf{v}\| = 20 \text{ and } \theta = 120°.$$

$$= 20\left(-\frac{1}{2}\right)\mathbf{i} + 20\left(\frac{\sqrt{3}}{2}\right)\mathbf{j} \qquad \cos 120° = -\frac{1}{2} \text{ and } \sin 120° = \frac{\sqrt{3}}{2}.$$

$$= -10\mathbf{i} + 10\sqrt{3}\mathbf{j} \qquad \text{Simplify.}$$

The wind's velocity can be expressed in terms of **i** and **j** as $\mathbf{v} = -10\mathbf{i} + 10\sqrt{3}\mathbf{j}$. • • •

FIGURE 4.31 Vector **v** represents a wind blowing at 20 miles per hour in the direction N30°W.

✓ Check Point **8** The jet stream is blowing at 60 miles per hour in the direction N45°E. Express its velocity as a vector **v** in terms of **i** and **j**.

7 Solve applied problems involving vectors.

Application

Many physical concepts can be represented by vectors. A vector that represents a pull or push of some type is called a **force vector**. If you are holding a 10-pound package, two force vectors are involved. The force of gravity is exerting a force of magnitude 10 pounds directly downward. This force is shown by vector $\mathbf{F}_1$ in **Figure 4.32**. Assuming there is no upward or downward movement of the package, you are exerting a force of magnitude 10 pounds directly upward. This force is shown by vector $\mathbf{F}_2$ in **Figure 4.32**. It has the same magnitude as the force exerted on your package by gravity, but it acts in the opposite direction.

If $\mathbf{F}_1$ and $\mathbf{F}_2$ are two forces acting on an object, the net effect is the same as if just the resultant force, $\mathbf{F}_1 + \mathbf{F}_2$, acted on the object. If the object is not moving, as is the case with your 10-pound package, the vector sum of all forces is the zero vector.

FIGURE 4.32 Force vectors

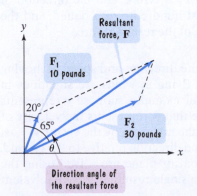

FIGURE 4.33

Finding the Resultant Force

Two forces, $\mathbf{F}_1$ and $\mathbf{F}_2$, of magnitude 10 and 30 pounds, respectively, act on an object. The direction of $\mathbf{F}_1$ is N20°E and the direction of $\mathbf{F}_2$ is N65°E. Find the magnitude and the direction of the resultant force. Express the magnitude to the nearest hundredth of a pound and the direction angle to the nearest tenth of a degree.

SOLUTION

The vectors $\mathbf{F}_1$ and $\mathbf{F}_2$ are shown in **Figure 4.33**. The direction angle for $\mathbf{F}_1$, from the positive x-axis to the vector, is $\theta_1 = 90° - 20°$, or 70°. We express $\mathbf{F}_1$ using the formula for a vector in terms of its magnitude and direction.

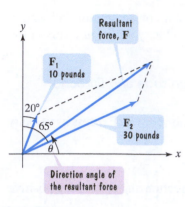

FIGURE 4.33 (repeated)

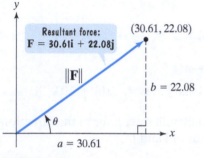

FIGURE 4.34

GREAT QUESTION!

In Figure 4.34, do I have to use the cosine or the sine to find the direction angle, θ?

No. If $\mathbf{F} = a\mathbf{i} + b\mathbf{j}$, the direction angle, θ, of $\mathbf{F}$ can also be found using

$$\tan \theta = \frac{b}{a}.$$

$$\mathbf{F}_1 = \|\mathbf{F}_1\| \cos \theta_1 \mathbf{i} + \|\mathbf{F}_1\| \sin \theta_1 \mathbf{j}$$
$$= 10 \cos 70°\mathbf{i} + 10 \sin 70°\mathbf{j} \qquad \text{\color{blue}{$\|\mathbf{F}_1\| = 10$ and $\theta_1 = 70°$.}}$$
$$\approx 3.42\mathbf{i} + 9.40\mathbf{j} \qquad \text{\color{blue}{Use a calculator.}}$$

Figure 4.33 illustrates that the direction angle for $\mathbf{F}_2$, from the positive x-axis to the vector, is $\theta_2 = 90° - 65°$, or $25°$. We express $\mathbf{F}_2$ using the formula for a vector in terms of its magnitude and direction.

$$\mathbf{F}_2 = \|\mathbf{F}_2\| \cos \theta_2 \mathbf{i} + \|\mathbf{F}_2\| \sin \theta_2 \mathbf{j}$$
$$= 30 \cos 25°\mathbf{i} + 30 \sin 25°\mathbf{j} \qquad \text{\color{blue}{$\|\mathbf{F}_2\| = 30$ and $\theta_2 = 25°$.}}$$
$$\approx 27.19\mathbf{i} + 12.68\mathbf{j} \qquad \text{\color{blue}{Use a calculator.}}$$

The resultant force, $\mathbf{F}$, is $\mathbf{F}_1 + \mathbf{F}_2$. Thus,

$$\mathbf{F} = \mathbf{F}_1 + \mathbf{F}_2$$
$$\approx (3.42\mathbf{i} + 9.40\mathbf{j}) + (27.19\mathbf{i} + 12.68\mathbf{j}) \qquad \text{\color{blue}{Use $\mathbf{F}_1$ and $\mathbf{F}_2$, found above.}}$$
$$= (3.42 + 27.19)\mathbf{i} + (9.40 + 12.68)\mathbf{j} \qquad \text{\color{blue}{Add the horizontal components.}}$$
$$\text{\color{blue}{Add the vertical components.}}$$
$$= 30.61\mathbf{i} + 22.08\mathbf{j}. \qquad \text{\color{blue}{Simplify.}}$$

Figure 4.34 shows the resultant force, $\mathbf{F}$, without showing $\mathbf{F}_1$ and $\mathbf{F}_2$.

Now that we have the resultant force vector, $\mathbf{F}$, we can find its magnitude.

$$\|\mathbf{F}\| = \sqrt{a^2 + b^2} = \sqrt{(30.61)^2 + (22.08)^2} \approx 37.74$$

The magnitude of the resultant force is approximately 37.74 pounds.

To find θ, the direction angle of the resultant force, we can use

$$\cos \theta = \frac{a}{\|\mathbf{F}\|} \quad \text{or} \quad \sin \theta = \frac{b}{\|\mathbf{F}\|}.$$

These ratios are illustrated for the right triangle in **Figure 4.34**.

Using the first formula, we obtain

$$\cos \theta = \frac{a}{\|\mathbf{F}\|} \approx \frac{30.61}{37.74}.$$

Thus,

$$\theta = \cos^{-1}\left(\frac{30.61}{37.74}\right) \approx 35.8°. \qquad \text{\color{blue}{Use a calculator.}}$$

The direction angle of the resultant force is approximately $35.8°$.

In summary, the two given forces are equivalent to a single force of approximately 37.74 pounds with a direction angle of approximately $35.8°$. (Answers may vary due to rounding.)   • • •

☑ **Check Point 9** Two forces, $\mathbf{F}_1$ and $\mathbf{F}_2$, of magnitude 30 and 60 pounds, respectively, act on an object. The direction of $\mathbf{F}_1$ is N10°E and the direction of $\mathbf{F}_2$ is N60°E. Find the magnitude, to the nearest hundredth of a pound, and the direction angle, to the nearest tenth of a degree, of the resultant force.

We have seen that velocity vectors represent the direction and speed of moving objects. Boats moving in currents and airplanes flying in winds are situations in which two velocity vectors act simultaneously. For example, suppose $\mathbf{v}$ represents the velocity of a plane in still air. Further suppose that $\mathbf{w}$ represents the velocity of the wind. The actual speed and direction of the plane are given by the vector $\mathbf{v} + \mathbf{w}$. This resultant vector describes the plane's speed and direction relative to the ground. Problems involving the resultant velocity of a boat or plane are solved using the same method that we used in Example 9 to find a single resultant force equivalent to two given forces.

CONCEPT AND VOCABULARY CHECK

Fill in each blank so that the resulting statement is true.

1. A quantity that has both magnitude and direction is called a/an _____.
2. A quantity that has magnitude but no direction is called a/an _____.

In Exercises 3–5, refer to the vectors shown below.

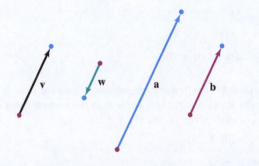

3. The vectors that appear to be equal are ____ and ____.
4. The vector that appears to be a scalar multiple of **v** is ____, where the scalar is positive and not 1.
5. The vector that appears to be a scalar multiple of **v** is ____, where the scalar is negative.
6. The vectors **i** and **j** both have magnitudes of 1 and are called _____ vectors. The direction of vector **i** is along the positive _____ -axis. The direction of vector **j** is along the positive _____ -axis.

7. Consider vector **v** from $(0, 0)$ to (a, b):
$$\mathbf{v} = a\mathbf{i} + b\mathbf{j}.$$
The horizontal component of **v** is ____. The vertical component of **v** is ____. The magnitude of **v** is given by $\|\mathbf{v}\| =$ _____.

8. A vector whose initial point is at the origin is called a/an _____ vector.

9. Vector **v** with initial point $P_1 = (x_1, y_1)$ and terminal point $P_2 = (x_2, y_2)$ is equal to the vector
$$\mathbf{v} = (\underline{\quad})\mathbf{i} + (\underline{\quad})\mathbf{j}.$$

10. If $\mathbf{v} = a_1\mathbf{i} + b_1\mathbf{j}$ and $\mathbf{w} = a_2\mathbf{i} + b_2\mathbf{j}$, then
$$\mathbf{v} + \mathbf{w} = (\underline{\quad})\mathbf{i} + (\underline{\quad})\mathbf{j}$$
$$\mathbf{v} - \mathbf{w} = (\underline{\quad})\mathbf{i} + (\underline{\quad})\mathbf{j}$$
$$k\mathbf{v} = (\underline{\quad})\mathbf{i} + (\underline{\quad})\mathbf{j}.$$

11. For any nonzero vector **v**, the unit vector that has the same direction as **v** is _____. To find this vector, divide **v** by its _____.

12. Let **v** be a nonzero vector. If θ is the direction angle measured from the positive x-axis to **v**, then the vector can be expressed in terms of its magnitude and direction angle as
$$\mathbf{v} = \|\mathbf{v}\| \underline{\quad} \mathbf{i} + \|\mathbf{v}\| \underline{\quad} \mathbf{j}.$$

13. If $\mathbf{F}_1$ and $\mathbf{F}_2$ are two forces acting on an object, the vector sum $\mathbf{F}_1 + \mathbf{F}_2$ is called the _____ force.

EXERCISE SET 4.3

Practice Exercises

*In Exercises 1–4, **u** and **v** have the same direction. In each exercise:*

 a. Find $\|\mathbf{u}\|$. **b.** Find $\|\mathbf{v}\|$. **c.** Is $\mathbf{u} = \mathbf{v}$? Explain.

1.

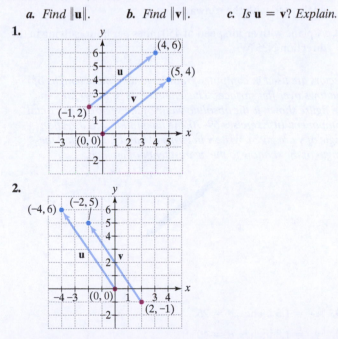

2.

3.

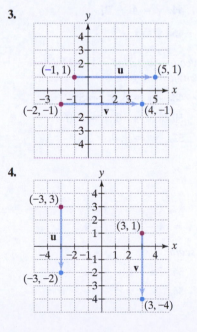

4.

In Exercises 5–12, sketch each vector as a position vector and find its magnitude.

5. $\mathbf{v} = 3\mathbf{i} + \mathbf{j}$
6. $\mathbf{v} = 2\mathbf{i} + 3\mathbf{j}$
7. $\mathbf{v} = \mathbf{i} - \mathbf{j}$
8. $\mathbf{v} = -\mathbf{i} - \mathbf{j}$
9. $\mathbf{v} = -6\mathbf{i} - 2\mathbf{j}$
10. $\mathbf{v} = 5\mathbf{i} - 2\mathbf{j}$
11. $\mathbf{v} = -4\mathbf{i}$
12. $\mathbf{v} = -5\mathbf{j}$

In Exercises 13–20, let $\mathbf{v}$ *be the vector from initial point* P_1 *to terminal point* P_2. *Write* $\mathbf{v}$ *in terms of* $\mathbf{i}$ *and* $\mathbf{j}$.

13. $P_1 = (-4, -4), P_2 = (6, 2)$
14. $P_1 = (2, -5), P_2 = (-6, 6)$
15. $P_1 = (-8, 6), P_2 = (-2, 3)$
16. $P_1 = (-7, -4), P_2 = (0, -2)$
17. $P_1 = (-1, 7), P_2 = (-7, -7)$
18. $P_1 = (-1, 6), P_2 = (7, -5)$
19. $P_1 = (-3, 4), P_2 = (6, 4)$
20. $P_1 = (4, -5), P_2 = (4, 3)$

In Exercises 21–38, let

$$\mathbf{u} = 2\mathbf{i} - 5\mathbf{j}, \mathbf{v} = -3\mathbf{i} + 7\mathbf{j}, \text{ and } \mathbf{w} = -\mathbf{i} - 6\mathbf{j}.$$

Find each specified vector or scalar.

21. $\mathbf{u} + \mathbf{v}$
22. $\mathbf{v} + \mathbf{w}$
23. $\mathbf{u} - \mathbf{v}$
24. $\mathbf{v} - \mathbf{w}$
25. $\mathbf{v} - \mathbf{u}$
26. $\mathbf{w} - \mathbf{v}$
27. $5\mathbf{v}$
28. $6\mathbf{v}$
29. $-4\mathbf{w}$
30. $-7\mathbf{w}$
31. $3\mathbf{w} + 2\mathbf{v}$
32. $3\mathbf{u} + 4\mathbf{v}$
33. $3\mathbf{v} - 4\mathbf{w}$
34. $4\mathbf{w} - 3\mathbf{v}$
35. $\|2\mathbf{u}\|$
36. $\|-2\mathbf{u}\|$
37. $\|\mathbf{w} - \mathbf{u}\|$
38. $\|\mathbf{u} - \mathbf{w}\|$

In Exercises 39–46, find the unit vector that has the same direction as the vector $\mathbf{v}$.

39. $\mathbf{v} = 6\mathbf{i}$
40. $\mathbf{v} = -5\mathbf{j}$
41. $\mathbf{v} = 3\mathbf{i} - 4\mathbf{j}$
42. $\mathbf{v} = 8\mathbf{i} - 6\mathbf{j}$
43. $\mathbf{v} = 3\mathbf{i} - 2\mathbf{j}$
44. $\mathbf{v} = 4\mathbf{i} - 2\mathbf{j}$
45. $\mathbf{v} = \mathbf{i} + \mathbf{j}$
46. $\mathbf{v} = \mathbf{i} - \mathbf{j}$

In Exercises 47–52, write the vector $\mathbf{v}$ *in terms of* $\mathbf{i}$ *and* $\mathbf{j}$ *whose magnitude* $\|\mathbf{v}\|$ *and direction angle* θ *are given.*

47. $\|\mathbf{v}\| = 6, \theta = 30°$
48. $\|\mathbf{v}\| = 8, \theta = 45°$
49. $\|\mathbf{v}\| = 12, \theta = 225°$
50. $\|\mathbf{v}\| = 10, \theta = 330°$
51. $\|\mathbf{v}\| = \frac{1}{2}, \theta = 113°$
52. $\|\mathbf{v}\| = \frac{1}{4}, \theta = 200°$

Practice Plus

In Exercises 53–56, let

$$\mathbf{u} = -2\mathbf{i} + 3\mathbf{j}, \mathbf{v} = 6\mathbf{i} - \mathbf{j}, \mathbf{w} = -3\mathbf{i}.$$

Find each specified vector or scalar.

53. $4\mathbf{u} - (2\mathbf{v} - \mathbf{w})$
54. $3\mathbf{u} - (4\mathbf{v} - \mathbf{w})$
55. $\|\mathbf{u} + \mathbf{v}\|^2 - \|\mathbf{u} - \mathbf{v}\|^2$
56. $\|\mathbf{v} + \mathbf{w}\|^2 - \|\mathbf{v} - \mathbf{w}\|^2$

In Exercises 57–60, let

$$\mathbf{u} = a_1\mathbf{i} + b_1\mathbf{j}$$
$$\mathbf{v} = a_2\mathbf{i} + b_2\mathbf{j}$$
$$\mathbf{w} = a_3\mathbf{i} + b_3\mathbf{j}.$$

Prove each property by obtaining the vector on each side of the equation. Have you proved a distributive, associative, or commutative property of vectors?

57. $\mathbf{u} + \mathbf{v} = \mathbf{v} + \mathbf{u}$
58. $(\mathbf{u} + \mathbf{v}) + \mathbf{w} = \mathbf{u} + (\mathbf{v} + \mathbf{w})$
59. $c(\mathbf{u} + \mathbf{v}) = c\mathbf{u} + c\mathbf{v}$
60. $(c + d)\mathbf{u} = c\mathbf{u} + d\mathbf{u}$

In Exercises 61–64, find the magnitude $\|\mathbf{v}\|$, *to the nearest hundredth, and the direction angle* θ, *to the nearest tenth of a degree, for each given vector* $\mathbf{v}$.

61. $\mathbf{v} = -10\mathbf{i} + 15\mathbf{j}$
62. $\mathbf{v} = 2\mathbf{i} - 8\mathbf{j}$
63. $\mathbf{v} = (4\mathbf{i} - 2\mathbf{j}) - (4\mathbf{i} - 8\mathbf{j})$
64. $\mathbf{v} = (7\mathbf{i} - 3\mathbf{j}) - (10\mathbf{i} - 3\mathbf{j})$

Application Exercises

In Exercises 65–68, a vector is described. Express the vector in terms of $\mathbf{i}$ *and* $\mathbf{j}$. *If exact values are not possible, round components to the nearest tenth.*

65. A quarterback releases a football with a speed of 44 feet per second at an angle of 30° with the horizontal.

66. A child pulls a sled along level ground by exerting a force of 30 pounds on a handle that makes an angle of 45° with the ground.

67. A plane approaches a runway at 150 miles per hour at an angle of 8° with the runway.

68. A plane with an airspeed of 450 miles per hour is flying in the direction N35°W.

Vectors are used in computer graphics to determine lengths of shadows over flat surfaces. The length of the shadow for $\mathbf{v}$ *in the figure shown is the absolute value of the vector's horizontal component. In Exercises 69–70, the magnitude and direction angle of* $\mathbf{v}$ *are given. Write* $\mathbf{v}$ *in terms of* $\mathbf{i}$ *and* $\mathbf{j}$. *Then find the length of the shadow to the nearest tenth of an inch.*

69. $\|\mathbf{v}\| = 1.5$ inches, $\theta = 25°$

70. $\|\mathbf{v}\| = 1.8$ inches, $\theta = 40°$

71. The magnitude and direction of two forces acting on an object are 70 pounds, S56°E, and 50 pounds, N72°E, respectively. Find the magnitude, to the nearest hundredth of a pound, and the direction angle, to the nearest tenth of a degree, of the resultant force.

72. The magnitude and direction exerted by two tugboats towing a ship are 4200 pounds, N65°E, and 3000 pounds, S58°E, respectively. Find the magnitude, to the nearest pound, and the direction angle, to the nearest tenth of a degree, of the resultant force.

73. The magnitude and direction exerted by two tugboats towing a ship are 1610 kilograms, N35°W, and 1250 kilograms, S55°W, respectively. Find the magnitude, to the nearest kilogram, and the direction angle, to the nearest tenth of a degree, of the resultant force.

74. The magnitude and direction of two forces acting on an object are 64 kilograms, N39°W, and 48 kilograms, S59°W, respectively. Find the magnitude, to the nearest hundredth of a kilogram, and the direction angle, to the nearest tenth of a degree, of the resultant force.

The figure shows a box being pulled up a ramp inclined at 18° from the horizontal.

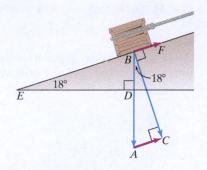

Use the following information to solve Exercises 75–76.

$\overrightarrow{BA}$ = force of gravity

$\|\overrightarrow{BA}\|$ = weight of the box

$\|\overrightarrow{AC}\|$ = magnitude of the force needed to pull the box up the ramp

$\|\overrightarrow{BC}\|$ = magnitude of the force of the box against the ramp

75. If the box weighs 100 pounds, find the magnitude of the force needed to pull it up the ramp.

76. If a force of 30 pounds is needed to pull the box up the ramp, find the weight of the box.

In Exercises 77–78, round answers to the nearest pound.

77. a. Find the magnitude of the force required to keep a 3500-pound car from sliding down a hill inclined at 5.5° from the horizontal.

b. Find the magnitude of the force of the car against the hill.

78. a. Find the magnitude of the force required to keep a 280-pound barrel from sliding down a ramp inclined at 12.5° from the horizontal.

b. Find the magnitude of the force of the barrel against the ramp.

The forces $\mathbf{F}_1, \mathbf{F}_2, \mathbf{F}_3, \ldots, \mathbf{F}_n$ *acting on an object are in* **equilibrium** *if the resultant force is the zero vector:*

$$\mathbf{F}_1 + \mathbf{F}_2 + \mathbf{F}_3 + \cdots + \mathbf{F}_n = \mathbf{0}.$$

In Exercises 79–82, the given forces are acting on an object.

a. *Find the resultant force.*

b. *What additional force is required for the given forces to be in equilibrium?*

79. $\mathbf{F}_1 = 3\mathbf{i} - 5\mathbf{j}, \quad \mathbf{F}_2 = 6\mathbf{i} + 2\mathbf{j}$

80. $\mathbf{F}_1 = -2\mathbf{i} + 3\mathbf{j}, \quad \mathbf{F}_2 = \mathbf{i} - \mathbf{j}, \quad \mathbf{F}_3 = 5\mathbf{i} - 12\mathbf{j}$

81.

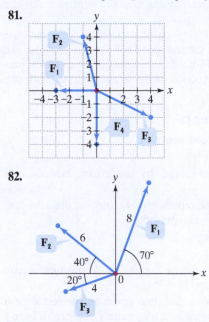

82.

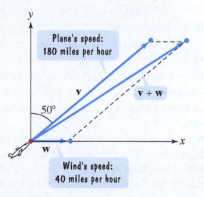

83. The figure shows a small plane flying at a speed of 180 miles per hour on a bearing of N50°E. The wind is blowing from west to east at 40 miles per hour. The figure indicates that **v** represents the velocity of the plane in still air and **w** represents the velocity of the wind.

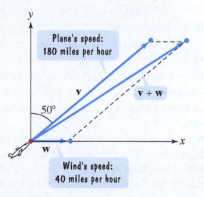

a. Express **v** and **w** in terms of their magnitudes and direction angles.

b. Find the resultant vector, **v** + **w**.

c. The magnitude of **v** + **w**, called the **ground speed** of the plane, gives its speed relative to the ground. Approximate the ground speed to the nearest mile per hour.

d. The direction angle of **v** + **w** gives the plane's true course relative to the ground. Approximate the true course to the nearest tenth of a degree. What is the plane's true bearing?

84. Use the procedure outlined in Exercise 83 to solve this exercise. A plane is flying at a speed of 400 miles per hour on a bearing of N50°W. The wind is blowing at 30 miles per hour on a bearing of N25°E.

  a. Approximate the plane's ground speed to the nearest mile per hour.

  b. Approximate the plane's true course to the nearest tenth of a degree. What is its true bearing?

85. A plane is flying at a speed of 320 miles per hour on a bearing of N70°E. Its ground speed is 370 miles per hour and its true course is 30°. Find the speed, to the nearest mile per hour, and the direction angle, to the nearest tenth of a degree, of the wind.

86. A plane is flying at a speed of 540 miles per hour on a bearing of S36°E. Its ground speed is 500 miles per hour and its true bearing is S44°E. Find the speed, to the nearest mile per hour, and the direction angle, to the nearest tenth of a degree, of the wind.

Explaining the Concepts

87. What is a directed line segment?

88. What are equal vectors?

89. If vector **v** is represented by an arrow, how is −3**v** represented?

90. If vectors **u** and **v** are represented by arrows, describe how the vector sum **u** + **v** is represented.

91. What is the vector **i**?

92. What is the vector **j**?

93. What is a position vector? How is a position vector represented using **i** and **j**?

94. If **v** is a vector between any two points in the rectangular coordinate system, explain how to write **v** in terms of **i** and **j**.

95. If two vectors are expressed in terms of **i** and **j**, explain how to find their sum.

96. If two vectors are expressed in terms of **i** and **j**, explain how to find their difference.

97. If a vector is expressed in terms of **i** and **j**, explain how to find the scalar multiplication of the vector and a given scalar *k*.

98. What is the zero vector?

99. Describe one similarity between the zero vector and the number 0.

100. Explain how to find the unit vector in the direction of any given vector **v**.

101. Explain how to write a vector in terms of its magnitude and direction.

102. You are on an airplane. The pilot announces the plane's speed over the intercom. Which speed do you think is being reported: the speed of the plane in still air or the speed after the effect of the wind has been accounted for? Explain your answer.

103. Use vectors to explain why it is difficult to hold a heavy stack of books perfectly still for a long period of time. As you become exhausted, what eventually happens? What does this mean in terms of the forces acting on the books?

Critical Thinking Exercises

Make Sense?   *In Exercises 104–107, determine whether each statement makes sense or does not make sense, and explain your reasoning.*

104. I used a vector to represent a wind velocity of 13 miles per hour from the west.

105. I used a vector to represent the average yearly rate of change in a man's height between ages 13 and 18.

106. Once I've found a unit vector **u**, the vector −**u** must also be a unit vector.

107. The resultant force of two forces that each have a magnitude of one pound is a vector whose magnitude is two pounds.

In Exercises 108–111, use the figure shown to determine whether each statement is true or false. If the statement is false, make the necessary change(s) to produce a true statement.

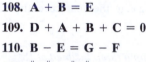

108. **A** + **B** = **E**

109. **D** + **A** + **B** + **C** = **0**

110. **B** − **E** = **G** − **F**

111. ‖**A**‖ = ‖**C**‖

112. Let **v** = *a***i** + *b***j**. Show that $\frac{\mathbf{v}}{\|\mathbf{v}\|}$ is a unit vector in the direction of **v**.

In Exercises 113–114, refer to the navigational compass shown in the figure. The compass is marked clockwise in degrees that start at north 0°.

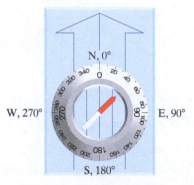

113. An airplane has an airspeed of 240 miles per hour and a compass heading of 280°. A steady wind of 30 miles per hour is blowing in the direction of 265°. What is the plane's true speed relative to the ground? What is its compass heading relative to the ground?

114. Two tugboats are pulling on a large ship that has gone aground. One tug pulls with a force of 2500 pounds in a compass direction of 55°. The second tug pulls with a force of 2000 pounds in a compass direction of 95°. Find the magnitude and the compass direction of the resultant force.

115. You want to fly your small plane due north, but there is a 75-kilometer wind blowing from west to east.

  a. Find the direction angle for where you should head the plane if your speed relative to the ground is 310 kilometers per hour.

  b. If you increase your airspeed, should the direction angle in part (a) increase or decrease? Explain your answer.

Retaining the Concepts

116. Determine the amplitude, period, and phase shift of $y = 3 \cos(2x + \pi)$. Then graph one period of the function. (Section 2.1, Example 6)

117. Verify the identity:

$$\frac{1 + \sin x}{1 - \sin x} - \frac{1 - \sin x}{1 + \sin x} = 4 \tan x \sec x.$$

(Section 3.1, Example 5)

118. Solve: $\cos 2x - \sin x = 0, 0 \le x < 2\pi$.

(Section 3.5, Example 8)

Preview Exercises

Exercises 119–121 will help you prepare for the material covered in the next section.

119. Find the obtuse angle θ, rounded to the nearest tenth of a degree, satisfying

$$\cos \theta = \frac{3(-1) + (-2)(4)}{\|\mathbf{v}\|\|\mathbf{w}\|},$$

where $\mathbf{v} = 3\mathbf{i} - 2\mathbf{j}$ and $\mathbf{w} = -\mathbf{i} + 4\mathbf{j}$.

120. If $\mathbf{w} = -2\mathbf{i} + 6\mathbf{j}$, find the following vector:

$$\frac{2(-2) + 4(-6)}{\|\mathbf{w}\|^2}\mathbf{w}.$$

121. Consider the triangle formed by vectors $\mathbf{u}$, $\mathbf{v}$, and $\mathbf{w}$.

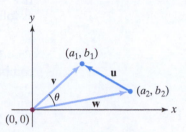

a. Use the magnitudes of the three vectors to write the Law of Cosines for the triangle shown in the figure: $\|\mathbf{u}\|^2 = ?$.

b. Use the coordinates of the points shown in the figure to write algebraic expressions for $\|\mathbf{u}\|$, $\|\mathbf{u}\|^2$, $\|\mathbf{v}\|$, $\|\mathbf{v}\|^2$, $\|\mathbf{w}\|$, and $\|\mathbf{w}\|^2$.

Section 4.4	The Dot Product

What am I supposed to learn?

After studying this section, you should be able to:

1. Find the dot product of two vectors.

2. Find the angle between two vectors.

3. Use the dot product to determine if two vectors are orthogonal.

4. Find the projection of a vector onto another vector.

5. Express a vector as the sum of two orthogonal vectors.

6. Compute work.

Talk about hard work! I can see the weightlifter's muscles quivering from the exertion of holding the barbell in a stationary position above her head. Still, I'm not sure if she's doing as much work as I am, sitting at my desk with my brain quivering from studying trigonometric functions and their applications.

Would it surprise you to know that neither you nor the weightlifter are doing any work at all? The definition of work in physics and mathematics is not the same as what we mean by "work" in everyday use. To understand what is involved in real work, we turn to a new vector operation called the dot product.

① Find the dot product of two vectors.

The Dot Product of Two Vectors

The operations of vector addition and scalar multiplication result in vectors. By contrast, the *dot product* of two vectors results in a scalar (a real number), rather than a vector.

Definition of the Dot Product

If $\mathbf{v} = a_1\mathbf{i} + b_1\mathbf{j}$ and $\mathbf{w} = a_2\mathbf{i} + b_2\mathbf{j}$ are vectors, the **dot product $\mathbf{v} \cdot \mathbf{w}$** is defined as follows:

$$\mathbf{v} \cdot \mathbf{w} = a_1a_2 + b_1b_2.$$

The dot product of two vectors is the sum of the products of their horizontal components and their vertical components.

EXAMPLE 1　Finding Dot Products

If $\mathbf{v} = 5\mathbf{i} - 2\mathbf{j}$ and $\mathbf{w} = -3\mathbf{i} + 4\mathbf{j}$, find each of the following dot products:

a. v · w　　　**b. w · v**　　　**c. v · v.**

SOLUTION

To find each dot product, multiply the two horizontal components, and then multiply the two vertical components. Finally, add the two products.

a. $\mathbf{v} \cdot \mathbf{w} = 5(-3) + (-2)(4) = -15 - 8 = -23$

> Multiply the horizontal components and multiply the vertical components of $\mathbf{v} = 5\mathbf{i} - 2\mathbf{j}$ and $\mathbf{w} = -3\mathbf{i} + 4\mathbf{j}$.

b. $\mathbf{w} \cdot \mathbf{v} = -3(5) + 4(-2) = -15 - 8 = -23$

> Multiply the horizontal components and multiply the vertical components of $\mathbf{w} = -3\mathbf{i} + 4\mathbf{j}$ and $\mathbf{v} = 5\mathbf{i} - 2\mathbf{j}$.

c. $\mathbf{v} \cdot \mathbf{v} = 5(5) + (-2)(-2) = 25 + 4 = 29$

> Multiply the horizontal components and multiply the vertical components of $\mathbf{v} = 5\mathbf{i} - 2\mathbf{j}$ and $\mathbf{v} = 5\mathbf{i} - 2\mathbf{j}$.

● ● ●

✓ Check Point 1 If $\mathbf{v} = 7\mathbf{i} - 4\mathbf{j}$ and $\mathbf{w} = 2\mathbf{i} - \mathbf{j}$, find each of the following dot products:

a. v · w　　　**b. w · v**　　　**c. w · w.**

In Example 1 and Check Point 1, did you notice that $\mathbf{v} \cdot \mathbf{w}$ and $\mathbf{w} \cdot \mathbf{v}$ produced the same scalar? The fact that $\mathbf{v} \cdot \mathbf{w} = \mathbf{w} \cdot \mathbf{v}$ follows from the definition of the dot product. Properties of the dot product are given in the following box. Proofs for some of these properties are given in the appendix.

Properties of the Dot Product

If $\mathbf{u}, \mathbf{v}$, and $\mathbf{w}$ are vectors, and c is a scalar, then

1. $\mathbf{u} \cdot \mathbf{v} = \mathbf{v} \cdot \mathbf{u}$
2. $\mathbf{u} \cdot (\mathbf{v} + \mathbf{w}) = \mathbf{u} \cdot \mathbf{v} + \mathbf{u} \cdot \mathbf{w}$
3. $\mathbf{0} \cdot \mathbf{v} = 0$
4. $\mathbf{v} \cdot \mathbf{v} = \|\mathbf{v}\|^2$
5. $(c\mathbf{u}) \cdot \mathbf{v} = c(\mathbf{u} \cdot \mathbf{v}) = \mathbf{u} \cdot (c\mathbf{v})$

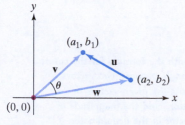

FIGURE 4.35

The Angle between Two Vectors

The Law of Cosines can be used to derive another formula for the dot product. This formula will give us a way to find the angle between two vectors.

Figure 4.35 shows vectors $\mathbf{v} = a_1\mathbf{i} + b_1\mathbf{j}$ and $\mathbf{w} = a_2\mathbf{i} + b_2\mathbf{j}$. By the definition of the dot product, we know that $\mathbf{v} \cdot \mathbf{w} = a_1a_2 + b_1b_2$. Our new formula for the dot product involves the angle between the vectors, shown as θ in the figure. Apply the Law of Cosines to the triangle shown in the figure.

$$\|\mathbf{u}\|^2 = \|\mathbf{v}\|^2 + \|\mathbf{w}\|^2 - 2\|\mathbf{v}\|\,\|\mathbf{w}\|\cos\theta \qquad \text{Use the Law of Cosines.}$$

$\mathbf{u} = (a_1 - a_2)\mathbf{i} + (b_1 - b_2)\mathbf{j}$	$\mathbf{v} = a_1\mathbf{i} + b_1\mathbf{j}$	$\mathbf{w} = a_2\mathbf{i} + b_2\mathbf{j}$
$\|\mathbf{u}\| = \sqrt{(a_1 - a_2)^2 + (b_1 - b_2)^2}$	$\|\mathbf{v}\| = \sqrt{a_1^2 + b_1^2}$	$\|\mathbf{w}\| = \sqrt{a_2^2 + b_2^2}$

$$(a_1 - a_2)^2 + (b_1 - b_2)^2 = (a_1^2 + b_1^2) + (a_2^2 + b_2^2) - 2\|\mathbf{v}\|\|\mathbf{w}\|\cos\theta$$

Substitute the squares of the magnitudes of vectors u, v, and w into the Law of Cosines.

$$a_1^2 - 2a_1a_2 + a_2^2 + b_1^2 - 2b_1b_2 + b_2^2 = a_1^2 + b_1^2 + a_2^2 + b_2^2 - 2\|\mathbf{v}\|\|\mathbf{w}\|\cos\theta$$

Square the binomials using $(A - B)^2 = A^2 - 2AB + B^2$.

$$-2a_1a_2 - 2b_1b_2 = -2\|\mathbf{v}\|\|\mathbf{w}\|\cos\theta$$

Subtract a_1^2, a_2^2, b_1^2, and b_2^2 from both sides of the equation.

$$a_1a_2 + b_1b_2 = \|\mathbf{v}\|\,\|\mathbf{w}\|\cos\theta$$

Divide both sides by -2.

By definition,
$$\mathbf{v} \cdot \mathbf{w} = a_1a_2 + b_1b_2.$$

$$\mathbf{v} \cdot \mathbf{w} = \|\mathbf{v}\|\|\mathbf{w}\|\cos\theta$$

Substitute v · w for the expression on the left side of the equation.

Alternative Formula for the Dot Product

If $\mathbf{v}$ and $\mathbf{w}$ are two nonzero vectors and θ is the smallest nonnegative angle between them, then

$$\mathbf{v} \cdot \mathbf{w} = \|\mathbf{v}\|\|\mathbf{w}\|\cos\theta.$$

 Find the angle between two vectors.

Solving the formula in the box for $\cos\theta$ gives us a formula for finding the angle between two vectors:

Formula for the Angle between Two Vectors

If $\mathbf{v}$ and $\mathbf{w}$ are two nonzero vectors and θ is the smallest nonnegative angle between $\mathbf{v}$ and $\mathbf{w}$, then

$$\cos\theta = \frac{\mathbf{v} \cdot \mathbf{w}}{\|\mathbf{v}\|\|\mathbf{w}\|} \quad \text{and} \quad \theta = \cos^{-1}\left(\frac{\mathbf{v} \cdot \mathbf{w}}{\|\mathbf{v}\|\|\mathbf{w}\|}\right).$$

EXAMPLE 2   Finding the Angle between Two Vectors

Find the angle θ between the vectors $\mathbf{v} = 3\mathbf{i} - 2\mathbf{j}$ and $\mathbf{w} = -\mathbf{i} + 4\mathbf{j}$, shown in **Figure 4.36** at the top of the next page. Round to the nearest tenth of a degree.

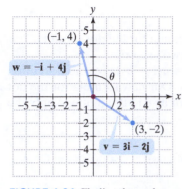

FIGURE 4.36 Finding the angle between two vectors

SOLUTION

Use the formula for the angle between two vectors.

$$\cos \theta = \frac{\mathbf{v} \cdot \mathbf{w}}{\|\mathbf{v}\| \|\mathbf{w}\|}$$

Begin with the formula for the cosine of the angle between two vectors.

$$= \frac{(3\mathbf{i} - 2\mathbf{j}) \cdot (-\mathbf{i} + 4\mathbf{j})}{\sqrt{3^2 + (-2)^2} \sqrt{(-1)^2 + 4^2}}$$

Substitute the given vectors in the numerator. Find the magnitude of each vector in the denominator.

$$= \frac{3(-1) + (-2)(4)}{\sqrt{13} \sqrt{17}}$$

Find the dot product in the numerator. Simplify in the denominator.

$$= -\frac{11}{\sqrt{221}}$$

Perform the indicated operations.

The angle θ between the vectors is

$$\theta = \cos^{-1}\left(-\frac{11}{\sqrt{221}}\right) \approx 137.7°.$$ Use a calculator.   • • •

 Check Point 2 Find the angle between the vectors $\mathbf{v} = 4\mathbf{i} - 3\mathbf{j}$ and $\mathbf{w} = \mathbf{i} + 2\mathbf{j}$. Round to the nearest tenth of a degree.

③ Use the dot product to determine if two vectors are orthogonal.

Parallel and Orthogonal Vectors

Two vectors are **parallel** when the angle θ between the vectors is 0° or 180°. If $\theta = 0°$, the vectors point in the same direction. If $\theta = 180°$, the vectors point in opposite directions. **Figure 4.37** shows parallel vectors.

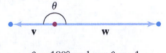

$\theta = 0°$ and $\cos \theta = 1$.
Vectors point in the same direction.

$\theta = 180°$ and $\cos \theta = -1$.
Vectors point in opposite directions.

FIGURE 4.37 Parallel vectors

FIGURE 4.38 Orthogonal vectors: $\theta = 90°$ and $\cos \theta = 0$

Two vectors are **orthogonal** when the angle between the vectors is 90°, shown in **Figure 4.38**. (The word *orthogonal*, rather than *perpendicular*, is used to describe vectors that meet at right angles.) We know that $\mathbf{v} \cdot \mathbf{w} = \|\mathbf{v}\| \|\mathbf{w}\| \cos \theta$. If $\mathbf{v}$ and $\mathbf{w}$ are orthogonal, then

$$\mathbf{v} \cdot \mathbf{w} = \|\mathbf{v}\| \|\mathbf{w}\| \cos 90° = \|\mathbf{v}\| \|\mathbf{w}\|(0) = 0.$$

Conversely, if $\mathbf{v}$ and $\mathbf{w}$ are vectors such that $\mathbf{v} \cdot \mathbf{w} = 0$, then $\|\mathbf{v}\| = 0$ or $\|\mathbf{w}\| = 0$ or $\cos \theta = 0$. If $\cos \theta = 0$, then $\theta = 90°$, so $\mathbf{v}$ and $\mathbf{w}$ are orthogonal.

The preceding discussion is summarized as follows:

> ### The Dot Product and Orthogonal Vectors
>
> Two nonzero vectors $\mathbf{v}$ and $\mathbf{w}$ are orthogonal if and only if $\mathbf{v} \cdot \mathbf{w} = 0$. Because $\mathbf{0} \cdot \mathbf{v} = 0$, the zero vector is orthogonal to every vector $\mathbf{v}$.

EXAMPLE 3   Determining Whether Vectors Are Orthogonal

Are the vectors $\mathbf{v} = 6\mathbf{i} - 3\mathbf{j}$ and $\mathbf{w} = \mathbf{i} + 2\mathbf{j}$ orthogonal?

SOLUTION

The vectors are orthogonal if their dot product is 0. Begin by finding $\mathbf{v} \cdot \mathbf{w}$.

$$\mathbf{v} \cdot \mathbf{w} = (6\mathbf{i} - 3\mathbf{j}) \cdot (\mathbf{i} + 2\mathbf{j}) = 6(1) + (-3)(2) = 6 - 6 = 0$$

The dot product is 0. Thus, the given vectors are orthogonal. They are shown in **Figure 4.39**.   • • •

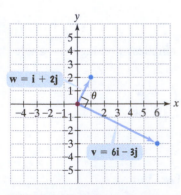

FIGURE 4.39 Orthogonal vectors

 Check Point 3 Are the vectors $\mathbf{v} = 2\mathbf{i} + 3\mathbf{j}$ and $\mathbf{w} = 6\mathbf{i} - 4\mathbf{j}$ orthogonal?

Find the projection of a vector onto another vector.

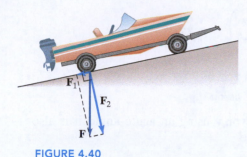

FIGURE 4.40

Projection of a Vector onto Another Vector

You know how to add two vectors to obtain a resultant vector. We now reverse this process by expressing a vector as the sum of two orthogonal vectors. By doing this, you can determine how much force is applied in a particular direction. For example, **Figure 4.40** shows a boat on a tilted ramp. The force due to gravity, **F**, is pulling straight down on the boat. Part of this force, **F₁**, is pushing the boat down the ramp. Another part of this force, **F₂**, is pressing the boat against the ramp, at a right angle to the incline. These two orthogonal vectors, **F₁** and **F₂**, are called the **vector components** of **F**. Notice that

$$\mathbf{F} = \mathbf{F}_1 + \mathbf{F}_2.$$

A method for finding **F₁** and **F₂** involves projecting a vector onto another vector.

Figure 4.41 shows two nonzero vectors, **v** and **w**, with the same initial point. The angle between the vectors, θ, is acute in **Figure 4.41(a)** and obtuse in **Figure 4.41(b)**. A third vector, called the **vector projection of v onto w**, is also shown in each figure, denoted by $\text{proj}_{\mathbf{w}}\mathbf{v}$.

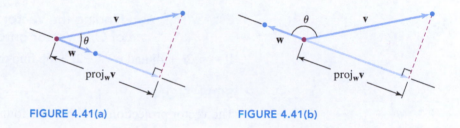

FIGURE 4.41(a)   FIGURE 4.41(b)

How is the vector projection of **v** onto **w** formed? Draw the line segment from the terminal point of **v** that forms a right angle with a line through **w**, shown in red. The projection of **v** onto **w** lies on a line through **w**, and is parallel to vector **w**. This vector begins at the common initial point of **v** and **w**. It ends at the point where the dashed red line segment intersects the line through **w**.

Our goal is to determine an expression for $\text{proj}_{\mathbf{w}}\mathbf{v}$. We begin with its magnitude. Refer to the acute angle θ in **Figure 4.41(a)**. By the definition of the cosine function,

$$\cos \theta = \frac{\|\text{proj}_{\mathbf{w}}\mathbf{v}\|}{\|\mathbf{v}\|}$$

> This is the magnitude of the vector projection of **v** onto **w**.

$$\|\mathbf{v}\| \cos \theta = \|\text{proj}_{\mathbf{w}}\mathbf{v}\| \qquad \text{Multiply both sides by } \|\mathbf{v}\|.$$

$$\|\text{proj}_{\mathbf{w}}\mathbf{v}\| = \|\mathbf{v}\| \cos \theta. \qquad \text{Reverse the two sides.}$$

We can rewrite the right side of this equation and obtain another expression for the magnitude of the vector projection of **v** onto **w**. To do so, use the alternate formula for the dot product, $\mathbf{v} \cdot \mathbf{w} = \|\mathbf{v}\| \|\mathbf{w}\| \cos \theta$.

Divide both sides of $\mathbf{v} \cdot \mathbf{w} = \|\mathbf{v}\| \|\mathbf{w}\| \cos \theta$ by $\|\mathbf{w}\|$:

$$\frac{\mathbf{v} \cdot \mathbf{w}}{\|\mathbf{w}\|} = \|\mathbf{v}\| \cos \theta.$$

The expression on the right side of this equation, $\|\mathbf{v}\| \cos \theta$, is the same expression that appears in the formula for $\|\text{proj}_{\mathbf{w}}\mathbf{v}\|$. Thus,

$$\|\text{proj}_{\mathbf{w}}\mathbf{v}\| = \|\mathbf{v}\| \cos \theta = \frac{\mathbf{v} \cdot \mathbf{w}}{\|\mathbf{w}\|}.$$

We use the formula for the magnitude of $proj_{\mathbf{w}}\mathbf{v}$ to find the vector itself. This is done by finding the scalar product of the magnitude and the unit vector in the direction of $\mathbf{w}$.

$$proj_{\mathbf{w}}\mathbf{v} = \left(\frac{\mathbf{v} \cdot \mathbf{w}}{\|\mathbf{w}\|}\right)\left(\frac{\mathbf{w}}{\|\mathbf{w}\|}\right) = \frac{\mathbf{v} \cdot \mathbf{w}}{\|\mathbf{w}\|^2}\mathbf{w}$$

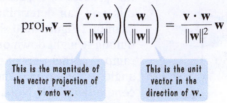

This is the magnitude of the vector projection of v onto w.

This is the unit vector in the direction of w.

Although we derived the expression for $proj_{\mathbf{w}}\mathbf{v}$ using an acute angle, it is also valid when the angle between $\mathbf{v}$ and $\mathbf{w}$ is obtuse.

The Vector Projection of $\mathbf{v}$ onto $\mathbf{w}$

If $\mathbf{v}$ and $\mathbf{w}$ are two nonzero vectors, the vector projection of $\mathbf{v}$ onto $\mathbf{w}$ is

$$proj_{\mathbf{w}}\mathbf{v} = \frac{\mathbf{v} \cdot \mathbf{w}}{\|\mathbf{w}\|^2}\mathbf{w}.$$

EXAMPLE 4   Finding the Vector Projection of One Vector onto Another

If $\mathbf{v} = 2\mathbf{i} + 4\mathbf{j}$ and $\mathbf{w} = -2\mathbf{i} + 6\mathbf{j}$, find the vector projection of $\mathbf{v}$ onto $\mathbf{w}$.

SOLUTION

The vector projection of $\mathbf{v}$ onto $\mathbf{w}$ is found using the formula for $proj_{\mathbf{w}}\mathbf{v}$.

$$proj_{\mathbf{w}}\mathbf{v} = \frac{\mathbf{v} \cdot \mathbf{w}}{\|\mathbf{w}\|^2}\mathbf{w} = \frac{(2\mathbf{i} + 4\mathbf{j}) \cdot (-2\mathbf{i} + 6\mathbf{j})}{\left(\sqrt{(-2)^2 + 6^2}\right)^2}\mathbf{w}$$

$$= \frac{2(-2) + 4(6)}{\left(\sqrt{40}\right)^2}\mathbf{w} = \frac{20}{40}\mathbf{w} = \tfrac{1}{2}(-2\mathbf{i} + 6\mathbf{j}) = -\mathbf{i} + 3\mathbf{j}$$

The three vectors, $\mathbf{v}$, $\mathbf{w}$, and $proj_{\mathbf{w}}\mathbf{v}$, are shown in **Figure 4.42**.   • • •

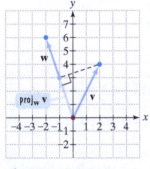

FIGURE 4.42 The vector projection of $\mathbf{v}$ onto $\mathbf{w}$

✓ Check Point **4** If $\mathbf{v} = 2\mathbf{i} - 5\mathbf{j}$ and $\mathbf{w} = \mathbf{i} - \mathbf{j}$, find the vector projection of $\mathbf{v}$ onto $\mathbf{w}$.

5 Express a vector as the sum of two orthogonal vectors.

We use the vector projection of $\mathbf{v}$ onto $\mathbf{w}$, $proj_{\mathbf{w}}\mathbf{v}$, to express $\mathbf{v}$ as the sum of two orthogonal vectors.

The Vector Components of $\mathbf{v}$

Let $\mathbf{v}$ and $\mathbf{w}$ be two nonzero vectors. Vector $\mathbf{v}$ can be expressed as the sum of two orthogonal vectors, $\mathbf{v}_1$ and $\mathbf{v}_2$, where $\mathbf{v}_1$ is parallel to $\mathbf{w}$ and $\mathbf{v}_2$ is orthogonal to $\mathbf{w}$.

$$\mathbf{v}_1 = proj_{\mathbf{w}}\mathbf{v} = \frac{\mathbf{v} \cdot \mathbf{w}}{\|\mathbf{w}\|^2}\mathbf{w}, \quad \mathbf{v}_2 = \mathbf{v} - \mathbf{v}_1$$

Thus, $\mathbf{v} = \mathbf{v}_1 + \mathbf{v}_2$. The vectors $\mathbf{v}_1$ and $\mathbf{v}_2$ are called the **vector components** of $\mathbf{v}$. The process of expressing $\mathbf{v}$ as $\mathbf{v}_1 + \mathbf{v}_2$ is called the **decomposition** of $\mathbf{v}$ into $\mathbf{v}_1$ and $\mathbf{v}_2$.

EXAMPLE 5   Decomposing a Vector into Two Orthogonal Vectors

Let $\mathbf{v} = 2\mathbf{i} + 4\mathbf{j}$ and $\mathbf{w} = -2\mathbf{i} + 6\mathbf{j}$. Decompose $\mathbf{v}$ into two vectors, $\mathbf{v}_1$ and $\mathbf{v}_2$, where $\mathbf{v}_1$ is parallel to $\mathbf{w}$ and $\mathbf{v}_2$ is orthogonal to $\mathbf{w}$.

SOLUTION

The vectors $v = 2i + 4j$ and $w = -2i + 6j$ are the vectors we worked with in Example 4. We use the formulas in the box on the preceding page.

$v_1 = \text{proj}_w v = -i + 3j$ We obtained this vector in Example 4.

$v_2 = v - v_1 = (2i + 4j) - (-i + 3j) = 3i + j$ • • •

✓ **Check Point 5** Let $v = 2i - 5j$ and $w = i - j$. (These are the vectors from Check Point 4.) Decompose v into two vectors, v_1 and v_2, where v_1 is parallel to w and v_2 is orthogonal to w.

⑥ Compute work.

Work: An Application of the Dot Product

The bad news: Your car just died. The good news: It died on a level road just 200 feet from a gas station. Exerting a constant force of 90 pounds, and not necessarily whistling as you work, you manage to push the car to the gas station.

Force: 90 pounds

A B

|— 200 feet —|

Although you did not whistle, you certainly did work pushing the car 200 feet from point A to point B. How much work did you do? If a constant force $\mathbf{F}$ is applied to an object, moving it from point A to point B in the direction of the force, the work, W, done is

$$W = (\text{magnitude of force})(\text{distance from } A \text{ to } B).$$

You pushed with a force of 90 pounds for a distance of 200 feet. The work done by your force is

$$W = (90 \text{ pounds})(200 \text{ feet})$$

or 18,000 foot-pounds. Work is often measured in foot-pounds or in newton-meters.

The photo on the left shows an adult pulling a small child in a wagon. Work is being done. However, the situation is not quite the same as pushing your car. Pushing the car, the force you applied was along the line of motion. By contrast, the force of the adult pulling the wagon is not applied along the line of the wagon's motion. In this case, the dot product is used to determine the work done by the force.

Definition of Work

The work, W, done by a force $\mathbf{F}$ moving an object from A to B is

$$W = \mathbf{F} \cdot \overrightarrow{AB}.$$

When computing work, it is often easier to use the alternative formula for the dot product. Thus,

$$W = \mathbf{F} \cdot \overrightarrow{AB} = \|\mathbf{F}\| \, \|\overrightarrow{AB}\| \cos \theta.$$

$\|\mathbf{F}\|$ is the magnitude of the force.

$\|\overrightarrow{AB}\|$ is the distance over which the constant force is applied.

θ is the angle between the force and the direction of motion.

It is correct to refer to W as either the work done or the work done by the force.

FIGURE 4.43 Computing work done pulling a sled 200 feet

EXAMPLE 6 Computing Work

A child pulls a sled along level ground by exerting a force of 30 pounds on a rope that makes an angle of 35° with the horizontal. How much work is done pulling the sled 200 feet?

SOLUTION

The situation is illustrated in **Figure 4.43**. The work done is

$$W = \|\mathbf{F}\| \|\overrightarrow{AB}\| \cos \theta = (30)(200) \cos 35° \approx 4915.$$

Magnitude of the force is 30 pounds.

Distance is 200 feet.

The angle between the force and the sled's motion is 35°.

Thus, the work done is approximately 4915 foot-pounds.   • • •

✓ **Check Point 6** A child pulls a wagon along level ground by exerting a force of 20 pounds on a handle that makes an angle of 30° with the horizontal. How much work is done pulling the wagon 150 feet?

CONCEPT AND VOCABULARY CHECK

Fill in each blank so that the resulting statement is true.

1. If $\mathbf{v} = a_1\mathbf{i} + b_1\mathbf{j}$ and $\mathbf{w} = a_2\mathbf{i} + b_2\mathbf{j}$ are vectors, the product $\mathbf{v} \cdot \mathbf{w}$, called the _____, is defined as $\mathbf{v} \cdot \mathbf{w} =$ _____.

2. If $\mathbf{v}$ and $\mathbf{w}$ are two nonzero vectors and θ is the smallest nonnegative angle between them, then $\mathbf{v} \cdot \mathbf{w} =$ _____.

3. If $\mathbf{v} \cdot \mathbf{w} = 0$, then the vectors $\mathbf{v}$ and $\mathbf{w}$ are _____.

4. True or false: Given two nonzero vectors $\mathbf{v}$ and $\mathbf{w}$, $\mathbf{v}$ can be decomposed into two vectors, one parallel to $\mathbf{w}$ and the other orthogonal to $\mathbf{w}$. _____

5. True or false: The definition of work indicates that work is a vector. _____

EXERCISE SET 4.4

Practice Exercises

In Exercises 1–8, use the given vectors to find $\mathbf{v} \cdot \mathbf{w}$ and $\mathbf{v} \cdot \mathbf{v}$.

1. $\mathbf{v} = 3\mathbf{i} + \mathbf{j}$, $\mathbf{w} = \mathbf{i} + 3\mathbf{j}$
2. $\mathbf{v} = 3\mathbf{i} + 3\mathbf{j}$, $\mathbf{w} = \mathbf{i} + 4\mathbf{j}$
3. $\mathbf{v} = 5\mathbf{i} - 4\mathbf{j}$, $\mathbf{w} = -2\mathbf{i} - \mathbf{j}$
4. $\mathbf{v} = 7\mathbf{i} - 2\mathbf{j}$, $\mathbf{w} = -3\mathbf{i} - \mathbf{j}$
5. $\mathbf{v} = -6\mathbf{i} - 5\mathbf{j}$, $\mathbf{w} = -10\mathbf{i} - 8\mathbf{j}$
6. $\mathbf{v} = -8\mathbf{i} - 3\mathbf{j}$, $\mathbf{w} = -10\mathbf{i} - 5\mathbf{j}$
7. $\mathbf{v} = 5\mathbf{i}$, $\mathbf{w} = \mathbf{j}$
8. $\mathbf{v} = \mathbf{i}$, $\mathbf{w} = -5\mathbf{j}$

In Exercises 9–16, let

$$\mathbf{u} = 2\mathbf{i} - \mathbf{j}, \quad \mathbf{v} = 3\mathbf{i} + \mathbf{j}, \quad and \quad \mathbf{w} = \mathbf{i} + 4\mathbf{j}.$$

Find each specified scalar.

9. $\mathbf{u} \cdot (\mathbf{v} + \mathbf{w})$
10. $\mathbf{v} \cdot (\mathbf{u} + \mathbf{w})$
11. $\mathbf{u} \cdot \mathbf{v} + \mathbf{u} \cdot \mathbf{w}$
12. $\mathbf{v} \cdot \mathbf{u} + \mathbf{v} \cdot \mathbf{w}$
13. $(4\mathbf{u}) \cdot \mathbf{v}$
14. $(5\mathbf{v}) \cdot \mathbf{w}$
15. $4(\mathbf{u} \cdot \mathbf{v})$
16. $5(\mathbf{v} \cdot \mathbf{w})$

In Exercises 17–22, find the angle between $\mathbf{v}$ and $\mathbf{w}$. Round to the nearest tenth of a degree.

17. $\mathbf{v} = 2\mathbf{i} - \mathbf{j}$, $\mathbf{w} = 3\mathbf{i} + 4\mathbf{j}$
18. $\mathbf{v} = -2\mathbf{i} + 5\mathbf{j}$, $\mathbf{w} = 3\mathbf{i} + 6\mathbf{j}$
19. $\mathbf{v} = -3\mathbf{i} + 2\mathbf{j}$, $\mathbf{w} = 4\mathbf{i} - \mathbf{j}$
20. $\mathbf{v} = \mathbf{i} + 2\mathbf{j}$, $\mathbf{w} = 4\mathbf{i} - 3\mathbf{j}$
21. $\mathbf{v} = 6\mathbf{i}$, $\mathbf{w} = 5\mathbf{i} + 4\mathbf{j}$
22. $\mathbf{v} = 3\mathbf{j}$, $\mathbf{w} = 4\mathbf{i} + 5\mathbf{j}$

In Exercises 23–32, use the dot product to determine whether $\mathbf{v}$ and $\mathbf{w}$ are orthogonal.

23. $\mathbf{v} = \mathbf{i} + \mathbf{j}$, $\mathbf{w} = \mathbf{i} - \mathbf{j}$
24. $\mathbf{v} = \mathbf{i} + \mathbf{j}$, $\mathbf{w} = -\mathbf{i} + \mathbf{j}$
25. $\mathbf{v} = 2\mathbf{i} + 8\mathbf{j}$, $\mathbf{w} = 4\mathbf{i} - \mathbf{j}$
26. $\mathbf{v} = 8\mathbf{i} - 4\mathbf{j}$, $\mathbf{w} = -6\mathbf{i} - 12\mathbf{j}$
27. $\mathbf{v} = 2\mathbf{i} - 2\mathbf{j}$, $\mathbf{w} = -\mathbf{i} + \mathbf{j}$
28. $\mathbf{v} = 5\mathbf{i} - 5\mathbf{j}$, $\mathbf{w} = \mathbf{i} - \mathbf{j}$
29. $\mathbf{v} = 3\mathbf{i}$, $\mathbf{w} = -4\mathbf{i}$
30. $\mathbf{v} = 5\mathbf{i}$, $\mathbf{w} = -6\mathbf{i}$
31. $\mathbf{v} = 3\mathbf{i}$, $\mathbf{w} = -4\mathbf{j}$
32. $\mathbf{v} = 5\mathbf{i}$, $\mathbf{w} = -6\mathbf{j}$

In Exercises 33–38, find $\text{proj}_\mathbf{w}\mathbf{v}$. *Then decompose* **v** *into two vectors,* $\mathbf{v}_1$ *and* $\mathbf{v}_2$, *where* $\mathbf{v}_1$ *is parallel to* **w** *and* $\mathbf{v}_2$ *is orthogonal to* **w**.

33. $\mathbf{v} = 3\mathbf{i} - 2\mathbf{j}, \quad \mathbf{w} = \mathbf{i} - \mathbf{j}$

34. $\mathbf{v} = 3\mathbf{i} - 2\mathbf{j}, \quad \mathbf{w} = 2\mathbf{i} + \mathbf{j}$

35. $\mathbf{v} = \mathbf{i} + 3\mathbf{j}, \quad \mathbf{w} = -2\mathbf{i} + 5\mathbf{j}$

36. $\mathbf{v} = 2\mathbf{i} + 4\mathbf{j}, \quad \mathbf{w} = -3\mathbf{i} + 6\mathbf{j}$

37. $\mathbf{v} = \mathbf{i} + 2\mathbf{j}, \quad \mathbf{w} = 3\mathbf{i} + 6\mathbf{j}$

38. $\mathbf{v} = 2\mathbf{i} + \mathbf{j}, \quad \mathbf{w} = 6\mathbf{i} + 3\mathbf{j}$

Practice Plus

In Exercises 39–42, let

$$\mathbf{u} = -\mathbf{i} + \mathbf{j}, \quad \mathbf{v} = 3\mathbf{i} - 2\mathbf{j}, \quad \text{and} \quad \mathbf{w} = -5\mathbf{j}.$$

Find each specified scalar or vector.

39. $5\mathbf{u} \cdot (3\mathbf{v} - 4\mathbf{w})$

40. $4\mathbf{u} \cdot (5\mathbf{v} - 3\mathbf{w})$

41. $\text{proj}_\mathbf{u}(\mathbf{v} + \mathbf{w})$

42. $\text{proj}_\mathbf{u}(\mathbf{v} - \mathbf{w})$

In Exercises 43–44, find the angle, in degrees, between **v** *and* **w**.

43. $\mathbf{v} = 2\cos\dfrac{4\pi}{3}\mathbf{i} + 2\sin\dfrac{4\pi}{3}\mathbf{j}, \quad \mathbf{w} = 3\cos\dfrac{3\pi}{2}\mathbf{i} + 3\sin\dfrac{3\pi}{2}\mathbf{j}$

44. $\mathbf{v} = 3\cos\dfrac{5\pi}{3}\mathbf{i} + 3\sin\dfrac{5\pi}{3}\mathbf{j}, \quad \mathbf{w} = 2\cos\pi\mathbf{i} + 2\sin\pi\mathbf{j}$

In Exercises 45–50, determine whether **v** *and* **w** *are parallel, orthogonal, or neither.*

45. $\mathbf{v} = 3\mathbf{i} - 5\mathbf{j}, \quad \mathbf{w} = 6\mathbf{i} - 10\mathbf{j}$

46. $\mathbf{v} = -2\mathbf{i} + 3\mathbf{j}, \quad \mathbf{w} = -6\mathbf{i} + 9\mathbf{j}$

47. $\mathbf{v} = 3\mathbf{i} - 5\mathbf{j}, \quad \mathbf{w} = 6\mathbf{i} + 10\mathbf{j}$

48. $\mathbf{v} = -2\mathbf{i} + 3\mathbf{j}, \quad \mathbf{w} = -6\mathbf{i} - 9\mathbf{j}$

49. $\mathbf{v} = 3\mathbf{i} - 5\mathbf{j}, \quad \mathbf{w} = 6\mathbf{i} + \dfrac{18}{5}\mathbf{j}$

50. $\mathbf{v} = -2\mathbf{i} + 3\mathbf{j}, \quad \mathbf{w} = -6\mathbf{i} - 4\mathbf{j}$

Application Exercises

51. The components of $\mathbf{v} = 240\mathbf{i} + 300\mathbf{j}$ represent the respective number of gallons of regular and premium gas sold at a station. The components of $\mathbf{w} = 2.90\mathbf{i} + 3.07\mathbf{j}$ represent the respective prices per gallon for each kind of gas. Find $\mathbf{v} \cdot \mathbf{w}$ and describe what the answer means in practical terms.

52. The components of $\mathbf{v} = 180\mathbf{i} + 450\mathbf{j}$ represent the respective number of one-day and three-day videos rented from a video store. The components of $\mathbf{w} = 3\mathbf{i} + 2\mathbf{j}$ represent the prices to rent the one-day and three-day videos, respectively. Find $\mathbf{v} \cdot \mathbf{w}$ and describe what the answer means in practical terms.

53. Find the work done in pushing a car along a level road from point A to point B, 80 feet from A, while exerting a constant force of 95 pounds. Round to the nearest foot-pound.

54. Find the work done when a crane lifts a 6000-pound boulder through a vertical distance of 12 feet. Round to the nearest foot-pound.

55. A wagon is pulled along level ground by exerting a force of 40 pounds on a handle that makes an angle of 32° with the horizontal. How much work is done pulling the wagon 100 feet? Round to the nearest foot-pound.

56. A wagon is pulled along level ground by exerting a force of 25 pounds on a handle that makes an angle of 38° with the horizontal. How much work is done pulling the wagon 100 feet? Round to the nearest foot-pound.

57. A force of 60 pounds on a rope is used to pull a box up a ramp inclined at 12° from the horizontal. The figure shows that the rope forms an angle of 38° with the horizontal. How much work is done pulling the box 20 feet along the ramp?

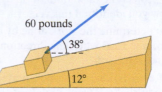

60 pounds
38°
12°

58. A force of 80 pounds on a rope is used to pull a box up a ramp inclined at 10° from the horizontal. The rope forms an angle of 33° with the horizontal. How much work is done pulling the box 25 feet along the ramp?

59. A force is given by the vector $\mathbf{F} = 3\mathbf{i} + 2\mathbf{j}$. The force moves an object along a straight line from the point $(4, 9)$ to the point $(10, 20)$. Find the work done if the distance is measured in feet and the force is measured in pounds.

60. A force is given by the vector $\mathbf{F} = 5\mathbf{i} + 7\mathbf{j}$. The force moves an object along a straight line from the point $(8, 11)$ to the point $(18, 20)$. Find the work done if the distance is measured in meters and the force is measured in newtons.

61. A force of 4 pounds acts in the direction of 50° to the horizontal. The force moves an object along a straight line from the point $(3, 7)$ to the point $(8, 10)$, with distance measured in feet. Find the work done by the force.

62. A force of 6 pounds acts in the direction of 40° to the horizontal. The force moves an object along a straight line from the point $(5, 9)$ to the point $(8, 20)$, with the distance measured in feet. Find the work done by the force.

63. Refer to **Figure 4.40** on page 253. Suppose a boat weighs 700 pounds and is on a ramp inclined at 30°. Represent the force due to gravity, **F**, using

$$\mathbf{F} = -700\mathbf{j}.$$

a. Write a unit vector along the ramp in the upward direction.

b. Find the vector projection of **F** onto the unit vector from part (a).

c. What is the magnitude of the vector projection in part (b)? What does this represent?

64. Refer to **Figure 4.40** on page 253. Suppose a boat weighs 650 pounds and is on a ramp inclined at 30°. Represent the force due to gravity, **F**, using

$$\mathbf{F} = -650\mathbf{j}.$$

a. Write a unit vector along the ramp in the upward direction.

b. Find the vector projection of **F** onto the unit vector from part (a).

c. What is the magnitude of the vector projection in part (b)? What does this represent?

Explaining the Concepts

65. Explain how to find the dot product of two vectors.

66. Using words and no symbols, describe how to find the dot product of two vectors with the alternative formula

$$\mathbf{v} \cdot \mathbf{w} = \|\mathbf{v}\|\|\mathbf{w}\|\cos \theta.$$

67. Describe how to find the angle between two vectors.

68. What are parallel vectors?

69. What are orthogonal vectors?

70. How do you determine if two vectors are orthogonal?

71. Draw two vectors, $\mathbf{v}$ and $\mathbf{w}$, with the same initial point. Show the vector projection of $\mathbf{v}$ onto $\mathbf{w}$ in your diagram. Then describe how you identified this vector.

72. How do you determine the work done by a force $\mathbf{F}$ in moving an object from A to B when the direction of the force is not along the line of motion?

73. A weightlifter is holding a barbell perfectly still above his head, his body shaking from the effort. How much work is the weightlifter doing? Explain your answer.

74. Describe one way in which the everyday use of the word *work* is different from the definition of work given in this section.

Critical Thinking Exercises

Make Sense? *In Exercises 75–78, determine whether each statement makes sense or does not make sense, and explain your reasoning.*

75. Although I expected vector operations to produce another vector, the dot product of two vectors is not a vector, but a real number.

76. I've noticed that whenever the dot product is negative, the angle between the two vectors is obtuse.

77. I'm working with a unit vector, so its dot product with itself must be 1.

78. The weightlifter does more work in raising 300 kilograms above her head than Atlas, who is supporting the entire world.

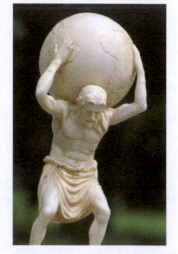

In Exercises 79–81, use the vectors

$$\mathbf{u} = a_1\mathbf{i} + b_1\mathbf{j}, \quad \mathbf{v} = a_2\mathbf{i} + b_2\mathbf{j}, \quad \text{and} \quad \mathbf{w} = a_3\mathbf{i} + b_3\mathbf{j},$$

to prove the given property.

79. $\mathbf{u} \cdot \mathbf{v} = \mathbf{v} \cdot \mathbf{u}$

80. $(c\mathbf{u}) \cdot \mathbf{v} = c(\mathbf{u} \cdot \mathbf{v})$

81. $\mathbf{u} \cdot (\mathbf{v} + \mathbf{w}) = \mathbf{u} \cdot \mathbf{v} + \mathbf{u} \cdot \mathbf{w}$

82. If $\mathbf{v} = -2\mathbf{i} + 5\mathbf{j}$, find a vector orthogonal to $\mathbf{v}$.

83. Find a value of b so that $15\mathbf{i} - 3\mathbf{j}$ and $-4\mathbf{i} + b\mathbf{j}$ are orthogonal.

84. Prove that the projection of $\mathbf{v}$ onto $\mathbf{i}$ is $(\mathbf{v} \cdot \mathbf{i})\mathbf{i}$.

85. Find two vectors $\mathbf{v}$ and $\mathbf{w}$ such that the projection of $\mathbf{v}$ onto $\mathbf{w}$ is $\mathbf{v}$.

Group Exercise

86. Group members should research and present a report on unusual and interesting applications of vectors.

Retaining the Concepts

87. Solve: $2\sin^2 x - 1 = 0, 0 \le x < 2\pi$.

(Section 3.5, Example 5)

88. Verify the identity:

$$\sin 2x = \frac{2\tan x}{1 + \tan^2 x}.$$

(Section 3.3, Examples 3 and 6)

89. Use a right triangle to write $\cos(\tan^{-1} x)$ as an algebraic expression. Assume that x is positive and that the given inverse trigonometric function is defined for the expression in x. (Section 2.3, Example 9)

Preview Exercises

Exercises 90–92 will help you prepare for the material covered in the first section of the next chapter.

90. Multiply: $(7 - 3x)(-2 - 5x)$.

91. Simplify: $\sqrt{18} - \sqrt{8}$.

92. Rationalize the denominator: $\dfrac{7 + 4\sqrt{2}}{2 - 5\sqrt{2}}$.

CHAPTER 4   Summary, Review, and Test

SUMMARY

DEFINITIONS AND CONCEPTS	EXAMPLES

4.1 and 4.2   The Law of Sines; The Law of Cosines

a. The Law of Sines

$$\frac{a}{\sin A} = \frac{b}{\sin B} = \frac{c}{\sin C}$$

Ex. 1, p. 213;
Ex. 2, p. 214;
Ex. 3, p. 215;
Ex. 4, p. 216

b. The Law of Sines is used to solve SAA, ASA, and SSA (the ambiguous case) triangles. The ambiguous case may result in no triangle, one triangle, or two triangles; see the box on page 215.

Ex. 5, p. 217

c. The area of a triangle equals one-half the product of the lengths of two sides times the sine of their included angle.

Ex. 6, p. 218

d. The Law of Cosines

$$a^2 = b^2 + c^2 - 2bc \cos A$$
$$b^2 = a^2 + c^2 - 2ac \cos B$$
$$c^2 = a^2 + b^2 - 2ab \cos C$$

e. The Law of Cosines is used to find the side opposite the given angle in an SAS triangle; see the box at the bottom of the page on page 225. The Law of Cosines is also used to find the angle opposite the longest side in an SSS triangle; see the box on page 226.

Ex. 1, p. 226;
Ex. 2, p. 227

f. Heron's Formula for the Area of a Triangle
The area of a triangle with sides $a, b,$ and c is $\sqrt{s(s - a)(s - b)(s - c)}$, where s is one-half its perimeter: $s = \frac{1}{2}(a + b + c)$.

Ex. 4, p. 228

4.3   Vectors

a. A vector is a directed line segment.

b. Equal vectors have the same magnitude and the same direction.

Ex. 1, p. 235

c. The vector $k\mathbf{v}$, the scalar multiple of the vector $\mathbf{v}$ and the scalar k, has magnitude $|k|\|\mathbf{v}\|$. The direction of $k\mathbf{v}$ is the same as that of $\mathbf{v}$ if $k > 0$ and opposite $\mathbf{v}$ if $k < 0$.

Figure 4.23, p. 236

d. The sum $\mathbf{u} + \mathbf{v}$, called the resultant vector, can be expressed geometrically. Position $\mathbf{u}$ and $\mathbf{v}$ so that the terminal point of $\mathbf{u}$ coincides with the initial point of $\mathbf{v}$. The vector $\mathbf{u} + \mathbf{v}$ extends from the initial point of $\mathbf{u}$ to the terminal point of $\mathbf{v}$.

Figure 4.24, p. 236

e. The difference of two vectors, $\mathbf{u} - \mathbf{v}$, is defined as $\mathbf{u} + (-\mathbf{v})$.

Figure 4.25, p. 237

f. The vector $\mathbf{i}$ is the unit vector whose direction is along the positive x-axis. The vector $\mathbf{j}$ is the unit vector whose direction is along the positive y-axis.

g. Vector $\mathbf{v}$, from $(0, 0)$ to (a, b), called a position vector, is represented as $\mathbf{v} = a\mathbf{i} + b\mathbf{j}$, where a is the horizontal component and b is the vertical component. The magnitude of $\mathbf{v}$ is given by $\|\mathbf{v}\| = \sqrt{a^2 + b^2}$.

Ex. 2, p. 238

h. Vector $\mathbf{v}$ from (x_1, y_1) to (x_2, y_2) is equal to the position vector $\mathbf{v} = (x_2 - x_1)\mathbf{i} + (y_2 - y_1)\mathbf{j}$. In rectangular coordinates, the term "vector" refers to the position vector in terms of $\mathbf{i}$ and $\mathbf{j}$ that is equal to it.

Ex. 3, p. 239

DEFINITIONS AND CONCEPTS

i. Operations with Vectors in Terms of $\mathbf{i}$ and $\mathbf{j}$

If $\mathbf{v} = a_1\mathbf{i} + b_1\mathbf{j}$ and $\mathbf{w} = a_2\mathbf{i} + b_2\mathbf{j}$, then

- $\mathbf{v} + \mathbf{w} = (a_1 + a_2)\mathbf{i} + (b_1 + b_2)\mathbf{j}$
- $\mathbf{v} - \mathbf{w} = (a_1 - a_2)\mathbf{i} + (b_1 - b_2)\mathbf{j}$
- $k\mathbf{v} = (ka_1)\mathbf{i} + (kb_1)\mathbf{j}$

Ex. 4, p. 240;
Ex. 5, p. 240;
Ex. 6, p. 241

j. The zero vector $\mathbf{0}$ is the vector whose magnitude is 0 and is assigned no direction. Many properties of vector addition and scalar multiplication involve the zero vector. Some of these properties are listed in the box on page 241.

k. The vector $\dfrac{\mathbf{v}}{\|\mathbf{v}\|}$ is the unit vector that has the same direction as $\mathbf{v}$.

Ex. 7, p. 242

l. A vector with magnitude $\|\mathbf{v}\|$ and direction angle θ, the angle that $\mathbf{v}$ makes with the positive x-axis, can be expressed in terms of its magnitude and direction angle as

$$\mathbf{v} = \|\mathbf{v}\| \cos \theta \mathbf{i} + \|\mathbf{v}\| \sin \theta \mathbf{j}.$$

Ex. 8, p. 243;
Ex. 9, p. 243

4.4 The Dot Product

a. Definition of the Dot Product

If $\mathbf{v} = a_1\mathbf{i} + b_1\mathbf{j}$ and $\mathbf{w} = a_2\mathbf{i} + b_2\mathbf{j}$, the dot product of $\mathbf{v}$ and $\mathbf{w}$ is defined by $\mathbf{v} \cdot \mathbf{w} = a_1 a_2 + b_1 b_2$.

Ex. 1, p. 250

b. Alternative Formula for the Dot Product: $\mathbf{v} \cdot \mathbf{w} = \|\mathbf{v}\| \|\mathbf{w}\| \cos \theta$, where θ is the smallest nonnegative angle between $\mathbf{v}$ and $\mathbf{w}$

c. Angle between Two Vectors

$$\cos \theta = \frac{\mathbf{v} \cdot \mathbf{w}}{\|\mathbf{v}\| \|\mathbf{w}\|} \quad \text{and} \quad \theta = \cos^{-1}\left(\frac{\mathbf{v} \cdot \mathbf{w}}{\|\mathbf{v}\| \|\mathbf{w}\|}\right)$$

Ex. 2, p. 251

d. Two vectors are orthogonal when the angle between them is 90°. To show that two vectors are orthogonal, show that their dot product is zero.

Ex. 3, p. 252

e. The vector projection of $\mathbf{v}$ onto $\mathbf{w}$ is given by

$$\text{proj}_{\mathbf{w}}\mathbf{v} = \frac{\mathbf{v} \cdot \mathbf{w}}{\|\mathbf{w}\|^2}\mathbf{w}.$$

Ex. 4, p. 254

f. A vector may be expressed as the sum of two orthogonal vectors, called the vector components. See the box at the bottom of the page on page 254.

Ex. 5, p. 254

g. The work, W, done by a force $\mathbf{F}$ moving an object from A to B is $W = \mathbf{F} \cdot \overrightarrow{AB}$.

Thus, $W = \|\mathbf{F}\| \|\overrightarrow{AB}\| \cos \theta$, where θ is the angle between the force and the direction of motion.

Ex. 6, p. 256

REVIEW EXERCISES

4.1 and 4.2

In Exercises 1–12, solve each triangle. Round lengths to the nearest tenth and angle measures to the nearest degree. If no triangle exists, state "no triangle." If two triangles exist, solve each triangle.

1. $A = 70°$, $B = 55°$, $a = 12$
2. $B = 107°$, $C = 30°$, $c = 126$
3. $B = 66°$, $a = 17$, $c = 12$
4. $a = 117$, $b = 66$, $c = 142$
5. $A = 35°$, $B = 25°$, $c = 68$
6. $A = 39°$, $a = 20$, $b = 26$
7. $C = 50°$, $a = 3$, $c = 1$
8. $A = 162°$, $b = 11.2$, $c = 48.2$
9. $a = 26.1$, $b = 40.2$, $c = 36.5$
10. $A = 40°$, $a = 6$, $b = 4$
11. $B = 37°$, $a = 12.4$, $b = 8.7$
12. $A = 23°$, $a = 54.3$, $b = 22.1$

In Exercises 13–16, find the area of the triangle having the given measurements. Round to the nearest square unit.

13. $C = 42°$, $a = 4$ feet, $b = 6$ feet
14. $A = 22°$, $b = 4$ feet, $c = 5$ feet
15. $a = 2$ meters, $b = 4$ meters, $c = 5$ meters
16. $a = 2$ meters, $b = 2$ meters, $c = 2$ meters
17. The A-frame cabin shown below is 35 feet wide. The roof of the cabin makes a 55° angle with the cabin's base. Find the length of one side of the roof from its ground level to the peak. Round to the nearest tenth of a foot.

18. Two cars leave a city at the same time and travel along straight highways that differ in direction by 80°. One car averages 60 miles per hour and the other averages 50 miles per hour. How far apart will the cars be after 30 minutes? Round to the nearest tenth of a mile.
19. Two airplanes leave an airport at the same time on different runways. One flies on a bearing of N66.5°W at 325 miles per hour. The other airplane flies on a bearing of S26.5°W at 300 miles per hour. How far apart will the airplanes be after two hours?
20. The figure shows three roads that intersect to bound a triangular piece of land. Find the lengths of the other two sides of the land to the nearest foot.

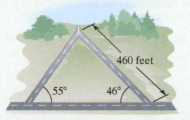

21. A commercial piece of real estate is priced at $5.25 per square foot. Find the cost, to the nearest dollar, of a triangular lot measuring 260 feet by 320 feet by 450 feet.

4.3

In Exercises 22–24, sketch each vector as a position vector and find its magnitude.

22. $\mathbf{v} = -3\mathbf{i} - 4\mathbf{j}$ 23. $\mathbf{v} = 5\mathbf{i} - 2\mathbf{j}$
24. $\mathbf{v} = -3\mathbf{j}$

In Exercises 25–26, let $\mathbf{v}$ *be the vector from initial point* P_1 *to terminal point* P_2. *Write* $\mathbf{v}$ *in terms of* $\mathbf{i}$ *and* $\mathbf{j}$.

25. $P_1 = (2, -1)$, $P_2 = (5, -3)$
26. $P_1 = (-3, 0)$, $P_2 = (-2, -2)$

In Exercises 27–30, let

$$\mathbf{v} = \mathbf{i} - 5\mathbf{j} \quad \text{and} \quad \mathbf{w} = -2\mathbf{i} + 7\mathbf{j}.$$

Find each specified vector or scalar.

27. $\mathbf{v} + \mathbf{w}$ 28. $\mathbf{w} - \mathbf{v}$
29. $6\mathbf{v} - 3\mathbf{w}$ 30. $\|-2\mathbf{v}\|$

In Exercises 31–32, find the unit vector that has the same direction as the vector $\mathbf{v}$.

31. $\mathbf{v} = 8\mathbf{i} - 6\mathbf{j}$
32. $\mathbf{v} = -\mathbf{i} + 2\mathbf{j}$
33. The magnitude and direction angle of $\mathbf{v}$ are $\|\mathbf{v}\| = 12$ and $\theta = 60°$. Express $\mathbf{v}$ in terms of $\mathbf{i}$ and $\mathbf{j}$.
34. The magnitude and direction of two forces acting on an object are 100 pounds, N25°E, and 200 pounds, N80°E, respectively. Find the magnitude, to the nearest pound, and the direction angle, to the nearest tenth of a degree, of the resultant force.
35. Your boat is moving at a speed of 15 miles per hour at an angle of 25° upstream on a river flowing at 4 miles per hour. The situation is illustrated in the figure below.

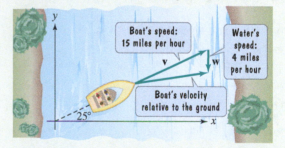

   a. Find the vector representing your boat's velocity relative to the ground.
   b. What is the speed of your boat, to the nearest mile per hour, relative to the ground?
   c. What is the boat's direction angle, to the nearest tenth of a degree, relative to the ground?

4.4

36. If $\mathbf{u} = 5\mathbf{i} + 2\mathbf{j}$, $\mathbf{v} = \mathbf{i} - \mathbf{j}$, and $\mathbf{w} = 3\mathbf{i} - 7\mathbf{j}$, find $\mathbf{u} \cdot (\mathbf{v} + \mathbf{w})$.

In Exercises 37–39, find the dot product $\mathbf{v} \cdot \mathbf{w}$. *Then find the angle between* $\mathbf{v}$ *and* $\mathbf{w}$ *to the nearest tenth of a degree.*

37. $\mathbf{v} = 2\mathbf{i} + 3\mathbf{j}$, $\mathbf{w} = 7\mathbf{i} - 4\mathbf{j}$
38. $\mathbf{v} = 2\mathbf{i} + 4\mathbf{j}$, $\mathbf{w} = 6\mathbf{i} - 11\mathbf{j}$
39. $\mathbf{v} = 2\mathbf{i} + \mathbf{j}$, $\mathbf{w} = \mathbf{i} - \mathbf{j}$

In Exercises 40–41, use the dot product to determine whether **v** *and* **w** *are orthogonal.*

40. $\mathbf{v} = 12\mathbf{i} - 8\mathbf{j}$, $\mathbf{w} = 2\mathbf{i} + 3\mathbf{j}$

41. $\mathbf{v} = \mathbf{i} + 3\mathbf{j}$, $\mathbf{w} = -3\mathbf{i} - \mathbf{j}$

In Exercises 42–43, find $\text{proj}_\mathbf{w}\,\mathbf{v}$. *Then decompose* **v** *into two vectors,* $\mathbf{v}_1$ *and* $\mathbf{v}_2$, *where* $\mathbf{v}_1$ *is parallel to* **w** *and* $\mathbf{v}_2$ *is orthogonal to* **w**.

42. $\mathbf{v} = -2\mathbf{i} + 5\mathbf{j}$, $\mathbf{w} = 5\mathbf{i} + 4\mathbf{j}$

43. $\mathbf{v} = -\mathbf{i} + 2\mathbf{j}$, $\mathbf{w} = 3\mathbf{i} - \mathbf{j}$

44. A heavy crate is dragged 50 feet along a level floor. Find the work done if a force of 30 pounds at an angle of 42° is used.

CHAPTER 4 TEST

1. In oblique triangle ABC, $A = 34°$, $B = 68°$, and $a = 4.8$. Find b to the nearest tenth.

2. In oblique triangle ABC, $C = 68°$, $a = 5$, and $b = 6$. Find c to the nearest tenth.

3. In oblique triangle ABC, $a = 17$ inches, $b = 45$ inches, and $c = 32$ inches. Find the area of the triangle to the nearest square inch.

4. If $P_1 = (-2, 3)$, $P_2 = (-1, 5)$, and **v** is the vector from P_1 to P_2,

    a. Write **v** in terms of **i** and **j**.

    b. Find $\|\mathbf{v}\|$.

In Exercises 5–8, let

$$\mathbf{v} = -5\mathbf{i} + 2\mathbf{j} \quad \text{and} \quad \mathbf{w} = 2\mathbf{i} - 4\mathbf{j}.$$

Find the specified vector, scalar, or angle.

5. $3\mathbf{v} - 4\mathbf{w}$

6. $\mathbf{v} \cdot \mathbf{w}$

7. the angle between **v** and **w**, to the nearest degree

8. $\text{proj}_\mathbf{w}\mathbf{v}$

9. A small fire is sighted from ranger stations A and B. Station B is 1.6 miles due east of station A. The bearing of the fire from station A is N35°E and the bearing of the fire from station B is N50°W. How far, to the nearest tenth of a mile, is the fire from station A?

10. The magnitude and direction of two forces acting on an object are 250 pounds, N60°E, and 150 pounds, S45°E. Find the magnitude, to the nearest pound, and the direction angle, to the nearest tenth of a degree, of the resultant force.

11. A child is pulling a wagon with a force of 40 pounds. How much work is done in moving the wagon 60 feet if the handle makes an angle of 35° with the ground? Round to the nearest foot-pound.

CUMULATIVE REVIEW EXERCISES (CHAPTERS 1–4)

In Exercises 1–2, solve each equation.

1. $2\sin^2\theta - 3\sin\theta + 1 = 0$, $0 \le \theta < 2\pi$

2. $\sin\theta\cos\theta = -\frac{1}{2}$, $0 \le \theta < 2\pi$

In Exercises 3–4, graph one complete cycle.

3. $y = 3\sin(2x - \pi)$

4. $y = -4\cos\pi x$

In Exercises 5–6, verify each identity.

5. $\sin\theta\csc\theta - \cos^2\theta = \sin^2\theta$

6. $\cos\left(\theta + \dfrac{3\pi}{2}\right) = \sin\theta$

7. A circle has a radius of 20 inches. Find the length of the arc intercepted by a central angle of 45°. Express arc length in terms of π. Then round your answer to two decimal places.

In Exercises 8–9, find the exact value of each expression.

8. $2\sin\dfrac{\pi}{3} - 3\tan\dfrac{\pi}{6}$

9. $\sin\left(\tan^{-1}\frac{1}{2}\right)$

10. Sighting the top of a building, a surveyor measured the angle of elevation to be 27°. The transit is 5 feet above the ground and 325 feet from the building. Find the building's height to the nearest foot.

11. If $\tan\theta = -\frac{5}{2}$ and $\sin\theta > 0$, find the exact value of $\cos\theta$ and $\sin\theta$.

12. In oblique triangle ABC, $A = 12°$, $B = 75°$, and $a = 20$. Find b to the nearest tenth.

13. In oblique triangle ABC, $a = 25$, $b = 32$, and $C = 30°$. Solve the triangle. Round lengths to the nearest tenth and angle measures to the nearest degree.

14. Convert $-720°$ to radians.

15. Convert $\dfrac{14\pi}{9}$ radians to degrees.

16. The point $(6, -3)$ is on the terminal side of angle θ in standard position. Find the exact value of $\cot\theta$.

17. A ship is due west of a lighthouse. A second ship is 14 miles south of the first ship. The bearing from the second ship to the lighthouse is N 58° E. How far, to the nearest tenth of a mile, is the first ship from the lighthouse?

18. An object moves in simple harmonic motion described by $d = 4\sin 5t$, where t is measured in seconds and d in meters. Find **a.** the maximum displacement; **b.** the frequency; and **c.** the time required for one cycle.

19. Use a half-angle formula to find the exact value of $\cos 22.5°$.

20. If $\mathbf{v} = 2\mathbf{i} + 7\mathbf{j}$ and $\mathbf{w} = \mathbf{i} - 2\mathbf{j}$, find **a.** $3\mathbf{v} - \mathbf{w}$ and **b.** $\mathbf{v} \cdot \mathbf{w}$.

CHAPTER 5

Complex Numbers, Polar Coordinates, and Parametric Equations

Can mathematical models be created for events that appear to involve random behavior, such as stock market fluctuations or air turbulence? Chaos theory, a new frontier of mathematics, offers models and computer-generated images that reveal order and underlying patterns where only the erratic and the unpredictable had been observed. Because most behavior is chaotic, the computer has become a canvas that looks more like the real world than anything previously seen. Magnified portions of these computer images yield repetitions of the original structure, as well as new and unexpected patterns. The computer generates these visualizations of chaos by plotting large numbers of points for functions whose domains are complex numbers involving the square root of negative one.

HERE'S WHERE YOU'LL FIND THIS APPLICATION:

We present a trigonometric approach to $\sqrt{-1}$, or i, in Section 5.2, hinting at chaotic possibilities in the section opener and the Blitzer Bonus on page 281. If you are intrigued by how the trigonometry of complex numbers in Section 5.2 reveals that the world is not random (rather, the underlying patterns are far more intricate than we had previously assumed), we suggest reading *Chaos* by James Gleick, published by Penguin Books.

A Brief Review • The Set of Real Numbers

- The sets that make up the real numbers are summarized in the following table. We refer to these sets as subsets of the real numbers, meaning that all elements in each subset are also elements in the set of real numbers.

Important Subsets of the Real Numbers

Name	Description	Examples
Natural numbers	$\{1, 2, 3, 4, 5, \ldots\}$ These are the numbers that we use for counting.	$2, 3, 5, 17$
Whole numbers	$\{0, 1, 2, 3, 4, 5, \ldots\}$ The set of whole numbers includes 0 and the natural numbers.	$0, 2, 3, 5, 17$
Integers	$\{\ldots, -5, -4, -3, -2, -1, 0, 1, 2, 3, 4, 5, \ldots\}$ The set of integers includes the negatives of the natural numbers and the whole numbers.	$-17, -5, -3, -2, 0, 2, 3, 5, 17$
Rational numbers	$\left\{ \dfrac{a}{b} \middle| a \text{ and } b \text{ are integers and } b \neq 0 \right\}$ The set of rational numbers is the set of all numbers that can be expressed as a quotient of two integers, with the denominator not 0. Rational numbers can be expressed as terminating or repeating decimals.	$-17 = \frac{-17}{1}, -5 = \frac{-5}{1}, -3, -2,$ $0, 2, 3, 5, 17,$ $\frac{2}{5} = 0.4,$ $\frac{-2}{3} = -0.6666\ldots = -0.\overline{6}$
Irrational numbers	The set of irrational numbers is the set of all numbers whose decimal representations are neither terminating nor repeating. Irrational numbers cannot be expressed as a quotient of integers.	$\sqrt{2} \approx 1.414214$ $-\sqrt{3} \approx -1.73205$ $\pi \approx 3.142$ $-\frac{\pi}{2} \approx -1.571$

The set of real numbers, shown in the figure on the right, is the set of numbers that are either rational or irrational.

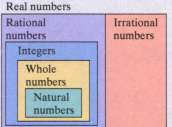

Section 5.1

Complex Numbers

Who is this kid warning us about our eyeballs turning black if we attempt to find the square root of −9? Don't believe what you hear on the street. Although square roots of negative numbers are not real numbers, they do play a significant role in trigonometry. In this section, we move beyond the real numbers and discuss square roots with negative radicands.

The Imaginary Unit *i*

We begin with a simple-looking quadratic equation: $x^2 = -1$. Because the square of a real number is never negative, there is no real number x such that $x^2 = -1$. To provide a setting in which such equations have solutions, mathematicians have invented an expanded system of numbers, the complex numbers. The *imaginary number i*, defined to be a solution of the equation $x^2 = -1$, is the basis of this new number system.

© Roz Chast/The New Yorker Collection/Cartoonbank

> **The Imaginary Unit *i***
>
> The **imaginary unit *i*** is defined as
> $$i = \sqrt{-1}, \quad \text{where } i^2 = -1.$$

Using the imaginary unit *i*, we can express the square root of any negative number as a real multiple of *i*. For example,
$$\sqrt{-25} = \sqrt{-1}\sqrt{25} = i\sqrt{25} = 5i.$$
We can check this result by squaring $5i$ and obtaining -25.
$$(5i)^2 = 5^2 i^2 = 25(-1) = -25$$

A new system of numbers, called *complex numbers*, is based on adding multiples of *i*, such as $5i$, to real numbers.

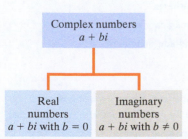

FIGURE 5.1 The complex number system

> **Complex Numbers and Imaginary Numbers**
>
> The set of all numbers in the form
> $$a + bi,$$
> with real numbers a and b, and i, the imaginary unit, is called the set of **complex numbers**. The real number a is called the **real part** and the real number b is called the **imaginary part** of the complex number $a + bi$. If $b \neq 0$, then the complex number is called an **imaginary number** (**Figure 5.1**). An imaginary number in the form bi is called a **pure imaginary number**.

Here are some examples of complex numbers. Each number can be written in the form $a + bi$.

$$-4 + 6i \qquad\qquad 2i = 0 + 2i \qquad\qquad 3 = 3 + 0i$$

| a, the real part, is −4. | b, the imaginary part, is 6. | | a, the real part, is 0. | b, the imaginary part, is 2. | | a, the real part, is 3. | b, the imaginary part, is 0. |

Can you see that b, the imaginary part, is not zero in the first two complex numbers? Because $b \neq 0$, these complex numbers are imaginary numbers. Furthermore, the imaginary number $2i$ is a pure imaginary number. By contrast, the imaginary part of the complex number on the right is zero. This complex number is not an imaginary number. The number 3, or $3 + 0i$, is a real number.

A complex number is said to be **simplified** if it is expressed in the **standard form** $a + bi$. If b contains a radical, we usually write i before the radical. For example, we write $7 + 3i\sqrt{5}$ rather than $7 + 3\sqrt{5}i$, which could easily be confused with $7 + 3\sqrt{5i}$.

Expressed in standard form, two complex numbers are equal if and only if their real parts are equal and their imaginary parts are equal.

Equality of Complex Numbers

$a + bi = c + di$ if and only if $a = c$ and $b = d$.

1 Add and subtract complex numbers.

Operations with Complex Numbers

The form of a complex number $a + bi$ is like the binomial $a + bx$. Consequently, we can add, subtract, and multiply complex numbers using the same methods we used for binomials, remembering that $i^2 = -1$.

Adding and Subtracting Complex Numbers

1. $(a + bi) + (c + di) = (a + c) + (b + d)i$

In words, this says that you add complex numbers by adding their real parts, adding their imaginary parts, and expressing the sum as a complex number.

2. $(a + bi) - (c + di) = (a - c) + (b - d)i$

In words, this says that you subtract complex numbers by subtracting their real parts, subtracting their imaginary parts, and expressing the difference as a complex number.

EXAMPLE 1   Adding and Subtracting Complex Numbers

Perform the indicated operations, writing the result in standard form:

a. $(5 - 11i) + (7 + 4i)$ **b.** $(-5 + i) - (-11 - 6i)$.

GREAT QUESTION!

Are operations with complex numbers similar to operations with polynomials?

Yes. The following examples, using the same integers as in Example 1, show how operations with complex numbers are just like operations with polynomials.

a. $(5 - 11x) + (7 + 4x)$
$= 12 - 7x$

b. $(-5 + x) - (-11 - 6x)$
$= -5 + x + 11 + 6x$
$= 6 + 7x$

SOLUTION

a. $(5 - 11i) + (7 + 4i)$

$= 5 - 11i + 7 + 4i$ Remove the parentheses.

$= 5 + 7 - 11i + 4i$ Group real and imaginary terms.

$= (5 + 7) + (-11 + 4)i$ Add real parts and add imaginary parts.

$= 12 - 7i$ Simplify.

b. $(-5 + i) - (-11 - 6i)$

$= -5 + i + 11 + 6i$ Remove the parentheses. Change signs of real and imaginary parts in the complex number being subtracted.

$= -5 + 11 + i + 6i$ Group real and imaginary terms.

$= (-5 + 11) + (1 + 6)i$ Add real parts and add imaginary parts.

$= 6 + 7i$ Simplify.

• • •

Check Point 1 Perform the indicated operations, writing the result in standard form:

    a. $(5 - 2i) + (3 + 3i)$ **b.** $(2 + 6i) - (12 - i)$.

Multiplication of complex numbers is performed the same way as multiplication of polynomials, using the distributive property and the FOIL method. After completing the multiplication, we replace any occurrences of i^2 with -1. This idea is illustrated in the next example.

② Multiply complex numbers.

EXAMPLE 2   Multiplying Complex Numbers

Find the products:

    a. $4i(3 - 5i)$ **b.** $(7 - 3i)(-2 - 5i)$.

SOLUTION

a. $4i(3 - 5i)$

$= 4i \cdot 3 - 4i \cdot 5i$ Distribute $4i$ throughout the parentheses.

$= 12i - 20i^2$ Multiply.

$= 12i - 20(-1)$ Replace i^2 with -1.

$= 20 + 12i$ Simplify to $12i + 20$ and write in standard form.

b. $(7 - 3i)(-2 - 5i)$

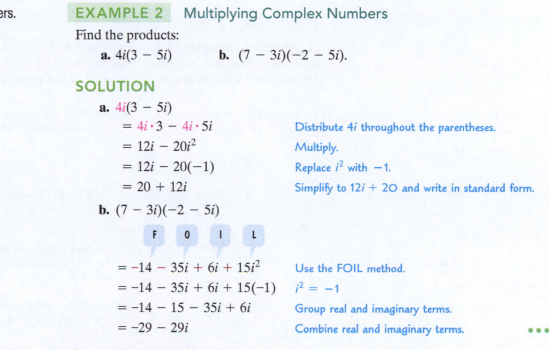

$= -14 - 35i + 6i + 15i^2$ Use the FOIL method.

$= -14 - 35i + 6i + 15(-1)$ $i^2 = -1$

$= -14 - 15 - 35i + 6i$ Group real and imaginary terms.

$= -29 - 29i$ Combine real and imaginary terms.   ● ● ●

 Check Point 2 Find the products:

    a. $7i(2 - 9i)$ **b.** $(5 + 4i)(6 - 7i)$.

③ Divide complex numbers.

Complex Conjugates and Division

It is possible to multiply imaginary numbers and obtain a real number. This occurs when we multiply $a + bi$ and $a - bi$.

$$(a + bi)(a - bi) = a^2 - abi + abi - b^2i^2$$ Use the FOIL method.

$$= a^2 - b^2(-1)$$ $i^2 = -1$

$$= a^2 + b^2$$ Notice that this product eliminates i.

For the complex number $a + bi$, we define its *complex conjugate* to be $a - bi$. The multiplication of complex conjugates results in a real number.

> **Conjugate of a Complex Number**
>
> The **complex conjugate** of the number $a + bi$ is $a - bi$, and the complex conjugate of $a - bi$ is $a + bi$. The multiplication of complex conjugates gives a real number.
>
> $$(a + bi)(a - bi) = a^2 + b^2$$
> $$(a - bi)(a + bi) = a^2 + b^2$$

Complex conjugates are used to divide complex numbers. The goal of the division procedure is to obtain a real number in the denominator. This real number becomes the denominator of a and b in the quotient $a + bi$. By multiplying the numerator and the denominator of the division by the complex conjugate of the denominator, you will obtain this real number in the denominator.

> **EXAMPLE 3** Using Complex Conjugates to Divide Complex Numbers

Divide and express the result in standard form: $\dfrac{7 + 4i}{2 - 5i}$.

SOLUTION

The complex conjugate of the denominator, $2 - 5i$, is $2 + 5i$. Multiplication of both the numerator and the denominator by $2 + 5i$ will eliminate i from the denominator while maintaining the value of the expression.

$$\frac{7 + 4i}{2 - 5i} = \frac{(7 + 4i)}{(2 - 5i)} \cdot \frac{(2 + 5i)}{(2 + 5i)}$$

Multiply the numerator and the denominator by the complex conjugate of the denominator.

$$= \frac{14 + 35i + 8i + 20i^2}{2^2 + 5^2}$$

Use the FOIL method in the numerator and $(a - bi)(a + bi) = a^2 + b^2$ in the denominator.

$$= \frac{14 + 43i + 20(-1)}{29}$$

In the numerator, combine imaginary terms and replace i^2 with -1. In the denominator, $2^2 + 5^2 = 4 + 25 = 29$.

$$= \frac{-6 + 43i}{29}$$

Combine real terms in the numerator: $14 + 20(-1) = 14 - 20 = -6$.

$$= -\frac{6}{29} + \frac{43}{29}i$$

Express the answer in standard form.

Observe that the quotient is expressed in the standard form $a + bi$, with $a = -\frac{6}{29}$ and $b = \frac{43}{29}$. • • •

✅ **Check Point 3** Divide and express the result in standard form: $\dfrac{5 + 4i}{4 - i}$.

④ Perform operations with square roots of negative numbers.

Roots of Negative Numbers

The square of $4i$ and the square of $-4i$ both result in -16:

$$(4i)^2 = 16i^2 = 16(-1) = -16 \qquad (-4i)^2 = 16i^2 = 16(-1) = -16.$$

Consequently, in the complex number system -16 has two square roots, namely, $4i$ and $-4i$. We call $4i$ the **principal square root** of -16.

> #### Principal Square Root of a Negative Number
>
> For any positive real number b, the **principal square root** of the negative number $-b$ is defined by
> $$\sqrt{-b} = i\sqrt{b}.$$

GREAT QUESTION!

In the definition $\sqrt{-b} = i\sqrt{b}$, why did you write i in the front? Could I write $\sqrt{-b} = \sqrt{b}i$?

Yes, $\sqrt{-b} = \sqrt{b}i$. However, it's tempting to write $\sqrt{-b} = \sqrt{bi}$, which is incorrect. In order to avoid writing i under a radical, let's agree to write i before any radical.

Consider the multiplication problem

$$5i \cdot 2i = 10i^2 = 10(-1) = -10.$$

This problem can also be given in terms of principal square roots of negative numbers:

$$\sqrt{-25} \cdot \sqrt{-4}.$$

Because the product rule for radicals only applies to real numbers, multiplying radicands is incorrect. **When performing operations with square roots of negative numbers, begin by expressing all square roots in terms of i.** Then perform the indicated operation.

Correct:	Incorrect:
$\sqrt{-25} \cdot \sqrt{-4} = i\sqrt{25} \cdot i\sqrt{4}$	$\sqrt{-25} \cdot \sqrt{-4} = \sqrt{(-25)(-4)}$
$\qquad = 5i \cdot 2i$	$\qquad = \sqrt{100}$
$\qquad = 10i^2 = 10(-1) = -10$	$\qquad = 10$

EXAMPLE 4 Operations Involving Square Roots of Negative Numbers

Perform the indicated operations and write the result in standard form:

a. $\sqrt{-18} - \sqrt{-8}$ **b.** $\left(-1 + \sqrt{-5}\right)^2$ **c.** $\dfrac{-25 + \sqrt{-50}}{15}$.

SOLUTION

Begin by expressing all square roots of negative numbers in terms of i.

a. $\sqrt{-18} - \sqrt{-8} = i\sqrt{18} - i\sqrt{8} = i\sqrt{9 \cdot 2} - i\sqrt{4 \cdot 2}$
$\qquad\qquad\qquad\quad = 3i\sqrt{2} - 2i\sqrt{2} = i\sqrt{2}$

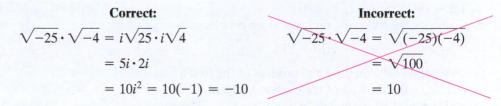
$$(A + B)^2 = A^2 + 2 \cdot A \cdot B + B^2$$

b. $\left(-1 + \sqrt{-5}\right)^2 = \left(-1 + i\sqrt{5}\right)^2 = (-1)^2 + 2(-1)(i\sqrt{5}) + (i\sqrt{5})^2$
$\qquad\qquad\qquad\qquad = 1 - 2i\sqrt{5} + 5i^2$
$\qquad\qquad\qquad\qquad = 1 - 2i\sqrt{5} + 5(-1)$
$\qquad\qquad\qquad\qquad = -4 - 2i\sqrt{5}$

c. $\dfrac{-25 + \sqrt{-50}}{15}$

$= \dfrac{-25 + i\sqrt{50}}{15}$ $\qquad$ $\sqrt{-b} = i\sqrt{b}$

$= \dfrac{-25 + 5i\sqrt{2}}{15}$ $\qquad$ $\sqrt{50} = \sqrt{25 \cdot 2} = 5\sqrt{2}$

$= \dfrac{-25}{15} + \dfrac{5i\sqrt{2}}{15}$ $\qquad$ Write the complex number in standard form.

$= -\dfrac{5}{3} + i\dfrac{\sqrt{2}}{3}$ $\qquad$ Simplify. $\qquad\qquad$ • • •

✓ **Check Point 4** Perform the indicated operations and write the result in standard form:

a. $\sqrt{-27} + \sqrt{-48}$ **b.** $\left(-2 + \sqrt{-3}\right)^2$ **c.** $\dfrac{-14 + \sqrt{-12}}{2}$.

CONCEPT AND VOCABULARY CHECK

Fill in each blank so that the resulting statement is true.

1. The imaginary unit i is defined as $i = $ _____, where $i^2 = $ ____.

2. The set of all numbers in the form $a + bi$ is called the set of _____ numbers. If $b \neq 0$, then the number is also called a/an _____ number. If $b = 0$, then the number is also called a/an ____ number.

3. $-9i + 3i = $ ____

4. $10i - (-4i) = $ ____

5. Consider the following multiplication problem:

$$(3 + 2i)(6 - 5i).$$

Using the FOIL method, the product of the first terms is _____, the product of the outside terms is _____, and the product of the inside terms is _____. The product of the last terms in terms of i^2 is _____, which simplifies to _____.

6. The conjugate of $2 - 9i$ is _____.

7. The division

$$\frac{7 + 4i}{2 - 5i}$$

is performed by multiplying the numerator and denominator by _____.

8. $\sqrt{-20} = $ ___$\sqrt{20} = $ ___$\sqrt{4 \cdot 5} = $ _____

EXERCISE SET 5.1

Practice Exercises

In Exercises 1–8, add or subtract as indicated and write the result in standard form.

1. $(7 + 2i) + (1 - 4i)$

2. $(-2 + 6i) + (4 - i)$

3. $(3 + 2i) - (5 - 7i)$

4. $(-7 + 5i) - (-9 - 11i)$

5. $6 - (-5 + 4i) - (-13 - i)$

6. $7 - (-9 + 2i) - (-17 - i)$

7. $8i - (14 - 9i)$

8. $15i - (12 - 11i)$

In Exercises 9–20, find each product and write the result in standard form.

9. $-3i(7i - 5)$

10. $-8i(2i - 7)$

11. $(-5 + 4i)(3 + i)$

12. $(-4 - 8i)(3 + i)$

13. $(7 - 5i)(-2 - 3i)$

14. $(8 - 4i)(-3 + 9i)$

15. $(3 + 5i)(3 - 5i)$

16. $(2 + 7i)(2 - 7i)$

17. $(-5 + i)(-5 - i)$

18. $(-7 - i)(-7 + i)$

19. $(2 + 3i)^2$

20. $(5 - 2i)^2$

In Exercises 21–28, divide and express the result in standard form.

21. $\dfrac{2}{3 - i}$

22. $\dfrac{3}{4 + i}$

23. $\dfrac{2i}{1 + i}$

24. $\dfrac{5i}{2 - i}$

25. $\dfrac{8i}{4 - 3i}$

26. $\dfrac{-6i}{3 + 2i}$

27. $\dfrac{2 + 3i}{2 + i}$

28. $\dfrac{3 - 4i}{4 + 3i}$

In Exercises 29–44, perform the indicated operations and write the result in standard form.

29. $\sqrt{-64} - \sqrt{-25}$

30. $\sqrt{-81} - \sqrt{-144}$

31. $5\sqrt{-16} + 3\sqrt{-81}$

32. $5\sqrt{-8} + 3\sqrt{-18}$

33. $\left(-2 + \sqrt{-4}\right)^2$

34. $\left(-5 - \sqrt{-9}\right)^2$

35. $\left(-3 - \sqrt{-7}\right)^2$

36. $\left(-2 + \sqrt{-11}\right)^2$

37. $\dfrac{-8 + \sqrt{-32}}{24}$

38. $\dfrac{-12 + \sqrt{-28}}{32}$

39. $\dfrac{-6 - \sqrt{-12}}{48}$

40. $\dfrac{-15 - \sqrt{-18}}{33}$

41. $\sqrt{-8}\left(\sqrt{-3} - \sqrt{5}\right)$

42. $\sqrt{-12}\left(\sqrt{-4} - \sqrt{2}\right)$

43. $\left(3\sqrt{-5}\right)\left(-4\sqrt{-12}\right)$

44. $\left(3\sqrt{-7}\right)\left(2\sqrt{-8}\right)$

Practice Plus

In Exercises 45–50, perform the indicated operation(s) and write the result in standard form.

45. $(2 - 3i)(1 - i) - (3 - i)(3 + i)$
46. $(8 + 9i)(2 - i) - (1 - i)(1 + i)$
47. $(2 + i)^2 - (3 - i)^2$
48. $(4 - i)^2 - (1 + 2i)^2$
49. $5\sqrt{-16} + 3\sqrt{-81}$
50. $5\sqrt{-8} + 3\sqrt{-18}$
51. Evaluate $x^2 - 2x + 2$ for $x = 1 + i$.
52. Evaluate $x^2 - 2x + 5$ for $x = 1 - 2i$.
53. Evaluate $\dfrac{x^2 + 19}{2 - x}$ for $x = 3i$.
54. Evaluate $\dfrac{x^2 + 11}{3 - x}$ for $x = 4i$.

Application Exercises

Complex numbers are used in electronics to describe the current in an electric circuit. Ohm's law relates the current in a circuit, I, in amperes, the voltage of the circuit, E, in volts, and the resistance of the circuit, R, in ohms, by the formula $E = IR$. Use this formula to solve Exercises 55–56.

55. Find E, the voltage of a circuit, if $I = (4 - 5i)$ amperes and $R = (3 + 7i)$ ohms.
56. Find E, the voltage of a circuit, if $I = (2 - 3i)$ amperes and $R = (3 + 5i)$ ohms.
57. The mathematician Girolamo Cardano is credited with the first use (in 1545) of negative square roots in solving the now-famous problem, "Find two numbers whose sum is 10 and whose product is 40." Show that the complex numbers $5 + i\sqrt{15}$ and $5 - i\sqrt{15}$ satisfy the conditions of the problem. (Cardano did not use the symbolism $i\sqrt{15}$ or even $\sqrt{-15}$. He wrote R.m 15 for $\sqrt{-15}$, meaning "radix minus 15." He regarded the numbers $5 + $ R.m 15 and $5 - $ R.m 15 as "fictitious" or "ghost numbers," and considered the problem "manifestly impossible." But in a mathematically adventurous spirit, he exclaimed, "Nevertheless, we will operate.")

Explaining the Concepts

58. What is i?
59. Explain how to add complex numbers. Provide an example with your explanation.
60. Explain how to multiply complex numbers and give an example.
61. What is the complex conjugate of $2 + 3i$? What happens when you multiply this complex number by its complex conjugate?
62. Explain how to divide complex numbers. Provide an example with your explanation.
63. Explain each of the three jokes in the cartoon on page 265.
64. A stand-up comedian uses algebra in some jokes, including one about a telephone recording that announces "You have just reached an imaginary number. Please multiply by i and dial again." Explain the joke.

Explain the error in Exercises 65–66.

65. $\sqrt{-9} + \sqrt{-16} = \sqrt{-25} = i\sqrt{25} = 5i$
66. $\left(\sqrt{-9}\right)^2 = \sqrt{-9} \cdot \sqrt{-9} = \sqrt{81} = 9$

Critical Thinking Exercises

Make Sense? *In Exercises 67–70, determine whether each statement makes sense or does not make sense, and explain your reasoning.*

67. The humor in the cartoon is based on the fact that "rational" and "real" have different meanings in mathematics and in everyday speech.

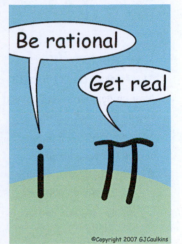

Be rational

Get real

©Copyright 2007 GJCaulkins

© 2007 GJ Caulkins

68. The word *imaginary* in imaginary numbers tells me that these numbers are undefined.
69. By writing the imaginary number $5i$, I can immediately see that 5 is the constant and i is the variable.
70. When I add or subtract complex numbers, I am basically combining like terms.

In Exercises 71–74, determine whether each statement is true or false. If the statement is false, make the necessary change(s) to produce a true statement.

71. Some irrational numbers are not complex numbers.
72. $(3 + 7i)(3 - 7i)$ is an imaginary number.
73. $\dfrac{7 + 3i}{5 + 3i} = \dfrac{7}{5}$
74. In the complex number system, $x^2 + y^2$ (the sum of two squares) can be factored as $(x + yi)(x - yi)$.

In Exercises 75–77, perform the indicated operations and write the result in standard form.

75. $\dfrac{4}{(2 + i)(3 - i)}$
76. $\dfrac{1 + i}{1 + 2i} + \dfrac{1 - i}{1 - 2i}$
77. $\dfrac{8}{1 + \dfrac{2}{i}}$

Retaining the Concepts

78. Verify the identity:
    $$\sin^2 x \tan^2 x + \cos^2 x \tan^2 x = \sec^2 x - 1.$$
    (Section 3.1, Example 3)
79. Solve: $2\cos^2 x + 3\sin x - 3 = 0, \quad 0 \le x < 2\pi$.
    (Section 3.5, Example 7)
80. Determine the amplitude, period, and phase shift of $y = \frac{1}{2}\cos(3x + \frac{\pi}{2})$. Then graph one period of the function. (Section 2.1, Example 6)

Preview Exercises

Exercises 81–83 will help you prepare for the material covered in the next section. In each exercise, perform the indicated operation and write the result in the standard form $a + bi$.

81. $(1 + i)(2 + 2i)$
82. $(-1 + i\sqrt{3})(-1 + i\sqrt{3})(-1 + i\sqrt{3})$
83. $\dfrac{2 + 2i}{1 + i}$

Section 5.2

Section 5.2 Complex Numbers in Polar Form; DeMoivre's Theorem

What am I supposed to learn?

After studying this section, you should be able to:

1. Plot complex numbers in the complex plane.
2. Find the absolute value of a complex number.
3. Write complex numbers in polar form.
4. Convert a complex number from polar to rectangular form.
5. Find products of complex numbers in polar form.
6. Find quotients of complex numbers in polar form.
7. Find powers of complex numbers in polar form.
8. Find roots of complex numbers in polar form.

One of the new frontiers of mathematics suggests that there is an underlying order in things that appear to be random, such as the hiss and crackle of background noises as you tune a radio. Irregularities in the heartbeat, some of them severe enough to cause a heart attack, or irregularities in our sleeping patterns, such as insomnia, are examples of chaotic behavior. Chaos in the mathematical sense does not mean a complete lack of form or arrangement. In mathematics, chaos is used to describe something that appears to be random but is not actually random. The patterns of chaos appear in images like the one shown here, called the Mandelbrot set. Magnified portions of this image yield repetitions of the original structure, as well as new and unexpected patterns. The Mandelbrot set transforms the hidden structure of chaotic events into a source of wonder and inspiration.

The Mandelbrot set is made possible by opening up graphing to include complex numbers in the form $a + bi$, where $i = \sqrt{-1}$. In this section, you will learn how to graph complex numbers and write them in terms of trigonometric functions.

A magnification of the Mandelbrot set

The Complex Plane

① Plot complex numbers in the complex plane.

We know that a real number can be represented as a point on a number line. By contrast, a complex number $z = a + bi$ is represented as a point (a, b) in a coordinate plane, as shown in **Figure 5.2**. The horizontal axis of the coordinate plane is called the **real axis**. The vertical axis is called the **imaginary axis**. The coordinate system is called the **complex plane**. Every complex number corresponds to a point in the complex plane and every point in the complex plane corresponds to a complex number. When we represent a complex number as a point in the complex plane, we say that we are **plotting the complex number**.

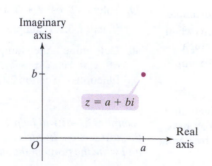

FIGURE 5.2 Plotting $z = a + bi$ in the complex plane

EXAMPLE 1   Plotting Complex Numbers

Plot each complex number in the complex plane:

a. $z = 3 + 4i$ **b.** $z = -1 - 2i$ **c.** $z = -3$ **d.** $z = -4i$.

SOLUTION

See **Figure 5.3**.

a. We plot the complex number $z = 3 + 4i$ the same way we plot $(3, 4)$ in the rectangular coordinate system. We move three units to the right on the real axis and four units up parallel to the imaginary axis.

b. The complex number $z = -1 - 2i$ corresponds to the point $(-1, -2)$ in the rectangular coordinate system. Plot the complex number by moving one unit to the left on the real axis and two units down parallel to the imaginary axis.

c. Because $z = -3 = -3 + 0i$, this complex number corresponds to the point $(-3, 0)$. We plot -3 by moving three units to the left on the real axis.

d. Because $z = -4i = 0 - 4i$, this number corresponds to the point $(0, -4)$. We plot the complex number by moving four units down on the imaginary axis. ● ● ●

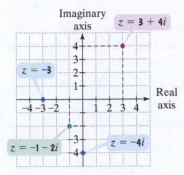

FIGURE 5.3 Plotting complex numbers

✓ **Check Point 1** Plot each complex number in the complex plane:

a. $z = 2 + 3i$ **b.** $z = -3 - 5i$ **c.** $z = -4$ **d.** $z = -i$.

② Find the absolute value of a complex number.

Recall that the absolute value of a real number is its distance from 0 on a number line. The **absolute value of the complex number** $z = a + bi$, denoted by $|z|$, is the distance from the origin to the point z in the complex plane. **Figure 5.4** illustrates that we can use the Pythagorean Theorem to represent $|z|$ in terms of a and b: $|z| = \sqrt{a^2 + b^2}$.

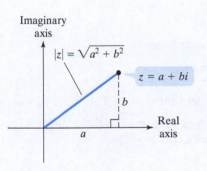

FIGURE 5.4

> ### The Absolute Value of a Complex Number
>
> The **absolute value** of the complex number $a + bi$ is
> $$|z| = |a + bi| = \sqrt{a^2 + b^2}.$$

EXAMPLE 2   Finding the Absolute Value of a Complex Number

Determine the absolute value of each of the following complex numbers:

a. $z = 3 + 4i$ **b.** $z = -1 - 2i$.

SOLUTION

a. The absolute value of $z = 3 + 4i$ is found using $a = 3$ and $b = 4$.

$$|z| = \sqrt{3^2 + 4^2} = \sqrt{9 + 16} = \sqrt{25} = 5$$ Use $|z| = \sqrt{a^2 + b^2}$ with $a = 3$ and $b = 4$.

Thus, the distance from the origin to the point $z = 3 + 4i$, shown in quadrant I in **Figure 5.5**, is five units.

b. The absolute value of $z = -1 - 2i$ is found using $a = -1$ and $b = -2$.

$$|z| = \sqrt{(-1)^2 + (-2)^2} = \sqrt{1 + 4} = \sqrt{5}$$ Use $|z| = \sqrt{a^2 + b^2}$ with $a = -1$ and $b = -2$.

Thus, the distance from the origin to the point $z = -1 - 2i$, shown in quadrant III in **Figure 5.5**, is $\sqrt{5}$ units. ● ● ●

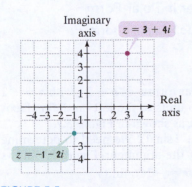

FIGURE 5.5

✓ **Check Point 2** Determine the absolute value of each of the following complex numbers:

    a. $z = 5 + 12i$ **b.** $2 - 3i$.

3   Write complex numbers in polar form.

Polar Form of a Complex Number

A complex number in the form $z = a + bi$ is said to be in **rectangular form**. Suppose that its absolute value is r. In **Figure 5.6**, we let θ be an angle in standard position whose terminal side passes through the point (a, b). From the figure, we see that

$$r = \sqrt{a^2 + b^2}.$$

Likewise, according to the definitions of the trigonometric functions,

$$\cos \theta = \frac{a}{r} \qquad\qquad \sin \theta = \frac{b}{r} \qquad\qquad \tan \theta = \frac{b}{a}.$$

$$a = r \cos \theta \qquad\qquad b = r \sin \theta$$

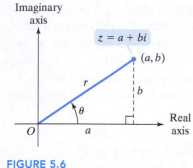

FIGURE 5.6

By substituting the expressions for a and b into $z = a + bi$, we write the complex number in terms of trigonometric functions.

$$z = a + bi = r \cos \theta + (r \sin \theta)i = r(\cos \theta + i \sin \theta)$$

    $a = r \cos \theta$ and $b = r \sin \theta$.

    Factor out r from each of the two previous terms.

The expression $z = r(\cos \theta + i \sin \theta)$ is called the **polar form of a complex number**.

Polar Form of a Complex Number

The complex number $z = a + bi$ is written in **polar form** as

$$z = r(\cos \theta + i \sin \theta),$$

where $a = r \cos \theta$, $b = r \sin \theta$, $r = \sqrt{a^2 + b^2}$, and $\tan \theta = \dfrac{b}{a}$. The value of r is called the **modulus** (plural: moduli) of the complex number z and the angle θ is called the **argument** of the complex number z with $0 \leq \theta < 2\pi$.

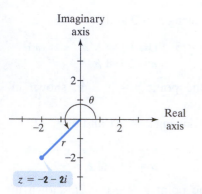

FIGURE 5.7 Plotting $z = -2 - 2i$ and writing the number in polar form

EXAMPLE 3   Writing a Complex Number in Polar Form

Plot $z = -2 - 2i$ in the complex plane. Then write z in polar form.

SOLUTION

The complex number $z = -2 - 2i$ is in rectangular form $z = a + bi$, with $a = -2$ and $b = -2$. We plot the number by moving two units to the left on the real axis and two units down parallel to the imaginary axis, as shown in **Figure 5.7**.

    By definition, the polar form of z is $r(\cos \theta + i \sin \theta)$. We need to determine the value for r, the modulus, and the value for θ, the argument. **Figure 5.7** shows r and θ. We use $r = \sqrt{a^2 + b^2}$ with $a = -2$ and $b = -2$ to find r.

$$r = \sqrt{a^2 + b^2} = \sqrt{(-2)^2 + (-2)^2} = \sqrt{4 + 4} = \sqrt{8} = \sqrt{4 \cdot 2} = 2\sqrt{2}$$

We use $\tan \theta = \dfrac{b}{a}$ with $a = -2$ and $b = -2$ to find θ.

$$\tan \theta = \frac{b}{a} = \frac{-2}{-2} = 1$$

We know that $\tan \dfrac{\pi}{4} = 1$. **Figure 5.7** shows that the argument, θ, satisfying $\tan \theta = 1$ lies in quadrant III. Thus,

$$\theta = \pi + \frac{\pi}{4} = \frac{4\pi}{4} + \frac{\pi}{4} = \frac{5\pi}{4}.$$

We use $r = 2\sqrt{2}$ and $\theta = \dfrac{5\pi}{4}$ to write the polar form. The polar form of $z = -2 - 2i$ is

$$z = r(\cos \theta + i \sin \theta) = 2\sqrt{2}\left(\cos \frac{5\pi}{4} + i \sin \frac{5\pi}{4}\right).$$ • • •

⊘ Check Point **3**   Plot $z = -1 - i\sqrt{3}$ in the complex plane. Then write z in polar form. Express the argument in radians. (We write $-1 - i\sqrt{3}$, rather than $-1 - \sqrt{3}i$, which could easily be confused with $-1 - \sqrt{3i}$.)

4   Convert a complex number from polar to rectangular form.

EXAMPLE 4   Writing a Complex Number in Rectangular Form

Write $z = 2(\cos 60° + i \sin 60°)$ in rectangular form.

SOLUTION

The complex number $z = 2(\cos 60° + i \sin 60°)$ is in polar form, with $r = 2$ and $\theta = 60°$. We use exact values for $\cos 60°$ and $\sin 60°$ to write the number in rectangular form.

$$2(\cos 60° + i \sin 60°) = 2\left(\frac{1}{2} + i\frac{\sqrt{3}}{2}\right) = 1 + i\sqrt{3}$$

The rectangular form of $z = 2(\cos 60° + i \sin 60°)$ is

$$z = 1 + i\sqrt{3}.$$

Write i before the radical to avoid confusion with $\sqrt{3i}$.   • • •

⊘ Check Point **4**   Write $z = 4(\cos 30° + i \sin 30°)$ in rectangular form.

5   Find products of complex numbers in polar form.

Products and Quotients in Polar Form

We can multiply and divide complex numbers fairly quickly if the numbers are expressed in polar form.

Product of Two Complex Numbers in Polar Form

Let $z_1 = r_1(\cos \theta_1 + i \sin \theta_1)$ and $z_2 = r_2(\cos \theta_2 + i \sin \theta_2)$ be two complex numbers in polar form. Their product, $z_1 z_2$, is

$$z_1 z_2 = r_1 r_2[\cos(\theta_1 + \theta_2) + i \sin(\theta_1 + \theta_2)].$$

To multiply two complex numbers, multiply moduli and add arguments.

To prove that $z_1z_2 = r_1r_2[\cos(\theta_1 + \theta_2) + i\sin(\theta_1 + \theta_2)]$ we begin by multiplying z_1 and z_2 using the FOIL method. Then we simplify the product using the sum formulas for sine and cosine.

$z_1z_2 = [r_1(\cos\theta_1 + i\sin\theta_1)][r_2(\cos\theta_2 + i\sin\theta_2)]$

$\quad = r_1r_2(\cos\theta_1 + i\sin\theta_1)(\cos\theta_2 + i\sin\theta_2)$ Rearrange factors.

F   O   I   L

$\quad = r_1r_2(\cos\theta_1\cos\theta_2 + i\cos\theta_1\sin\theta_2 + i\sin\theta_1\cos\theta_2 + i^2\sin\theta_1\sin\theta_2)$ Use the FOIL method.

$\quad = r_1r_2[\cos\theta_1\cos\theta_2 + i(\cos\theta_1\sin\theta_2 + \sin\theta_1\cos\theta_2) + i^2\sin\theta_1\sin\theta_2]$ Factor i from the second and third terms.

$\quad = r_1r_2[\cos\theta_1\cos\theta_2 + i(\cos\theta_1\sin\theta_2 + \sin\theta_1\cos\theta_2) - \sin\theta_1\sin\theta_2]$ $i^2 = -1$

$\quad = r_1r_2[(\cos\theta_1\cos\theta_2 - \sin\theta_1\sin\theta_2) + i(\sin\theta_1\cos\theta_2 + \cos\theta_1\sin\theta_2)]$ Rearrange terms.

This is $\cos(\theta_1 + \theta_2)$.   This is $\sin(\theta_1 + \theta_2)$.

$\quad = r_1r_2[\cos(\theta_1 + \theta_2) + i\sin(\theta_1 + \theta_2)]$

This result gives a rule for finding the product of two complex numbers in polar form. The two parts to the rule are shown in the following voice balloons.

$$z_1z_2 = r_1r_2[\cos(\theta_1 + \theta_2) + i\sin(\theta_1 + \theta_2)]$$

Multiply moduli.   Add arguments.

EXAMPLE 5   Finding Products of Complex Numbers in Polar Form

Find the product of the complex numbers. Leave the answer in polar form.

$$z_1 = 4(\cos 50° + i\sin 50°) \qquad z_2 = 7(\cos 100° + i\sin 100°)$$

SOLUTION

z_1z_2

$\quad = [4(\cos 50° + i\sin 50°)][7(\cos 100° + i\sin 100°)]$ Form the product of the given numbers.

$\quad = (4 \cdot 7)[\cos(50° + 100°) + i\sin(50° + 100°)]$ Multiply moduli and add arguments.

$\quad = 28(\cos 150° + i\sin 150°)$ Simplify.   • • •

 Check Point 5 Find the product of the complex numbers. Leave the answer in polar form.

$$z_1 = 6(\cos 40° + i\sin 40°) \qquad z_2 = 5(\cos 20° + i\sin 20°)$$

6 Find quotients of complex numbers in polar form.

Using algebraic methods for dividing complex numbers and the difference formulas for sine and cosine, we can obtain a rule for dividing complex numbers in polar form. The proof of this rule can be found in the appendix. You can derive the rule on your own by working Exercise 110 in this section's Exercise Set.

Quotient of Two Complex Numbers in Polar Form

Let $z_1 = r_1(\cos\theta_1 + i\sin\theta_1)$ and $z_2 = r_2(\cos\theta_2 + i\sin\theta_2)$ be two complex numbers in polar form. Their quotient, $\dfrac{z_1}{z_2}$, is

$$\frac{z_1}{z_2} = \frac{r_1}{r_2}[\cos(\theta_1 - \theta_2) + i\sin(\theta_1 - \theta_2)].$$

To divide two complex numbers, divide moduli and subtract arguments.

EXAMPLE 6   Finding Quotients of Complex Numbers in Polar Form

Find the quotient $\dfrac{z_1}{z_2}$ of the complex numbers. Leave the answer in polar form.

$$z_1 = 12\left(\cos \frac{3\pi}{4} + i \sin \frac{3\pi}{4}\right) \qquad z_2 = 4\left(\cos \frac{\pi}{4} + i \sin \frac{\pi}{4}\right)$$

SOLUTION

$$\frac{z_1}{z_2} = \frac{12\left(\cos \dfrac{3\pi}{4} + i \sin \dfrac{3\pi}{4}\right)}{4\left(\cos \dfrac{\pi}{4} + i \sin \dfrac{\pi}{4}\right)} \qquad \text{Form the quotient of the given numbers.}$$

$$= \frac{12}{4}\left[\cos\left(\frac{3\pi}{4} - \frac{\pi}{4}\right) + i \sin\left(\frac{3\pi}{4} - \frac{\pi}{4}\right)\right] \qquad \text{Divide moduli and subtract arguments.}$$

$$= 3\left(\cos \frac{\pi}{2} + i \sin \frac{\pi}{2}\right) \qquad \text{Simplify: } \frac{3\pi}{4} - \frac{\pi}{4} = \frac{2\pi}{4} = \frac{\pi}{2}. \quad \bullet\bullet\bullet$$

✓ Check Point 6 Find the quotient $\dfrac{z_1}{z_2}$ of the complex numbers. Leave the answer in polar form.

$$z_1 = 50\left(\cos \frac{4\pi}{3} + i \sin \frac{4\pi}{3}\right) \qquad z_2 = 5\left(\cos \frac{\pi}{3} + i \sin \frac{\pi}{3}\right)$$

⑦ Find powers of complex numbers in polar form.

Powers of Complex Numbers in Polar Form

We can use a formula to find powers of complex numbers if the complex numbers are expressed in polar form. This formula can be illustrated by repeatedly multiplying by $r(\cos \theta + i \sin \theta)$.

$z = r(\cos \theta + i \sin \theta)$	Start with z.
$z \cdot z = r(\cos \theta + i \sin \theta)r(\cos \theta + i \sin \theta)$	Multiply z by $z = r(\cos \theta + i \sin \theta)$.
$z^2 = r^2(\cos 2\theta + i \sin 2\theta)$	Multiply moduli: $r \cdot r = r^2$. Add arguments: $\theta + \theta = 2\theta$.
$z^2 \cdot z = r^2(\cos 2\theta + i \sin 2\theta)r(\cos \theta + i \sin \theta)$	Multiply z^2 by $z = r(\cos \theta + i \sin \theta)$.
$z^3 = r^3(\cos 3\theta + i \sin 3\theta)$	Multiply moduli: $r^2 \cdot r = r^3$. Add arguments: $2\theta + \theta = 3\theta$.
$z^3 \cdot z = r^3(\cos 3\theta + i \sin 3\theta)r(\cos \theta + i \sin \theta)$	Multiply z^3 by $z = r(\cos \theta + i \sin \theta)$.
$z^4 = r^4(\cos 4\theta + i \sin 4\theta)$	Multiply moduli: $r^3 \cdot r = r^4$. Add arguments: $3\theta + \theta = 4\theta$.

Do you see a pattern forming? If n is a positive integer, it appears that z^n is obtained by raising the modulus to the nth power and multiplying the argument by n. The formula for the nth power of a complex number is known as **DeMoivre's Theorem** in honor of the French mathematician Abraham DeMoivre (1667–1754).

> **DeMoivre's Theorem**
>
> Let $z = r(\cos \theta + i \sin \theta)$ be a complex number in polar form. If n is a positive integer, then z to the nth power, z^n, is
>
> $$z^n = [r(\cos \theta + i \sin \theta)]^n = r^n(\cos n\theta + i \sin n\theta).$$

EXAMPLE 7 Finding the Power of a Complex Number

Find $[2(\cos 20° + i \sin 20°)]^6$. Write the answer in rectangular form, $a + bi$.

SOLUTION

We begin by applying DeMoivre's Theorem.

$$[2(\cos 20° + i \sin 20°)]^6$$

$$= 2^6[\cos(6 \cdot 20°) + i \sin(6 \cdot 20°)] \quad \text{Raise the modulus to the 6th power and multiply the argument by 6.}$$

$$= 64(\cos 120° + i \sin 120°) \quad \text{Simplify.}$$

$$= 64\left(-\frac{1}{2} + i\frac{\sqrt{3}}{2}\right) \quad \text{Write the answer in rectangular form.}$$

$$= -32 + 32i\sqrt{3} \quad \text{Multiply and express the answer in } a + bi \text{ form.} \quad \bullet\bullet\bullet$$

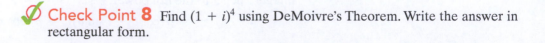 **Check Point 7** Find $[2(\cos 30° + i \sin 30°)]^5$. Write the answer in rectangular form.

EXAMPLE 8 Finding the Power of a Complex Number

Find $(1 + i)^8$ using DeMoivre's Theorem. Write the answer in rectangular form, $a + bi$.

SOLUTION

DeMoivre's Theorem applies to complex numbers in polar form. Thus, we must first write $1 + i$ in $r(\cos \theta + i \sin \theta)$ form. Then we can use DeMoivre's Theorem. The complex number $1 + i$ is plotted in **Figure 5.8**. From the figure, we obtain values for r and θ.

$$r = \sqrt{a^2 + b^2} = \sqrt{1^2 + 1^2} = \sqrt{2}$$

$$\tan \theta = \frac{b}{a} = \frac{1}{1} = 1 \text{ and } \theta = \frac{\pi}{4} \text{ because } \theta \text{ lies in quadrant I.}$$

Using these values,

$$1 + i = r(\cos \theta + i \sin \theta) = \sqrt{2}\left(\cos \frac{\pi}{4} + i \sin \frac{\pi}{4}\right).$$

Now we use DeMoivre's Theorem to raise $1 + i$ to the 8th power.

$(1 + i)^8$

$$= \left[\sqrt{2}\left(\cos \frac{\pi}{4} + i \sin \frac{\pi}{4}\right)\right]^8 \quad \text{Work with the polar form of } 1 + i.$$

$$= (\sqrt{2})^8\left[\cos\left(8 \cdot \frac{\pi}{4}\right) + i \sin\left(8 \cdot \frac{\pi}{4}\right)\right] \quad \text{Apply DeMoivre's Theorem. Raise the modulus to the 8th power and multiply the argument by 8.}$$

$$= 16(\cos 2\pi + i \sin 2\pi) \quad \text{Simplify: } (\sqrt{2})^8 = (2^{1/2})^8 = 2^4 = 16.$$

$$= 16(1 + 0i) \quad \cos 2\pi = 1 \text{ and } \sin 2\pi = 0.$$

$$= 16 \text{ or } 16 + 0i \quad \text{Simplify.} \quad \bullet\bullet\bullet$$

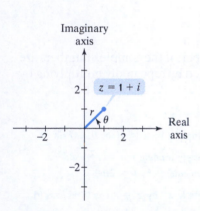

Imaginary axis

$z = 1 + i$

Real axis

FIGURE 5.8 Plotting $1 + i$ and writing the number in polar form

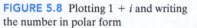 **Check Point 8** Find $(1 + i)^4$ using DeMoivre's Theorem. Write the answer in rectangular form.

8 Find roots of complex numbers in polar form.

Roots of Complex Numbers in Polar Form

In Example 7, we showed that

$$[2(\cos 20° + i \sin 20°)]^6 = 64(\cos 120° + i \sin 120°).$$

We say that $2(\cos 20° + i \sin 20°)$ is a **complex sixth root** of $64(\cos 120° + i \sin 120°)$. It is one of six distinct complex sixth roots of $64(\cos 120° + i \sin 120°)$.

In general, if a complex number z satisfies the equation

$$z^n = w,$$

we say that z is a **complex nth root** of w. It is one of n distinct nth complex roots that can be found using the following theorem:

DeMoivre's Theorem for Finding Complex Roots

Let $w = r(\cos \theta + i \sin \theta)$ be a complex number in polar form. If $w \neq 0$, w has n distinct complex nth roots given by the formula

$$z_k = \sqrt[n]{r}\left[\cos\left(\frac{\theta + 2\pi k}{n}\right) + i \sin\left(\frac{\theta + 2\pi k}{n}\right)\right] \quad \text{(radians)}$$

$$\text{or} \quad z_k = \sqrt[n]{r}\left[\cos\left(\frac{\theta + 360°k}{n}\right) + i \sin\left(\frac{\theta + 360°k}{n}\right)\right] \quad \text{(degrees)},$$

where $k = 0, 1, 2, \ldots, n - 1$.

By raising the radian or degree formula for z_k to the nth power, you can use DeMoivre's Theorem for powers to show that $z_k^n = w$. Thus, each z_k is a complex nth root of w.

DeMoivre's Theorem for finding complex roots states that every complex number has two distinct complex square roots, three distinct complex cube roots, four distinct complex fourth roots, and so on. Each root has the same modulus, $\sqrt[n]{r}$.

Successive roots have arguments that differ by the same amount, $\dfrac{2\pi}{n}$ or $\dfrac{360°}{n}$. This means that if you plot all the complex roots of any number, they will be equally spaced on a circle centered at the origin, with radius $\sqrt[n]{r}$.

EXAMPLE 9   Finding the Roots of a Complex Number

Find all the complex fourth roots of $16(\cos 120° + i \sin 120°)$. Write roots in polar form, with θ in degrees.

SOLUTION

There are exactly four fourth roots of the given complex number. From DeMoivre's Theorem for finding complex roots, the fourth roots of $16(\cos 120° + i \sin 120°)$ are

$$z_k = \sqrt[4]{16}\left[\cos\left(\frac{120° + 360°k}{4}\right) + i \sin\left(\frac{120° + 360°k}{4}\right)\right], k = 0, 1, 2, 3.$$

Use $z_k = \sqrt[n]{r}\left[\cos\left(\frac{\theta + 360°k}{n}\right) + i \sin\left(\frac{\theta + 360°k}{n}\right)\right]$.
In $16(\cos 120° + i \sin 120°)$, $r = 16$ and $\theta = 120°$.
Because we are finding fourth roots, $n = 4$.

The four fourth roots are found by substituting $0, 1, 2,$ and 3 for k in the expression for z_k, repeated in the margin. Thus, the four complex fourth roots are as follows:

$z_k = \sqrt[4]{16}\left[\cos\left(\dfrac{120° + 360°k}{4}\right) + i\sin\left(\dfrac{120° + 360°k}{4}\right)\right]$

The formula for the four fourth roots of $16(\cos 120° + i\sin 120°), k = 0, 1, 2, 3$ (repeated)

$$z_0 = \sqrt[4]{16}\left[\cos\left(\dfrac{120° + 360° \cdot 0}{4}\right) + i\sin\left(\dfrac{120° + 360° \cdot 0}{4}\right)\right]$$

$$= \sqrt[4]{16}\left(\cos\dfrac{120°}{4} + i\sin\dfrac{120°}{4}\right) = 2(\cos 30° + i\sin 30°)$$

$$z_1 = \sqrt[4]{16}\left[\cos\left(\dfrac{120° + 360° \cdot 1}{4}\right) + i\sin\left(\dfrac{120° + 360° \cdot 1}{4}\right)\right]$$

$$= \sqrt[4]{16}\left(\cos\dfrac{480°}{4} + i\sin\dfrac{480°}{4}\right) = 2(\cos 120° + i\sin 120°)$$

$$z_2 = \sqrt[4]{16}\left[\cos\left(\dfrac{120° + 360° \cdot 2}{4}\right) + i\sin\left(\dfrac{120° + 360° \cdot 2}{4}\right)\right]$$

$$= \sqrt[4]{16}\left(\cos\dfrac{840°}{4} + i\sin\dfrac{840°}{4}\right) = 2(\cos 210° + i\sin 210°)$$

$$z_3 = \sqrt[4]{16}\left[\cos\left(\dfrac{120° + 360° \cdot 3}{4}\right) + i\sin\left(\dfrac{120° + 360° \cdot 3}{4}\right)\right]$$

$$= \sqrt[4]{16}\left(\cos\dfrac{1200°}{4} + i\sin\dfrac{1200°}{4}\right) = 2(\cos 300° + i\sin 300°). \quad \bullet\bullet\bullet$$

In **Figure 5.9**, we have plotted each of the four fourth roots of $16(\cos 120° + i\sin 120°)$. Notice that they are equally spaced at $90°$ intervals on a circle with radius 2.

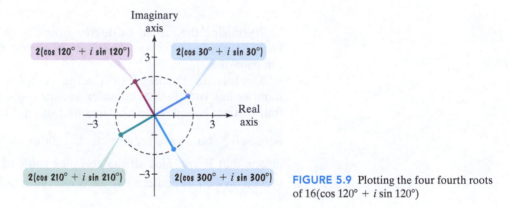

FIGURE 5.9 Plotting the four fourth roots of $16(\cos 120° + i\sin 120°)$

✓ **Check Point 9** Find all the complex fourth roots of $16(\cos 60° + i\sin 60°)$. Write roots in polar form, with θ in degrees.

EXAMPLE 10   Finding the Roots of a Complex Number

Find all the cube roots of 8. Write roots in rectangular form.

SOLUTION

DeMoivre's Theorem for roots applies to complex numbers in polar form. Thus, we will first write 8, or $8 + 0i$, in polar form. We express θ in radians, although degrees could also be used.

$$8 = r(\cos\theta + i\sin\theta) = 8(\cos 0 + i\sin 0)$$

There are exactly three cube roots of 8. From DeMoivre's Theorem for finding complex roots, the cube roots of 8 are

$$z_k = \sqrt[3]{8}\left[\cos\left(\frac{0 + 2\pi k}{3}\right) + i\sin\left(\frac{0 + 2\pi k}{3}\right)\right], k = 0, 1, 2.$$

> Use $z_k = \sqrt[n]{r}\left[\cos\left(\frac{\theta + 2\pi k}{n}\right) + i\sin\left(\frac{\theta + 2\pi k}{n}\right)\right]$.
> In $8(\cos 0 + i \sin 0)$, $r = 8$ and $\theta = 0$.
> Because we are finding cube roots, $n = 3$.

The three cube roots of 8 are found by substituting 0, 1, and 2 for k in the expression for z_k above the voice balloon. Thus, the three cube roots of 8 are

$$z_0 = \sqrt[3]{8}\left[\cos\left(\frac{0 + 2\pi \cdot 0}{3}\right) + i\sin\left(\frac{0 + 2\pi \cdot 0}{3}\right)\right]$$

$$= 2(\cos 0 + i \sin 0) = 2(1 + i \cdot 0) = 2$$

$$z_1 = \sqrt[3]{8}\left[\cos\left(\frac{0 + 2\pi \cdot 1}{3}\right) + i\sin\left(\frac{0 + 2\pi \cdot 1}{3}\right)\right]$$

$$= 2\left(\cos\frac{2\pi}{3} + i\sin\frac{2\pi}{3}\right) = 2\left(-\frac{1}{2} + i \cdot \frac{\sqrt{3}}{2}\right) = -1 + i\sqrt{3}$$

$$z_2 = \sqrt[3]{8}\left[\cos\left(\frac{0 + 2\pi \cdot 2}{3}\right) + i\sin\left(\frac{0 + 2\pi \cdot 2}{3}\right)\right]$$

$$= 2\left(\cos\frac{4\pi}{3} + i\sin\frac{4\pi}{3}\right) = 2\left(-\frac{1}{2} + i \cdot \left(-\frac{\sqrt{3}}{2}\right)\right) = -1 - i\sqrt{3}.$$

The three cube roots of 8 are plotted in **Figure 5.10**.   • • •

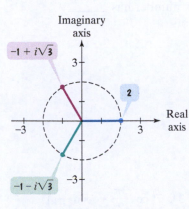

FIGURE 5.10 The three cube roots of 8 are equally spaced at intervals of $\frac{2\pi}{3}$ on a circle with radius 2.

DISCOVERY

Use DeMoivre's Theorem to cube $-1 + i\sqrt{3}$ or $-1 - i\sqrt{3}$ and obtain 8.

⊘ **Check Point 10** Find all the cube roots of 27. Write roots in rectangular form.

Blitzer Bonus ‖ The Mandelbrot Set

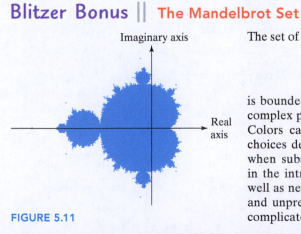

FIGURE 5.11

The set of all complex numbers for which the sequence

$$z, z^2 + z, (z^2 + z)^2 + z, [(z^2 + z)^2 + z]^2 + z, \ldots$$

is bounded is called the **Mandelbrot set**. Plotting these complex numbers in the complex plane results in a graph that is "buglike" in shape, shown in **Figure 5.11**. Colors can be added to the boundary of the graph. At the boundary, color choices depend on how quickly the numbers in the boundary approach infinity when substituted into the sequence shown. The magnified boundary is shown in the introduction to this section. It includes the original buglike structure, as well as new and interesting patterns. With each level of magnification, repetition and unpredictable formations interact to create what has been called the most complicated mathematical object ever known.

CONCEPT AND VOCABULARY CHECK

Fill in each blank so that the resulting statement is true.

1. In the complex plane, the horizontal axis is called the _____ axis and the vertical axis is called the _____ axis.

2. The value $\sqrt{a^2 + b^2}$ is the _____ of the complex number $a + bi$.

3. In the polar form of a complex number, $r(\cos \theta + i \sin \theta)$, r is called the _____ and θ is called the _____, $0 \le \theta < 2\pi$.

4. To convert a complex number from rectangular form, $z = a + bi$, to polar form, $z = r(\cos \theta + i \sin \theta)$, we use the relationships $r =$ _____ and

   $\tan \theta =$ _____, noting the quadrant in which the graph of z lies.

5. $r_1(\cos \theta_1 + i \sin \theta_1) \cdot r_2(\cos \theta_2 + i \sin \theta_2) =$ _____ $[\cos(\underline{\quad}) + i \sin (\underline{\quad})]$
   The product of two complex numbers in polar form is found by _____ their moduli and _____ their arguments.

6. $\dfrac{r_1(\cos \theta_1 + i \sin \theta_1)}{r_2(\cos \theta_2 + i \sin \theta_2)}$

   $=$ _____ $[\cos(\underline{\quad}) + i \sin (\underline{\quad})]$
   The quotient of two complex numbers in polar form is found by _____ their moduli and _____ their arguments.

7. DeMoivre's Theorem states that
   $[r(\cos \theta + i \sin \theta)]^n$

   $=$ _____ $(\cos \underline{\quad} + i \sin \underline{\quad})$.

8. Every nonzero complex number has _____ distinct complex nth roots.

EXERCISE SET 5.2

Practice Exercises

In Exercises 1–10, plot each complex number and find its absolute value.

1. $z = 4i$
2. $z = 3i$
3. $z = 3$
4. $z = 4$
5. $z = 3 + 2i$
6. $z = 2 + 5i$
7. $z = 3 - i$
8. $z = 4 - i$
9. $z = -3 + 4i$
10. $z = -3 - 4i$

In Exercises 11–26, plot each complex number. Then write the complex number in polar form. You may express the argument in degrees or radians.

11. $2 + 2i$
12. $1 + i\sqrt{3}$
13. $-1 - i$
14. $2 - 2i$
15. $-4i$
16. $-3i$
17. $2\sqrt{3} - 2i$
18. $-2 + 2i\sqrt{3}$
19. -3
20. -4
21. $-3\sqrt{2} - 3i\sqrt{3}$
22. $3\sqrt{2} - 3i\sqrt{2}$
23. $-3 + 4i$
24. $-2 + 3i$
25. $2 - i\sqrt{3}$
26. $1 - i\sqrt{5}$

In Exercises 27–36, write each complex number in rectangular form. If necessary, round to the nearest tenth.

27. $6(\cos 30° + i \sin 30°)$
28. $12(\cos 60° + i \sin 60°)$
29. $4(\cos 240° + i \sin 240°)$
30. $10(\cos 210° + i \sin 210°)$
31. $8\left(\cos \dfrac{7\pi}{4} + i \sin \dfrac{7\pi}{4}\right)$
32. $4\left(\cos \dfrac{5\pi}{6} + i \sin \dfrac{5\pi}{6}\right)$

33. $5\left(\cos \dfrac{\pi}{2} + i \sin \dfrac{\pi}{2}\right)$
34. $7\left(\cos \dfrac{3\pi}{2} + i \sin \dfrac{3\pi}{2}\right)$
35. $20(\cos 205° + i \sin 205°)$
36. $30(\cos 2.3 + i \sin 2.3)$

In Exercises 37–44, find the product of the complex numbers. Leave answers in polar form.

37. $z_1 = 6(\cos 20° + i \sin 20°)$
    $z_2 = 5(\cos 50° + i \sin 50°)$

38. $z_1 = 4(\cos 15° + i \sin 15°)$
    $z_2 = 7(\cos 25° + i \sin 25°)$

39. $z_1 = 3\left(\cos \dfrac{\pi}{5} + i \sin \dfrac{\pi}{5}\right)$
    $z_2 = 4\left(\cos \dfrac{\pi}{10} + i \sin \dfrac{\pi}{10}\right)$

40. $z_1 = 3\left(\cos \dfrac{5\pi}{8} + i \sin \dfrac{5\pi}{8}\right)$
    $z_2 = 10\left(\cos \dfrac{\pi}{16} + i \sin \dfrac{\pi}{16}\right)$

41. $z_1 = \cos \dfrac{\pi}{4} + i \sin \dfrac{\pi}{4}$
    $z_2 = \cos \dfrac{\pi}{3} + i \sin \dfrac{\pi}{3}$

42. $z_1 = \cos \dfrac{\pi}{6} + i \sin \dfrac{\pi}{6}$
    $z_2 = \cos \dfrac{\pi}{4} + i \sin \dfrac{\pi}{4}$

43. $z_1 = 1 + i$
    $z_2 = -1 + i$

44. $z_1 = 1 + i$
    $z_2 = 2 + 2i$

In Exercises 45–52, find the quotient $\dfrac{z_1}{z_2}$ of the complex numbers.

Leave answers in polar form. In Exercises 49–50, express the argument as an angle between 0° and 360°.

45. $z_1 = 20(\cos 75° + i \sin 75°)$
$z_2 = 4(\cos 25° + i \sin 25°)$

46. $z_1 = 50(\cos 80° + i \sin 80°)$
$z_2 = 10(\cos 20° + i \sin 20°)$

47. $z_1 = 3\left(\cos \dfrac{\pi}{5} + i \sin \dfrac{\pi}{5}\right)$

$z_2 = 4\left(\cos \dfrac{\pi}{10} + i \sin \dfrac{\pi}{10}\right)$

48. $z_1 = 3\left(\cos \dfrac{5\pi}{18} + i \sin \dfrac{5\pi}{18}\right)$

$z_2 = 10\left(\cos \dfrac{\pi}{16} + i \sin \dfrac{\pi}{16}\right)$

49. $z_1 = \cos 80° + i \sin 80°$
$z_2 = \cos 200° + i \sin 200°$

50. $z_1 = \cos 70° + i \sin 70°$
$z_2 = \cos 230° + i \sin 230°$

51. $z_1 = 2 + 2i$
$z_2 = 1 + i$

52. $z_1 = 2 - 2i$
$z_2 = 1 - i$

In Exercises 53–64, use DeMoivre's Theorem to find the indicated power of the complex number. Write answers in rectangular form.

53. $[4(\cos 15° + i \sin 15°)]^3$

54. $[2(\cos 10° + i \sin 10°)]^3$

55. $[2(\cos 80° + i \sin 80°)]^3$

56. $[2(\cos 40° + i \sin 40°)]^3$

57. $\left[\dfrac{1}{2}\left(\cos \dfrac{\pi}{12} + i \sin \dfrac{\pi}{12}\right)\right]^6$

58. $\left[\dfrac{1}{2}\left(\cos\dfrac{\pi}{10} + i \sin \dfrac{\pi}{10}\right)\right]^5$

59. $\left[\sqrt{2}\left(\cos\dfrac{5\pi}{6} + i \sin \dfrac{5\pi}{6}\right)\right]^4$

60. $\left[\sqrt{3}\left(\cos\dfrac{5\pi}{18} + i \sin\dfrac{5\pi}{18}\right)\right]^6$

61. $(1 + i)^5$

62. $(1 - i)^5$

63. $\left(\sqrt{3} - i\right)^6$

64. $\left(\sqrt{2} - i\right)^4$

In Exercises 65–68, find all the complex roots. Write roots in polar form with θ in degrees.

65. The complex square roots of $9(\cos 30° + i \sin 30°)$

66. The complex square roots of $25(\cos 210° + i \sin 210°)$

67. The complex cube roots of $8(\cos 210° + i \sin 210°)$

68. The complex cube roots of $27(\cos 306° + i \sin 306°)$

In Exercises 69–76, find all the complex roots. Write roots in rectangular form. If necessary, round to the nearest tenth.

69. The complex fourth roots of $81\left(\cos \dfrac{4\pi}{3} + i \sin \dfrac{4\pi}{3}\right)$

70. The complex fifth roots of $32\left(\cos \dfrac{5\pi}{3} + i \sin \dfrac{5\pi}{3}\right)$

71. The complex fifth roots of 32

72. The complex sixth roots of 64

73. The complex cube roots of 1

74. The complex cube roots of i

75. The complex fourth roots of $1 + i$

76. The complex fifth roots of $-1 + i$

Practice Plus

In Exercises 77–80, convert to polar form and then perform the indicated operations. Express answers in polar and rectangular form.

77. $i(2 + 2i)\left(-\sqrt{3} + i\right)$

78. $(1 + i)\left(1 - i\sqrt{3}\right)\left(-\sqrt{3} + i\right)$

79. $\dfrac{\left(1 + i\sqrt{3}\right)(1 - i)}{2\sqrt{3} - 2i}$

80. $\dfrac{\left(-1 + i\sqrt{3}\right)\left(2 - 2i\sqrt{3}\right)}{4\sqrt{3} - 4i}$

In Exercises 81–86, solve each equation in the complex number system. Express solutions in polar and rectangular form.

81. $x^6 - 1 = 0$

82. $x^6 + 1 = 0$

83. $x^4 + 16i = 0$

84. $x^5 - 32i = 0$

85. $x^3 - \left(1 + i\sqrt{3}\right) = 0$

86. $x^3 - \left(1 - i\sqrt{3}\right) = 0$

In calculus, it can be shown that

$$e^{i\theta} = \cos\theta + i \sin\theta.$$

In Exercises 87–90, use this result to plot each complex number.

87. $e^{\frac{\pi i}{4}}$ **88.** $e^{\frac{\pi i}{6}}$ **89.** $-e^{-\pi i}$ **90.** $-2e^{-2\pi i}$

Application Exercises

In Exercises 91–92, show that the given complex number z plots as a point in the Mandelbrot set.

   a. Write the first six terms of the sequence

$$z_1, z_2, z_3, z_4, z_5, z_6, \ldots$$

   where

   $z_1 = z$: *Write the given number.*
   $z_2 = z^2 + z$: *Square z_1 and add the given number.*
   $z_3 = (z^2 + z)^2 + z$: *Square z_2 and add the given number.*
   $z_4 = [(z^2 + z)^2 + z]^2 + z$: *Square z_3 and add the given number.*
   z_5: *Square z_4 and add the given number.*
   z_6: *Square z_5 and add the given number.*

   b. If the sequence that you began writing in part (a) is bounded, the given complex number belongs to the Mandelbrot set. Show that the sequence is bounded by writing two complex numbers. One complex number should be greater in absolute value than the absolute values of the terms in the sequence. The second complex number should be less in absolute value than the absolute values of the terms in the sequence.

91. $z = i$ **92.** $z = -i$

Explaining the Concepts

93. Explain how to plot a complex number in the complex plane. Provide an example with your explanation.

94. How do you determine the absolute value of a complex number?

95. What is the polar form of a complex number?

96. If you are given a complex number in rectangular form, how do you write it in polar form?

97. If you are given a complex number in polar form, how do you write it in rectangular form?

98. Explain how to find the product of two complex numbers in polar form.

99. Explain how to find the quotient of two complex numbers in polar form.

100. Explain how to find the power of a complex number in polar form.

101. Explain how to use DeMoivre's Theorem for finding complex roots to find the two square roots of 9.

102. Describe the graph of all complex numbers with an absolute value of 6.

103. The image of the Mandelbrot set in the section opener exhibits self-similarity: Magnified portions repeat much of the pattern of the whole structure, as well as new and unexpected patterns. Describe an object in nature that exhibits self-similarity.

Technology Exercises

104. Use the rectangular-to-polar feature on a graphing utility to verify any four of your answers in Exercises 11–26. Be aware that you may have to adjust the angle for the correct quadrant.

105. Use the polar-to-rectangular feature on a graphing utility to verify any four of your answers in Exercises 27–36.

Critical Thinking Exercises

Make Sense? *In Exercises 106–109, determine whether each statement makes sense or does not make sense, and explain your reasoning.*

106. A complex number $a + bi$ can be interpreted geometrically as the point (a, b) in the xy-plane.

107. I multiplied two complex numbers in polar form by first multiplying the moduli and then multiplying the arguments.

108. The proof of the formula for the product of two complex numbers in polar form uses the sum formulas for cosines and sines that I studied in a previous chapter.

109. My work with complex numbers verified that the only possible cube root of 8 is 2.

110. Prove the rule for finding the quotient of two complex numbers in polar form. Begin the proof as follows, using the conjugate of the denominator's second factor:

$$\frac{r_1(\cos\theta_1 + i\sin\theta_1)}{r_2(\cos\theta_2 + i\sin\theta_2)} = \frac{r_1(\cos\theta_1 + i\sin\theta_1)}{r_2(\cos\theta_2 + i\sin\theta_2)} \cdot \frac{(\cos\theta_2 - i\sin\theta_2)}{(\cos\theta_2 - i\sin\theta_2)}.$$

Perform the indicated multiplications. Then use the difference formulas for sine and cosine.

111. Plot each of the complex fourth roots of 1.

Group Exercise

112. Group members should prepare and present a seminar on mathematical chaos. Include one or more of the following topics in your presentation: fractal images, the role of complex numbers in generating fractal images, algorithms, iterations, iteration number, and fractals in nature. Be sure to include visual images that will intrigue your audience.

Retaining the Concepts

113. Verify the identity:

$$\frac{1}{\sin x \cos x} - \frac{\cos x}{\sin x} = \tan x.$$

(Section 3.1, Example 6)

114. Use the Law of Sines to solve the triangle shown in the figure with $B = 32°$, $C = 39°$, and $c = 40$. Round lengths of sides to the nearest tenth. (Section 4.1, Example 1)

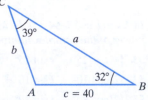

115. Find the exact value of $\cos 75°$ using $\cos 75° = \cos(120° - 45°)$ and the difference formula for cosines. (Section 3.2, Example 1)

Preview Exercises

Exercises 116–118 will help you prepare for the material covered in the next section.

116. Graph: $y = 3$.

117. Graph: $x^2 + (y - 1)^2 = 1$.

118. Complete the square and write the equation in standard form: $x^2 + 6x + y^2 = 0$. Then give the center and radius of the circle, and graph the equation.

A Brief Review • Equations of Lines and Circles in Rectangular Coordinates

- The general form of the equation of a line is $Ax + By + C = 0$, where A and B are not both zero.
- A horizontal line is given by an equation of the form $y = b$. A vertical line is given by an equation of the form $x = a$.

EXAMPLE

The graph of $y = -4$

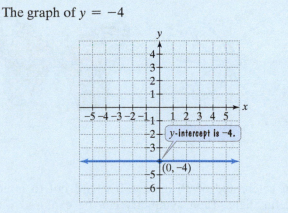

The graph of $x = 2$

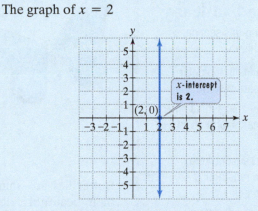

- The standard form of the equation of a circle with center (h, k) and radius r is $(x - h)^2 + (y - k)^2 = r^2$.
- The general form of the equation of a circle is $x^2 + y^2 + Dx + Ey + F = 0$.

We can convert the general form of the equation of a circle to the standard form by completing the square on x and y. To complete the square on $x^2 + bx$, take half the coefficient of x. Then square this number. By adding the square of half the coefficient of x, a perfect square trinomial will result. Factor this trinomial.

EXAMPLE

Write in standard form and graph: $x^2 + y^2 + 4x - 6y - 23 = 0$.

Because we plan to complete the square on both x and y, let's rearrange the terms so that x-terms are arranged in descending order, y-terms are arranged in descending order, and the constant term appears on the right.

$$x^2 + y^2 + 4x - 6y - 23 = 0$$ This is the given equation.

$$(x^2 + 4x \quad) + (y^2 - 6y \quad) = 23$$ Rewrite in anticipation of completing the square.

$$(x^2 + 4x + 4) + (y^2 - 6y + 9) = 23 + 4 + 9$$ Complete the square on x: $\frac{1}{2} \cdot 4 = 2$ and $2^2 = 4$, so add 4 to both sides. Complete the square on y: $\frac{1}{2}(-6) = -3$ and $(-3)^2 = 9$, so add 9 to both sides..

> Remember that numbers added on the left side must also be added on the right side.

$$(x + 2)^2 + (y - 3)^2 = 36$$ Factor on the left and add on the right.

This last equation, $(x + 2)^2 + (y - 3)^2 = 36$, is in standard form. We can identify the circle's center and radius by comparing this equation to the standard form of the equation of a circle, $(x - h)^2 + (y - k)^2 = r^2$.

$$(x + 2)^2 + (y - 3)^2 = 36$$
$$(x - (-2))^2 + (y - 3)^2 = 6^2$$

> This is $(x - h)^2$, with $h = -2$.
> This is $(y - k)^2$, with $k = 3$.
> This is r^2, with $r = 6$.

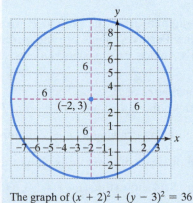

The graph of $(x + 2)^2 + (y - 3)^2 = 36$

We use the center, $(h, k) = (-2, 3)$, and the radius, $r = 6$, to graph the circle. The graph is shown on the left.

Section 5.3   Polar Coordinates

What am I supposed to learn?

After studying this section, you should be able to:

1. Plot points in the polar coordinate system.

2. Find multiple sets of polar coordinates for a given point.

3. Convert a point from polar to rectangular coordinates.

4. Convert a point from rectangular to polar coordinates.

5. Convert an equation from rectangular to polar coordinates.

6. Convert an equation from polar to rectangular coordinates.

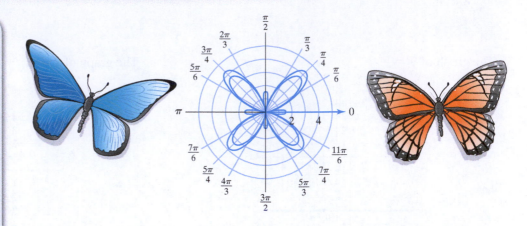

Butterflies are among the most celebrated of all insects. It's hard not to notice their beautiful colors and graceful flight. Their symmetry can be explored with trigonometric functions and a system for plotting points called the *polar coordinate system*. In many cases, polar coordinates are simpler and easier to use than rectangular coordinates.

Plotting Points in the Polar Coordinate System

The foundation of the polar coordinate system is a horizontal ray that extends to the right. The ray is called the **polar axis** and is shown in **Figure 5.12**. The endpoint of the ray is called the **pole**.

A point P in the polar coordinate system is represented by an ordered pair of numbers (r, θ). **Figure 5.13** shows $P = (r, \theta)$ in the polar coordinate system.

1. Plot points in the polar coordinate system.

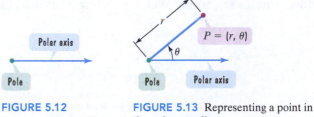

FIGURE 5.12

FIGURE 5.13 Representing a point in the polar coordinate system

- r is a directed distance from the pole to P. (We shall see that r can be positive, negative, or zero.)
- θ is an angle from the polar axis to the line segment from the pole to P. This angle can be measured in degrees or radians. Positive angles are measured counterclockwise from the polar axis. Negative angles are measured clockwise from the polar axis.

We refer to the ordered pair (r, θ) as the **polar coordinates** of P.

Let's look at a specific example. Suppose that the polar coordinates of a point P are $\left(3, \dfrac{\pi}{4}\right)$. Because θ is positive, we locate this point by drawing $\theta = \dfrac{\pi}{4}$ counter-clockwise from the polar axis. Then we count out a distance of three units along the terminal side of the angle to reach the point P. **Figure 5.14** shows that $(r, \theta) = \left(3, \dfrac{\pi}{4}\right)$ lies three units from the pole on the terminal side of the angle $\theta = \dfrac{\pi}{4}$.

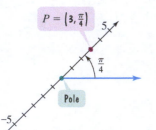

FIGURE 5.14 Locating a point in polar coordinates

The sign of r is important in locating $P = (r, \theta)$ in polar coordinates.

> **The Sign of r and a Point's Location in Polar Coordinates**
>
> The point $P = (r, \theta)$ is located $|r|$ units from the pole. If $r > 0$, the point lies on the terminal side of θ. If $r < 0$, the point lies along the ray opposite the terminal side of θ. If $r = 0$, the point lies at the pole, regardless of the value of θ.

EXAMPLE 1 Plotting Points in a Polar Coordinate System

Plot the points with the following polar coordinates:

    a. $(2, 135°)$ **b.** $\left(-3, \dfrac{3\pi}{2}\right)$ **c.** $\left(-1, -\dfrac{\pi}{4}\right)$.

SOLUTION

    a. To plot the point $(r, \theta) = (2, 135°)$, begin with the $135°$ angle. Because $135°$ is a positive angle, draw $\theta = 135°$ counterclockwise from the polar axis. Now consider $r = 2$. Because $r > 0$, plot the point by going out two units on the terminal side of θ. **Figure 5.15(a)** shows the point. The concentric circles in the figure are drawn to help plot the point at the appropriate distance from the pole.

    b. To plot the point $(r, \theta) = \left(-3, \dfrac{3\pi}{2}\right)$, begin with the $\dfrac{3\pi}{2}$ angle. Because $\dfrac{3\pi}{2}$ is a positive angle, we draw $\theta = \dfrac{3\pi}{2}$ counterclockwise from the polar axis. Now consider $r = -3$. Because $r < 0$, plot the point by going out three units along the ray *opposite* the terminal side of θ. **Figure 5.15(b)** shows the point.

    c. To plot the point $(r, \theta) = \left(-1, -\dfrac{\pi}{4}\right)$, begin with the $-\dfrac{\pi}{4}$ angle. Because $-\dfrac{\pi}{4}$ is a negative angle, draw $\theta = -\dfrac{\pi}{4}$ clockwise from the polar axis. Now consider $r = -1$. Because $r < 0$, plot the point by going out one unit along the ray *opposite* the terminal side of θ. **Figure 5.15(c)** shows the point.   •••

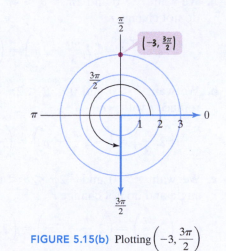

FIGURE 5.15(a) Plotting $(2, 135°)$

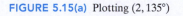

FIGURE 5.15(b) Plotting $\left(-3, \dfrac{3\pi}{2}\right)$

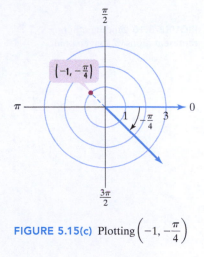

FIGURE 5.15(c) Plotting $\left(-1, -\dfrac{\pi}{4}\right)$

✓ **Check Point 1** Plot the points with the following polar coordinates:

    a. $(3, 315°)$ **b.** $(-2, \pi)$ **c.** $\left(-1, -\dfrac{\pi}{2}\right)$.

② Find multiple sets of polar coordinates for a given point.

Multiple Representations of Points in the Polar Coordinate System

In rectangular coordinates, each point (x, y) has exactly one representation. By contrast, any point in polar coordinates can be represented in infinitely many ways. For example,

$$(r, \theta) = (r, \theta + 2\pi) \qquad \text{and} \qquad (r, \theta) = (-r, \theta + \pi).$$

Adding 1 revolution, or 2π radians, to the angle does not change the point's location.

Adding $\frac{1}{2}$ revolution, or π radians, to the angle and replacing r with $-r$ does not change the point's location.

DISCOVERY

Illustrate the statements in the voice balloons by plotting the points with the following polar coordinates:

a. $\left(1, \dfrac{\pi}{2}\right)$ and $\left(1, \dfrac{5\pi}{2}\right)$

b. $\left(3, \dfrac{\pi}{4}\right)$ and $\left(-3, \dfrac{5\pi}{4}\right)$.

Thus, to find two other representations for the point (r, θ),

- Add 2π to the angle and do not change r.
- Add π to the angle and replace r with $-r$.

Continually adding or subtracting 2π in either of these representations does not change the point's location.

> ### Multiple Representations of Points
>
> If n is any integer, the point (r, θ) can be represented as
>
> $$(r, \theta) = (r, \theta + 2n\pi) \quad \text{or} \quad (r, \theta) = (-r, \theta + \pi + 2n\pi).$$

EXAMPLE 2 Finding Other Polar Coordinates for a Given Point

The point $\left(2, \dfrac{\pi}{3}\right)$ is plotted in **Figure 5.16**. Find another representation of this point in which

a. r is positive and $2\pi < \theta < 4\pi$.
b. r is negative and $0 < \theta < 2\pi$.
c. r is positive and $-2\pi < \theta < 0$.

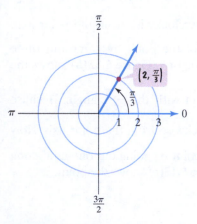

FIGURE 5.16 Finding other representations of a given point

SOLUTION

a. We want $r > 0$ and $2\pi < \theta < 4\pi$. Using $\left(2, \dfrac{\pi}{3}\right)$, add 2π to the angle and do not change r.

$$\left(2, \frac{\pi}{3}\right) = \left(2, \frac{\pi}{3} + 2\pi\right) = \left(2, \frac{\pi}{3} + \frac{6\pi}{3}\right) = \left(2, \frac{7\pi}{3}\right)$$

b. We want $r < 0$ and $0 < \theta < 2\pi$. Using $\left(2, \dfrac{\pi}{3}\right)$, add π to the angle and replace r with $-r$.

$$\left(2, \frac{\pi}{3}\right) = \left(-2, \frac{\pi}{3} + \pi\right) = \left(-2, \frac{\pi}{3} + \frac{3\pi}{3}\right) = \left(-2, \frac{4\pi}{3}\right)$$

c. We want $r > 0$ and $-2\pi < \theta < 0$. Using $\left(2, \dfrac{\pi}{3}\right)$, subtract 2π from the angle and do not change r.

$$\left(2, \frac{\pi}{3}\right) = \left(2, \frac{\pi}{3} - 2\pi\right) = \left(2, \frac{\pi}{3} - \frac{6\pi}{3}\right) = \left(2, -\frac{5\pi}{3}\right) \qquad \bullet\bullet\bullet$$

✓ **Check Point 2** Find another representation of $\left(5, \dfrac{\pi}{4}\right)$ in which

a. r is positive and $2\pi < \theta < 4\pi$.

b. r is negative and $0 < \theta < 2\pi$.

c. r is positive and $-2\pi < \theta < 0$.

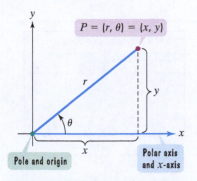

FIGURE 5.17 Polar and rectangular coordinate systems

Relations between Polar and Rectangular Coordinates

We now consider both polar and rectangular coordinates simultaneously. **Figure 5.17** shows the two coordinate systems. The polar axis coincides with the positive x-axis and the pole coincides with the origin. A point P, other than the origin, has rectangular coordinates (x, y) and polar coordinates (r, θ), as indicated in the figure. We wish to find equations relating the two sets of coordinates. From the figure, we see that

$$x^2 + y^2 = r^2$$

$$\sin \theta = \frac{y}{r} \qquad \cos \theta = \frac{x}{r} \qquad \tan \theta = \frac{y}{x}.$$

These relationships hold when P is in any quadrant and when $r > 0$ or $r < 0$.

Relations between Polar and Rectangular Coordinates

$$x = r \cos \theta$$
$$y = r \sin \theta$$
$$x^2 + y^2 = r^2$$
$$\tan \theta = \frac{y}{x}$$

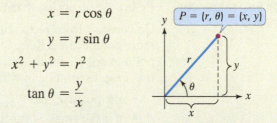

③ Convert a point from polar to rectangular coordinates.

Point Conversion from Polar to Rectangular Coordinates

To convert a point from polar coordinates (r, θ) to rectangular coordinates (x, y), use the formulas $x = r \cos \theta$ and $y = r \sin \theta$.

EXAMPLE 3   Polar-to-Rectangular Point Conversion

Find the rectangular coordinates of the points with the following polar coordinates:

a. $\left(2, \dfrac{3\pi}{2}\right)$ **b.** $\left(-8, \dfrac{\pi}{3}\right)$.

SOLUTION
We find (x, y) by substituting the given values for r and θ into $x = r \cos \theta$ and $y = r \sin \theta$.

a. We begin with the rectangular coordinates of the point $(r, \theta) = \left(2, \dfrac{3\pi}{2}\right)$.

$$x = r \cos \theta = 2 \cos \frac{3\pi}{2} = 2 \cdot 0 = 0$$

$$y = r \sin \theta = 2 \sin \frac{3\pi}{2} = 2(-1) = -2$$

The rectangular coordinates of $\left(2, \dfrac{3\pi}{2}\right)$ are $(0, -2)$. See **Figure 5.18**.

b. We now find the rectangular coordinates of the point $(r, \theta) = \left(-8, \dfrac{\pi}{3}\right)$.

$$x = r \cos \theta = -8 \cos \frac{\pi}{3} = -8\left(\frac{1}{2}\right) = -4$$

$$y = r \sin \theta = -8 \sin \frac{\pi}{3} = -8\left(\frac{\sqrt{3}}{2}\right) = -4\sqrt{3}$$

The rectangular coordinates of $\left(-8, \dfrac{\pi}{3}\right)$ are $\left(-4, -4\sqrt{3}\right)$.   • • •

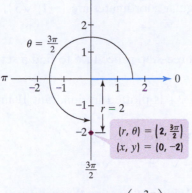

FIGURE 5.18 Converting $\left(2, \dfrac{3\pi}{2}\right)$ to rectangular coordinates

Some graphing utilities can convert a point from polar coordinates to rectangular coordinates. Consult your manual. The screen on the right verifies the polar-rectangular conversion in Example 3(a). It shows that the rectangular coordinates of

$(r, \theta) = \left(2, \dfrac{3\pi}{2}\right)$ are $(0, -2)$. Notice that the

x- and y-coordinates are displayed separately.

```
P▸Rx(2,3π/2)
                              0.
P▸Ry(2,3π/2)
                              -2.
```

⊘ **Check Point 3** Find the rectangular coordinates of the points with the following polar coordinates:

    a. $(3, \pi)$ **b.** $\left(-10, \dfrac{\pi}{6}\right)$.

④ Convert a point from rectangular to polar coordinates.

Point Conversion from Rectangular to Polar Coordinates

Conversion from rectangular coordinates (x, y) to polar coordinates (r, θ) is a bit more complicated. Keep in mind that there are infinitely many representations for a point in polar coordinates. If the point (x, y) lies in one of the four quadrants, we will use a representation in which

- r is positive, and
- θ is the smallest positive angle with the terminal side passing through (x, y).

These conventions provide the following procedure:

> **Converting a Point from Rectangular to Polar Coordinates**
> $(r > 0 \text{ and } 0 \leq \theta < 2\pi)$
>
> **1.** Plot the point (x, y).
> **2.** Find r by computing the distance from the origin to (x, y): $r = \sqrt{x^2 + y^2}$.
> **3.** Find θ using $\tan \theta = \dfrac{y}{x}$ with the terminal side of θ passing through (x, y).

EXAMPLE 4   Rectangular-to-Polar Point Conversion

Find polar coordinates of the point whose rectangular coordinates are $\left(-1, \sqrt{3}\right)$.

SOLUTION

We begin with $(x, y) = \left(-1, \sqrt{3}\right)$ and use our three-step procedure to find a set of polar coordinates (r, θ).

Step 1 **Plot the point (x, y).** The point $\left(-1, \sqrt{3}\right)$ is plotted in quadrant II in **Figure 5.19**.

Step 2 **Find r by computing the distance from the origin to (x, y).**

$$r = \sqrt{x^2 + y^2} = \sqrt{(-1)^2 + \left(\sqrt{3}\right)^2} = \sqrt{1 + 3} = \sqrt{4} = 2$$

Step 3 **Find θ using $\tan \theta = \dfrac{y}{x}$ with the terminal side of θ passing through (x, y).**

$$\tan \theta = \frac{y}{x} = \frac{\sqrt{3}}{-1} = -\sqrt{3}$$

We know that $\tan \dfrac{\pi}{3} = \sqrt{3}$. Because θ lies in quadrant II,

$$\theta = \pi - \frac{\pi}{3} = \frac{3\pi}{3} - \frac{\pi}{3} = \frac{2\pi}{3}.$$

One representation of $\left(-1, \sqrt{3}\right)$ in polar coordinates is $(r, \theta) = \left(2, \dfrac{2\pi}{3}\right).$ • • •

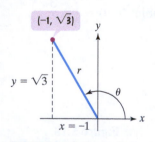

FIGURE 5.19 Converting $\left(-1, \sqrt{3}\right)$ to polar coordinates

TECHNOLOGY

The screen shows the rectangular-polar conversion for $(-1, \sqrt{3})$ on a graphing utility. In Example 4, we showed that $(x, y) = (-1, \sqrt{3})$ can be represented in polar coordinates as $(r, \theta) = \left(2, \dfrac{2\pi}{3}\right)$.

Using $\dfrac{2\pi}{3} \approx 2.094395102$ verifies that our conversion is correct. Notice that the r- and (approximate) θ-coordinates are displayed separately.

```
R▶Pr(-1,√3)
                            2.
R▶Pθ(-1,√3)
                   2.094395102
■
```

✓ **Check Point 4** Find polar coordinates of the point whose rectangular coordinates are $(1, -\sqrt{3})$.

If a point (x, y) lies on a positive or negative axis, we use a representation in which

- r is positive, and
- θ is the smallest quadrantal angle that lies on the same positive or negative axis as (x, y).

In these cases, you can find r and θ by plotting (x, y) and inspecting the figure. Let's see how this is done.

EXAMPLE 5   Rectangular-to-Polar Point Conversion

Find polar coordinates of the point whose rectangular coordinates are $(-2, 0)$.

SOLUTION

We begin with $(x, y) = (-2, 0)$ and find a set of polar coordinates (r, θ).

Step 1   Plot the point (x, y). The point $(-2, 0)$ is plotted in **Figure 5.20**.

Step 2   Find r, the distance from the origin to (x, y). Can you tell by looking at **Figure 5.20** that this distance is 2?

$$r = \sqrt{x^2 + y^2} = \sqrt{(-2)^2 + 0^2} = \sqrt{4} = 2$$

Step 3   Find θ with θ lying on the same positive or negative axis as (x, y). The point $(-2, 0)$ is on the negative x-axis. Thus, θ lies on the negative x-axis and $\theta = \pi$. One representation of $(-2, 0)$ in polar coordinates is $(2, \pi)$.   •••

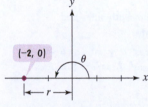

FIGURE 5.20 Converting $(-2, 0)$ to polar coordinates

✓ **Check Point 5** Find polar coordinates of the point whose rectangular coordinates are $(0, -4)$. Express θ in radians.

⑤ Convert an equation from rectangular to polar coordinates.

Equation Conversion from Rectangular to Polar Coordinates

A **polar equation** is an equation whose variables are r and θ. Two examples of polar equations are

$$r = \frac{5}{\cos \theta + \sin \theta} \qquad \text{and} \qquad r = 3 \csc \theta.$$

To convert a rectangular equation in x and y to a polar equation in r and θ, replace x with $r \cos \theta$ and y with $r \sin \theta$.

EXAMPLE 6 Converting Equations from Rectangular to Polar Coordinates

Convert each rectangular equation to a polar equation that expresses r in terms of θ:

a. $x + y = 5$ **b.** $(x - 1)^2 + y^2 = 1$.

SOLUTION

Our goal is to obtain equations in which the variables are r and θ rather than x and y. We use $x = r \cos \theta$ and $y = r \sin \theta$. We then solve the equations for r, obtaining equivalent equations that give r in terms of θ.

a.

$x + y = 5$	This is the given equation in rectangular coordinates. The graph is a line passing through (5, O) and (O, 5).
$r \cos \theta + r \sin \theta = 5$	Replace x with $r \cos \theta$ and y with $r \sin \theta$.
$r(\cos \theta + \sin \theta) = 5$	Factor out r.
$r = \dfrac{5}{\cos \theta + \sin \theta}$	Divide both sides of the equation by $\cos \theta + \sin \theta$ and solve for r.

Thus, the polar equation for $x + y = 5$ is $r = \dfrac{5}{\cos \theta + \sin \theta}$.

b.

$(x - 1)^2 + y^2 = 1$	This is the given equation in rectangular coordinates. The graph is a circle with radius 1 and center at $(h, k) = (1, O)$.

The standard form of a circle's equation is $(x - h)^2 + (y - k)^2 = r^2$, with radius r and center at (h, k).

$(r \cos \theta - 1)^2 + (r \sin \theta)^2 = 1$	Replace x with $r \cos \theta$ and y with $r \sin \theta$.
$r^2 \cos^2 \theta - 2r \cos \theta + 1 + r^2 \sin^2 \theta = 1$	Use $(A - B)^2 = A^2 - 2AB + B^2$ to square $r \cos \theta - 1$.
$r^2 \cos^2 \theta + r^2 \sin^2 \theta - 2r \cos \theta = 0$	Subtract 1 from both sides and rearrange terms.
$r^2 - 2r \cos \theta = 0$	Simplify: $r^2 \cos^2 \theta + r^2 \sin^2 \theta = r^2(\cos^2 \theta + \sin^2 \theta) = r^2 \cdot 1 = r^2$.
$r(r - 2 \cos \theta) = 0$	Factor out r.
$r = 0$ or $r - 2 \cos \theta = 0$	Set each factor equal to O.
$r = 2 \cos \theta$	Solve for r.

The graph of $r = 0$ is a single point, the pole. Because the pole also satisfies the equation $r = 2 \cos \theta$ (for $\theta = \frac{\pi}{2}$, $r = 0$), it is not necessary to include the equation $r = 0$. Thus, the polar equation for $(x - 1)^2 + y^2 = 1$ is $r = 2 \cos \theta$. **• • •**

✓ **Check Point 6** Convert each rectangular equation to a polar equation that expresses r in terms of θ:

a. $3x - y = 6$ **b.** $x^2 + (y + 1)^2 = 1$.

⑥ Convert an equation from polar to rectangular coordinates.

Equation Conversion from Polar to Rectangular Coordinates

When we convert an equation from polar to rectangular coordinates, our goal is to obtain an equation in which the variables are x and y rather than r and θ. We use one or more of the following equations:

$$r^2 = x^2 + y^2 \qquad r \cos \theta = x \qquad r \sin \theta = y \qquad \tan \theta = \frac{y}{x}.$$

To use these equations, it is sometimes necessary to do something to the given polar equation. This could include squaring both sides, using an identity, taking the tangent of both sides, or multiplying both sides by r.

EXAMPLE 7   Converting Equations from Polar to Rectangular Form

Convert each polar equation to a rectangular equation in x and y:

     a. $r = 5$ **b.** $\theta = \dfrac{\pi}{4}$ **c.** $r = 3 \csc \theta$ **d.** $r = -6 \cos \theta$.

SOLUTION

In each case, let's express the rectangular equation in a form that enables us to recognize its graph.

a. We use $r^2 = x^2 + y^2$ to convert the polar equation $r = 5$ to a rectangular equation.

$r = 5$	This is the given polar equation.
$r^2 = 25$	Square both sides.
$x^2 + y^2 = 25$	Use $r^2 = x^2 + y^2$ on the left side.

The rectangular equation for $r = 5$ is $x^2 + y^2 = 25$. The graph is a circle with center at $(0, 0)$ and radius 5.

b. We use $\tan \theta = \dfrac{y}{x}$ to convert the polar equation $\theta = \dfrac{\pi}{4}$ to a rectangular equation in x and y.

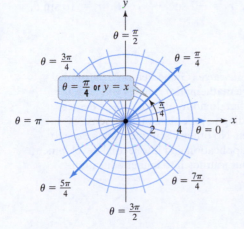

$\theta = \dfrac{\pi}{4}$	This is the given polar equation.
$\tan \theta = \tan \dfrac{\pi}{4}$	Take the tangent of both sides.
$\tan \theta = 1$	$\tan \dfrac{\pi}{4} = 1$
$\dfrac{y}{x} = 1$	Use $\tan \theta = \dfrac{y}{x}$ on the left side.
$y = x$	Multiply both sides by x.

FIGURE 5.21

The rectangular equation for $\theta = \dfrac{\pi}{4}$ is $y = x$. The graph is a line that bisects quadrants I and III. **Figure 5.21** shows the line drawn in a polar coordinate system.

c. We use $r \sin \theta = y$ to convert the polar equation $r = 3 \csc \theta$ to a rectangular equation. To do this, we express the cosecant in terms of the sine.

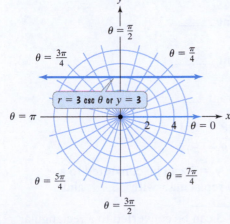

$r = 3 \csc \theta$	This is the given polar equation.
$r = \dfrac{3}{\sin \theta}$	$\csc \theta = \dfrac{1}{\sin \theta}$
$r \sin \theta = 3$	Multiply both sides by $\sin \theta$.
$y = 3$	Use $r \sin \theta = y$ on the left side.

The rectangular equation for $r = 3 \csc \theta$ is $y = 3$. The graph is a horizontal line three units above the x-axis. **Figure 5.22** shows the line drawn in a polar coordinate system.

FIGURE 5.22

d. To convert $r = -6 \cos \theta$ to rectangular coordinates, we multiply both sides by r. Then we use $r^2 = x^2 + y^2$ on the left side and $r \cos \theta = x$ on the right side.

$$r = -6 \cos \theta \qquad \text{This is the given polar equation.}$$
$$r^2 = -6r \cos \theta \qquad \text{Multiply both sides by } r.$$
$$x^2 + y^2 = -6x \qquad \text{Convert to rectangular coordinates:}$$
$$ \qquad r^2 = x^2 + y^2 \text{ and } r \cos \theta = x.$$
$$x^2 + 6x + y^2 = 0 \qquad \text{Add } 6x \text{ to both sides.}$$
$$x^2 + 6x + 9 + y^2 = 9 \qquad \text{Complete the square on } x: \frac{1}{2} \cdot 6 = 3 \text{ and}$$
$$ \qquad 3^2 = 9.$$
$$(x + 3)^2 + y^2 = 9 \qquad \text{Factor.}$$

The rectangular equation for $r = -6 \cos \theta$ is $(x + 3)^2 + y^2 = 9$. This last equation is the standard form of the equation of a circle, $(x - h)^2 + (y - k)^2 = r^2$, with radius r and center at (h, k). Thus, the graph of $(x + 3)^2 + y^2 = 9$ is a circle with center at $(-3, 0)$ and radius 3. • • •

Converting a polar equation to a rectangular equation may be a useful way to develop or check a graph. For example, the graph of the polar equation $r = 5$ consists of all points that are five units from the pole. Thus, the graph is a circle centered at the pole with radius 5. The rectangular equation for $r = 5$, namely, $x^2 + y^2 = 25$, has precisely the same graph (see **Figure 5.23**). We will discuss graphs of polar equations in the next section.

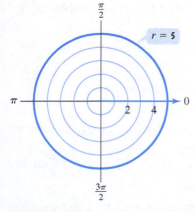

FIGURE 5.23 The equations $r = 5$ and $x^2 + y^2 = 25$ have the same graph.

✓ **Check Point 7** Convert each polar equation to a rectangular equation in x and y:

a. $r = 4$ **b.** $\theta = \dfrac{3\pi}{4}$ **c.** $r = -2 \sec \theta$ **d.** $r = 10 \sin \theta$.

ACHIEVING SUCCESS

A recent government study cited in *Math: A Rich Heritage* (Globe Fearon Educational Publisher) found this simple fact: **The more college mathematics courses you take, the greater your earning potential will be.** Even jobs that do not require a college degree require mathematical thinking that involves attending to precision, making sense of complex problems, and persevering in solving them. No other discipline comes close to math in offering a more extensive set of tools for application and intellectual development. Take as much math as possible as you continue your journey into higher education.

CONCEPT AND VOCABULARY CHECK

Fill in each blank so that the resulting statement is true.

1. The foundation of the polar coordinate system consists of a point, called the _____, and a ray extending out from it, called the _____.

2. The origin in the rectangular coordinate system coincides with the _____ in polar coordinates. The positive x-axis in rectangular coordinates coincides with the _____ in polar coordinates.

For each point with the given polar coordinates in Exercises 3–8 determine the quadrant in which the point lies if it is graphed in a rectangular coordinate system.

3. $(4, 135°)$; quadrant _____

4. $(-4, 135°)$; quadrant _____

5. $\left(2, \dfrac{5\pi}{3}\right)$; quadrant _____

6. $\left(-3, \dfrac{\pi}{4}\right)$; quadrant _____

7. $\left(5, -\dfrac{\pi}{4}\right)$; quadrant _____

8. $\left(-2, -\dfrac{\pi}{4}\right)$; quadrant _____

9. $(r, \theta) = (\text{_____}, \theta + 2\pi)$

10. $(r, \theta) = (\text{_____}, \theta + \pi)$

11. The equation $x + y = 7$ can be converted to a polar equation by replacing x with _____ and replacing y with _____.

12. The equation $r = 3$ can be converted to a rectangular equation by _____ both sides and then replacing r^2 with _____.

13. The equation $\theta = \dfrac{5\pi}{4}$ can be converted to a rectangular equation by taking the _____ of both sides and then replacing $\tan\theta$ with _____.

14. The equation $r = 4\sin\theta$ can be converted to a rectangular equation by _____ both sides by _____ and then replacing r^2 with _____ and $r\sin\theta$ with _____.

EXERCISE SET 5.3

Practice Exercises

In Exercises 1–10, indicate if the point with the given polar coordinates is represented by A, B, C, or D on the graph.

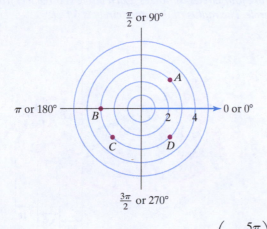

1. $(3, 225°)$ **2.** $(3, 315°)$ **3.** $\left(-3, \dfrac{5\pi}{4}\right)$

4. $\left(-3, \dfrac{\pi}{4}\right)$ **5.** $(3, \pi)$ **6.** $(-3, 0)$

7. $(3, -135°)$ **8.** $(3, -315°)$ **9.** $\left(-3, -\dfrac{3\pi}{4}\right)$

10. $\left(-3, -\dfrac{5\pi}{4}\right)$

In Exercises 11–20, use a polar coordinate system like the one shown for Exercises 1–10 to plot each point with the given polar coordinates.

11. $(2, 45°)$ **12.** $(1, 45°)$ **13.** $(3, 90°)$

14. $(2, 270°)$ **15.** $\left(3, \dfrac{4\pi}{3}\right)$ **16.** $\left(3, \dfrac{7\pi}{6}\right)$

17. $(-1, \pi)$ **18.** $\left(-1, \dfrac{3\pi}{2}\right)$ **19.** $\left(-2, -\dfrac{\pi}{2}\right)$

20. $(-3, -\pi)$

In Exercises 21–26, use a polar coordinate system like the one shown for Exercises 1–10 to plot each point with the given polar coordinates. Then find another representation (r, θ) of this point in which

  a. $r > 0$, $2\pi < \theta < 4\pi$.
  b. $r < 0$, $0 < \theta < 2\pi$.
  c. $r > 0$, $-2\pi < \theta < 0$.

21. $\left(5, \dfrac{\pi}{6}\right)$ **22.** $\left(8, \dfrac{\pi}{6}\right)$ **23.** $\left(10, \dfrac{3\pi}{4}\right)$

24. $\left(12, \dfrac{2\pi}{3}\right)$ **25.** $\left(4, \dfrac{\pi}{2}\right)$ **26.** $\left(6, \dfrac{\pi}{2}\right)$

In Exercises 27–32, select the representations that do not change the location of the given point.

27. $(7, 140°)$
  a. $(-7, 320°)$ **b.** $(-7, -40°)$
  c. $(-7, 220°)$ **d.** $(7, -220°)$

28. $(4, 120°)$
  a. $(-4, 300°)$ **b.** $(-4, -240°)$
  c. $(4, -240°)$ **d.** $(4, 480°)$

29. $\left(2, -\dfrac{3\pi}{4}\right)$
  a. $\left(2, -\dfrac{7\pi}{4}\right)$ **b.** $\left(2, \dfrac{5\pi}{4}\right)$
  c. $\left(-2, -\dfrac{\pi}{4}\right)$ **d.** $\left(-2, -\dfrac{7\pi}{4}\right)$

30. $\left(-2, \dfrac{7\pi}{6}\right)$
  a. $\left(-2, -\dfrac{5\pi}{6}\right)$ **b.** $\left(-2, -\dfrac{\pi}{6}\right)$
  c. $\left(2, -\dfrac{\pi}{6}\right)$ **d.** $\left(2, \dfrac{\pi}{6}\right)$

31. $\left(-5, -\dfrac{\pi}{4}\right)$
  a. $\left(-5, \dfrac{7\pi}{4}\right)$ **b.** $\left(5, -\dfrac{5\pi}{4}\right)$
  c. $\left(-5, \dfrac{11\pi}{4}\right)$ **d.** $\left(5, \dfrac{\pi}{4}\right)$

32. $(-6, 3\pi)$
  a. $(6, 2\pi)$ **b.** $(6, -\pi)$
  c. $(-6, \pi)$ **d.** $(-6, -2\pi)$

In Exercises 33–40, polar coordinates of a point are given. Find the rectangular coordinates of each point.

33. $(4, 90°)$ **34.** $(6, 180°)$

35. $\left(2, \dfrac{\pi}{3}\right)$ **36.** $\left(2, \dfrac{\pi}{6}\right)$

37. $\left(-4, \dfrac{\pi}{2}\right)$ **38.** $\left(-6, \dfrac{3\pi}{2}\right)$

39. $(7.4, 2.5)$ **40.** $(8.3, 4.6)$

In Exercises 41–48, the rectangular coordinates of a point are given. Find polar coordinates of each point. Express θ in radians.

41. $(-2, 2)$ **42.** $(2, -2)$

43. $(2, -2\sqrt{3})$ **44.** $(-2\sqrt{3}, 2)$

45. $\left(-\sqrt{3}, -1\right)$

46. $\left(-1, -\sqrt{3}\right)$

47. $(5, 0)$

48. $(0, -6)$

In Exercises 49–58, convert each rectangular equation to a polar equation that expresses r in terms of θ.

49. $3x + y = 7$

50. $x + 5y = 8$

51. $x = 7$

52. $y = 3$

53. $x^2 + y^2 = 9$

54. $x^2 + y^2 = 16$

55. $(x - 2)^2 + y^2 = 4$

56. $x^2 + (y + 3)^2 = 9$

57. $y^2 = 6x$

58. $x^2 = 6y$

In Exercises 59–74, convert each polar equation to a rectangular equation. Then use a rectangular coordinate system to graph the rectangular equation.

59. $r = 8$

60. $r = 10$

61. $\theta = \dfrac{\pi}{2}$

62. $\theta = \dfrac{\pi}{3}$

63. $r \sin \theta = 3$

64. $r \cos \theta = 7$

65. $r = 4 \csc \theta$

66. $r = 6 \sec \theta$

67. $r = \sin \theta$

68. $r = \cos \theta$

69. $r = 12 \cos \theta$

70. $r = -4 \sin \theta$

71. $r = 6 \cos \theta + 4 \sin \theta$

72. $r = 8 \cos \theta + 2 \sin \theta$

73. $r^2 \sin 2\theta = 2$

74. $r^2 \sin 2\theta = 4$

Practice Plus

In Exercises 75–78, show that each statement is true by converting the given polar equation to a rectangular equation.

75. Show that the graph of $r = a \sec \theta$ is a vertical line a units to the right of the y-axis if $a > 0$ and $|a|$ units to the left of the y-axis if $a < 0$.

76. Show that the graph of $r = a \csc \theta$ is a horizontal line a units above the x-axis if $a > 0$ and $|a|$ units below the x-axis if $a < 0$.

77. Show that the graph of $r = a \sin \theta$ is a circle with center at $\left(0, \dfrac{a}{2}\right)$ and radius $\dfrac{a}{2}$.

78. Show that the graph of $r = a \cos \theta$ is a circle with center at $\left(\dfrac{a}{2}, 0\right)$ and radius $\dfrac{a}{2}$.

In Exercises 79–80, convert each polar equation to a rectangular equation. Then determine the graph's slope and y-intercept.

79. $r \sin\left(\theta - \dfrac{\pi}{4}\right) = 2$

80. $r \cos\left(\theta + \dfrac{\pi}{6}\right) = 8$

In Exercises 81–82, find the rectangular coordinates of each pair of points. Then find the distance, in simplified radical form, between the points.

81. $\left(2, \dfrac{2\pi}{3}\right)$ and $\left(4, \dfrac{\pi}{6}\right)$

82. $(6, \pi)$ and $\left(5, \dfrac{7\pi}{4}\right)$

Application Exercises

Use the figure of the merry-go-round to solve Exercises 83–84. There are four circles of horses. Each circle is three feet from the next circle. The radius of the inner circle is 6 feet.

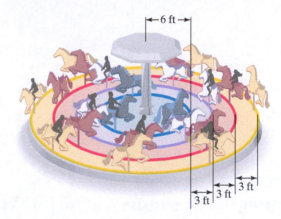

83. If a horse in the outer circle is $\frac{2}{3}$ of the way around the merry-go-round, give its polar coordinates.

84. If a horse in the inner circle is $\frac{5}{6}$ of the way around the merry-go-round, give its polar coordinates.

The wind is blowing at 10 knots. Sailboat racers look for a sailing angle to the 10-knot wind that produces maximum sailing speed. In this application, (r, θ) describes the sailing speed, r, in knots, at an angle θ to the 10-knot wind. Use this information to solve Exercises 85–87.

85. Interpret the polar coordinates: $(6.3, 50°)$.

86. Interpret the polar coordinates: $(7.4, 85°)$.

87. Four points in this 10-knot-wind situation are $(6.3, 50°)$, $(7.4, 85°)$, $(7.5, 105°)$, and $(7.3, 135°)$. Based on these points, which sailing angle to the 10-knot wind would you recommend to a serious sailboat racer? What sailing speed is achieved at this angle?

Explaining the Concepts

88. Explain how to plot (r, θ) if $r > 0$ and $\theta > 0$.

89. Explain how to plot (r, θ) if $r < 0$ and $\theta > 0$.

90. If you are given polar coordinates of a point, explain how to find two additional sets of polar coordinates for the point.

91. Explain how to convert a point from polar to rectangular coordinates. Provide an example with your explanation.

92. Explain how to convert a point from rectangular to polar coordinates. Provide an example with your explanation.

93. Explain how to convert from a rectangular equation to a polar equation.

94. In converting $r = 5$ from a polar equation to a rectangular equation, describe what should be done to both sides of the equation and why this should be done.

95. In converting $r = \sin \theta$ from a polar equation to a rectangular equation, describe what should be done to both sides of the equation and why this should be done.

96. Suppose that (r, θ) describes the sailing speed, r, in knots, at an angle θ to a wind blowing at 20 knots. You have a list of all ordered pairs (r, θ) for integral angles from $\theta = 0°$ to $\theta = 180°$. Describe a way to present this information so that a serious sailboat racer can visualize sailing speeds at different sailing angles to the wind.

Technology Exercises

In Exercises 97–99, polar coordinates of a point are given. Use a graphing utility to find the rectangular coordinates of each point to three decimal places.

97. $\left(4, \dfrac{2\pi}{3}\right)$

98. $(5.2, 1.7)$

99. $(-4, 1.088)$

In Exercises 100–102, the rectangular coordinates of a point are given. Use a graphing utility in radian mode to find polar coordinates of each point to three decimal places.

100. $(-5, 2)$

101. $(\sqrt{5}, 2)$

102. $(-4.308, -7.529)$

Critical Thinking Exercises

Make Sense? *In Exercises 103–106, determine whether each statement makes sense or does not make sense, and explain your reasoning.*

103. I must have made a mistake because my polar representation of a given point is not the same as the answer in the back of the book.

104. When converting a point from polar coordinates to rectangular coordinates, there are infinitely many possible rectangular coordinate pairs.

105. After plotting the point with rectangular coordinates $(0, -4)$, I found polar coordinates without having to show any work.

106. When I convert an equation from polar form to rectangular form, the rectangular equation might not define y as a function of x.

107. Prove that the distance, d, between two points with polar coordinates (r_1, θ_1) and (r_2, θ_2) is

$$d = \sqrt{r_1^2 + r_2^2 - 2r_1 r_2 \cos(\theta_2 - \theta_1)}.$$

108. Use the formula in Exercise 107 to find the distance between $\left(2, \dfrac{5\pi}{6}\right)$ and $\left(4, \dfrac{\pi}{6}\right)$. Express the answer in simplified radical form.

Retaining the Concepts

109. Solve: $\cos x \tan^2 x = 3 \cos x, \quad 0 \le x < 2\pi.$

(Section 3.5, Example 6)

110. Solve triangle ABC with $A = 20°$, $b = 60$, $c = 68$. Round lengths of sides to the nearest tenth and angle measures to the nearest degree.

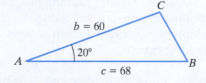

(Section 4.2, Example 1)

111. Verify the identity:

$$\sin 2x + 1 = (\sin x + \cos x)^2.$$

(Section 3.3, Examples 3 and 6)

Preview Exercises

Exercises 112–114 will help you prepare for the material covered in the next section. In each exercise, use a calculator to complete the table of coordinates. Where necessary, round to two decimal places. Then plot the resulting points, (r, θ), using a polar coordinate system.

112.

θ	0	$\dfrac{\pi}{6}$	$\dfrac{\pi}{3}$	$\dfrac{\pi}{2}$	$\dfrac{2\pi}{3}$	$\dfrac{5\pi}{6}$	π
$r = 1 - \cos \theta$							

113.

θ	0	$\dfrac{\pi}{6}$	$\dfrac{\pi}{3}$	$\dfrac{\pi}{2}$	$\dfrac{2\pi}{3}$	$\dfrac{5\pi}{6}$	π	$\dfrac{7\pi}{6}$	$\dfrac{4\pi}{3}$	$\dfrac{3\pi}{2}$
$r = 1 + 2 \sin \theta$										

114.

θ	0	$\dfrac{\pi}{6}$	$\dfrac{\pi}{4}$	$\dfrac{\pi}{3}$	$\dfrac{\pi}{2}$	$\dfrac{2\pi}{3}$	$\dfrac{3\pi}{4}$	$\dfrac{5\pi}{6}$	π
$r = 4 \sin 2\theta$									

CHAPTER 5	Mid-Chapter Check Point

WHAT YOU KNOW: We defined the imaginary unit i as $i = \sqrt{-1}$, where $i^2 = -1$. The set of numbers in the form $a + bi$ is the set of complex numbers, where a is the real part and b is the imaginary part. We performed operations with complex numbers, including division, where we multiplied the numerator and the denominator by the complex conjugate of the denominator. We saw that the polar form of $z = a + bi$ is $z = r(\cos\theta + i\sin\theta)$, where $a = r\cos\theta$, $b = r\sin\theta$, $r = \sqrt{a^2 + b^2}$, and $\tan\theta = \dfrac{b}{a}$.

We called r the modulus and θ the argument of z, with $0 \le \theta < 2\pi$. We multiplied numbers in polar form by multiplying moduli and adding arguments. Division was performed by dividing moduli and subtracting arguments. We used DeMoivre's Theorem to find powers of complex numbers in polar form:

$$[r(\cos\theta + i\sin\theta)]^n = r^n(\cos n\theta + i\sin n\theta)$$

We also used the theorem to find the n distinct nth roots of $r(\cos\theta + i\sin\theta)$.

We then turned our attention to the polar coordinate system, where points are represented by (r, θ), where r is the directed distance from the pole to the point and θ is the angle from the polar axis to line segment OP. We saw that points have multiple representations:

$$(r, \theta) = (r, \theta + 2n\pi) \text{ or } (r, \theta) = (-r, \theta + \pi + 2n\pi),$$
n is any integer.

We used the relations between polar and rectangular coordinates

$$x = r\cos\theta, y = r\sin\theta, x^2 + y^2 = r^2, \tan\theta = \frac{y}{x},$$

to convert both points and equations from polar coordinates to rectangular coordinates, and vice-versa.

In Exercises 1–6, perform the indicated operations and write the result in standard form.

1. $(6 - 2i) - (7 - i)$ **2.** $3i(2 + i)$

3. $(1 + i)(4 - 3i)$ **4.** $\dfrac{1 + i}{1 - i}$

5. $\sqrt{-75} - \sqrt{-12}$ **6.** $\left(2 - \sqrt{-3}\right)^2$

7. Plot $-3 - 3i$. Then write the complex number in polar form, expressing the argument in degrees or radians.

8. Write $12\left(\cos\dfrac{2\pi}{3} + i\sin\dfrac{2\pi}{3}\right)$ in rectangular form.

In Exercises 9–11, perform the indicated operations. Leave answers in polar form.

9. $4(\cos 15° + i\sin 15°) \cdot 10(\cos 3° + i\sin 3°)$

10. $\dfrac{18\left(\cos\dfrac{\pi}{2} + i\sin\dfrac{\pi}{2}\right)}{6\left(\cos\dfrac{\pi}{3} + i\sin\dfrac{\pi}{3}\right)}$

11. $[3(\cos 10° + i\sin 10°)]^4$

12. Find the three cube roots of $8(\cos 210° + i\sin 210°)$. Leave answers in polar form.

13. Find the three cube roots of 1. Write roots in rectangular form.

In Exercises 14–17, convert the given coordinates to the indicated ordered pair.

14. $\left(-3, \dfrac{5\pi}{4}\right)$ to (x, y)

15. $\left(6, -\dfrac{\pi}{2}\right)$ to (x, y)

16. $\left(2, -2\sqrt{3}\right)$ to (r, θ)

17. $(-6, 0)$ to (r, θ)

In Exercises 18–19, plot each point in polar coordinates. Then find another representation (r, θ) of this point in which:

     a. $r > 0$, $2\pi < \theta < 4\pi$.

     b. $r < 0$, $0 < \theta < 2\pi$.

     c. $r > 0$, $-2\pi < \theta < 0$.

18. $\left(4, \dfrac{3\pi}{4}\right)$ **19.** $\left(\dfrac{5}{2}, \dfrac{\pi}{2}\right)$

In Exercises 20–22, convert each rectangular equation to a polar equation that expresses r in terms of θ.

20. $5x - y = 7$

21. $y = -7$

22. $(x + 1)^2 + y^2 = 1$

In Exercises 23–26, convert each polar equation to a rectangular equation. Then use your knowledge of the rectangular equation to graph the polar equation in a polar coordinate system.

23. $r = 6$

24. $\theta = \dfrac{\pi}{3}$

25. $r = -3\csc\theta$

26. $r = -10\cos\theta$

Section 5.4   Graphs of Polar Equations

What am I supposed to learn?

After studying this section, you should be able to:

1. Use point plotting to graph polar equations.

2. Use symmetry to graph polar equations.

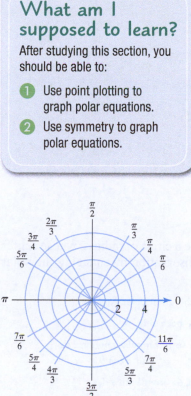

FIGURE 5.24 A polar coordinate grid

1. Use point plotting to graph polar equations.

The America's Cup is the supreme event in ocean sailing. Competition is fierce and the costs are huge. Competitors look to mathematics to provide the critical innovation that can make the difference between winning and losing. In this section's Exercise Set, you will see how graphs of polar equations play a role in sailing faster using mathematics.

Using Polar Grids to Graph Polar Equations

Recall that a **polar equation** is an equation whose variables are r and θ. The **graph of a polar equation** is the set of all points whose polar coordinates satisfy the equation. We use **polar grids** like the one shown in **Figure 5.24** to graph polar equations. The grid consists of circles with centers at the pole. This polar grid shows five such circles. A polar grid also shows lines passing through the pole. In this grid, each line represents an angle for which we know the exact values of the trigonometric functions.

Many polar coordinate grids show more circles and more lines through the pole than in **Figure 5.24**. See if your campus bookstore has paper with polar grids and use the polar graph paper throughout this section.

Graphing a Polar Equation by Point Plotting

One method for graphing a polar equation such as $r = 4 \cos \theta$ is the **point-plotting method**. First, we make a table of values that satisfy the equation. Next, we plot these ordered pairs as points in the polar coordinate system. Finally, we connect the points with a smooth curve. This often gives us a picture of all ordered pairs (r, θ) that satisfy the equation.

EXAMPLE 1   Graphing an Equation Using the Point-Plotting Method

Graph the polar equation $r = 4 \cos \theta$ with θ in radians.

SOLUTION

We construct a partial table of coordinates for $r = 4 \cos \theta$ using multiples of $\frac{\pi}{6}$. Then we plot the points and join them with a smooth curve, as shown in **Figure 5.25** on the next page.

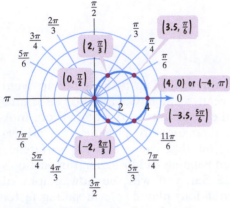

FIGURE 5.25 The graph of $r = 4 \cos \theta$

θ	$r = 4 \cos \theta$	(r, θ)
0	$4 \cos 0 = 4 \cdot 1 = 4$	$(4, 0)$
$\dfrac{\pi}{6}$	$4 \cos \dfrac{\pi}{6} = 4 \cdot \dfrac{\sqrt{3}}{2} = 2\sqrt{3} \approx 3.5$	$\left(3.5, \dfrac{\pi}{6}\right)$
$\dfrac{\pi}{3}$	$4 \cos \dfrac{\pi}{3} = 4 \cdot \dfrac{1}{2} = 2$	$\left(2, \dfrac{\pi}{3}\right)$
$\dfrac{\pi}{2}$	$4 \cos \dfrac{\pi}{2} = 4 \cdot 0 = 0$	$\left(0, \dfrac{\pi}{2}\right)$
$\dfrac{2\pi}{3}$	$4 \cos \dfrac{2\pi}{3} = 4\left(-\dfrac{1}{2}\right) = -2$	$\left(-2, \dfrac{2\pi}{3}\right)$
$\dfrac{5\pi}{6}$	$4 \cos \dfrac{5\pi}{6} = 4\left(-\dfrac{\sqrt{3}}{2}\right) = -2\sqrt{3} \approx -3.5$	$\left(-3.5, \dfrac{5\pi}{6}\right)$
π	$4 \cos \pi = 4(-1) = -4$	$(-4, \pi)$

The points repeat.

•••

The graph of $r = 4 \cos \theta$ in **Figure 5.25** looks like a circle of radius 2 whose center is at the point $(x, y) = (2, 0)$. We can verify this observation by changing the polar equation to a rectangular equation.

$r = 4 \cos \theta$ This is the given polar equation.

$r^2 = 4r \cos \theta$ Multiply both sides by r.

$x^2 + y^2 = 4x$ Convert to rectangular coordinates: $r^2 = x^2 + y^2$ and $r \cos \theta = x$.

$x^2 - 4x + y^2 = 0$ Subtract $4x$ from both sides.

$x^2 - 4x + 4 + y^2 = 4$ Complete the square on x: $\frac{1}{2}(-4) = -2$ and $(-2)^2 = 4$. Add 4 to both sides.

$(x - 2)^2 + y^2 = 2^2$ Factor.

This last equation is the standard form of the equation of a circle, $(x - h)^2 + (y - k)^2 = r^2$, with radius r and center at (h, k). Thus, the radius is 2 and the center is at $(h, k) = (2, 0)$.

In general, circles have simpler equations in polar form than in rectangular form.

TECHNOLOGY

A graphing utility can be used to obtain the graph of a polar equation. Use the polar mode with angle measure in radians. You must enter the minimum and maximum values for θ and an increment setting for θ, called θ step. θ step determines the number of points that the graphing utility will plot. Make θ step relatively small so that a significant number of points are plotted.

Shown is the graph of $r = 4 \cos \theta$ in a $[-8, 8, 1]$ by $[-5, 5, 1]$ viewing rectangle with

$\theta \min = 0$

$\theta \max = 2\pi$

$\theta \text{ step} = \dfrac{\pi}{48}.$

A square setting was used.

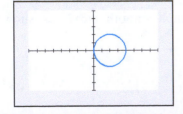

> ### Circles in Polar Coordinates
>
> The graphs of
>
> $$r = a \cos \theta \quad \text{and} \quad r = a \sin \theta, \, a > 0,$$
>
> are circles.
>
>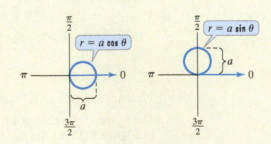

✅ **Check Point 1** Graph the equation $r = 4 \sin \theta$ with θ in radians. Use multiples of $\dfrac{\pi}{6}$ from 0 to π to generate coordinates for points (r, θ).

② Use symmetry to graph polar equations.

Graphing a Polar Equation Using Symmetry

If the graph of a polar equation exhibits symmetry, you may be able to graph it more quickly. Three types of symmetry can be helpful.

Tests for Symmetry in Polar Coordinates

Symmetry with Respect to the Polar Axis (x-Axis)	Symmetry with Respect to the Line $\theta = \frac{\pi}{2}$ (y-Axis)	Symmetry with Respect to the Pole (Origin)
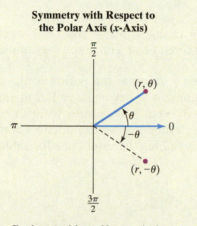	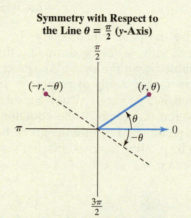	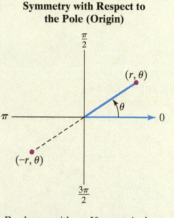
Replace θ with $-\theta$. If an equivalent equation results, the graph is symmetric with respect to the polar axis.	Replace (r, θ) with $(-r, -\theta)$. If an equivalent equation results, the graph is symmetric with respect to $\theta = \frac{\pi}{2}$.	Replace r with $-r$. If an equivalent equation results, the graph is symmetric with respect to the pole.

If a polar equation passes a symmetry test, then its graph exhibits that symmetry. By contrast, if a polar equation fails a symmetry test, then its graph *may or may not* have that kind of symmetry. Thus, the graph of a polar equation may have a symmetry even if it fails a test for that particular symmetry. Nevertheless, the symmetry tests are useful. If we detect symmetry, we can obtain a graph of the equation by plotting fewer points.

EXAMPLE 2 Graphing a Polar Equation Using Symmetry

Check for symmetry and then graph the polar equation:

$$r = 1 - \cos\theta.$$

SOLUTION

We apply each of the tests for symmetry.

Polar Axis: Replace θ with $-\theta$ in $r = 1 - \cos\theta$.

$r = 1 - \cos(-\theta)$ Replace θ with $-\theta$ in $r = 1 - \cos\theta$.

$r = 1 - \cos\theta$ The cosine function is even: $\cos(-\theta) = \cos\theta$.

Because the polar equation does not change when θ is replaced with $-\theta$, the graph is symmetric with respect to the polar axis.

The Line $\theta = \dfrac{\pi}{2}$: Replace (r, θ) with $(-r, -\theta)$ in $r = 1 - \cos\theta$.

$-r = 1 - \cos(-\theta)$ Replace r with $-r$ and θ with $-\theta$ in $r = 1 - \cos\theta$.

$-r = 1 - \cos\theta$ $\cos(-\theta) = \cos\theta$.

$r = \cos\theta - 1$ Multiply both sides by -1.

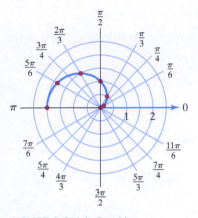

FIGURE 5.26(a) Graphing $r = 1 - \cos\theta$ for $0 \le \theta \le \pi$

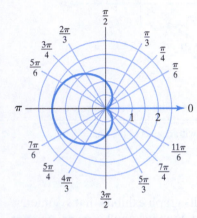

FIGURE 5.26(b) A complete graph of $r = 1 - \cos\theta$

Because the polar equation $r = 1 - \cos\theta$ changes to $r = \cos\theta - 1$ when (r, θ) is replaced with $(-r, -\theta)$, the equation fails this symmetry test. The graph may or may not be symmetric with respect to the line $\theta = \dfrac{\pi}{2}$.

The Pole: Replace r with $-r$ in $r = 1 - \cos\theta$.

$$-r = 1 - \cos\theta \quad \text{Replace } r \text{ with } -r \text{ in } r = 1 - \cos\theta.$$
$$r = \cos\theta - 1 \quad \text{Multiply both sides by } -1.$$

Because the polar equation $r = 1 - \cos\theta$ changes to $r = \cos\theta - 1$ when r is replaced with $-r$, the equation fails this symmetry test. The graph may or may not be symmetric with respect to the pole.

Now we are ready to graph $r = 1 - \cos\theta$. Because the period of the cosine function is 2π, we need not consider values of θ beyond 2π. Recall that we discovered the graph of the equation $r = 1 - \cos\theta$ has symmetry with respect to the polar axis. Because the graph has this symmetry, we can obtain a complete graph by plotting fewer points. Let's start by finding the values of r for values of θ from 0 to π.

θ	0	$\dfrac{\pi}{6}$	$\dfrac{\pi}{3}$	$\dfrac{\pi}{2}$	$\dfrac{2\pi}{3}$	$\dfrac{5\pi}{6}$	π
r	0	0.13	0.5	1	1.5	1.87	2

The values for r and θ are shown in the table. These values can be obtained using your calculator or possibly with the $\boxed{\text{TABLE}}$ feature on some graphing calculators. The points in the table are plotted in **Figure 5.26(a)**. Examine the graph. Keep in mind that the graph must be symmetric with respect to the polar axis. Thus, if we reflect the graph in **Figure 5.26(a)** about the polar axis, we will obtain a complete graph of $r = 1 - \cos\theta$. This graph is shown in **Figure 5.26(b)**. •••

✓ **Check Point 2** Check for symmetry and then graph the polar equation:
$$r = 1 + \cos\theta.$$

EXAMPLE 3 Graphing a Polar Equation

Graph the polar equation: $r = 1 + 2\sin\theta$.

SOLUTION

We first check for symmetry.
$$r = 1 + 2\sin\theta$$

Polar Axis	**The Line $\theta = \frac{\pi}{2}$**	**The Pole**
Replace θ with $-\theta$.	Replace (r, θ) with $(-r, -\theta)$.	Replace r with $-r$.
$r = 1 + 2\sin(-\theta)$	$-r = 1 + 2\sin(-\theta)$	$-r = 1 + 2\sin\theta$
$r = 1 + 2(-\sin\theta)$	$-r = 1 - 2\sin\theta$	$r = -1 - 2\sin\theta$
$r = 1 - 2\sin\theta$	$r = -1 + 2\sin\theta$	

None of these equations are equivalent to $r = 1 + 2\sin\theta$. Thus, the graph may or may not have each of these kinds of symmetry.

Now we are ready to graph $r = 1 + 2 \sin \theta$. Because the period of the sine function is 2π, we need not consider values of θ beyond 2π. We identify points on the graph of $r = 1 + 2 \sin \theta$ by assigning values to θ and calculating the corresponding values of r. The values for r and θ are in the tables above **Figure 5.27(a)**, **Figure 5.27(b)**, and **Figure 5.27(c)**. The complete graph of $r = 1 + 2 \sin \theta$ is shown in **Figure 5.27(c)**. The inner loop indicates that the graph passes through the pole twice.

θ	0	$\dfrac{\pi}{6}$	$\dfrac{\pi}{3}$	$\dfrac{\pi}{2}$	$\dfrac{2\pi}{3}$	$\dfrac{5\pi}{6}$	π
r	1	2	2.73	3	2.73	2	1

θ	$\dfrac{7\pi}{6}$	$\dfrac{4\pi}{3}$	$\dfrac{3\pi}{2}$
r	0	-0.73	-1

θ	$\dfrac{5\pi}{3}$	$\dfrac{11\pi}{6}$	2π
r	-0.73	0	1

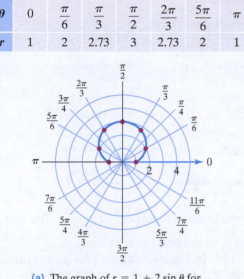

(a) The graph of $r = 1 + 2 \sin \theta$ for $0 \le \theta \le \pi$

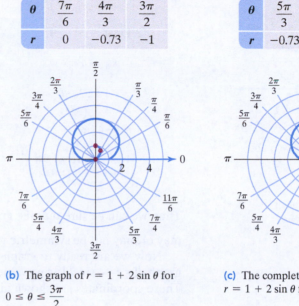

(b) The graph of $r = 1 + 2 \sin \theta$ for $0 \le \theta \le \dfrac{3\pi}{2}$

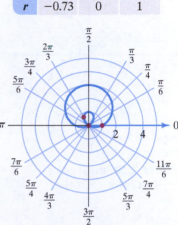

(c) The complete graph of $r = 1 + 2 \sin \theta$ for $0 \le \theta \le 2\pi$

FIGURE 5.27 Graphing $r = 1 + 2 \sin \theta$

Although the polar equation $r = 1 + 2 \sin \theta$ failed the test for symmetry with respect to the line $\theta = \frac{\pi}{2}$ (the y-axis), its graph in **Figure 5.27(c)** reveals this kind of symmetry.   • • •

We're not quite sure if the polar graph in **Figure 5.27(c)** looks like a snail. However, the graph is called a limaçon, pronounced "LEE-ma-sohn," which is a French word for snail. Limaçons come with and without inner loops.

Limaçons

The graphs of

$$r = a + b \sin \theta, \quad r = a - b \sin \theta,$$
$$r = a + b \cos \theta, \quad r = a - b \cos \theta, \quad a > 0, b > 0$$

are called **limaçons**. The ratio $\dfrac{a}{b}$ determines a limaçon's shape.

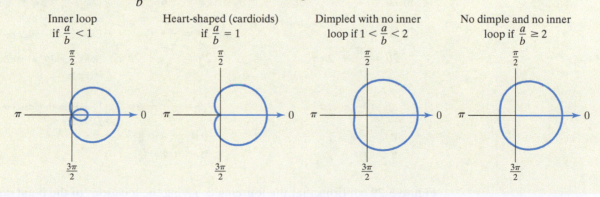

Inner loop if $\dfrac{a}{b} < 1$ Heart-shaped (cardioids) if $\dfrac{a}{b} = 1$ Dimpled with no inner loop if $1 < \dfrac{a}{b} < 2$ No dimple and no inner loop if $\dfrac{a}{b} \ge 2$

✓ **Check Point 3** Graph the polar equation: $r = 1 - 2 \sin \theta$.

Graphing a Polar Equation

Graph the polar equation: $r = 4 \sin 2\theta$.

SOLUTION

We first check for symmetry.

$$r = 4 \sin 2\theta$$

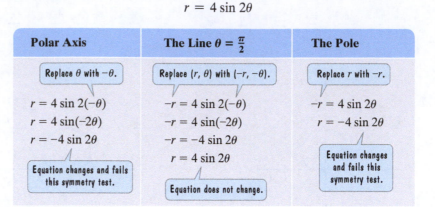

Polar Axis	The Line $\theta = \frac{\pi}{2}$	The Pole
Replace θ with $-\theta$.	Replace (r, θ) with $(-r, -\theta)$.	Replace r with $-r$.
$r = 4 \sin 2(-\theta)$ $r = 4 \sin(-2\theta)$ $r = -4 \sin 2\theta$	$-r = 4 \sin 2(-\theta)$ $-r = 4 \sin(-2\theta)$ $-r = -4 \sin 2\theta$ $r = 4 \sin 2\theta$	$-r = 4 \sin 2\theta$ $r = -4 \sin 2\theta$
Equation changes and fails this symmetry test.	Equation does not change.	Equation changes and fails this symmetry test.

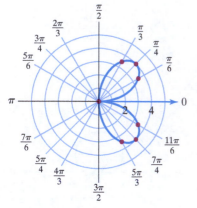

FIGURE 5.28 The graph of $r = 4 \sin 2\theta$ for $0 \le \theta \le \pi$

Thus, we can be sure that the graph is symmetric with respect to $\theta = \frac{\pi}{2}$. The graph may or may not be symmetric with respect to the polar axis or the pole.

Now we are ready to graph $r = 4 \sin 2\theta$. In **Figure 5.28**, we plot points on the graph of $r = 4 \sin 2\theta$ using values of θ from 0 to π and the corresponding values of r. These coordinates are shown in the table below.

θ	0	$\frac{\pi}{6}$	$\frac{\pi}{4}$	$\frac{\pi}{3}$	$\frac{\pi}{2}$	$\frac{2\pi}{3}$	$\frac{3\pi}{4}$	$\frac{5\pi}{6}$	π
r	0	3.46	4	3.46	0	-3.46	-4	-3.46	0

Now we can use symmetry with respect to the line $\theta = \frac{\pi}{2}$ (the y-axis) to complete the graph. By reflecting the graph in **Figure 5.28** about the y-axis, we obtain the complete graph of $r = 4 \sin 2\theta$ from 0 to 2π. The graph is shown in **Figure 5.29**.

Although the polar equation $r = 4 \sin 2\theta$ failed the tests for symmetry with respect to the polar axis (the x-axis) and the pole (the origin), its graph in **Figure 5.29** reveals all three types of symmetry.   • • •

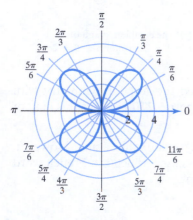

FIGURE 5.29 The graph of $r = 4 \sin 2\theta$ for $0 \le \theta \le 2\pi$

The curve in **Figure 5.29** is called a **rose with four petals**. We can use a trigonometric equation to confirm the four angles that give the location of the petal points. The petal points of $r = 4 \sin 2\theta$ are located at values of θ for which $r = 4$ or $r = -4$.

$4 \sin 2\theta = 4$ or	$4 \sin 2\theta = -4$	Use $r = 4 \sin 2\theta$ and set r equal to 4 or -4.
$\sin 2\theta = 1$	$\sin 2\theta = -1$	Divide both sides by 4.
$2\theta = \frac{\pi}{2} + 2n\pi$	$2\theta = \frac{3\pi}{2} + 2n\pi$	Solve for 2θ, where n is any integer.
$\theta = \frac{\pi}{4} + n\pi$	$\theta = \frac{3\pi}{4} + n\pi$	Divide both sides by 2 and solve for θ.
If $n = 0, \theta = \frac{\pi}{4}$. If $n = 1, \theta = \frac{5\pi}{4}$.	If $n = 0, \theta = \frac{3\pi}{4}$. If $n = 1, \theta = \frac{7\pi}{4}$.	

Figure 5.29 confirms that the four angles giving the locations of the petal points are $\frac{\pi}{4}, \frac{3\pi}{4}, \frac{5\pi}{4}$, and $\frac{7\pi}{4}$.

TECHNOLOGY

The graph of

$$r = 4 \sin 2\theta$$

was initially obtained using a $[-4, 4, 1]$ by $[-4, 4, 1]$ viewing rectangle and

$$\theta \min = 0, \quad \theta \max = 2\pi,$$

$$\theta \text{ step} = \frac{\pi}{48}.$$

The graph is distorted. Use the ZSquare option in the Zoom menu to see the true shape of the graph.

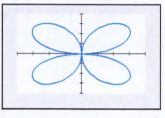
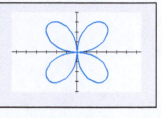

Rose Curves

The graphs of

$$r = a \sin n\theta \quad \text{and} \quad r = a \cos n\theta, \quad a \neq 0,$$

are called **rose curves**. If n is even, the rose has $2n$ petals. If n is odd, the rose has n petals.

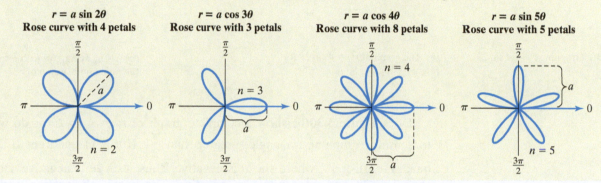

| $r = a \sin 2\theta$
Rose curve with 4 petals | $r = a \cos 3\theta$
Rose curve with 3 petals | $r = a \cos 4\theta$
Rose curve with 8 petals | $r = a \sin 5\theta$
Rose curve with 5 petals |

✅ **Check Point 4** Graph the polar equation: $r = 3 \cos 2\theta$.

EXAMPLE 5   Graphing a Polar Equation

Graph the polar equation: $r^2 = 4 \sin 2\theta$.

SOLUTION

We first check for symmetry.

$$r^2 = 4 \sin 2\theta$$

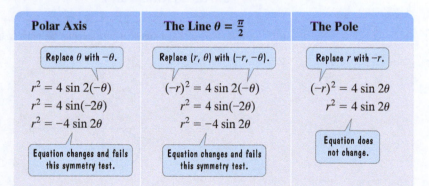

Polar Axis	The Line $\theta = \frac{\pi}{2}$	The Pole
Replace θ with $-\theta$.	Replace (r, θ) with $(-r, -\theta)$.	Replace r with $-r$.
$r^2 = 4 \sin 2(-\theta)$ $r^2 = 4 \sin(-2\theta)$ $r^2 = -4 \sin 2\theta$	$(-r)^2 = 4 \sin 2(-\theta)$ $r^2 = 4 \sin(-2\theta)$ $r^2 = -4 \sin 2\theta$	$(-r)^2 = 4 \sin 2\theta$ $r^2 = 4 \sin 2\theta$
Equation changes and fails this symmetry test.	Equation changes and fails this symmetry test.	Equation does not change.

Thus, we can be sure that the graph is symmetric with respect to the pole. The graph may or may not be symmetric with respect to the polar axis or the line $\theta = \frac{\pi}{2}$.

Now we are ready to graph $r^2 = 4 \sin 2\theta$. In **Figure 5.30(a)**, we plot points on the graph by using values of θ from 0 to $\dfrac{\pi}{2}$ and the corresponding values of r. These coordinates are shown in the table to the left of **Figure 5.30(a)**. Notice that the points in **Figure 5.30(a)** are shown for $r \geq 0$. Because the graph is symmetric with respect to the pole, we can reflect the graph in **Figure 5.30(a)** about the pole and obtain the graph in **Figure 5.30(b)**.

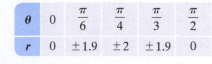

θ	0	$\dfrac{\pi}{6}$	$\dfrac{\pi}{4}$	$\dfrac{\pi}{3}$	$\dfrac{\pi}{2}$
r	0	± 1.9	± 2	± 1.9	0

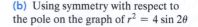

FIGURE 5.30 Graphing $r^2 = 4 \sin 2\theta$

(a) The graph of $r^2 = 4 \sin 2\theta$ for $0 \leq \theta \leq \dfrac{\pi}{2}$ and $r \geq 0$

(b) Using symmetry with respect to the pole on the graph of $r^2 = 4 \sin 2\theta$

Does **Figure 5.30(b)** show a complete graph of $r^2 = 4 \sin 2\theta$ or do we need to continue graphing for angles greater than $\dfrac{\pi}{2}$? If θ is in quadrant II, 2θ is in quadrant III or IV, where $\sin 2\theta$ is negative. Thus, $4 \sin 2\theta$ is negative. However, $r^2 = 4 \sin 2\theta$ and r^2 cannot be negative. The same observation applies to quadrant IV. This means that there are no points on the graph in quadrants II or IV. Thus, **Figure 5.30(b)** shows the complete graph of $r^2 = 4 \sin 2\theta$. ●●●

The curve in **Figure 5.30(b)** is shaped like a propeller and is called a *lemniscate*.

Lemniscates

The graphs of

$$r^2 = a^2 \sin 2\theta \quad \text{and} \quad r^2 = a^2 \cos 2\theta, \quad a \neq 0$$

are called **lemniscates**.

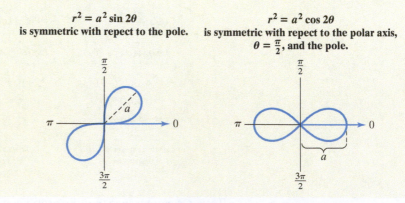

$r^2 = a^2 \sin 2\theta$
is symmetric with repect to the pole.

$r^2 = a^2 \cos 2\theta$
is symmetric with repect to the polar axis, $\theta = \dfrac{\pi}{2}$, and the pole.

✓ **Check Point 5** Graph the polar equation: $r^2 = 4 \cos 2\theta$.

CONCEPT AND VOCABULARY CHECK

Fill in each blank so that the resulting statement is true.

1. One method for graphing a polar equation is the point-plotting method. We substitute convenient values of _____ into the equation and then determine the values for _____.

2. In polar coordinates, the graphs of $r = a \cos \theta$ and $r = a \sin \theta$ are _____.

3. To test whether the graph of a polar equation may be symmetric with respect to the polar axis (x-axis), replace _____ with _____.

4. To test whether the graph of a polar equation may be symmetric with respect to the line $\theta = \dfrac{\pi}{2}$ (y-axis), replace _____ with _____.

5. To test whether the graph of a polar equation may be symmetric with respect to the pole (origin), replace _____ with _____.

6. True or false: The graph of a polar equation may have symmetry even if it fails a test for that particular symmetry. _____

7. The graphs of $r = a + b \sin \theta$, $r = a - b \sin \theta$, $r = a + b \cos \theta$, and $r = a - b \cos \theta$, $a > 0$, $b > 0$, are called _____, a French word for snail. The ratio $\dfrac{a}{b}$ determines the graph's shape. If $\dfrac{a}{b} = 1$, the graph is shaped like a heart and called a/an _____. If $\dfrac{a}{b} < 1$, the graph has an inner _____.

8. The graphs of $r = a \sin n\theta$ and $r = a \cos n\theta$, $a \neq 0$, are called rose curves. If n is even, the rose has _____ petals. If n is odd, the rose has _____ petals.

9. The graphs of $r^2 = a^2 \sin 2\theta$ and $r = a^2 \cos 2\theta$, $a \neq 0$, are shaped like propellers and called _____. The graph of $r^2 = a^2 \sin 2\theta$ is symmetric with respect to the _____. The graph of $r^2 = a^2 \cos 2\theta$ is symmetric with respect to the _____, the _____, and _____.

EXERCISE SET 5.4

Practice Exercises

In Exercises 1–6, the graph of a polar equation is given. Select the polar equation for each graph from the following options.

$$r = 2 \sin \theta, \quad r = 2 \cos \theta, \quad r = 1 + \sin \theta,$$
$$r = 1 - \sin \theta, \quad r = 3 \sin 2\theta, \quad r = 3 \sin 3\theta$$

1.

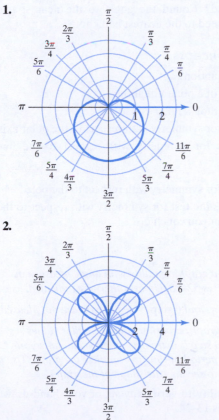

2.

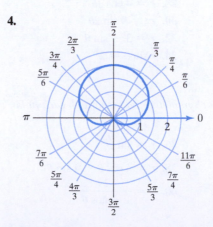

3.

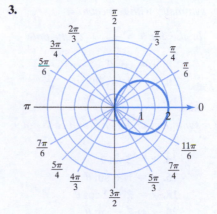

4.

5.

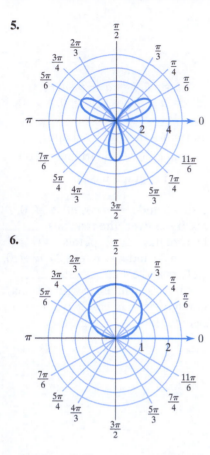

6.

In Exercises 7–12, test for symmetry with respect to

    ***a.** the polar axis.*   ***b.** the line $\theta = \dfrac{\pi}{2}$.*   ***c.** the pole.*

7. $r = \sin \theta$ **8.** $r = \cos \theta$
9. $r = 4 + 3 \cos \theta$ **10.** $r = 2 \cos 2\theta$
11. $r^2 = 16 \cos 2\theta$ **12.** $r^2 = 16 \sin 2\theta$

In Exercises 13–34, test for symmetry and then graph each polar equation.

13. $r = 2 \cos \theta$ **14.** $r = 2 \sin \theta$
15. $r = 1 - \sin \theta$ **16.** $r = 1 + \sin \theta$
17. $r = 2 + 2 \cos \theta$ **18.** $r = 2 - 2 \cos \theta$
19. $r = 2 + \cos \theta$ **20.** $r = 2 - \sin \theta$
21. $r = 1 + 2 \cos \theta$ **22.** $r = 1 - 2 \cos \theta$
23. $r = 2 - 3 \sin \theta$ **24.** $r = 2 + 4 \sin \theta$
25. $r = 2 \cos 2\theta$ **26.** $r = 2 \sin 2\theta$
27. $r = 4 \sin 3\theta$ **28.** $r = 4 \cos 3\theta$
29. $r^2 = 9 \cos 2\theta$ **30.** $r^2 = 9 \sin 2\theta$
31. $r = 1 - 3 \sin \theta$ **32.** $r = 3 + \sin \theta$
33. $r \cos \theta = -3$ **34.** $r \sin \theta = 2$

Practice Plus

In Exercises 35–44, test for symmetry and then graph each polar equation.

35. $r = \cos \dfrac{\theta}{2}$ **36.** $r = \sin \dfrac{\theta}{2}$
37. $r = \sin \theta + \cos \theta$ **38.** $r = 4 \cos \theta + 4 \sin \theta$
39. $r = \dfrac{1}{1 - \cos \theta}$ **40.** $r = \dfrac{2}{1 - \cos \theta}$

41. $r = \sin \theta \cos^2 \theta$ **42.** $r = \dfrac{3 \sin 2\theta}{\sin^3 \theta + \cos^3 \theta}$
43. $r = 2 + 3 \sin 2\theta$ **44.** $r = 2 - 4 \cos 2\theta$

Application Exercises

In Exercise Set 5.3, we considered an application in which sailboat racers look for a sailing angle to a 10-knot wind that produces maximum sailing speed. This situation is now represented by the polar graph in the figure shown. Each point (r, θ) on the graph gives the sailing speed, r, in knots, at an angle θ to the 10-knot wind. Use this information to solve Exercises 45–49.

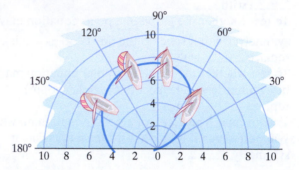

45. What is the speed, to the nearest knot, of a sailboat sailing at a 60° angle to the wind?

46. What is the speed, to the nearest knot, of a sailboat sailing at a 120° angle to the wind?

47. What is the speed, to the nearest knot, of a sailboat sailing at a 90° angle to the wind?

48. What is the speed, to the nearest knot, of a sailboat sailing at a 180° angle to the wind?

49. What angle to the wind produces the maximum sailing speed? What is the speed? Round the angle to the nearest five degrees and the speed to the nearest half knot.

Explaining the Concepts

50. What is a polar equation?

51. What is the graph of a polar equation?

52. Describe how to graph a polar equation.

53. Describe the test for symmetry with respect to the polar axis.

54. Describe the test for symmetry with respect to the line $\theta = \dfrac{\pi}{2}$.

55. Describe the test for symmetry with respect to the pole.

56. If an equation fails the test for symmetry with respect to the polar axis, what can you conclude?

Technology Exercises

Use the polar mode of a graphing utility with angle measure in radians to solve Exercises 57–88. Unless otherwise indicated, use $\theta \min = 0$, $\theta \max = 2\pi$, and $\theta \text{ step} = \dfrac{\pi}{48}$. If you are not pleased with the quality of the graph, experiment with smaller values for θ step. However, if θ step is extremely small, it can take your graphing utility a long period of time to complete the graph. Use a square window setting to see the true shape of the graph.

57. Use a graphing utility to verify any six of your hand-drawn graphs in Exercises 13–34.

In Exercises 58–75, use a graphing utility to graph the polar equation.

58. $r = 4 \cos 5\theta$

59. $r = 4 \sin 5\theta$

60. $r = 4 \cos 6\theta$

61. $r = 4 \sin 6\theta$

62. $r = 2 + 2 \cos \theta$

63. $r = 2 + 2 \sin \theta$

64. $r = 4 + 2 \cos \theta$

65. $r = 4 + 2 \sin \theta$

66. $r = 2 + 4 \cos \theta$

67. $r = 2 + 4 \sin \theta$

68. $r = \dfrac{3}{\sin \theta}$

69. $r = \dfrac{3}{\cos \theta}$

70. $r = \cos \dfrac{3}{2}\theta$

71. $r = \cos \dfrac{5}{2}\theta$

72. $r = 3 \sin\left(\theta + \dfrac{\pi}{4}\right)$

73. $r = 2 \cos\left(\theta - \dfrac{\pi}{4}\right)$

74. $r = \dfrac{1}{1 - \sin \theta}$

75. $r = \dfrac{1}{3 - 2 \sin \theta}$

In Exercises 76–78, find the smallest interval for θ starting with θ min $= 0$ so that your graphing utility graphs the given polar equation exactly once without retracing any portion of it.

76. $r = 4 \sin \theta$

77. $r = 4 \sin 2\theta$

78. $r^2 = 4 \sin 2\theta$

In Exercises 79–82, use a graphing utility to graph each butterfly curve. Experiment with the range setting, particularly θ step, to produce a butterfly of the best possible quality.

79. $r = \cos^2 5\theta + \sin 3\theta + 0.3$

80. $r = \sin^4 4\theta + \cos 3\theta$

81. $r = \sin^5 \theta + 8 \sin \theta \cos^3 \theta$

82. $r = 1.5^{\sin \theta} - 2.5 \cos 4\theta + \sin^7 \dfrac{\theta}{15}$ (Use θ min $= 0$ and

θ max $= 20\pi$.)

83. Use a graphing utility to graph $r = \sin n\theta$ for $n = 1, 2, 3, 4, 5,$ and 6. Use a separate viewing screen for each of the six graphs. What is the pattern for the number of loops that occur corresponding to each value of n? What is happening to the shape of the graphs as n increases? For each graph, what is the smallest interval for θ so that the graph is traced only once?

84. Repeat Exercise 83 for $r = \cos n\theta$. Are your conclusions the same as they were in Exercise 83?

85. Use a graphing utility to graph $r = 1 + 2 \sin n\theta$ for $n = 1, 2, 3, 4, 5,$ and 6. Use a separate viewing screen for each of the six graphs. What is the pattern for the number of large and small petals that occur corresponding to each value of n? How are the large and small petals related when n is odd and when n is even?

86. Repeat Exercise 85 for $r = 1 + 2 \cos n\theta$. Are your conclusions the same as they were in Exercise 85?

87. Graph the spiral $r = \theta$. Use a $[-48, 48, 6]$ by $[-30, 30, 6]$ viewing rectangle. Let θ min $= 0$ and θ max $= 2\pi$, then θ min $= 0$ and θ max $= 4\pi$, and finally θ min $= 0$ and θ max $= 8\pi$.

88. Graph the spiral $r = \dfrac{1}{\theta}$. Use a $[-1.6, 1.6, 1]$ by $[-1, 1, 1]$ viewing rectangle. Let θ min $= 0$ and θ max $= 2\pi$, then θ min $= 0$ and θ max $= 4\pi$, and finally θ min $= 0$ and θ max $= 8\pi$.

Critical Thinking Exercises

Make Sense? *In Exercises 89–92, determine whether each statement makes sense or does not make sense, and explain your reasoning.*

89. I'm working with a polar equation that failed the symmetry test with respect to $\theta = \dfrac{\pi}{2}$, so my graph will not have this kind of symmetry.

90. The graph of my limaçon exhibits none of the three kinds of symmetry discussed in this section.

91. There are no points on my graph of $r^2 = 9 \cos 2\theta$ for which $\dfrac{\pi}{4} < \theta < \dfrac{3\pi}{4}$.

92. I'm graphing a polar equation in which for every value of θ there is exactly one corresponding value of r, yet my polar coordinate graph fails the vertical line for functions.

In Exercises 93–94, graph r_1 and r_2 in the same polar coordinate system. What is the relationship between the two graphs?

93. $r_1 = 4 \cos 2\theta$, $r_2 = 4 \cos 2\left(\theta - \dfrac{\pi}{4}\right)$

94. $r_1 = 2 \sin 3\theta$, $r_2 = 2 \sin 3\left(\theta + \dfrac{\pi}{6}\right)$

95. Describe a test for symmetry with respect to the line $\theta = \dfrac{\pi}{2}$ in which r is not replaced.

Retaining the Concepts

96. Determine the amplitude and period of $y = -4 \cos \dfrac{\pi}{2} x$. Then graph the function for $-4 \leq x \leq 4$. (Section 2.1, Example 5)

97. Verify the identity:

$$\cot x + \tan x = \csc x \sec x.$$

(Section 3.1, Example 2)

98. Two fire-lookout stations are 10 miles apart with station B directly east of station A. Both stations spot a fire on a mountain to the north. The bearing from station A to the fire is N39°E (39° east of north). The bearing from station B to the fire is N42°W (42° west of north). How far, to the nearest tenth of a mile, is the fire from station A? (Section 4.1, Example 7)

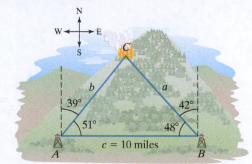

Preview Exercises

Exercises 99–101 will help you prepare for the material covered in the next section. In each exercise, graph the equation in a rectangular coordinate system.

99. $y^2 = 4(x + 1)$

100. $y = \dfrac{1}{2}x^2 + 1, \quad x \geq 0$

101. $\dfrac{x^2}{25} + \dfrac{y^2}{4} = 1$

A Brief Review • The Ellipse

- An ellipse is the set of all points in a plane the sum of whose distances from two fixed points, the foci (plural of focus), is constant. The midpoint of the segment connecting the foci is the center of the ellipse.
- An ellipse can be elongated in any direction. The figure shows ellipses that are elongated horizontally or vertically. The line through the foci intersects the ellipse at two points, the vertices (singular: vertex). The line segment that joins the vertices is the major axis. The line segment whose endpoints are on the ellipse and that is perpendicular to the major axis at the center is called the minor axis of the ellipse.

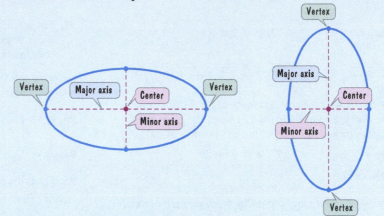

Horizontal and vertical elongations of an ellipse

- Standard forms of the equations of an ellipse with center at the origin, $a^2 > b^2$, are

$$\frac{x^2}{a^2} + \frac{y^2}{b^2} = 1 \quad \text{and} \quad \frac{x^2}{b^2} + \frac{y^2}{a^2} = 1$$

| Endpoints of major axis are a units left and right of center. | Endpoints of minor axis are b units above and below center. | Endpoints of minor axis are b units left and right of center. | Endpoints of major axis are a units above and below center. |

EXAMPLE

Graph:

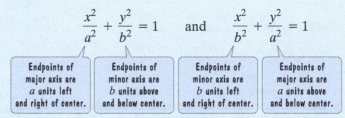

$$\frac{x^2}{9} + \frac{y^2}{4} = 1.$$

| $a^2 = 9$: Endpoints of major axis are **3 units left and right of center.** | $b^2 = 4$: Endpoints of minor axis are **2 units above and below center.** |

- Standard forms of the equations of an ellipse with center at (h, k), $a^2 > b^2$, are

$$\frac{(x-h)^2}{a^2} + \frac{(y-k)^2}{b^2} = 1 \quad \text{and} \quad \frac{(x-h)^2}{b^2} + \frac{(y-k)^2}{a^2} = 1$$

| Endpoints of major axis are a units left and right of center. | Endpoints of minor axis are b units above and below center. | Endpoints of minor axis are b units left and right of center. | Endpoints of major axis are a units above and below center. |

Section 5.5   Parametric Equations

What am I supposed to learn?

After studying this section, you should be able to:

① Use point plotting to graph plane curves described by parametric equations.

② Eliminate the parameter.

③ Find parametric equations for functions.

④ Understand the advantages of parametric representations.

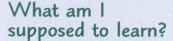

What a baseball game! You got to see the great Albert Pujols of the Los Angeles Angels blast a powerful homer. In less than 8 seconds, the parabolic path of his home run took the ball a horizontal distance of over 1000 feet. Is there a way to model this path that gives both the ball's location and the time that it is in each of its positions? In this section, we look at ways of describing curves that reveal the where and the when of motion.

Plane Curves and Parametric Equations

You throw a ball from a height of 6 feet, with an initial velocity of 90 feet per second and at an angle of 40° with the horizontal. After t seconds, the location of the ball can be described by

$$x = (90 \cos 40°)t \quad \text{and} \quad y = 6 + (90 \sin 40°)t - 16t^2.$$

This is the ball's horizontal distance, in feet.   This is the ball's vertical height, in feet.

Because we can use these equations to calculate the location of the ball at any time t, we can describe the path of the ball. For example, to determine the location when $t = 1$ second, substitute 1 for t in each equation:

$$x = (90 \cos 40°)t = (90 \cos 40°)(1) \approx 68.9 \text{ feet}$$
$$y = 6 + (90 \sin 40°)t - 16t^2 = 6 + (90 \sin 40°)(1) - 16(1)^2 \approx 47.9 \text{ feet}.$$

This tells us that after one second, the ball has traveled a horizontal distance of approximately 68.9 feet, and the height of the ball is approximately 47.9 feet. **Figure 5.31** displays this information and the results for calculations corresponding to $t = 2$ seconds and $t = 3$ seconds.

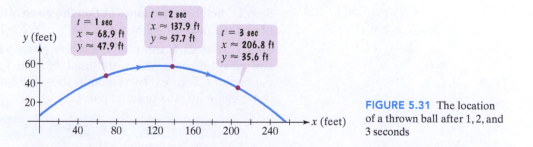

FIGURE 5.31 The location of a thrown ball after 1, 2, and 3 seconds

The voice balloons in **Figure 5.31** tell where the ball is located and when the ball is at a given point (x, y) on its path. The variable t, called a **parameter**, gives the various times for the ball's location. The equations that describe where the ball is located express both x and y as functions of t and are called **parametric equations**.

$$x = (90 \cos 40°)t \qquad y = 6 + (90 \sin 40°)t - 16t^2$$

This is the parametric equation for x.

This is the parametric equation for y.

The collection of points (x, y) in **Figure 5.31** on the previous page is called a **plane curve**.

Plane Curves and Parametric Equations

Suppose that t is a number in an interval I. A **plane curve** is the set of ordered pairs (x, y), where

$$x = f(t), \quad y = g(t) \quad \text{for } t \text{ in interval } I.$$

The variable t is called a **parameter**, and the equations $x = f(t)$ and $y = g(t)$ are called **parametric equations** for the curve.

① Use point plotting to graph plane curves described by parametric equations.

Graphing Plane Curves

Graphing a plane curve represented by parametric equations involves plotting points in the rectangular coordinate system and connecting them with a smooth curve.

Graphing a Plane Curve Described by Parametric Equations

1. Select some values of t on the given interval.
2. For each value of t, use the given parametric equations to compute x and y.
3. Plot the points (x, y) in the order of increasing t and connect them with a smooth curve.

Take a second look at **Figure 5.31**. Do you notice arrows along the curve? These arrows show the direction, or **orientation**, along the curve as t increases. After graphing a plane curve described by parametric equations, use arrows between the points to show the orientation of the curve corresponding to increasing values of t.

EXAMPLE 1   Graphing a Curve Defined by Parametric Equations

Graph the plane curve defined by the parametric equations:

$$x = t^2 - 1, \quad y = 2t, \quad -2 \le t \le 2.$$

SOLUTION

Step 1   Select some values of t on the given interval. We will select integral values of t on the interval $-2 \le t \le 2$. Let $t = -2, -1, 0, 1,$ and 2.

Step 2   For each value of t, use the given parametric equations to compute x and y. We organize our work in a table. The first column lists the choices for the parameter t. The next two columns show the corresponding values for x and y. The last column lists the ordered pair (x, y).

t	$x = t^2 - 1$	$y = 2t$	(x, y)
-2	$(-2)^2 - 1 = 4 - 1 = 3$	$2(-2) = -4$	$(3, -4)$
-1	$(-1)^2 - 1 = 1 - 1 = 0$	$2(-1) = -2$	$(0, -2)$
0	$0^2 - 1 = -1$	$2(0) = 0$	$(-1, 0)$
1	$1^2 - 1 = 0$	$2(1) = 2$	$(0, 2)$
2	$2^2 - 1 = 4 - 1 = 3$	$2(2) = 4$	$(3, 4)$

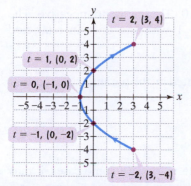

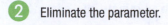

FIGURE 5.32 The plane curve defined by $x = t^2 - 1$, $y = 2t$, $-2 \leq t \leq 2$

② Eliminate the parameter.

TECHNOLOGY

A graphing utility can be used to obtain a plane curve represented by parametric equations. Set the mode to parametric and enter the equations. You must enter the minimum and maximum values for t and an increment setting for t ($t$step). The setting $t$step determines the number of points the graphing utility will plot.

Shown below is the plane curve for

$$x = t^2 - 1$$
$$y = 2t$$

in a $[-8, 8, 1]$ by $[-5, 5, 1]$ viewing rectangle with $t$min $= -2$, $t$max $= 2$, and $t$step $= 0.01$.

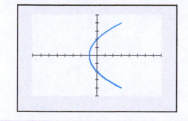

Step 3 **Plot the points (x, y) in the order of increasing t and connect them with a smooth curve.** The plane curve defined by the parametric equations on the given interval is shown in **Figure 5.32**. The arrows show the direction, or orientation, along the curve as t varies from -2 to 2. •••

✓ **Check Point 1** Graph the plane curve defined by the parametric equations:

$$x = t^2 + 1, \qquad y = 3t, \qquad -2 \leq t \leq 2.$$

Eliminating the Parameter

The graph in **Figure 5.32** shows the plane curve for $x = t^2 - 1$, $y = 2t$, $-2 \leq t \leq 2$. Even if we examine the parametric equations carefully, we may not be able to tell that the corresponding plane curve is a portion of a parabola. By **eliminating the parameter**, we can write one equation in x and y that is equivalent to the two parametric equations. The voice balloons illustrate this process.

Begin with the parametric equations.	Solve for t in one of the equations.	Substitute the expression for t in the other parametric equation.

$$x = t^2 - 1 \qquad\qquad \text{Using } y = 2t, \qquad\qquad \text{Using } t = \frac{y}{2} \text{ and } x = t^2 - 1,$$
$$y = 2t \qquad\qquad\qquad t = \frac{y}{2}. \qquad\qquad\qquad x = \left(\frac{y}{2}\right)^2 - 1.$$

The rectangular equation (the equation in x and y), $x = \frac{y^2}{4} - 1$, can be written as $y^2 = 4(x + 1)$. This is the standard form of the equation of a parabola with vertex at $(-1, 0)$ and axis of symmetry along the x-axis. Because the parameter t is restricted to the interval $[-2, 2]$, the plane curve in **Figure 5.32** and the technology box on the left shows only a part of the parabola.

Our discussion illustrates a second method for graphing a plane curve described by parametric equations. Eliminate the parameter t and graph the resulting rectangular equation in x and y. However, **you may need to change the domain of the rectangular equation to be consistent with the domain for the parametric equation in x.** This situation is illustrated in Example 2.

EXAMPLE 2 Finding and Graphing the Rectangular Equation of a Curve Defined Parametrically

Sketch the plane curve represented by the parametric equations

$$x = \sqrt{t} \quad \text{and} \quad y = \tfrac{1}{2}t + 1$$

by eliminating the parameter.

SOLUTION

We eliminate the parameter t and then graph the resulting rectangular equation.

Begin with the parametric equations.	Solve for t in one of the equations.	Substitute the expression for t in the other parametric equation.

$$x = \sqrt{t} \qquad\qquad \text{Using } x = \sqrt{t} \text{ and } \qquad \text{Using } t = x^2 \text{ and } y = \tfrac{1}{2}t + 1,$$
$$y = \tfrac{1}{2}t + 1 \qquad\qquad \text{squaring both sides,} \qquad\qquad y = \tfrac{1}{2}x^2 + 1.$$
$$t = x^2.$$

Because t is not limited to a closed interval, you might be tempted to graph the entire bowl-shaped parabola whose equation is $y = \tfrac{1}{2}x^2 + 1$. However, take a second look at the parametric equation for x:

$$x = \sqrt{t}.$$

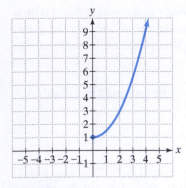

FIGURE 5.33 The plane curve for $x = \sqrt{t}$ and $y = \frac{1}{2}t + 1$, or $y = \frac{1}{2}x^2 + 1$, $x \geq 0$

The equation $x = \sqrt{t}$ is defined only when $t \geq 0$. Thus, x is nonnegative. The plane curve is the parabola given by $y = \frac{1}{2}x^2 + 1$ with the domain restricted to $x \geq 0$. The plane curve is shown in **Figure 5.33**. • • •

✅ **Check Point 2** Sketch the plane curve represented by the parametric equations

$$x = \sqrt{t} \quad \text{and} \quad y = 2t - 1$$

by eliminating the parameter.

Eliminating the parameter is not always a simple matter. In some cases, it may not be possible. When this occurs, you can use point plotting to obtain a plane curve.

Trigonometric identities can be helpful in eliminating the parameter. For example, consider the plane curve defined by the parametric equations

$$x = \sin t, \quad y = \cos t, \quad 0 \leq t < 2\pi.$$

We use the trigonometric identity $\sin^2 t + \cos^2 t = 1$ to eliminate the parameter. Square each side of each parametric equation and then add.

$$
\begin{aligned}
x^2 &= \sin^2 t \\
y^2 &= \cos^2 t \\
\hline
x^2 + y^2 &= \sin^2 t + \cos^2 t
\end{aligned}
$$

This is the sum of the two equations above the horizontal lines.

Using a Pythagorean identity, we write this equation as $x^2 + y^2 = 1$. The plane curve is a circle with center $(0, 0)$ and radius 1. It is shown in **Figure 5.34**.

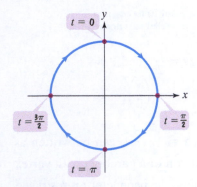

FIGURE 5.34 The plane curve defined by $x = \sin t, y = \cos t$, $0 \leq t < 2\pi$

EXAMPLE 3 Finding and Graphing the Rectangular Equation of a Curve Defined Parametrically

Sketch the plane curve represented by the parametric equations

$$x = 5 \cos t, \quad y = 2 \sin t, \quad 0 \leq t \leq \pi$$

by eliminating the parameter.

SOLUTION

We eliminate the parameter using the identity $\cos^2 t + \sin^2 t = 1$. To apply the identity, divide the parametric equation for x by 5 and the parametric equation for y by 2.

$$\frac{x}{5} = \cos t \quad \text{and} \quad \frac{y}{2} = \sin t$$

Square and add these two equations.

$$
\begin{aligned}
\frac{x^2}{25} &= \cos^2 t \\
\frac{y^2}{4} &= \sin^2 t \\
\hline
\frac{x^2}{25} + \frac{y^2}{4} &= \cos^2 t + \sin^2 t
\end{aligned}
$$

This is the sum of the two equations above the horizontal lines.

Using a Pythagorean identity, we write this equation as

$$\frac{x^2}{25} + \frac{y^2}{4} = 1.$$

This rectangular equation is the standard form of the equation for an ellipse centered at $(0, 0)$.

$$\frac{x^2}{25} + \frac{y^2}{4} = 1$$

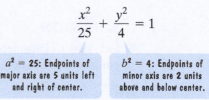

$a^2 = 25$: Endpoints of major axis are 5 units left and right of center.

$b^2 = 4$: Endpoints of minor axis are 2 units above and below center.

The ellipse is shown in **Figure 5.35(a)**. However, this is not the plane curve. We are given that $0 \le t \le \pi$. Because t is restricted to the interval $[0, \pi]$, the plane curve is only a portion of the ellipse. Use the starting and ending values for t, 0 and π, respectively, and a value of t in the interval $(0, \pi)$ to find which portion to include.

Begin at $t = 0$.

$x = 5 \cos t = 5 \cos 0 = 5 \cdot 1 = 5$

$y = 2 \sin t = 2 \sin 0 = 2 \cdot 0 = 0$

Increase to $t = \frac{\pi}{2}$.

$x = 5 \cos t = 5 \cos \frac{\pi}{2} = 5 \cdot 0 = 0$

$y = 2 \sin t = 2 \sin \frac{\pi}{2} = 2 \cdot 1 = 2$

End at $t = \pi$.

$x = 5 \cos t = 5 \cos \pi = 5(-1) = -5$

$y = 2 \sin t = 2 \sin \pi = 2(0) = 0$

Points on the plane curve include $(5, 0)$, which is the starting point, $(0, 2)$, and $(-5, 0)$, which is the ending point. The plane curve is the top half of the ellipse, shown in **Figure 5.35(b)**.

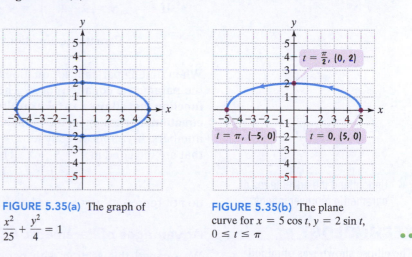

FIGURE 5.35(a) The graph of $\frac{x^2}{25} + \frac{y^2}{4} = 1$

FIGURE 5.35(b) The plane curve for $x = 5 \cos t$, $y = 2 \sin t$, $0 \le t \le \pi$

Check Point 3 Sketch the plane curve represented by the parametric equations $x = 6 \cos t$, $y = 4 \sin t$, $\pi \le t \le 2\pi$ by eliminating the parameter.

③ Find parametric equations for functions.

Finding Parametric Equations

Infinitely many pairs of parametric equations can represent the same plane curve. If the plane curve is defined by the function $y = f(x)$, here is a procedure for finding a set of parametric equations:

Parametric Equations for the Function $y = f(x)$

One set of parametric equations for the plane curve defined by $y = f(x)$ is

$$x = t \quad \text{and} \quad y = f(t),$$

in which t is in the domain of f.

EXAMPLE 4   Finding Parametric Equations

Find a set of parametric equations for the parabola whose equation is $y = 9 - x^2$.

SOLUTION

Let $x = t$. Parametric equations for $y = f(x)$ are $x = t$ and $y = f(t)$. Thus, parametric equations for $y = 9 - x^2$ are

$$x = t \quad \text{and} \quad y = 9 - t^2. \qquad \bullet\bullet\bullet$$

Check Point **4** Find a set of parametric equations for the parabola whose equation is $y = x^2 - 25$.

You can write other sets of parametric equations for $y = 9 - x^2$ by starting with a different parametric equation for x. Here are three more sets of parametric equations for

$$y = 9 - x^2:$$

- If $x = t^3, y = 9 - (t^3)^2 = 9 - t^6.$

  Parametric equations are $x = t^3$ and $y = 9 - t^6$.

- If $x = t + 1, y = 9 - (t + 1)^2 = 9 - (t^2 + 2t + 1) = 8 - t^2 - 2t.$

  Parametric equations are $x = t + 1$ and $y = 8 - t^2 - 2t$.

- If $x = \dfrac{t}{2}, y = 9 - \left(\dfrac{t}{2}\right)^2 = 9 - \dfrac{t^2}{4}.$

  Parametric equations are $x = \dfrac{t}{2}$ and $y = 9 - \dfrac{t^2}{4}$.

When finding parametric equations for $y = 9 - x^2$, can we start with any choice for the parametric equation for x? The answer is no. **The substitution for x must be a function that allows x to take on all the values in the domain of the given rectangular equation.** For example, the domain of the function $y = 9 - x^2$ is the set of all real numbers. If you incorrectly let $x = t^2$, these values of x exclude negative numbers that are included in $y = 9 - x^2$. The parametric equations

$$x = t^2 \quad \text{and} \quad y = 9 - (t^2)^2 = 9 - t^4$$

do not represent $y = 9 - x^2$ because only points for which $x \geq 0$ are obtained.

④ Understand the advantages of parametric representations.

Advantages of Parametric Equations over Rectangular Equations

We opened this section with parametric equations that described the horizontal distance and the vertical height of your thrown baseball after t seconds. Parametric equations are frequently used to represent the path of a moving object. If t represents time, parametric equations give the location of a moving object and tell when the object is located at each of its positions. Rectangular equations tell where the moving object is located but do not reveal when the object is in a particular position.

When using technology to obtain graphs, parametric equations that represent relations that are not functions are often easier to use than their corresponding rectangular equations. It is far easier to enter the equation of an ellipse given by the parametric equations

$$x = 2 + 3 \cos t \quad \text{and} \quad y = 3 + 2 \sin t$$

than to use the rectangular equivalent

$$\frac{(x - 2)^2}{9} + \frac{(y - 3)^2}{4} = 1.$$

The rectangular equation must first be solved for y and then entered as two separate equations before a graphing utility reveals the ellipse.

TECHNOLOGY

The ellipse shown was obtained using the parametric mode and the radian mode of a graphing utility.

$$x(t) = 2 + 3 \cos t$$
$$y(t) = 3 + 2 \sin t$$

We used a $[-3.6, 7.6, 1]$ by $[-1, 6, 1]$ viewing rectangle with $t\text{min} = 0$, $t\text{max} = 6.3$, and $t\text{step} = 0.1$.

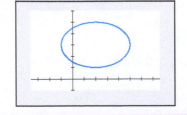

Blitzer Bonus || The Parametrization of DNA

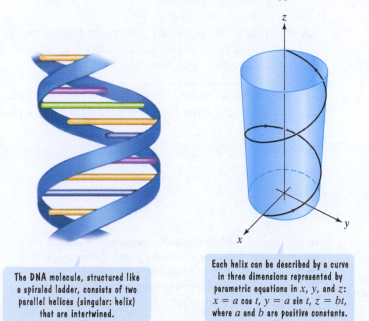

The DNA molecule, structured like a spiraled ladder, consists of two parallel helices (singular: helix) that are intertwined.

Each helix can be described by a curve in three dimensions represented by parametric equations in x, y, and z: $x = a \cos t$, $y = a \sin t$, $z = bt$, where a and b are positive constants.

DNA, the molecule of biological inheritance, is hip. At least that's what a new breed of marketers would like you to believe. For $2500, you can spit into a test tube and a Web-based company will tell you your risks for heart attack and other conditions.

It's been more than 60 years since James Watson and Francis Crick defined the structure, or shape, of DNA. A knowledge of how a molecule is structured does not always lead to an understanding of how it works, but it did in the case of DNA. The structure, which Watson and Crick announced in *Nature* in 1953, immediately suggested how the molecule could be reproduced and how it could contain biological information.

The structure of the DNA molecule reveals the vital role that trigonometric functions play in the genetic information and instruction codes necessary for the maintenance and continuation of life.

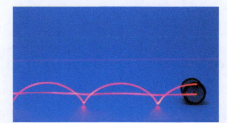

Linear functions and cycloids are used to describe rolling motion. The light at the rolling circle's center shows that it moves linearly. By contrast, the light at the circle's edge has rotational motion and traces out a cycloid.

A curve that is used in physics for much of the theory of light is called a **cycloid**. The path of a fixed point on the circumference of a circle as it rolls along a line is a cycloid. A point on the rim of a bicycle wheel traces out a cycloid curve, shown in **Figure 5.36**. If the radius of the circle is a, the parametric equations of the cycloid are

$$x = a(t - \sin t) \quad \text{and} \quad y = a(1 - \cos t).$$

It is an extremely complicated task to represent the cycloid in rectangular form.

Cycloids are used to solve problems that involve the "shortest time." For example, **Figure 5.37** shows a bead sliding down a wire. For the bead to travel along the wire in the shortest possible time, the shape of the wire should be that of an inverted cycloid.

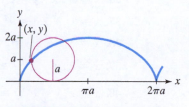

FIGURE 5.36 The curve traced by a fixed point on the circumference of a circle rolling along a straight line is a cycloid.

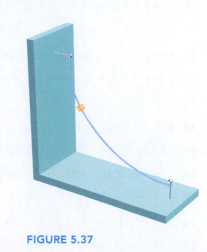

FIGURE 5.37

CONCEPT AND VOCABULARY CHECK

Fill in each blank so that the resulting statement is true.

1. The pair of equations $x = \sqrt{t}$ and $y = 2t - 1$ are called _____ equations and the common variable t is called the _____. The graph for this pair of equations is called a/an _____.

2. Eliminating the parameter from $x = f(t)$ and $y = g(t)$ means eliminating ____ from the pair of equations to obtain one equation in ____ and ____ only.

3. In order to eliminate the parameter from $x = 3 \sin t$ and $y = 2 \cos t$, isolate _____ and _____, square the two equations, and then use the identity _____.

4. True or false: There is more than one way for pairs of parametric equations to represent the same plane curve. _____

EXERCISE SET 5.5

Practice Exercises

In Exercises 1–8, parametric equations and a value for the parameter t are given. Find the coordinates of the point on the plane curve described by the parametric equations corresponding to the given value of t.

1. $x = 3 - 5t, y = 4 + 2t; t = 1$

2. $x = 7 - 4t, y = 5 + 6t; t = 1$

3. $x = t^2 + 1, y = 5 - t^3; t = 2$

4. $x = t^2 + 3, y = 6 - t^3; t = 2$

5. $x = 4 + 2 \cos t, y = 3 + 5 \sin t; t = \dfrac{\pi}{2}$

6. $x = 2 + 3 \cos t, y = 4 + 2 \sin t; t = \pi$

7. $x = (60 \cos 30°)t, y = 5 + (60 \sin 30°)t - 16t^2; t = 2$

8. $x = (80 \cos 45°)t, y = 6 + (80 \sin 45°)t - 16t^2; t = 2$

In Exercises 9–20, use point plotting to graph the plane curve described by the given parametric equations. Use arrows to show the orientation of the curve corresponding to increasing values of t.

9. $x = t + 2, y = t^2; -2 \le t \le 2$

10. $x = t - 1, y = t^2; -2 \le t \le 2$

11. $x = t - 2, y = 2t + 1; -2 \le t \le 3$

12. $x = t - 3, y = 2t + 2; -2 \le t \le 3$

13. $x = t + 1, y = \sqrt{t}; t \ge 0$

14. $x = \sqrt{t}, y = t - 1; t \ge 0$

15. $x = \cos t, y = \sin t; 0 \le t < 2\pi$

16. $x = -\sin t, y = -\cos t; 0 \le t < 2\pi$

17. $x = t^2, y = t^3; -\infty < t < \infty$

18. $x = t^2 + 1, y = t^3 - 1; -\infty < t < \infty$

19. $x = 2t, y = |t - 1|; -\infty < t < \infty$

20. $x = |t + 1|, y = t - 2; -\infty < t < \infty$

In Exercises 21–40, eliminate the parameter t. Then use the rectangular equation to sketch the plane curve represented by the given parametric equations. Use arrows to show the orientation of the curve corresponding to increasing values of t. (If an interval for t is not specified, assume that $-\infty < t < \infty$.)

21. $x = t, y = 2t$

22. $x = t, y = -2t$

23. $x = 2t - 4, y = 4t^2$

24. $x = t - 2, y = t^2$

25. $x = \sqrt{t}, y = t - 1$

26. $x = \sqrt{t}, y = t + 1$

27. $x = 2 \sin t, y = 2 \cos t; 0 \le t < 2\pi$

28. $x = 3 \sin t, y = 3 \cos t; 0 \le t < 2\pi$

29. $x = 1 + 3 \cos t, y = 2 + 3 \sin t; 0 \le t < 2\pi$

30. $x = -1 + 2 \cos t, y = 1 + 2 \sin t; 0 \le t < 2\pi$

31. $x = 2 \cos t, y = 3 \sin t; 0 \le t < 2\pi$

32. $x = 3 \cos t, y = 5 \sin t; 0 \le t < 2\pi$

33. $x = 1 + 3 \cos t, y = -1 + 2 \sin t; 0 \le t \le \pi$

34. $x = 2 + 4 \cos t, y = -1 + 3 \sin t; 0 \le t \le \pi$

35. $x = \sec t, y = \tan t$

36. $x = 5 \sec t, y = 3 \tan t$

37. $x = t^2 + 2, y = t^2 - 2$

38. $x = \sqrt{t} + 2, y = \sqrt{t} - 2$

39. $x = 2^t, y = 2^{-t}; t \ge 0$

40. $x = e^t, y = e^{-t}; t \ge 0$

In Exercises 41–43, eliminate the parameter. Write the resulting equation in standard form.

41. A circle: $x = h + r \cos t, y = k + r \sin t$

42. An ellipse: $x = h + a \cos t, y = k + b \sin t$

43. A hyperbola: $x = h + a \sec t, y = k + b \tan t$

44. The following are parametric equations of the line through (x_1, y_1) and (x_2, y_2):

$$x = x_1 + t(x_2 - x_1) \quad \text{and} \quad y = y_1 + t(y_2 - y_1).$$

Eliminate the parameter and write the resulting equation in point-slope form.

In Exercises 45–52, use your answers from Exercises 41–44 and the parametric equations given in Exercises 41–44 to find a set of parametric equations for the conic section or the line.

45. Circle: Center: $(3, 5)$; Radius: 6

46. Circle: Center: $(4, 6)$; Radius: 9

47. Ellipse: Center: $(-2, 3)$; Vertices: 5 units to the left and right of the center; Endpoints of Minor Axis: 2 units above and below the center

48. Ellipse: Center: $(4, -1)$; Vertices: 5 units above and below the center; Endpoints of Minor Axis: 3 units to the left and right of the center

49. Hyperbola: Vertices: $(4, 0)$ and $(-4, 0)$; Foci: $(6, 0)$ and $(-6, 0)$

50. Hyperbola: Vertices: $(0, 4)$ and $(0, -4)$; Foci: $(0, 5)$ and $(0, -5)$

51. Line: Passes through $(-2, 4)$ and $(1, 7)$

52. Line: Passes through $(3, -1)$ and $(9, 12)$

In Exercises 53–56, find two different sets of parametric equations for each rectangular equation.

53. $y = 4x - 3$

54. $y = 2x - 5$

55. $y = x^2 + 4$

56. $y = x^2 - 3$

In Exercises 57–58, the parametric equations of four plane curves are given. Graph each plane curve and determine how they differ from each other.

57. a. $x = t$ and $y = t^2 - 4$

    b. $x = t^2$ and $y = t^4 - 4$

    c. $x = \cos t$ and $y = \cos^2 t - 4$

    d. $x = e^t$ and $y = e^{2t} - 4$

58. a. $x = t, y = \sqrt{4 - t^2}; -2 \le t \le 2$

    b. $x = \sqrt{4 - t^2}, y = t; -2 \le t \le 2$

    c. $x = 2 \sin t, y = 2 \cos t; 0 \le t < 2\pi$

    d. $x = 2 \cos t, y = 2 \sin t; 0 \le t < 2\pi$

Practice Plus

In Exercises 59–62, sketch the plane curve represented by the given parametric equations. Then use interval notation to give each relation's domain and range.

59. $x = 4 \cos t + 2, y = 4 \cos t - 1$

60. $x = 2 \sin t - 3, y = 2 \sin t + 1$

61. $x = t^2 + t + 1, y = 2t$

62. $x = t^2 - t + 6, y = 3t$

In Exercises 63–68, sketch the function represented by the given parametric equations. Then use the graph to determine each of the following:

    a. *intervals, if any, on which the function is increasing and intervals, if any, on which the function is decreasing.*

    b. *the number, if any, at which the function has a maximum and this maximum value, or the number, if any, at which the function has a minimum and this minimum value.*

63. $x = 2^t, y = t$

64. $x = e^t, y = t$

65. $x = \dfrac{t}{2}, y = 2t^2 - 8t + 3$

66. $x = \dfrac{t}{2}, y = -2t^2 + 8t - 1$

67. $x = 2(t - \sin t), y = 2(1 - \cos t); 0 \le t \le 2\pi$

68. $x = 3(t - \sin t), y = 3(1 - \cos t); 0 \le t \le 2\pi$

Application Exercises

The path of a projectile that is launched h feet above the ground with an initial velocity of v_0 feet per second and at an angle θ with the horizontal is given by the parametric equations

$$x = (v_0 \cos \theta)t \quad \text{and} \quad y = h + (v_0 \sin \theta)t - 16t^2,$$

where t is the time, in seconds, after the projectile was launched. The parametric equation for x gives the projectile's horizontal distance, in feet. The parametric equation for y gives the projectile's height, in feet. Use these parametric equations to solve Exercises 69–70.

69. The figure shows the path for a baseball hit by Albert Pujols. The ball was hit with an initial velocity of 180 feet per second at an angle of 40° to the horizontal. The ball was hit at a height 3 feet off the ground.

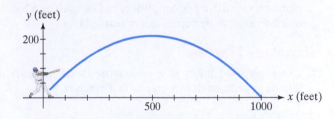

    a. Find the parametric equations that describe the position of the ball as a function of time.

    b. Describe the ball's position after 1, 2, and 3 seconds. Round to the nearest tenth of a foot. Locate your solutions on the plane curve.

    c. How long, to the nearest tenth of a second, is the ball in flight? What is the total horizontal distance that it travels before it lands? Is your answer consistent with the figure shown?

    d. You meet Albert Pujols and he asks you to tell him something interesting about the path of the baseball that he hit. Use the graph to respond to his request. Then verify your observation algebraically.

70. The figure shows the path for a baseball that was hit with an initial velocity of 150 feet per second at an angle of 35° to the horizontal. The ball was hit at a height of 3 feet off the ground.

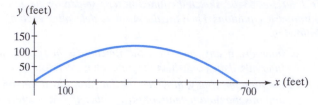

y (feet)

a. Find the parametric equations that describe the position of the ball as a function of time.

b. Describe the ball's position after 1, 2, and 3 seconds. Round to the nearest tenth of a foot. Locate your solutions on the plane curve.

c. How long is the ball in flight? (Round to the nearest tenth of a second.) What is the total horizontal distance that it travels, to the nearest tenth of a foot, before it lands? Is your answer consistent with the figure shown?

d. Use the graph to describe something about the path of the baseball that might be of interest to the player who hit the ball. Then verify your observation algebraically.

Explaining the Concepts

71. What are plane curves and parametric equations?

72. How is point plotting used to graph a plane curve described by parametric equations? Give an example with your description.

73. What is the significance of arrows along a plane curve?

74. What does it mean to eliminate the parameter? What useful information can be obtained by doing this?

75. Explain how the rectangular equation $y = 5x$ can have infinitely many sets of parametric equations.

76. Discuss how the parametric equations for the path of a projectile (see Exercises 69–70) and the ability to obtain plane curves with a graphing utility can be used by a baseball coach to analyze performances of team players.

Technology Exercises

77. Use a graphing utility in a parametric mode to verify any five of your hand-drawn graphs in Exercises 9–40.

In Exercises 78–82, use a graphing utility to obtain the plane curve represented by the given parametric equations.

78. Cycloid: $x = 3(t - \sin t), y = 3(1 - \cos t)$; $[0, 60, 5] \times [0, 8, 1], 0 \le t < 6\pi$

79. Cycloid: $x = 2(t - \sin t), y = 2(1 - \cos t)$; $[0, 60, 5] \times [0, 8, 1], 0 \le t < 6\pi$

80. Witch of Agnesi: $x = 2 \cot t, y = 2 \sin^2 t$; $[-6, 6, 1] \times [-4, 4, 1], 0 \le t < 2\pi$

81. Hypocycloid: $x = 4 \cos^3 t, y = 4 \sin^3 t$; $[-8, 8, 1] \times [-5, 5, 1], 0 \le t < 2\pi$

82. Lissajous Curve: $x = 2 \cos t, y = \sin 2t$; $[-3, 3, 1] \times [-2, 2, 1], 0 \le t < 2\pi$

Use the equations for the path of a projectile given prior to Exercises 69–70 to solve Exercises 83–85.

In Exercises 83–84, use a graphing utility to obtain the path of a projectile launched from the ground ($h = 0$) at the specified values of θ and v_0. In each exercise, use the graph to determine the maximum height and the time at which the projectile reaches its maximum height. Also use the graph to determine the range of the projectile and the time it hits the ground. Round all answers to the nearest tenth.

83. $\theta = 55°, v_0 = 200$ feet per second

84. $\theta = 35°, v_0 = 300$ feet per second

85. A baseball player throws a ball with an initial velocity of 140 feet per second at an angle of 22° to the horizontal. The ball leaves the player's hand at a height of 5 feet.

a. Write the parametric equations that describe the ball's position as a function of time.

b. Use a graphing utility to obtain the path of the baseball.

c. Find the ball's maximum height and the time at which it reaches this height. Round all answers to the nearest tenth.

d. How long is the ball in the air?

e. How far does the ball travel?

Critical Thinking Exercises

Make Sense? *In Exercises 86–89, determine whether each statement makes sense or does not make sense, and explain your reasoning.*

86. Parametric equations allow me to use functions to describe curves that are not graphs of functions.

87. Parametric equations let me think of a curve as a path traced out by a moving point.

88. I represented $y = x^2 - 9$ with the parametric equations $x = t^2$ and $y = t^4 - 9$.

89. I found alternate pairs of parametric equations for the same rectangular equation.

90. Eliminate the parameter: $x = \cos^3 t$ and $y = \sin^3 t$.

91. The plane curve described by the parametric equations $x = 3 \cos t$ and $y = 3 \sin t, 0 \le t < 2\pi$, has a counterclockwise orientation. Alter one or both parametric equations so that you obtain the same plane curve with the opposite orientation.

92. The figure shows a circle of radius a rolling along a horizontal line. Point P traces out a cycloid. Angle t, in radians, is the angle through which the circle has rolled. C is the center of the circle.

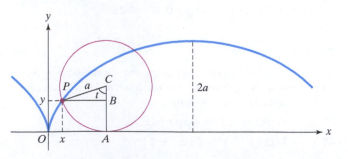

Refer to the figure at the bottom of the previous page. Use the suggestions in parts (a) and (b) to prove that the parametric equations of the cycloid are $x = a(t - \sin t)$ and $y = a(1 - \cos t)$.

a. Derive the parametric equation for x using the figure and
$$x = OA - xA.$$

b. Derive the parametric equation for y using the figure and
$$y = AC - BC.$$

Retaining the Concepts

93. Graph $y = 3\sin 2x$. Then use the graph to obtain the graph of $y = 3\csc 2x$.   (Section 2.2, Example 4)

94. Let $\mathbf{v} = 2\mathbf{i} - 5\mathbf{j}$ and $\mathbf{w} = -4\mathbf{i} + 3\mathbf{j}$. Find each of the following.

a. $\mathbf{v} + \mathbf{w}$ **b.** $2\mathbf{v} - \mathbf{w}$ **c.** $\mathbf{v} \cdot \mathbf{w}$ **d.** $\mathbf{v} \cdot \mathbf{v}$

(Section 4.3, Examples 4–6; Section 4.4, Example 1)

95. Solve triangle ABC with $A = 39°$, $b = 5$, and $c = 7$. Round lengths of sides to the nearest tenth and angle measures to the nearest degree.

(Section 4.2, Example 1)

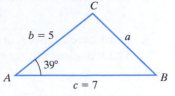

CHAPTER 5   **Summary, Review, and Test**

SUMMARY

DEFINITIONS AND CONCEPTS	**EXAMPLES**

5.1   Complex Numbers

a. The imaginary unit i is defined as
$$i = \sqrt{-1}, \text{ where } i^2 = -1.$$
The set of numbers in the form $a + bi$ is called the set of complex numbers; a is the real part and b is the imaginary part. If $b = 0$, the complex number is a real number. If $b \neq 0$, the complex number is an imaginary number. Complex numbers in the form bi are called pure imaginary numbers.

b. Rules for adding and subtracting complex numbers are given in the box on page 266.   Ex. 1, p. 266

c. To multiply complex numbers, multiply as if they are polynomials. After completing the multiplication, replace i^2 with -1 and simplify.   Ex. 2, p. 267

d. The complex conjugate of $a + bi$ is $a - bi$ and vice versa. The multiplication of complex conjugates gives a real number:
$$(a + bi)(a - bi) = a^2 + b^2.$$

e. To divide complex numbers, multiply the numerator and the denominator by the complex conjugate of the denominator.   Ex. 3, p. 268

f. When performing operations with square roots of negative numbers, begin by expressing all square roots in terms of i. The principal square root of $-b$ is defined by
$$\sqrt{-b} = i\sqrt{b}.$$
   Ex. 4, p. 269

5.2   Complex Numbers in Polar Form; DeMoivre's Theorem

a. The complex number $z = a + bi$ is represented as a point (a, b) in the complex plane, shown in Figure 5.2 on page 272.   Ex. 1, p. 273

b. The absolute value of $z = a + bi$ is $|z| = |a + bi| = \sqrt{a^2 + b^2}$.   Ex. 2, p. 273

c. The polar form of $z = a + bi$ is $z = r(\cos\theta + i\sin\theta)$, where $a = r\cos\theta$, $b = r\sin\theta$, $r = \sqrt{a^2 + b^2}$, and $\tan\theta = \dfrac{b}{a}$. We call r the modulus and θ the argument of z, with $0 \leq \theta < 2\pi$.   Ex. 3, p. 274; Ex. 4, p. 275

d. Multiplying Complex Numbers in Polar Form: Multiply moduli and add arguments. See the box on page 275.   Ex. 5, p. 276

DEFINITIONS AND CONCEPTS	**EXAMPLES**
e. Dividing Complex Numbers in Polar Form: Divide moduli and subtract arguments. See the box on page 276.	Ex. 6, p. 277
f. DeMoivre's Theorem is used to find powers of complex numbers in polar form. $$[r(\cos\theta + i\sin\theta)]^n = r^n(\cos n\theta + i\sin n\theta)$$	Ex. 7, p. 278; Ex. 8, p. 278
g. DeMoivre's Theorem can be used to find roots of complex numbers in polar form. The n distinct nth roots of $r(\cos\theta + i\sin\theta)$ are $$\sqrt[n]{r}\left[\cos\left(\frac{\theta + 2\pi k}{n}\right) + i\sin\left(\frac{\theta + 2\pi k}{n}\right)\right]$$ or $$\sqrt[n]{r}\left[\cos\left(\frac{\theta + 360°k}{n}\right) + i\sin\left(\frac{\theta + 360°k}{n}\right)\right],$$ where $k = 0, 1, 2, \ldots, n - 1$.	Ex. 9, p. 279; Ex. 10, p. 280

5.3 and 5.4 Polar Coordinates; Graphs of Polar Equations

a. A point P in the polar coordinate system is represented by (r, θ), where r is the directed distance from the pole to the point and θ is the angle from the polar axis to line segment OP. The elements of the ordered pair (r, θ) are called the polar coordinates of P. See Figure 5.13 on page 286. When r in (r, θ) is negative, a point is located $|r|$ units along the ray opposite the terminal side of θ. Important information about the sign of r and the location of the point (r, θ) is found in the box on page 287.	Ex. 1, p. 287
b. Multiple Representations of Points If n is any integer, $(r, \theta) = (r, \theta + 2n\pi)$ or $(r, \theta) = (-r, \theta + \pi + 2n\pi)$.	Ex. 2, p. 288
c. Relations between Polar and Rectangular Coordinates $$x = r\cos\theta, \quad y = r\sin\theta, \quad x^2 + y^2 = r^2, \quad \tan\theta = \frac{y}{x}$$	
d. To convert a point from polar coordinates (r, θ) to rectangular coordinates (x, y), use $x = r\cos\theta$ and $y = r\sin\theta$.	Ex. 3, p. 289
e. A point in rectangular coordinates (x, y) can be converted to polar coordinates (r, θ). Use the procedure in the box on page 290.	Ex. 4, p. 290; Ex. 5, p. 291
f. To convert a rectangular equation to a polar equation, replace x with $r\cos\theta$ and y with $r\sin\theta$.	Ex. 6, p. 292
g. To convert a polar equation to a rectangular equation, use one or more of $$r^2 = x^2 + y^2, \quad r\cos\theta = x, \quad r\sin\theta = y, \quad \text{and} \quad \tan\theta = \frac{y}{x}.$$ It is often necessary to do something to the given polar equation before using the preceding expressions.	Ex. 7, p. 293
h. A polar equation is an equation whose variables are r and θ. The graph of a polar equation is the set of all points whose polar coordinates satisfy the equation.	Ex. 1, p. 299
i. Polar equations can be graphed using point plotting and symmetry (see the box on page 301).	Ex. 2, p. 301
j. The graphs of $r = a\cos\theta$ and $r = a\sin\theta$ are circles. See the box on page 300. The graphs of $r = a \pm b\sin\theta$ and $r = a \pm b\cos\theta$ are called limaçons ($a > 0$ and $b > 0$), shown in the box on page 303. The graphs of $r = a\sin n\theta$ and $r = a\cos n\theta, a \neq 0$, are rose curves with $2n$ petals if n is even and n petals if n is odd. See the box on page 305. The graphs of $r^2 = a^2\sin 2\theta$ and $r^2 = a^2\cos 2\theta, a \neq 0$, are called lemniscates and are shown in the box on page 306.	Ex. 3, p. 302; Ex. 4, p. 304; Ex. 5, p. 305

DEFINITIONS AND CONCEPTS

EXAMPLES

5.5 Parametric Equations

a. The relationship between the parametric equations $x = f(t)$ and $y = g(t)$ and plane curves is described in the first box on page 312.

b. Point plotting can be used to graph a plane curve described by parametric equations. See the second box on page 312.

Ex. 1, p. 312

c. Plane curves can be sketched by eliminating the parameter t and graphing the resulting rectangular equation. It is sometimes necessary to change the domain of the rectangular equation to be consistent with the domain for the parametric equation in x.

Ex. 2, p. 313; Ex. 3, p. 314

d. Infinitely many pairs of parametric equations can represent the same plane curve. One pair for $y = f(x)$ is $x = t$ and $y = f(t)$, in which t is in the domain of f.

Ex. 4, p. 316

REVIEW EXERCISES

5.1

In Exercises 1–10, perform the indicated operations and write the result in standard form.

1. $(8 - 3i) - (17 - 7i)$

2. $4i(3i - 2)$

3. $(7 - i)(2 + 3i)$

4. $(3 - 4i)^2$

5. $(7 + 8i)(7 - 8i)$

6. $\dfrac{6}{5 + i}$

7. $\dfrac{3 + 4i}{4 - 2i}$

8. $\sqrt{-32} - \sqrt{-18}$

9. $\left(-2 + \sqrt{-100}\right)^2$

10. $\dfrac{4 + \sqrt{-8}}{2}$

5.2

In Exercises 11–14, plot each complex number. Then write the complex number in polar form. You may express the argument in degrees or radians.

11. $1 - i$

12. $-2\sqrt{3} + 2i$

13. $-3 - 4i$

14. $-5i$

In Exercises 15–18, write each complex number in rectangular form. If necessary, round to the nearest tenth.

15. $8(\cos 60° + i \sin 60°)$

16. $4(\cos 210° + i \sin 210°)$

17. $6\left(\cos \dfrac{2\pi}{3} + i \sin \dfrac{2\pi}{3}\right)$

18. $0.6(\cos 100° + i \sin 100°)$

In Exercises 19–21, find the product of the complex numbers. Leave answers in polar form.

19. $z_1 = 3(\cos 40° + i \sin 40°)$
$z_2 = 5(\cos 70° + i \sin 70°)$

20. $z_1 = \cos 210° + i \sin 210°$
$z_2 = \cos 55° + i \sin 55°$

21. $z_1 = 4\left(\cos \dfrac{3\pi}{7} + i \sin \dfrac{3\pi}{7}\right)$

$z_2 = 10\left(\cos \dfrac{4\pi}{7} + i \sin \dfrac{4\pi}{7}\right)$

In Exercises 22–24, find the quotient $\dfrac{z_1}{z_2}$ of the complex numbers. Leave answers in polar form.

22. $z_1 = 10(\cos 10° + i \sin 10°)$
$z_2 = 5(\cos 5° + i \sin 5°)$

23. $z_1 = 5\left(\cos \dfrac{4\pi}{3} + i \sin \dfrac{4\pi}{3}\right)$

$z_2 = 10\left(\cos \dfrac{\pi}{3} + i \sin \dfrac{\pi}{3}\right)$

24. $z_1 = 2\left(\cos \dfrac{5\pi}{3} + i \sin \dfrac{5\pi}{3}\right)$

$z_2 = \cos \dfrac{\pi}{2} + i \sin \dfrac{\pi}{2}$

In Exercises 25–29, use DeMoivre's Theorem to find the indicated power of the complex number. Write answers in rectangular form.

25. $[2(\cos 20° + i \sin 20°)]^3$

26. $[4(\cos 50° + i \sin 50°)]^3$

27. $\left[\dfrac{1}{2}\left(\cos \dfrac{\pi}{14} + i \sin \dfrac{\pi}{14}\right)\right]^7$

28. $\left(1 - i\sqrt{3}\right)^2$

29. $(-2 - 2i)^5$

In Exercises 30–31, find all the complex roots. Write roots in polar form with θ in degrees.

30. The complex square roots of $49(\cos 50° + i \sin 50°)$

31. The complex cube roots of $125(\cos 165° + i \sin 165°)$

In Exercises 32–35, find all the complex roots. Write roots in rectangular form.

32. The complex fourth roots of $16\left(\cos \dfrac{2\pi}{3} + i \sin \dfrac{2\pi}{3}\right)$

33. The complex cube roots of $8i$

34. The complex cube roots of -1

35. The complex fifth roots of $-1 - i$

5.3 and 5.4

In Exercises 36–41, plot each point in polar coordinates and find its rectangular coordinates.

36. $(4, 60°)$ **37.** $(3, 150°)$ **38.** $\left(-4, \dfrac{4\pi}{3}\right)$

39. $\left(-2, \dfrac{5\pi}{4}\right)$ **40.** $\left(-4, -\dfrac{\pi}{2}\right)$ **41.** $\left(-2, -\dfrac{\pi}{4}\right)$

In Exercises 42–44, plot each point in polar coordinates. Then find another representation (r, θ) of this point in which
   a. $r > 0, 2\pi < \theta < 4\pi$.
   b. $r < 0, 0 < \theta < 2\pi$.
   c. $r > 0, -2\pi < \theta < 0$.

42. $\left(3, \dfrac{\pi}{6}\right)$ **43.** $\left(2, \dfrac{2\pi}{3}\right)$ **44.** $\left(3.5, \dfrac{\pi}{2}\right)$

In Exercises 45–50, the rectangular coordinates of a point are given. Find polar coordinates of each point.

45. $(-4, 4)$ **46.** $(3, -3)$ **47.** $(5, 12)$

48. $(-3, 4)$ **49.** $(0, -5)$ **50.** $(1, 0)$

In Exercises 51–53, convert each rectangular equation to a polar equation that expresses r in terms of θ.

51. $2x + 3y = 8$

52. $x^2 + y^2 = 100$

53. $(x - 6)^2 + y^2 = 36$

In Exercises 54–60, convert each polar equation to a rectangular equation. Then use your knowledge of the rectangular equation to graph the polar equation in a polar coordinate system.

54. $r = 3$ **55.** $\theta = \dfrac{3\pi}{4}$

56. $r \cos \theta = -1$ **57.** $r = 5 \csc \theta$

58. $r = 3 \cos \theta$ **59.** $4r \cos \theta + r \sin \theta = 8$

60. $r^2 \sin 2\theta = -2$

In Exercises 61–63, test for symmetry with respect to
   a. the polar axis. *b. the line $\theta = \dfrac{\pi}{2}$.*
   c. the pole.

61. $r = 5 + 3 \cos \theta$ **62.** $r = 3 \sin \theta$

63. $r^2 = 9 \cos 2\theta$

In Exercises 64–70, graph each polar equation. Be sure to test for symmetry.

64. $r = 3 \cos \theta$ **65.** $r = 2 + 2 \sin \theta$

66. $r = \sin 2\theta$ **67.** $r = 2 + \cos \theta$

68. $r = 1 + 3 \sin \theta$ **69.** $r = 1 - 2 \cos \theta$

70. $r^2 = \cos 2\theta$

5.5

In Exercises 71–76, eliminate the parameter and graph the plane curve represented by the parametric equations. Use arrows to show the orientation of each plane curve.

71. $x = 2t - 1, y = 1 - t; -\infty < t < \infty$

72. $x = t^2, y = t - 1; -1 \le t \le 3$

73. $x = 4t^2, y = t + 1; -\infty < t < \infty$

74. $x = 4 \sin t, y = 3 \cos t; 0 \le t < \pi$

75. $x = 3 + 2 \cos t, y = 1 + 2 \sin t; 0 \le t < 2\pi$

76. $x = 3 \sec t, y = 3 \tan t; 0 \le t \le \dfrac{\pi}{4}$

77. Find two different sets of parametric equations for $y = x^2 + 6$.

78. The path of a projectile that is launched h feet above the ground with an initial velocity of v_0 feet per second and at an angle θ with the horizontal is given by the parametric equations

$$x = (v_0 \cos \theta)t \quad \text{and} \quad y = h + (v_0 \sin \theta)t - 16t^2,$$

where t is the time, in seconds, after the projectile was launched. A football player throws a football with an initial velocity of 100 feet per second at an angle of 40° to the horizontal. The ball leaves the player's hand at a height of 6 feet.
   a. Find the parametric equations that describe the position of the ball as a function of time.
   b. Describe the ball's position after 1, 2, and 3 seconds. Round to the nearest tenth of a foot.
   c. How long, to the nearest tenth of a second, is the ball in flight? What is the total horizontal distance that it travels before it lands?
   d. Graph the parametric equations in part (a) using a graphing utility. Use the graph to determine when the ball is at its maximum height. What is its maximum height? Round answers to the nearest tenth.

CHAPTER 5 TEST

In Exercises 1–3, perform the indicated operations and write the result in standard form.

1. $(6 - 7i)(2 + 5i)$ **2.** $\dfrac{5}{2 - i}$

3. $2\sqrt{-49} + 3\sqrt{-64}$

4. Write $-\sqrt{3} + i$ in polar form.

In Exercises 5–7, perform the indicated operation. Leave answers in polar form.

5. $5(\cos 15° + i \sin 15°) \cdot 10(\cos 5° + i \sin 5°)$

6. $\dfrac{2\left(\cos \dfrac{\pi}{2} + i \sin \dfrac{\pi}{2}\right)}{4\left(\cos \dfrac{\pi}{3} + i \sin \dfrac{\pi}{3}\right)}$

7. $[2(\cos 10° + i \sin 10°)]^5$

8. Find the three cube roots of 27. Write roots in rectangular form.

9. Plot $\left(4, \dfrac{5\pi}{4}\right)$ in the polar coordinate system. Then write two other ordered pairs (r, θ) that name this point.

10. If the rectangular coordinates of a point are $(1, -1)$, find polar coordinates of the point.

11. Convert $x^2 + (y + 8)^2 = 64$ to a polar equation that expresses r in terms of θ.

12. Convert to a rectangular equation and then graph: $r = -4 \sec \theta$.

In Exercises 13–14, graph each polar equation.

13. $r = 1 + \sin \theta$ 14. $r = 1 + 3 \cos \theta$

In Exercises 15–16, eliminate the parameter and graph the plane curve represented by the parametric equations. Use arrows to show the orientation of each plane curve.

15. $x = \sqrt{t}, y = t + 1; -\infty < t < \infty$

16. $x = 3 \sin t, y = 2 \cos t; 0 \le t < 2\pi$

CUMULATIVE REVIEW EXERCISES (CHAPTERS 1–5)

In Exercises 1–3, solve each equation on the interval $[0, 2\pi)$.

1. $2 \sin^2 x = 3 \cos x + 3$

2. $\sin 2x = \sin x$

3. $\tan 2x = 1$

In Exercises 4–7, graph each equation.

4. $y = 2 \sin 2\pi x, 0 \le x \le 2$

5. $y = 3 \cos(4x + \pi)$ (Graph one period.)

6. $y = 2 \tan \dfrac{x}{2}, -\pi < x < 3\pi$

7. $y = 4 \csc 2x,$ (Graph one period.)

In Exercises 8–10, verify each identity.

8. $\sec x - \cos x = \tan x \sin x$

9. $\tan(x + \pi) = \tan x$

10. $\dfrac{1}{1 + \sin x} + \dfrac{1}{1 - \sin x} = 2 \sec^2 x$

11. Use DeMoivre's Theorem to find
$$[\sqrt{2}(\cos 15° + i \sin 15°)]^4.$$
Write the answer in rectangular form.

12. Find the dot product $\mathbf{v} \cdot \mathbf{w}$ and the angle between $\mathbf{v}$ and $\mathbf{w}$:
$$\mathbf{v} = -2\mathbf{i} + \mathbf{j}, \mathbf{w} = 4\mathbf{i} - 3\mathbf{j}.$$

13. In oblique triangle ABC, $A = 64°$, $B = 72°$, and $a = 13.6$. Solve the triangle. Round lengths to the nearest tenth.

14. In oblique triangle ABC, $A = 120°$, $b = 7$, and $c = 8$. Solve the triangle. Round lengths to the nearest tenth and angle measures to the nearest degree.

15. If $\mathbf{v} = -6\mathbf{i} + 5\mathbf{j}$ and $\mathbf{w} = -7\mathbf{i} + 3\mathbf{j}$, find $4\mathbf{w} - 5\mathbf{v}$.

16. You are standing on level ground 400 feet from the center of a building. On the top of the building's flat roof is a statue. The angle of elevation to the bottom of the statue is 54° and the angle of elevation to the top is 57°. Find the height of the statue to the nearest tenth of a foot.

In Exercises 17–18, find the exact value of each expression. Do not use a calculator.

17. $\cos \dfrac{7\pi}{6}$ 18. $\sin \left[\tan^{-1} \left(-\dfrac{4}{3} \right) \right]$

19. Consider a circle of radius 6 inches with a 120° central angle. Find the length of the arc intercepted by the central angle and the area of the sector formed by the central angle. Express arc length and area in terms of π. Then round each answer to two decimal places.

20. Write $\cos \left(\sin^{-1} \dfrac{x}{2} \right)$ as an algebraic expression. Assume $x > 0$ and $\dfrac{x}{2}$ is in the domain of the inverse sine function.

21. A small fire is sighted from ranger stations A and B. Station B is 2.2 miles due east of station A. The bearing of the fire from station A is N32°E and the bearing of the fire from station B is N54°W. How far, to the nearest tenth of a mile, is the fire from station A?

22. A object moves in simple harmonic motion described by $d = -14 \cos \pi t$, where t is measured in seconds and d in inches. Find **a.** the maximum displacement, **b.** the frequency, and **c.** the time required for one oscillation.

23. Plot $\left(2, \dfrac{3\pi}{4} \right)$ in the polar coordinate system. Then write two other pairs (r, θ) that name this point.

24. Find polar coordinates of the point whose rectangular coordinates are $(\sqrt{3}, 1)$.

25. Convert $(x + 1)^2 + y^2 = 1$ to a polar equation that expresses r in terms of θ.

26. Convert to a rectangular equation and then graph: $r = 2 \sec \theta$.

27. Graph $r = 3 \cos 2\theta$ in polar coordinates.

28. Use the parametric equations $x = 7 \sin t$ and $y = 3 \cos t, 0 \le t \le \pi$, and eliminate the parameter. Graph the plane curve represented by the parametric equations. Use graph arrows to show the orientation of the curve.

Where Did That Come From?
Selected Proofs

Section 4.2	The Law of Cosines

Heron's Formula for the Area of a Triangle

The area of a triangle with sides $a, b,$ and c is

$$\text{Area} = \sqrt{s(s-a)(s-b)(s-c)},$$

where s is one-half its perimeter: $s = \frac{1}{2}(a + b + c)$.

Proof

The proof of Heron's formula begins with a half-angle formula and the Law of Cosines.

$$\cos\frac{C}{2} = \sqrt{\frac{1 + \cos C}{2}} = \sqrt{\frac{1 + \dfrac{a^2 + b^2 - c^2}{2ab}}{2}}$$

> This is the Law of Cosines $c^2 = a^2 + b^2 - 2ab\cos C$ solved for $\cos C$.

$$= \sqrt{\frac{a^2 + 2ab + b^2 - c^2}{4ab}} = \sqrt{\frac{(a+b)^2 - c^2}{4ab}} = \sqrt{\frac{(a+b+c)(a+b-c)}{4ab}}$$

> Multiply the numerator and denominator of the radicand by $2ab$.

> Factor $a^2 + 2ab + b^2$.

> Factor the numerator as the difference of two squares.

We now introduce the expression for one-half the perimeter: $s = \frac{1}{2}(a + b + c)$. We replace $a + b + c$ in the numerator by $2s$. We also find an expression for $a + b - c$ as follows:

$$a + b - c = a + b + c - 2c = 2s - 2c = 2(s - c).$$

Thus,

$$\cos\frac{C}{2} = \sqrt{\frac{(a+b+c)(a+b-c)}{4ab}} = \sqrt{\frac{2s\cdot 2(s-c)}{4ab}} = \sqrt{\frac{s(s-c)}{ab}}.$$

In a similar manner, we obtain

$$\sin\frac{C}{2} = \sqrt{\frac{1 - \cos C}{2}} = \sqrt{\frac{(s-a)(s-b)}{ab}}.$$

From our work in Section 4.1, we know that the area of a triangle is one-half the product of the length of two sides times the sine of their included angle.

$$\text{Area} = \frac{1}{2}ab\sin C$$

$$= \frac{1}{2}ab \cdot 2\sin\frac{C}{2}\cos\frac{C}{2}$$

> $\sin C = \sin 2\dfrac{C}{2} = 2\sin\dfrac{C}{2}\cos\dfrac{C}{2}$

$$= ab\sqrt{\frac{(s-a)(s-b)}{ab}}\sqrt{\frac{s(s-c)}{ab}}$$

> Use the preceding expressions for $\sin\dfrac{C}{2}$ and $\cos\dfrac{C}{2}$.

$$= ab \frac{\sqrt{s(s-a)(s-b)(s-c)}}{\sqrt{a^2 b^2}} \qquad \text{Multiply the radicands.}$$

$$= \sqrt{s(s-a)(s-b)(s-c)} \qquad \text{Simplify: } \frac{ab}{\sqrt{a^2 b^2}} = \frac{ab}{ab} = 1.$$

Section 4.4   The Dot Product

Properties of the Dot Product

If $\mathbf{u}$, $\mathbf{v}$, and $\mathbf{w}$ are vectors, and c is a scalar, then

1. $\mathbf{u} \cdot \mathbf{v} = \mathbf{v} \cdot \mathbf{u}$
2. $\mathbf{u} \cdot (\mathbf{v} + \mathbf{w}) = \mathbf{u} \cdot \mathbf{v} + \mathbf{u} \cdot \mathbf{w}$
3. $\mathbf{0} \cdot \mathbf{v} = 0$
4. $\mathbf{v} \cdot \mathbf{v} = \|\mathbf{v}\|^2$
5. $(c\mathbf{u}) \cdot \mathbf{v} = c(\mathbf{u} \cdot \mathbf{v}) = \mathbf{u} \cdot (c\mathbf{v})$

Proof

To prove the second property, let

$$\mathbf{u} = u_1 \mathbf{i} + u_2 \mathbf{j}, \quad \mathbf{v} = v_1 \mathbf{i} + v_2 \mathbf{j}, \quad \text{and} \quad \mathbf{w} = w_1 \mathbf{i} + w_2 \mathbf{j}.$$

Then

$$\mathbf{u} \cdot (\mathbf{v} + \mathbf{w}) = (u_1 \mathbf{i} + u_2 \mathbf{j}) \cdot [(v_1 \mathbf{i} + v_2 \mathbf{j}) + (w_1 \mathbf{i} + w_2 \mathbf{j})] \qquad \text{These are the given vectors.}$$

$$= (u_1 \mathbf{i} + u_2 \mathbf{j}) \cdot [(v_1 + w_1)\mathbf{i} + (v_2 + w_2)\mathbf{j}] \qquad \text{Add horizontal components and add vertical components.}$$

$$= u_1(v_1 + w_1) + u_2(v_2 + w_2) \qquad \text{Multiply horizontal components and multiply vertical components.}$$

$$= u_1 v_1 + u_1 w_1 + u_2 v_2 + u_2 w_2 \qquad \text{Use the distributive property.}$$

$$= \underbrace{u_1 v_1 + u_2 v_2}_{\text{This is the dot product of } \mathbf{u} \text{ and } \mathbf{v}.} + \underbrace{u_1 w_1 + u_2 w_2}_{\text{This is the dot product of } \mathbf{u} \text{ and } \mathbf{w}.} \qquad \text{Rearrange terms.}$$

$$= \mathbf{u} \cdot \mathbf{v} + \mathbf{u} \cdot \mathbf{w}.$$

To prove the third property, let

$$\mathbf{0} = 0\mathbf{i} + 0\mathbf{j} \quad \text{and} \quad \mathbf{v} = v_1 \mathbf{i} + v_2 \mathbf{j}.$$

Then

$$\mathbf{0} \cdot \mathbf{v} = (0\mathbf{i} + 0\mathbf{j}) \cdot (v_1 \mathbf{i} + v_2 \mathbf{j}) \qquad \text{These are the given vectors.}$$

$$= 0 \cdot v_1 + 0 \cdot v_2 \qquad \text{Multiply horizontal components and multiply vertical components.}$$

$$= 0 + 0$$

$$= 0.$$

To prove the first part of the fifth property, let

$$\mathbf{u} = u_1 \mathbf{i} + u_2 \mathbf{j} \quad \text{and} \quad \mathbf{v} = v_1 \mathbf{i} + v_2 \mathbf{j}.$$

Then

$$(c\mathbf{u}) \cdot \mathbf{v} = [c(u_1\mathbf{i} + u_2\mathbf{j})] \cdot (v_1\mathbf{i} + v_2\mathbf{j})$$ These are the given vectors.

$$= (cu_1\mathbf{i} + cu_2\mathbf{j}) \cdot (v_1\mathbf{i} + v_2\mathbf{j})$$ Multiply each component of $u_1\mathbf{i} + u_2\mathbf{j}$ by c.

$$= cu_1v_1 + cu_2v_2$$ Multiply horizontal components and multiply vertical components.

$$= c(u_1v_1 + u_2v_2)$$ Factor out c from both terms.

This is the dot product of **u** and **v**.

$$= c(\mathbf{u} \cdot \mathbf{v})$$

Section 5.2   Complex Numbers in Polar Form; DeMoivre's Theorem

The Quotient of Two Complex Numbers in Polar Form

Let $z_1 = r_1(\cos\theta_1 + i\sin\theta_1)$ and $z_2 = r_2(\cos\theta_2 + i\sin\theta_2)$ be two complex numbers in polar form. Their quotient, $\dfrac{z_1}{z_2}$, is

$$\frac{z_1}{z_2} = \frac{r_1}{r_2}[\cos(\theta_1 - \theta_2) + i\sin(\theta_1 - \theta_2)].$$

Proof

We begin by multiplying the numerator and denominator of the quotient, $\dfrac{z_1}{z_2}$, by the conjugate of the expression in parentheses in the denominator. Then we simplify the quotient using the difference formulas for sine and cosine.

$$\frac{z_1}{z_2} = \frac{r_1(\cos\theta_1 + i\sin\theta_1)}{r_2(\cos\theta_2 + i\sin\theta_2)}$$ This is the given quotient.

$$= \frac{r_1(\cos\theta_1 + i\sin\theta_1)(\cos\theta_2 - i\sin\theta_2)}{r_2(\cos\theta_2 + i\sin\theta_2)(\cos\theta_2 - i\sin\theta_2)}$$ Multiply the numerator and denominator by the conjugate of the expression in parentheses in the denominator. Recall that the conjugate of $a + bi$ is $a - bi$.

$$= \frac{r_1(\cos\theta_1 + i\sin\theta_1)(\cos\theta_2 - i\sin\theta_2)}{r_2(\cos^2\theta_2 + \sin^2\theta_2)}$$ Multiply the conjugates in the denominator.

$$= \frac{r_1(\cos\theta_1 + i\sin\theta_1)(\cos\theta_2 - i\sin\theta_2)}{r_2}$$ Use a Pythagorean identity: $\cos^2\theta_2 + \sin^2\theta_2 = 1$.

$$= \frac{r_1}{r_2}(\cos\theta_1\cos\theta_2 - i\cos\theta_1\sin\theta_2 + i\sin\theta_1\cos\theta_2 - i^2\sin\theta_1\sin\theta_2)$$ Use the FOIL method.

$$= \frac{r_1}{r_2}[\cos\theta_1\cos\theta_2 + i(\sin\theta_1\cos\theta_2 - \cos\theta_1\sin\theta_2) - i^2\sin\theta_1\sin\theta_2]$$ Factor i from the second and third terms.

$$= \frac{r_1}{r_2}[\cos\theta_1\cos\theta_2 + i(\sin\theta_1\cos\theta_2 - \cos\theta_1\sin\theta_2) - (-1)\sin\theta_1\sin\theta_2]$$ $i^2 = -1$

$$= \frac{r_1}{r_2}[(\cos\theta_1\cos\theta_2 + \sin\theta_1\sin\theta_2) + i(\sin\theta_1\cos\theta_2 - \cos\theta_1\sin\theta_2)]$$ Rearrange terms.

This is $\cos(\theta_1 - \theta_2)$.   This is $\sin(\theta_1 - \theta_2)$.

$$= \frac{r_1}{r_2}[\cos(\theta_1 - \theta_2) + i\sin(\theta_1 - \theta_2)]$$

ANSWERS TO SELECTED EXERCISES

CHAPTER 1

Section 1.1

Check Point Exercises

1. 3.5 radians   **2. a.** $\frac{\pi}{3}$ radians   **b.** $\frac{3\pi}{2}$ radians   **c.** $-\frac{5\pi}{3}$ radians   **3. a.** 45°   **b.** $-240°$ **c.** 343.8°   **d.** $-269.3°$

4. a.   **b.**   **c.**   **d.**   **5. a.** 40°   **b.** 225°   **6. a.** $\frac{3\pi}{5}$ **b.** $\frac{29\pi}{15}$ **7. a.** 135°

b. $\frac{5\pi}{3}$ **c.** $\frac{11\pi}{6}$ **d.** 4.8   **8.** $\frac{3\pi}{2}$ in. $\approx$ 4.71 in.

9. $15\pi \approx 47.12$ sq ft   **10.** 135π in./min ≈ 424 in./min

Concept and Vocabulary Check

1. origin; x-axis   **2.** counterclockwise; clockwise   **3.** acute; right; obtuse; straight   **4.** $\frac{s}{r}$ **5.** $\frac{\pi}{180°}$ **6.** $\frac{180°}{\pi}$ **7.** coterminal; 360°; 2π

8. $r\theta$ **9.** $\frac{1}{2}r^2\theta$ **10.** false   **11.** $r\omega$; angular

Exercise Set 1.1

1. obtuse   **3.** acute   **5.** straight   **7.** 4 radians   **9.** $\frac{4}{3}$ radians   **11.** 4 radians   **13.** $\frac{\pi}{4}$ radian   **15.** $\frac{3\pi}{4}$ radians   **17.** $\frac{5\pi}{3}$ radians

19. $-\frac{5\pi}{4}$ radians   **21.** 90°   **23.** 120°   **25.** 210°   **27.** $-540°$ **29.** 0.31 radian   **31.** -0.70 radian   **33.** 3.49 radians   **35.** 114.59°

37. 13.85°   **39.** $-275.02°$

41. ; quadrant III   **43.** ; quadrant II   **45.** ; quadrant III   **47.** ; quadrant II

49. ; quadrant III   **51.** ; quadrant II   **53.** ; quadrant II   **55.** ; quadrant I

57. 35°   **59.** 210°   **61.** 315°   **63.** $\frac{7\pi}{6}$ **65.** $\frac{3\pi}{5}$ **67.** $\frac{99\pi}{50}$ **69.** $\frac{11\pi}{7}$ **71.** 3π in. ≈ 9.42 in.   **73.** 10π ft ≈ 31.42 ft

75. $5\pi \approx 15.71$ sq m   **77.** $\frac{32\pi}{3} \approx 33.51$ sq in.   **79.** $\frac{12\pi \text{ radians}}{\text{second}}$ **81.** $-\frac{4\pi}{3}$ and $\frac{2\pi}{3}$ **83.** $-\frac{3\pi}{4}$ and $\frac{5\pi}{4}$ **85.** $-\frac{\pi}{2}$ and $\frac{3\pi}{2}$ **87.** $\frac{11\pi}{6}$

89. $\frac{22\pi}{3}$ **91.** $\frac{1}{2}$ radian   **93.** 60°; $\frac{\pi}{3}$ radians   **95.** $\frac{8\pi}{3}$ in. ≈ 8.38 in.   **97.** 12π in. ≈ 37.70 in.   **99.** $600\pi \approx 1884.96$ sq ft

101. 2 radians; 114.59°   **103.** 2094 mi   **105.** 1047 mph   **107.** 1508 ft/min   **121.** 30.25°   **123.** 30°25′12″   **125.** does not make sense

127. makes sense   **129.** smaller than a right angle   **131.** 89°   **133.** $\frac{5}{13}$ **134.** $\frac{\sqrt{2}}{2}$ **135.** 1

Section 1.2

Check Point Exercises

1. $\sin \theta = \frac{3}{5}$; $\cos \theta = \frac{4}{5}$; $\tan \theta = \frac{3}{4}$; $\csc \theta = \frac{5}{3}$; $\sec \theta = \frac{5}{4}$; $\cot \theta = \frac{4}{3}$ **2.** $\sin \theta = \frac{1}{5}$; $\cos \theta = \frac{2\sqrt{6}}{5}$; $\tan \theta = \frac{\sqrt{6}}{12}$; $\csc \theta = 5$; $\sec \theta = \frac{5\sqrt{6}}{12}$; $\cot \theta = 2\sqrt{6}$

3. $\sqrt{2}$; $\sqrt{2}$; 1   **4.** $\sqrt{3}$; $\frac{\sqrt{3}}{3}$ **5.** $\tan \theta = \frac{2\sqrt{5}}{5}$; $\csc \theta = \frac{3}{2}$; $\sec \theta = \frac{3\sqrt{5}}{5}$; $\cot \theta = \frac{\sqrt{5}}{2}$ **6.** $\frac{\sqrt{3}}{2}$ **7. a.** $\cos 44°$ **b.** $\tan \frac{5\pi}{12}$

8. a. 0.9553   **b.** 1.0025   **9.** 333.9 yd   **10.** 54°

Concept and Vocabulary Check

1. $\sin \theta = \frac{a}{c}$; $\cos \theta = \frac{b}{c}$; $\tan \theta = \frac{a}{b}$; $\csc \theta = \frac{c}{a}$; $\sec \theta = \frac{c}{b}$; $\cot \theta = \frac{b}{a}$ **2.** opposite; adjacent to; hypotenuse   **3.** true   **4.** $\sin \theta$; $\cos \theta$; $\tan \theta$
5. $\tan \theta$; $\cot \theta$ **6.** 1; $\sec^2 \theta$; $\csc^2 \theta$ **7.** $\sin \theta$; $\tan \theta$; $\sec \theta$

Exercise Set 1.2

1. 15; $\sin \theta = \frac{3}{5}$; $\cos \theta = \frac{4}{5}$; $\tan \theta = \frac{3}{4}$; $\csc \theta = \frac{5}{3}$; $\sec \theta = \frac{5}{4}$; $\cot \theta = \frac{4}{3}$ **3.** 20; $\sin \theta = \frac{20}{29}$; $\cos \theta = \frac{21}{29}$; $\tan \theta = \frac{20}{21}$; $\csc \theta = \frac{29}{20}$; $\sec \theta = \frac{29}{21}$; $\cot \theta = \frac{21}{20}$
5. 24; $\sin \theta = \frac{5}{13}$; $\cos \theta = \frac{12}{13}$; $\tan \theta = \frac{5}{12}$; $\csc \theta = \frac{13}{5}$; $\sec \theta = \frac{13}{12}$; $\cot \theta = \frac{12}{5}$ **7.** 28; $\sin \theta = \frac{4}{5}$; $\cos \theta = \frac{3}{5}$; $\tan \theta = \frac{4}{3}$; $\csc \theta = \frac{5}{4}$; $\sec \theta = \frac{5}{3}$; $\cot \theta = \frac{3}{4}$
9. $\frac{\sqrt{3}}{2}$ **11.** $\sqrt{2}$ **13.** $\sqrt{3}$ **15.** 0 **17.** $\tan \theta = \frac{8}{15}$; $\csc \theta = \frac{17}{8}$; $\sec \theta = \frac{17}{15}$; $\cot \theta = \frac{15}{8}$ **19.** $\tan \theta = \frac{\sqrt{2}}{4}$; $\csc \theta = 3$; $\sec \theta = \frac{3\sqrt{2}}{4}$; $\cot \theta = 2\sqrt{2}$
21. $\frac{\sqrt{13}}{7}$ **23.** $\frac{5}{8}$ **25.** 1 **27.** 1 **29.** 1 **31.** $\cos 83°$ **33.** $\sec 65°$ **35.** $\cot \frac{7\pi}{18}$ **37.** $\sin \frac{\pi}{10}$ **39.** 0.6157 **41.** 0.6420
43. 3.4203 **45.** 0.9511 **47.** 3.7321 **49.** 188 cm   **51.** 182 in.   **53.** 41 m   **55.** $17°$ **57.** $78°$ **59.** 1.147 radians   **61.** 0.395 radian
63. 0 **65.** 2 **67.** 1 **69.** $\frac{2\sqrt{3} - 1}{2}$ **71.** $\frac{1}{4}$ **73.** 529 yd   **75.** $36°$ **77.** 2879 ft   **79.** $37°$ **93.** $0.92106, -0.19735$; $0.95534, -0.148878$;
$0.98007, -0.099667$; $0.99500, -0.04996$; $0.99995, -0.005$; $0.9999995, -0.0005$; $0.999999995, -0.00005$; $0.99999999995, -0.000005$; $\frac{\cos \theta - 1}{\theta}$ approaches
0 as θ approaches 0.   **95.** does not make sense   **97.** makes sense   **99.** true   **101.** false   **103.** As θ approaches $90°$, $\tan \theta$ increases without
bound.   **105. a.** $\frac{y}{r}$ **b.** $\frac{4}{5}$; positive   **106. a.** $\frac{x}{r}$ **b.** $-\frac{3\sqrt{34}}{34}$; negative   **107. a.** $15°$ **b.** $\frac{\pi}{6}$

Section 1.3

Check Point Exercises

1. $\sin \theta = -\frac{3\sqrt{10}}{10}$; $\cos \theta = \frac{\sqrt{10}}{10}$; $\tan \theta = -3$; $\csc \theta = -\frac{\sqrt{10}}{3}$; $\sec \theta = \sqrt{10}$; $\cot \theta = -\frac{1}{3}$ **2. a.** 1; undefined   **b.** 0; 1 **c.** -1; undefined
d. 0; -1 **3.** quadrant III   **4.** $\frac{\sqrt{10}}{10}$, $-\frac{\sqrt{10}}{3}$ **5. a.** $30°$ **b.** $\frac{\pi}{4}$ **c.** $60°$ **d.** 0.46 **6. a.** $55°$ **b.** $\frac{\pi}{4}$ **c.** $\frac{\pi}{3}$
7. a. $-\frac{\sqrt{3}}{2}$ **b.** 1 **c.** $\frac{2\sqrt{3}}{3}$ **8. a.** $-\frac{\sqrt{3}}{2}$ **b.** $\frac{\sqrt{3}}{2}$

Concept and Vocabulary Check

1. $\sin \theta = \frac{y}{r}$; $\cos \theta = \frac{x}{r}$; $\tan \theta = \frac{y}{x}$; $\csc \theta = \frac{r}{y}$; $\sec \theta = \frac{r}{x}$; $\cot \theta = \frac{x}{y}$ **2.** undefined when $x = 0$: $\tan \theta$ and $\sec \theta$; undefined when
$y = 0$: $\cot \theta$ and $\csc \theta$; do not depend on r: $\tan \theta$ and $\cot \theta$ **3.** $\sin \theta$; $\csc \theta$ **4.** $\tan \theta$; $\cot \theta$ **5.** $\cos \theta$; $\sec \theta$ **6.** terminal; x
7. a. $180° - \theta$ **b.** $\theta - 180°$ **c.** $360° - \theta$

Exercise Set 1.3

1. $\sin \theta = \frac{3}{5}$; $\cos \theta = -\frac{4}{5}$; $\tan \theta = -\frac{3}{4}$; $\csc \theta = \frac{5}{3}$; $\sec \theta = -\frac{5}{4}$; $\cot \theta = -\frac{4}{3}$ **3.** $\sin \theta = \frac{3\sqrt{13}}{13}$; $\cos \theta = \frac{2\sqrt{13}}{13}$; $\tan \theta = \frac{3}{2}$; $\csc \theta = \frac{\sqrt{13}}{3}$; $\sec \theta = \frac{\sqrt{13}}{2}$;
$\cot \theta = \frac{2}{3}$ **5.** $\sin \theta = -\frac{\sqrt{2}}{2}$; $\cos \theta = \frac{\sqrt{2}}{2}$; $\tan \theta = -1$; $\csc \theta = -\sqrt{2}$; $\sec \theta = \sqrt{2}$; $\cot \theta = -1$ **7.** $\sin \theta = -\frac{5\sqrt{29}}{29}$; $\cos \theta = -\frac{2\sqrt{29}}{29}$; $\tan \theta = \frac{5}{2}$;
$\csc \theta = -\frac{\sqrt{29}}{5}$; $\sec \theta = -\frac{\sqrt{29}}{2}$; $\cot \theta = \frac{2}{5}$ **9.** -1 **11.** -1 **13.** undefined   **15.** 0 **17.** quadrant I   **19.** quadrant III
21. quadrant II   **23.** $\sin \theta = -\frac{4}{5}$; $\tan \theta = \frac{4}{3}$; $\csc \theta = -\frac{5}{4}$; $\sec \theta = -\frac{5}{3}$; $\cot \theta = \frac{3}{4}$
25. $\cos \theta = -\frac{12}{13}$; $\tan \theta = -\frac{5}{12}$; $\csc \theta = \frac{13}{5}$; $\sec \theta = -\frac{13}{12}$; $\cot \theta = -\frac{12}{5}$ **27.** $\sin \theta = -\frac{15}{17}$; $\tan \theta = -\frac{15}{8}$; $\csc \theta = -\frac{17}{15}$; $\sec \theta = \frac{17}{8}$; $\cot \theta = -\frac{8}{15}$
29. $\sin \theta = \frac{2\sqrt{13}}{13}$; $\cos \theta = -\frac{3\sqrt{13}}{13}$; $\csc \theta = \frac{\sqrt{13}}{2}$; $\sec \theta = -\frac{\sqrt{13}}{3}$; $\cot \theta = -\frac{3}{2}$ **31.** $\sin \theta = -\frac{4}{5}$; $\cos \theta = -\frac{3}{5}$; $\csc \theta = -\frac{5}{4}$; $\sec \theta = -\frac{5}{3}$; $\cot \theta = \frac{3}{4}$
33. $\sin \theta = -\frac{2\sqrt{2}}{3}$; $\cos \theta = -\frac{1}{3}$; $\tan \theta = 2\sqrt{2}$; $\csc \theta = -\frac{3\sqrt{2}}{4}$; $\cot \theta = \frac{\sqrt{2}}{4}$ **35.** $20°$ **37.** $25°$ **39.** $5°$ **41.** $\frac{\pi}{4}$ **43.** $\frac{\pi}{6}$ **45.** $30°$
47. $25°$ **49.** 1.56 **51.** $25°$ **53.** $\frac{\pi}{6}$ **55.** $\frac{\pi}{4}$ **57.** $\frac{\pi}{4}$ **59.** $\frac{\pi}{6}$ **61.** $-\frac{\sqrt{2}}{2}$ **63.** $\frac{\sqrt{3}}{3}$ **65.** $\sqrt{3}$ **67.** $\frac{\sqrt{3}}{2}$ **69.** -2 **71.** 1
73. $\frac{\sqrt{3}}{2}$ **75.** -1 **77.** $-\sqrt{2}$ **79.** $\sqrt{3}$ **81.** $\frac{\sqrt{2}}{2}$ **83.** $\frac{\sqrt{3}}{3}$ **85.** $\frac{\sqrt{3}}{2}$ **87.** $\frac{1 - \sqrt{3}}{2}$ **89.** $\frac{-\sqrt{6} - \sqrt{2}}{4}$ or $-\frac{\sqrt{6} + \sqrt{2}}{4}$
91. $-\frac{3}{2}$ **93.** $\frac{-1 - \sqrt{3}}{2}$ or $-\frac{1 + \sqrt{3}}{2}$ **95.** 1 **97.** $\frac{2\sqrt{2} - 4}{\pi}$ **99.** $\frac{\pi}{4}$ and $\frac{3\pi}{4}$ **101.** $\frac{5\pi}{4}$ and $\frac{7\pi}{4}$ **103.** $\frac{2\pi}{3}$ and $\frac{5\pi}{3}$
111. does not make sense   **113.** makes sense

114.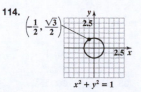

115. domain: $\{x|-1 \le x \le 1\}$ or $[-1, 1]$; range: $\{y|-1 \le y \le 1\}$ or $[-1, 1]$

116. a. $\frac{\sqrt{2}}{2}; -\frac{\sqrt{2}}{2}; \frac{\sqrt{3}}{2}; -\frac{\sqrt{3}}{2}$; no   **b.** $\frac{\sqrt{2}}{2}; \frac{\sqrt{2}}{2}; \frac{1}{2}; \frac{1}{2}$; no

Mid-Chapter 1 Check Point

1. $\frac{\pi}{18}$ **2.** $-\frac{7\pi}{12}$ **3.** $75°$ **4.** $-117°$

5. a. $\frac{5\pi}{3}$ **6. a.** $\frac{5\pi}{4}$ **7. a.** $150°$ **8.** $\sin \theta = \frac{5}{6}; \cos \theta = \frac{\sqrt{11}}{6}; \tan \theta = \frac{5\sqrt{11}}{11}; \csc \theta = \frac{6}{5}; \sec \theta = \frac{6\sqrt{11}}{11}; \cot \theta = \frac{\sqrt{11}}{5}$

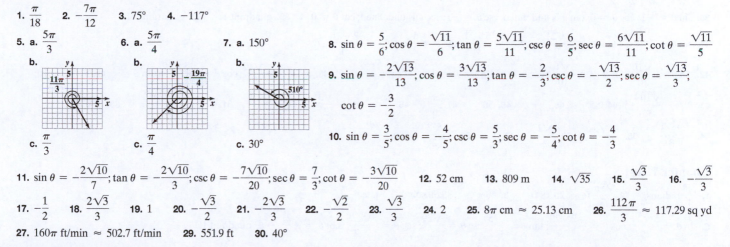

b.   **b.**   **b.**   **9.** $\sin \theta = -\frac{2\sqrt{13}}{13}; \cos \theta = \frac{3\sqrt{13}}{13}; \tan \theta = -\frac{2}{3}; \csc \theta = -\frac{\sqrt{13}}{2}; \sec \theta = \frac{\sqrt{13}}{3};$

$\cot \theta = -\frac{3}{2}$

c. $\frac{\pi}{3}$ **c.** $\frac{\pi}{4}$ **c.** $30°$ **10.** $\sin \theta = \frac{3}{5}; \cos \theta = -\frac{4}{5}; \csc \theta = \frac{5}{3}; \sec \theta = -\frac{5}{4}; \cot \theta = -\frac{4}{3}$

11. $\sin \theta = -\frac{2\sqrt{10}}{7}; \tan \theta = -\frac{2\sqrt{10}}{3}; \csc \theta = -\frac{7\sqrt{10}}{20}; \sec \theta = \frac{7}{3}; \cot \theta = -\frac{3\sqrt{10}}{20}$ **12.** 52 cm   **13.** 809 m   **14.** $\sqrt{35}$ **15.** $\frac{\sqrt{3}}{3}$ **16.** $-\frac{\sqrt{3}}{3}$

17. $-\frac{1}{2}$ **18.** $\frac{2\sqrt{3}}{3}$ **19.** 1 **20.** $-\frac{\sqrt{3}}{2}$ **21.** $-\frac{2\sqrt{3}}{3}$ **22.** $-\frac{\sqrt{2}}{2}$ **23.** $\frac{\sqrt{3}}{3}$ **24.** 2 **25.** 8π cm ≈ 25.13 cm   **26.** $\frac{112\pi}{3} \approx 117.29$ sq yd

27. 160π ft/min ≈ 502.7 ft/min   **29.** 551.9 ft   **30.** $40°$

Section 1.4

Check Point Exercises

1. $\sin t = \frac{1}{2}; \cos t = \frac{\sqrt{3}}{2}; \tan t = \frac{\sqrt{3}}{3}; \csc t = 2; \sec t = \frac{2\sqrt{3}}{3}; \cot t = \sqrt{3}$ **2.** $\sin \pi = 0; \cos \pi = -1; \tan \pi = 0; \csc \pi$ is undefined;

$\sec \pi = -1; \cot \pi$ is undefined   **3. a.** $\frac{1}{2}$ **b.** $-\frac{\sqrt{3}}{3}$ **4. a.** $\frac{\sqrt{2}}{2}$ **b.** $\frac{\sqrt{3}}{2}$

Concept and Vocabulary Check

1. intercepted arc   **2.** cosine; sine   **3.** sine; cosine; $(-\infty, \infty)$ **4.** $1; -1; [-1, 1]$ **5.** $\cos t; \sec t;$ even   **6.** $-\sin t; -\csc t; -\tan t; -\cot t;$ odd
7. periodic; period   **8.** $\sin t; \cos t;$ periodic; 2π **9.** $\tan t; \cot t;$ periodic; π

Exercise Set 1.4

1. $\sin t = \frac{8}{17}; \cos t = -\frac{15}{17}; \tan t = -\frac{8}{15}; \csc t = \frac{17}{8}; \sec t = -\frac{17}{15}; \cot t = -\frac{15}{8}$ **3.** $\sin t = -\frac{\sqrt{2}}{2}; \cos t = \frac{\sqrt{2}}{2}; \tan t = -1; \csc t = -\sqrt{2}; \sec t = \sqrt{2};$

$\cot t = -1$ **5.** $\frac{1}{2}$ **7.** $-\frac{\sqrt{3}}{2}$ **9.** 0 **11.** -2 **13.** $\frac{2\sqrt{3}}{3}$ **15.** -1 **17.** undefined   **19. a.** $\frac{\sqrt{3}}{2}$ **b.** $\frac{\sqrt{3}}{2}$ **21. a.** $\frac{1}{2}$ **b.** $-\frac{1}{2}$

23. a. $-\sqrt{3}$ **b.** $\sqrt{3}$ **25. a.** $\frac{\sqrt{2}}{2}$ **b.** $\frac{\sqrt{2}}{2}$ **27. a.** 0 **b.** 0 **29. a.** 0 **b.** 0 **31. a.** $-\frac{\sqrt{2}}{2}$ **b.** $-\frac{\sqrt{2}}{2}$ **33.** $-2a$ **35.** $3b$

37. $a - b + c$ **39.** $-a - b + c$ **41.** $3a + 2b - 2c$ **43. a.** 12 hr   **b.** 20.3 hr   **c.** 3.7 hr   **45. a.** $1; 0; -1; 0; 1$ **b.** 28 days

53. makes sense   **55.** does not make sense   **57.** 0 **59.** $-\frac{1}{4}$ **61.** $\frac{1}{2}; 0; -\frac{1}{2}; 0; \frac{1}{2}$ **62.** $0; 4; 0; -4; 0$

63. $0; \frac{3}{2}; 3; \frac{3}{2}; 0; -\frac{3}{2}; -3; -\frac{3}{2}; 0$

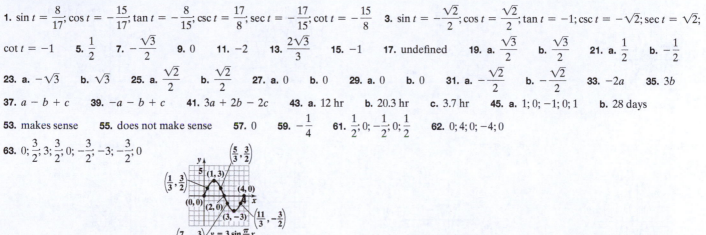

Chapter 1 Review Exercises

1. 4.5 radians   **2.** $\frac{\pi}{12}$ radian   **3.** $\frac{2\pi}{3}$ radians   **4.** $\frac{7\pi}{4}$ radians   **5.** $300°$ **6.** $252°$ **7.** $-150°$

8.   **9.**   **10.**   **11.**   **12.**

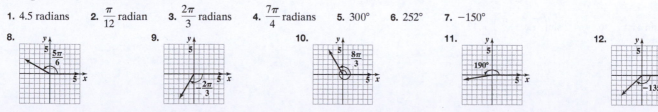

13. 40°   **14.** 275°   **15.** $\dfrac{5\pi}{4}$ **16.** $\dfrac{7\pi}{6}$ **17.** $\dfrac{4\pi}{3}$ **18.** $\dfrac{15\pi}{2}$ ft ≈ 23.56 ft   **19.** $3\pi \approx 9.42$ sq yd   **20.** 20.6π radians per min

21. 42,412 ft per min   **22.** $\sin\theta = \dfrac{5\sqrt{89}}{89}$; $\cos\theta = \dfrac{8\sqrt{89}}{89}$; $\tan\theta = \dfrac{5}{8}$; $\csc\theta = \dfrac{\sqrt{89}}{5}$; $\sec\theta = \dfrac{\sqrt{89}}{8}$; $\cot\theta = \dfrac{8}{5}$ **23.** $\dfrac{7}{2}$ **24.** $-\dfrac{1}{2}$ **25.** 1

26. 1   **27.** $\dfrac{\sqrt{21}}{7}$ **28.** $\cos 20°$ **29.** $\sin\dfrac{\pi}{6}$ **30.** 42 mm   **31.** 23 cm   **32.** 37 in.   **33.** $\sqrt{15}$ **34.** 772 ft   **35.** 31 m   **36.** 56°

37. $\sin\theta = -\dfrac{5\sqrt{26}}{26}$; $\cos\theta = -\dfrac{\sqrt{26}}{26}$; $\tan\theta = 5$; $\csc\theta = -\dfrac{\sqrt{26}}{5}$; $\sec\theta = -\sqrt{26}$; $\cot\theta = \dfrac{1}{5}$

38. $\sin\theta = -1$; $\cos\theta = 0$; $\tan\theta$ is undefined; $\csc\theta = -1$; $\sec\theta$ is undefined; $\cot\theta = 0$ **39.** quadrant I   **40.** quadrant III

41. $\sin\theta = -\dfrac{\sqrt{21}}{5}$; $\tan\theta = -\dfrac{\sqrt{21}}{2}$; $\csc\theta = -\dfrac{5\sqrt{21}}{21}$; $\sec\theta = \dfrac{5}{2}$; $\cot\theta = -\dfrac{2\sqrt{21}}{21}$

42. $\sin\theta = \dfrac{\sqrt{10}}{10}$; $\cos\theta = -\dfrac{3\sqrt{10}}{10}$; $\csc\theta = \sqrt{10}$; $\sec\theta = -\dfrac{\sqrt{10}}{3}$; $\cot\theta = -3$ **43.** $\sin\theta = -\dfrac{\sqrt{10}}{10}$; $\cos\theta = -\dfrac{3\sqrt{10}}{10}$; $\tan\theta = \dfrac{1}{3}$; $\csc\theta = -\sqrt{10}$;

$\sec\theta = -\dfrac{\sqrt{10}}{3}$ **44.** 85°   **45.** $\dfrac{3\pi}{8}$ **46.** 50°   **47.** $\dfrac{\pi}{6}$ **48.** $\dfrac{\pi}{3}$ **49.** $-\dfrac{\sqrt{3}}{2}$ **50.** $-\sqrt{3}$ **51.** $\sqrt{2}$ **52.** $\dfrac{\sqrt{3}}{2}$ **53.** $-\sqrt{3}$

54. $-\dfrac{2\sqrt{3}}{3}$ **55.** $-\dfrac{\sqrt{3}}{2}$ **56.** $\dfrac{\sqrt{2}}{2}$ **57.** 1   **58.** $-\dfrac{\sqrt{3}}{2}$ **59.** $\dfrac{\sqrt{3}}{2}$

Chapter 1 Test

1. $\dfrac{3\pi}{4}$ radians   **2.** $\dfrac{25\pi}{3}$ ft ≈ 26.18 ft   **3.** $60\pi \approx 188.50$ sq yd   **4. a.** $\dfrac{4\pi}{3}$ **b.** $\dfrac{\pi}{3}$

5. $\sin\theta = \dfrac{5\sqrt{29}}{29}$; $\cos\theta = -\dfrac{2\sqrt{29}}{29}$; $\tan\theta = -\dfrac{5}{2}$; $\csc\theta = \dfrac{\sqrt{29}}{5}$; $\sec\theta = -\dfrac{\sqrt{29}}{2}$; $\cot\theta = -\dfrac{2}{5}$ **6.** quadrant III

7. $\sin\theta = -\dfrac{2\sqrt{2}}{3}$; $\tan\theta = -2\sqrt{2}$; $\csc\theta = -\dfrac{3\sqrt{2}}{4}$; $\sec\theta = 3$; $\cot\theta = -\dfrac{\sqrt{2}}{4}$ **8.** $\dfrac{\sqrt{3}}{6}$ **9.** $-\sqrt{3}$ **10.** $-\dfrac{\sqrt{2}}{2}$ **11.** -2 **12.** $\dfrac{\sqrt{3}}{3}$

13. $\sqrt{3}$ **14. a.** $-a + b$ or $b - a$ **b.** $\dfrac{a}{b} - \dfrac{1}{b}$ or $\dfrac{a-1}{b}$ **15.** 23 yd   **16.** 36.1°   **17.** Trigonometric functions are periodic.

CHAPTER 2

Section 2.1
Check Point Exercises

1. 3   **2.** $\dfrac{1}{2}$ **3.** $2; 4\pi$ **4.** $3; \pi; \dfrac{\pi}{6}$

5. $4; 2$ **6.** $\dfrac{3}{2}; \pi; -\dfrac{\pi}{2}$ **7.**   **8.**

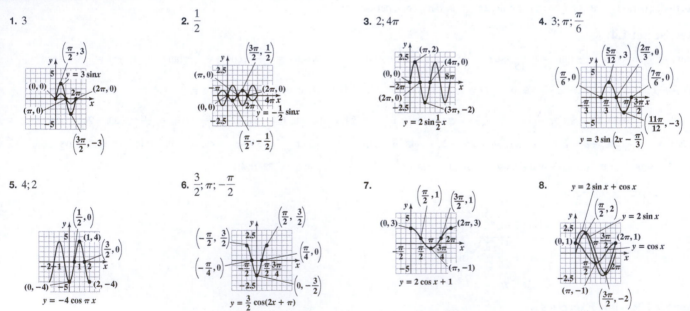

9. $y = 4\sin 4x$ **10.** $y = 2\sin\left(\dfrac{\pi}{6}x - \dfrac{\pi}{2}\right) + 12$

Concept and Vocabulary Check

1. $|A|$; $\dfrac{2\pi}{B}$ **2.** 3; 4π **3.** π, 0; $\dfrac{\pi}{4}$, $\dfrac{\pi}{2}$, $\dfrac{3\pi}{4}$, π **4.** $\dfrac{C}{B}$; right; left   **5.** $|A|$; $\dfrac{2\pi}{B}$ **6.** $\dfrac{1}{2}$; $\dfrac{2\pi}{3}$ **7.** false   **8.** true   **9.** true   **10.** true

Exercise Set 2.1

1. 4 **3.** $\dfrac{1}{3}$ **5.** 3 **7.** 1; π

9. 3; 4π **11.** 4; 2 **13.** 3; 1 **15.** 1; 3π

17. 1; 2π; π **19.** 1; π; $\dfrac{\pi}{2}$ **21.** 3; π; $\dfrac{\pi}{2}$ **23.** $\dfrac{1}{2}$; 2π; $-\dfrac{\pi}{2}$

25. 2; π; $-\dfrac{\pi}{4}$ **27.** 3; 2; $-\dfrac{2}{\pi}$ **29.** 2; 1; -2 **31.** 2

33. 2 **35.** 1; π **37.** 4; 1 **39.** 4; 4π

41. $\dfrac{1}{2}$; 6 **43.** 1; 2π, $\dfrac{\pi}{2}$ **45.** 3; π; $\dfrac{\pi}{2}$ **47.** $\dfrac{1}{2}$; $\dfrac{2\pi}{3}$; $-\dfrac{\pi}{6}$

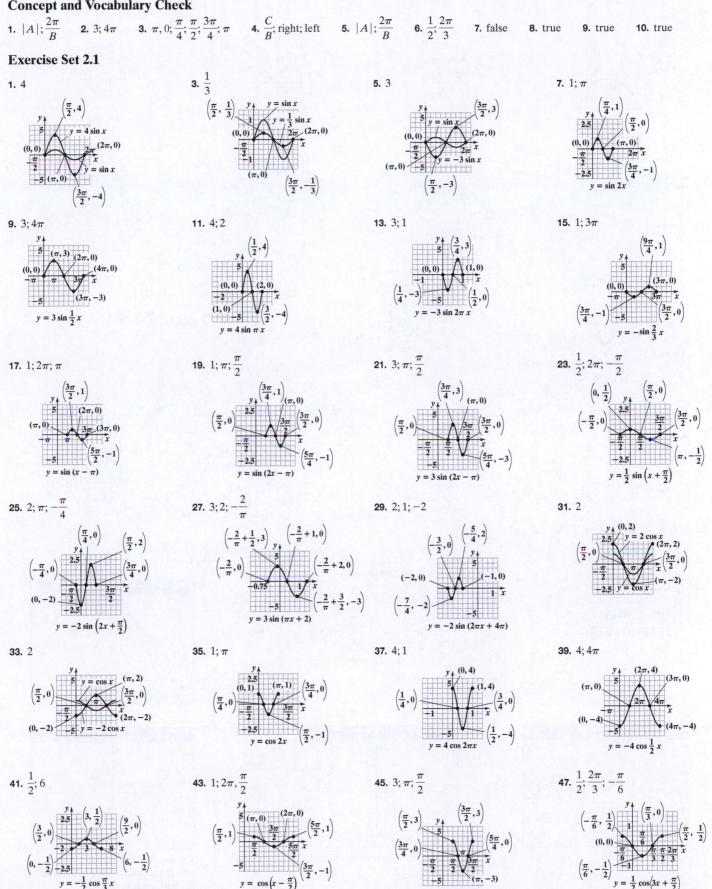

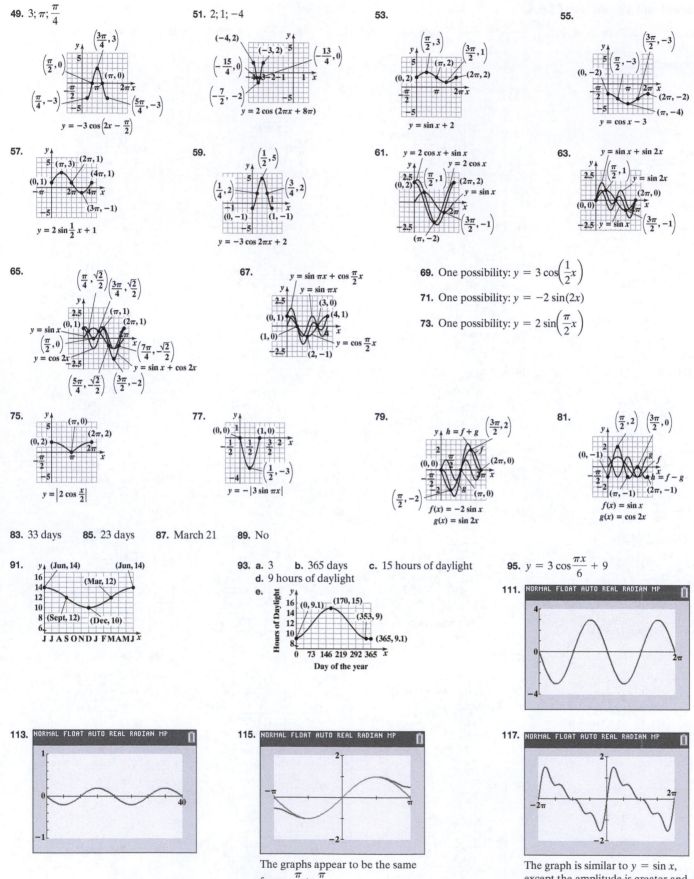

49. $3; \pi; \dfrac{\pi}{4}$

$y = -3 \cos\left(2x - \dfrac{\pi}{2}\right)$

51. $2; 1; -4$

$y = 2 \cos(2\pi x + 8\pi)$

53. $y = \sin x + 2$

55. $y = \cos x - 3$

57. $y = 2 \sin \dfrac{1}{2}x + 1$

59. $y = -3 \cos 2\pi x + 2$

61. $y = 2 \cos x + \sin x$

63. $y = \sin x + \sin 2x$

65. $y = \sin x + \cos 2x$

67. $y = \sin \pi x + \cos \dfrac{\pi}{2}x$

69. One possibility: $y = 3 \cos\left(\dfrac{1}{2}x\right)$

71. One possibility: $y = -2 \sin(2x)$

73. One possibility: $y = 2 \sin\left(\dfrac{\pi}{2}x\right)$

75. $y = \left|2 \cos \dfrac{x}{2}\right|$

77. $y = -|3 \sin \pi x|$

79. $f(x) = -2 \sin x$; $g(x) = \sin 2x$

81. $f(x) = \sin x$; $g(x) = \cos 2x$

83. 33 days   **85.** 23 days   **87.** March 21   **89.** No

91.

93. a. 3   **b.** 365 days   **c.** 15 hours of daylight
d. 9 hours of daylight
e.

95. $y = 3 \cos \dfrac{\pi x}{6} + 9$

111.

113.

115. The graphs appear to be the same from $-\dfrac{\pi}{2}$ to $\dfrac{\pi}{2}$.

117. The graph is similar to $y = \sin x$, except the amplitude is greater and the curve is less smooth.

119. a & c.

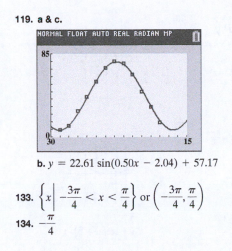

b. $y = 22.61 \sin(0.50x - 2.04) + 57.17$

121. makes sense   **123.** makes sense

125. a. range: $[-5, 1]$; $\left[-\dfrac{\pi}{6}, \dfrac{23\pi}{6}, \dfrac{\pi}{6}\right]$ by $[-5, 1, 1]$

b. range: $[-3, -1]$; $\left[-\dfrac{\pi}{6}, \dfrac{7\pi}{6}, \dfrac{\pi}{6}\right]$ by $[-3, -1, 1]$

127.

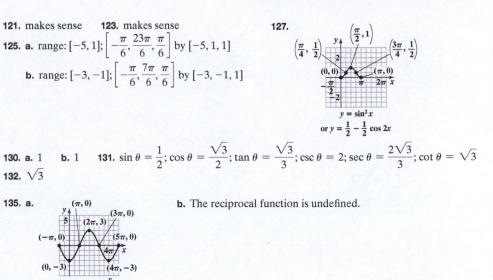

$y = \sin^2 x$
or $y = \dfrac{1}{2} - \dfrac{1}{2}\cos 2x$

130. a. 1   **b.** 1   **131.** $\sin\theta = \dfrac{1}{2}$; $\cos\theta = \dfrac{\sqrt{3}}{2}$; $\tan\theta = \dfrac{\sqrt{3}}{3}$; $\csc\theta = 2$; $\sec\theta = \dfrac{2\sqrt{3}}{3}$; $\cot\theta = \sqrt{3}$

132. $\sqrt{3}$

133. $\left\{x \,\middle|\, -\dfrac{3\pi}{4} < x < \dfrac{\pi}{4}\right\}$ or $\left(-\dfrac{3\pi}{4}, \dfrac{\pi}{4}\right)$

134. $-\dfrac{\pi}{4}$

135. a.

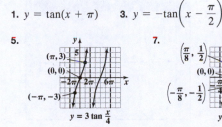

$y = -3\cos\dfrac{x}{2}$

b. The reciprocal function is undefined.

Section 2.2

Check Point Exercises

1. 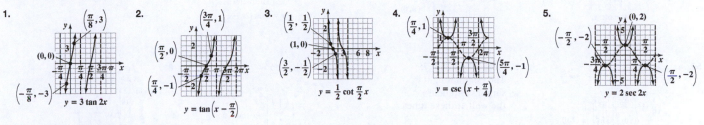 **2.** **3.** **4.** **5.**

$y = 3\tan 2x$ $y = \tan\left(x - \dfrac{\pi}{2}\right)$ $y = \dfrac{1}{2}\cot\dfrac{\pi}{2}x$ $y = \csc\left(x + \dfrac{\pi}{4}\right)$ $y = 2\sec 2x$

Concept and Vocabulary Check

1. $\left(-\dfrac{\pi}{4}, \dfrac{\pi}{4}\right)$; $-\dfrac{\pi}{4}, \dfrac{\pi}{4}$ **2.** $(0, \pi)$; 0; π **3.** $(0, 2)$; 0; 2 **4.** $\left(-\dfrac{\pi}{4}, \dfrac{3\pi}{4}\right)$; $-\dfrac{\pi}{4}, \dfrac{3\pi}{4}$ **5.** $3\sin 2x$ **6.** $y = 2\cos \pi x$ **7.** false   **8.** true

Exercise Set 2.2

1. $y = \tan(x + \pi)$ **3.** $y = -\tan\left(x - \dfrac{\pi}{2}\right)$

5. **7.** **9.** **11.** **13.** $y = -\cot x$

$y = 3\tan\dfrac{x}{4}$ $y = \dfrac{1}{2}\tan 2x$ $y = -2\tan\dfrac{1}{2}x$ $y = \tan(x - \pi)$

15. $y = \cot\left(x + \dfrac{\pi}{2}\right)$

17. **19.** **21.** **23.**

$y = 2\cot x$ $y = \dfrac{1}{2}\cot 2x$ $y = -3\cot\dfrac{\pi}{2}x$ $y = 3\cot\left(x + \dfrac{\pi}{2}\right)$

25. $y = -\dfrac{1}{2}\csc\dfrac{x}{2}$ **27.** $y = \dfrac{1}{2}\sec 2\pi x$ **29.** **31.**

$y = -\dfrac{1}{2}\csc\dfrac{x}{2}$ $y = \dfrac{1}{2}\sec 2\pi x$ $y = 3\sin x$ $y = \dfrac{1}{2}\sin\dfrac{x}{2}$

$y = 3\csc x$ $y = \dfrac{1}{2}\csc\dfrac{x}{2}$

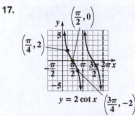
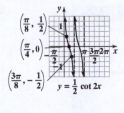
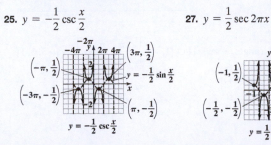
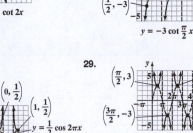
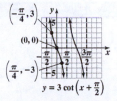
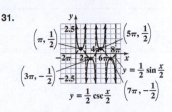
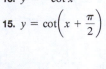
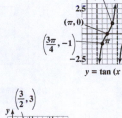
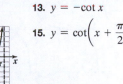

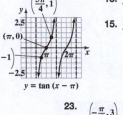
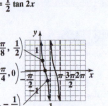

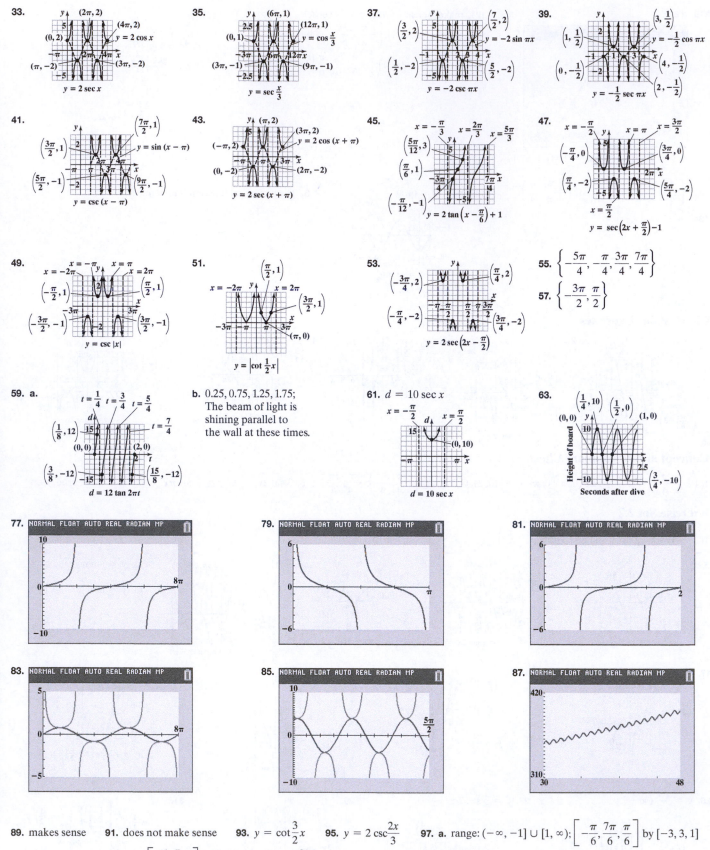

89. makes sense   **91.** does not make sense   **93.** $y = \cot \frac{3}{2}x$ **95.** $y = 2 \csc \frac{2x}{3}$ **97. a.** range: $(-\infty, -1] \cup [1, \infty)$; $\left[-\frac{\pi}{6}, \frac{7\pi}{6}, \frac{\pi}{6} \right]$ by $[-3, 3, 1]$

b. range: $(-\infty, -3] \cup [3, \infty)$; $\left[-\frac{1}{2}, \frac{7}{2}, 1 \right]$ by $[-6, 6, 1]$ **99.** $\frac{20\pi}{3} \approx 20.94$ in.   **100.** $\frac{\pi}{3}$ **101.** 84 m

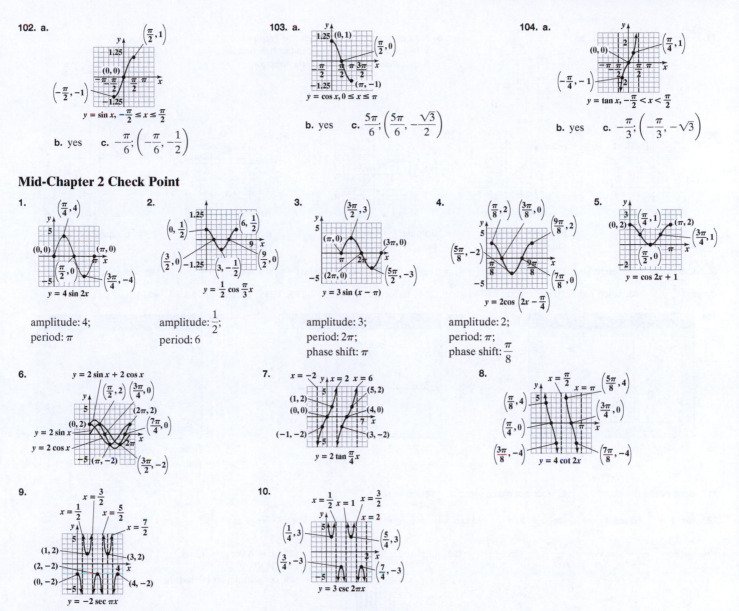

102. a.

$y = \sin x, -\frac{\pi}{2} \le x \le \frac{\pi}{2}$

b. yes   **c.** $-\frac{\pi}{6}; \left(-\frac{\pi}{6}, -\frac{1}{2}\right)$

103. a.

$y = \cos x, 0 \le x \le \pi$

b. yes   **c.** $\frac{5\pi}{6}; \left(\frac{5\pi}{6}, -\frac{\sqrt{3}}{2}\right)$

104. a.

$y = \tan x, -\frac{\pi}{2} < x < \frac{\pi}{2}$

b. yes   **c.** $-\frac{\pi}{3}; \left(-\frac{\pi}{3}, -\sqrt{3}\right)$

Mid-Chapter 2 Check Point

1.

$y = 4 \sin 2x$

amplitude: 4;
period: π

2.

$y = \frac{1}{2} \cos \frac{\pi}{3} x$

amplitude: $\frac{1}{2}$;
period: 6

3.

$y = 3 \sin (x - \pi)$

amplitude: 3;
period: 2π;
phase shift: π

4.

$y = 2\cos\left(2x - \frac{\pi}{4}\right)$

amplitude: 2;
period: π;
phase shift: $\frac{\pi}{8}$

5.

$y = \cos 2x + 1$

6.

$y = 2 \sin x + 2 \cos x$
$y = 2 \sin x$
$y = 2 \cos x$

7.

$y = 2 \tan \frac{\pi}{4} x$

8.

$y = 4 \cot 2x$

9.

$y = -2 \sec \pi x$

10.

$y = 3 \csc 2\pi x$

Section 2.3

Check Point Exercises

1. $\frac{\pi}{3}$ **2.** $-\frac{\pi}{4}$ **3.** $\frac{2\pi}{3}$ **4.** $-\frac{\pi}{4}$ **5. a.** 1.2310   **b.** -1.5429 **6. a.** 0.7   **b.** 0   **c.** not defined   **7.** $\frac{3}{5}$ **8.** $\frac{\sqrt{3}}{2}$ **9.** $\sqrt{x^2 + 1}$

Concept and Vocabulary Check

1. $-\frac{\pi}{2} \le x \le \frac{\pi}{2}; \sin^{-1} x$ **2.** $0 \le x \le \pi; \cos^{-1} x$ **3.** $-\frac{\pi}{2} < x < \frac{\pi}{2}; \tan^{-1} x$ **4.** $[-1, 1]; \left[-\frac{\pi}{2}, \frac{\pi}{2}\right]$ **5.** $[-1, 1]; [0, \pi]$ **6.** $(-\infty, \infty); \left(-\frac{\pi}{2}, \frac{\pi}{2}\right)$

7. $\left[-\frac{\pi}{2}, \frac{\pi}{2}\right]$ **8.** $[0, \pi]$ **9.** $\left(-\frac{\pi}{2}, \frac{\pi}{2}\right)$ **10.** false

Exercise Set 2.3

1. $\frac{\pi}{6}$ **3.** $\frac{\pi}{4}$ **5.** $-\frac{\pi}{6}$ **7.** $\frac{\pi}{6}$ **9.** $\frac{3\pi}{4}$ **11.** $\frac{\pi}{2}$ **13.** $\frac{\pi}{6}$ **15.** 0   **17.** $-\frac{\pi}{3}$ **19.** 0.30   **21.** -0.33 **23.** 1.19   **25.** 1.25

27. -1.52 **29.** -1.52 **31.** 0.9   **33.** $\frac{\pi}{3}$ **35.** $\frac{\pi}{6}$ **37.** 125   **39.** $-\frac{\pi}{6}$ **41.** $-\frac{\pi}{3}$ **43.** 0   **45.** not defined   **47.** $\frac{3}{5}$ **49.** $\frac{12}{5}$

51. $-\frac{3}{4}$ **53.** $\frac{\sqrt{2}}{2}$ **55.** $\frac{4\sqrt{15}}{15}$ **57.** $-2\sqrt{2}$ **59.** 2   **61.** $\frac{3\sqrt{13}}{13}$ **63.** $\frac{\sqrt{1 - x^2}}{x}$ **65.** $\sqrt{1 - 4x^2}$ **67.** $\frac{\sqrt{x^2 - 1}}{x}$ **69.** $\frac{\sqrt{3}}{x}$

71. $\dfrac{\sqrt{x^2+4}}{2}$

73. a.

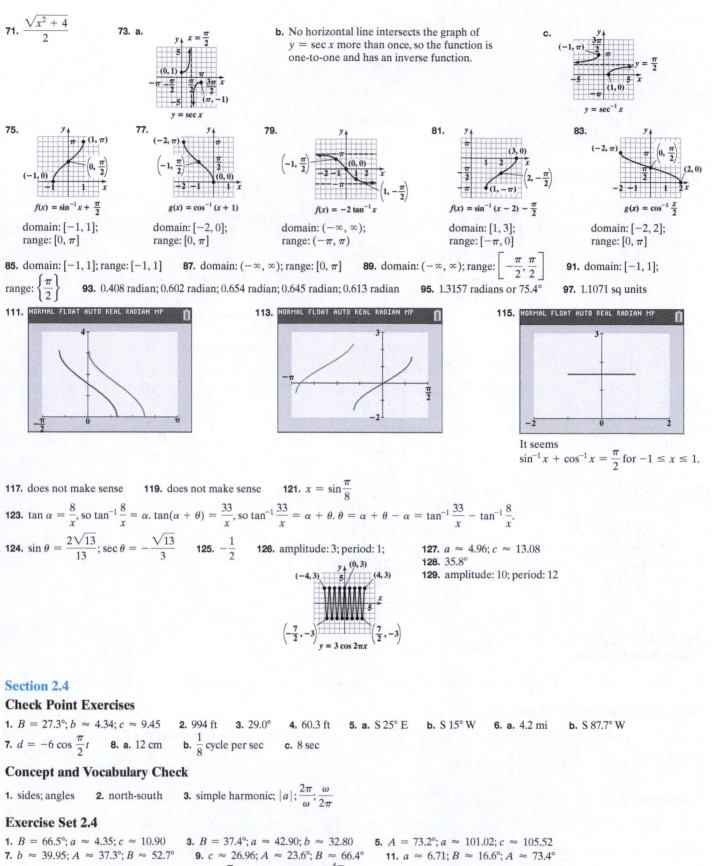

b. No horizontal line intersects the graph of $y = \sec x$ more than once, so the function is one-to-one and has an inverse function.

75. $f(x) = \sin^{-1} x + \dfrac{\pi}{2}$

domain: $[-1, 1]$; range: $[0, \pi]$

77. $g(x) = \cos^{-1}(x+1)$

domain: $[-2, 0]$; range: $[0, \pi]$

79. $f(x) = -2\tan^{-1} x$

domain: $(-\infty, \infty)$; range: $(-\pi, \pi)$

81. $f(x) = \sin^{-1}(x-2) - \dfrac{\pi}{2}$

domain: $[1, 3]$; range: $[-\pi, 0]$

83. $g(x) = \cos^{-1}\dfrac{x}{2}$

domain: $[-2, 2]$; range: $[0, \pi]$

85. domain: $[-1, 1]$; range: $[-1, 1]$ **87.** domain: $(-\infty, \infty)$; range: $[0, \pi]$ **89.** domain: $(-\infty, \infty)$; range: $\left[-\dfrac{\pi}{2}, \dfrac{\pi}{2}\right]$ **91.** domain: $[-1, 1]$;

range: $\left\{\dfrac{\pi}{2}\right\}$ **93.** 0.408 radian; 0.602 radian; 0.654 radian; 0.645 radian; 0.613 radian   **95.** 1.3157 radians or 75.4°   **97.** 1.1071 sq units

111.   **113.**   **115.**

It seems $\sin^{-1} x + \cos^{-1} x = \dfrac{\pi}{2}$ for $-1 \le x \le 1$.

117. does not make sense   **119.** does not make sense   **121.** $x = \sin\dfrac{\pi}{8}$

123. $\tan\alpha = \dfrac{8}{x}$, so $\tan^{-1}\dfrac{8}{x} = \alpha$. $\tan(\alpha + \theta) = \dfrac{33}{x}$, so $\tan^{-1}\dfrac{33}{x} = \alpha + \theta$. $\theta = \alpha + \theta - \alpha = \tan^{-1}\dfrac{33}{x} - \tan^{-1}\dfrac{8}{x}$.

124. $\sin\theta = \dfrac{2\sqrt{13}}{13}$; $\sec\theta = -\dfrac{\sqrt{13}}{3}$ **125.** $-\dfrac{1}{2}$ **126.** amplitude: 3; period: 1;

$y = 3\cos 2\pi x$

127. $a \approx 4.96$; $c \approx 13.08$
128. 35.8°
129. amplitude: 10; period: 12

Section 2.4

Check Point Exercises

1. $B = 27.3°$; $b \approx 4.34$; $c \approx 9.45$ **2.** 994 ft   **3.** 29.0°   **4.** 60.3 ft   **5. a.** S 25° E   **b.** S 15° W   **6. a.** 4.2 mi   **b.** S 87.7° W

7. $d = -6\cos\dfrac{\pi}{2}t$ **8. a.** 12 cm   **b.** $\dfrac{1}{8}$ cycle per sec   **c.** 8 sec

Concept and Vocabulary Check

1. sides; angles   **2.** north-south   **3.** simple harmonic; $|a|$; $\dfrac{2\pi}{\omega}$; $\dfrac{\omega}{2\pi}$

Exercise Set 2.4

1. $B = 66.5°$; $a \approx 4.35$; $c \approx 10.90$ **3.** $B = 37.4°$; $a \approx 42.90$; $b \approx 32.80$ **5.** $A = 73.2°$; $a \approx 101.02$; $c \approx 105.52$
7. $b \approx 39.95$; $A \approx 37.3°$; $B \approx 52.7°$ **9.** $c \approx 26.96$; $A \approx 23.6°$; $B \approx 66.4°$ **11.** $a \approx 6.71$; $B \approx 16.6°$; $A \approx 73.4°$
13. N 15° E   **15.** S 80° W   **17.** $d = -6\cos\dfrac{\pi}{2}t$ **19.** $d = -3\sin\dfrac{4\pi}{3}t$
21. a. 5 in.   **b.** $\dfrac{1}{4}$ cycle per sec   **c.** 4 sec   **23. a.** 6 in.   **b.** 1 cycle per sec   **c.** 1 sec   **25. a.** $\dfrac{1}{2}$ in.   **b.** $\dfrac{1}{\pi}$ or 0.32 cycle per sec   **c.** π or 3.14 sec
27. a. 5 in.   **b.** $\dfrac{1}{3}$ cycle per sec   **c.** 3 sec   **29.** 653 units   **31.** 39 units   **33.** 298 units   **35.** 257 units

37.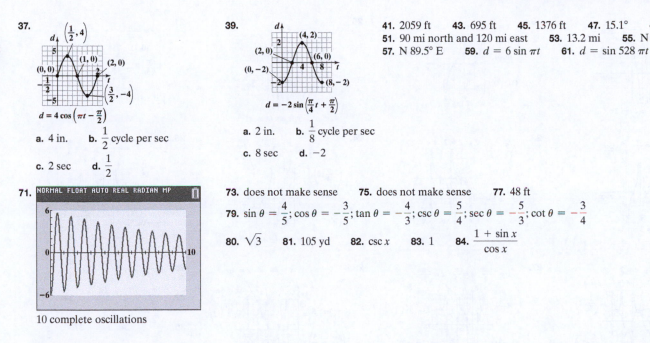

$$d = 4\cos\left(\pi t - \frac{\pi}{2}\right)$$

a. 4 in.   **b.** $\frac{1}{2}$ cycle per sec

c. 2 sec   **d.** $\frac{1}{2}$

39.

$$d = -2\sin\left(\frac{\pi}{4}t + \frac{\pi}{2}\right)$$

a. 2 in.   **b.** $\frac{1}{8}$ cycle per sec

c. 8 sec   **d.** -2

41. 2059 ft   **43.** 695 ft   **45.** 1376 ft   **47.** 15.1°   **49.** 33.7 ft
51. 90 mi north and 120 mi east   **53.** 13.2 mi   **55.** N 53° W
57. N 89.5° E   **59.** $d = 6\sin \pi t$ **61.** $d = \sin 528\pi t$

71.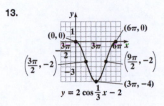

10 complete oscillations

73. does not make sense   **75.** does not make sense   **77.** 48 ft

79. $\sin\theta = \frac{4}{5}$; $\cos\theta = -\frac{3}{5}$; $\tan\theta = -\frac{4}{3}$; $\csc\theta = \frac{5}{4}$; $\sec\theta = -\frac{5}{3}$; $\cot\theta = -\frac{3}{4}$

80. $\sqrt{3}$ **81.** 105 yd   **82.** $\csc x$ **83.** 1   **84.** $\dfrac{1 + \sin x}{\cos x}$

Chapter 2 Review Exercises

1. 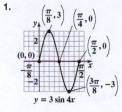 $y = 3\sin 4x$

2. 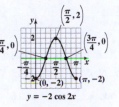 $y = -2\cos 2x$

3. 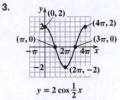 $y = 2\cos\frac{1}{2}x$

4. 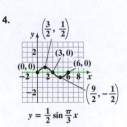 $y = \frac{1}{2}\sin\frac{\pi}{3}x$

5. 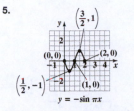 $y = -\sin \pi x$

6. 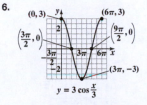 $y = 3\cos\frac{x}{3}$

7. 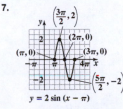 $y = 2\sin(x-\pi)$

8. 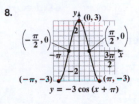 $y = -3\cos(x+\pi)$

9. 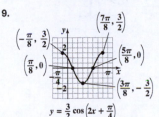 $y = \frac{3}{2}\cos\left(2x+\frac{\pi}{4}\right)$

10. 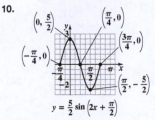 $y = \frac{5}{2}\sin\left(2x+\frac{\pi}{2}\right)$

11. 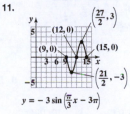 $y = -3\sin\left(\frac{\pi}{3}x - 3\pi\right)$

12. 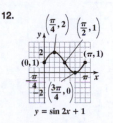 $y = \sin 2x + 1$

13. 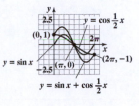 $y = 2\cos\frac{1}{3}x - 2$

14.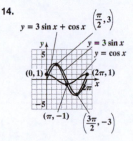

15.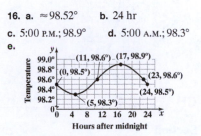

16. a. $\approx 98.52°$ **b.** 24 hr   **c.** 5:00 P.M.; 98.9°   **d.** 5:00 A.M.; 98.3°   **e.**

17. blue: $y = \sin\frac{\pi}{240}x$; red: $y = \sin\frac{\pi}{320}x$

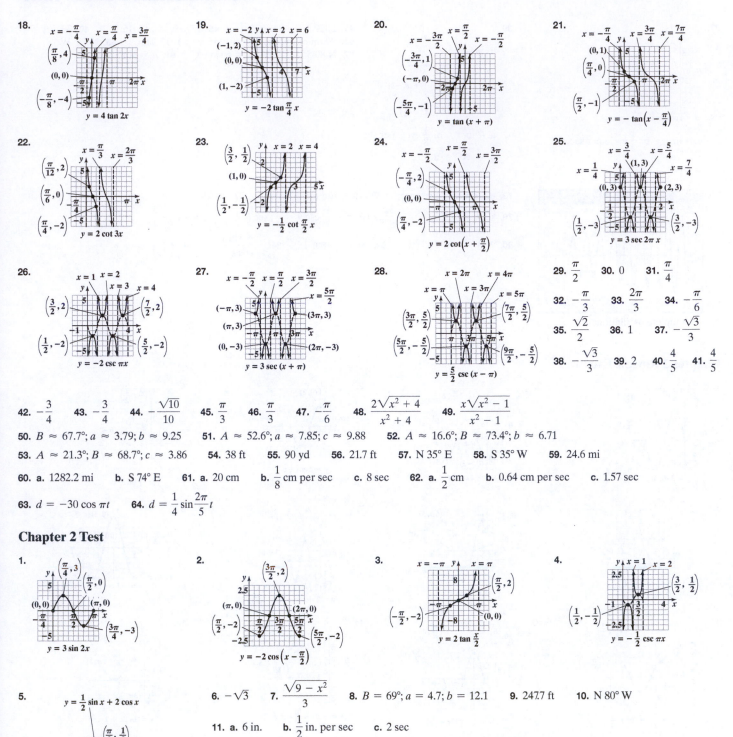

18. $y = 4 \tan 2x$

19. $y = -2 \tan \frac{\pi}{4} x$

20. $y = \tan(x + \pi)$

21. $y = -\tan\left(x - \frac{\pi}{4}\right)$

22. $y = 2 \cot 3x$

23. $y = -\frac{1}{2} \cot \frac{\pi}{2} x$

24. $y = 2 \cot\left(x + \frac{\pi}{2}\right)$

25. $y = 3 \sec 2\pi x$

26. $y = -2 \csc \pi x$

27. $y = 3 \sec(x + \pi)$

28. $y = \frac{5}{2} \csc(x - \pi)$

29. $\dfrac{\pi}{2}$ **30.** 0 **31.** $\dfrac{\pi}{4}$

32. $-\dfrac{\pi}{3}$ **33.** $\dfrac{2\pi}{3}$ **34.** $-\dfrac{\pi}{6}$

35. $\dfrac{\sqrt{2}}{2}$ **36.** 1 **37.** $-\dfrac{\sqrt{3}}{3}$

38. $-\dfrac{\sqrt{3}}{3}$ **39.** 2 **40.** $\dfrac{4}{5}$ **41.** $\dfrac{4}{5}$

42. $-\dfrac{3}{4}$ **43.** $-\dfrac{3}{4}$ **44.** $-\dfrac{\sqrt{10}}{10}$ **45.** $\dfrac{\pi}{3}$ **46.** $\dfrac{\pi}{3}$ **47.** $-\dfrac{\pi}{6}$ **48.** $\dfrac{2\sqrt{x^2 + 4}}{x^2 + 4}$ **49.** $\dfrac{x\sqrt{x^2 - 1}}{x^2 - 1}$

50. $B \approx 67.7°; a \approx 3.79; b \approx 9.25$ **51.** $A \approx 52.6°; a \approx 7.85; c \approx 9.88$ **52.** $A \approx 16.6°; B \approx 73.4°; b \approx 6.71$

53. $A \approx 21.3°; B \approx 68.7°; c \approx 3.86$ **54.** 38 ft **55.** 90 yd **56.** 21.7 ft **57.** N 35° E **58.** S 35° W **59.** 24.6 mi

60. a. 1282.2 mi **b.** S 74° E **61. a.** 20 cm **b.** $\dfrac{1}{8}$ cm per sec **c.** 8 sec **62. a.** $\dfrac{1}{2}$ cm **b.** 0.64 cm per sec **c.** 1.57 sec

63. $d = -30 \cos \pi t$ **64.** $d = \dfrac{1}{4} \sin \dfrac{2\pi}{5} t$

Chapter 2 Test

1. $y = 3 \sin 2x$

2. $y = -2 \cos\left(x - \frac{\pi}{2}\right)$

3. $y = 2 \tan \frac{x}{2}$

4. $y = -\frac{1}{2} \csc \pi x$

5. $y = \frac{1}{2} \sin x + 2 \cos x$

6. $-\sqrt{3}$ **7.** $\dfrac{\sqrt{9 - x^2}}{3}$ **8.** $B = 69°; a = 4.7; b = 12.1$ **9.** 247.7 ft **10.** N 80° W

11. a. 6 in. **b.** $\dfrac{1}{2}$ in. per sec **c.** 2 sec

CHAPTER 3

Section 3.1

Check Point Exercises

1. $\csc x \tan x = \dfrac{1}{\sin x} \cdot \dfrac{\sin x}{\cos x} = \dfrac{1}{\cos x} = \sec x$

2. $\cos x \cot x + \sin x = \cos x \cdot \dfrac{\cos x}{\sin x} + \sin x = \dfrac{\cos^2 x}{\sin x} + \sin x \cdot \dfrac{\sin x}{\sin x} = \dfrac{\cos^2 x + \sin^2 x}{\sin x} = \dfrac{1}{\sin x} = \csc x$

3. $\sin x - \sin x \cos^2 x = \sin x(1 - \cos^2 x) = \sin x \cdot \sin^2 x = \sin^3 x$

4. $\dfrac{1 + \cos \theta}{\sin \theta} = \dfrac{1}{\sin \theta} + \dfrac{\cos \theta}{\sin \theta} = \csc \theta + \cot \theta$

5. $\dfrac{\sin x}{1 + \cos x} + \dfrac{1 + \cos x}{\sin x} = \dfrac{\sin x(\sin x)}{(1 + \cos x)(\sin x)} + \dfrac{(1 + \cos x)(1 + \cos x)}{(\sin x)(1 + \cos x)} = \dfrac{\sin^2 x + 1 + 2\cos x + \cos^2 x}{(1 + \cos x)(\sin x)}$

$= \dfrac{\sin^2 x + \cos^2 x + 1 + 2\cos x}{(1 + \cos x)(\sin x)} = \dfrac{1 + 1 + 2\cos x}{(1 + \cos x)(\sin x)} = \dfrac{2 + 2\cos x}{(1 + \cos x)(\sin x)} = \dfrac{2(1 + \cos x)}{(1 + \cos x)(\sin x)} = \dfrac{2}{\sin x} = 2\csc x$

6. $\dfrac{\cos x}{1 + \sin x} = \dfrac{\cos x(1 - \sin x)}{(1 + \sin x)(1 - \sin x)} = \dfrac{\cos x(1 - \sin x)}{1 - \sin^2 x} = \dfrac{\cos x(1 - \sin x)}{\cos^2 x} = \dfrac{1 - \sin x}{\cos x}$ **7.** $\dfrac{\sec x + \csc(-x)}{\sec x \csc x} = \dfrac{\sec x - \csc x}{\sec x \csc x}$

$= \dfrac{\dfrac{1}{\cos x} - \dfrac{1}{\sin x}}{\dfrac{1}{\cos x} \cdot \dfrac{1}{\sin x}} = \dfrac{\dfrac{\sin x}{\cos x \cdot \sin x} - \dfrac{\cos x}{\cos x \cdot \sin x}}{\dfrac{1}{\cos x \cdot \sin x}} = \dfrac{\dfrac{\sin x - \cos x}{\cos x \cdot \sin x}}{\dfrac{1}{\cos x \cdot \sin x}} = \dfrac{\sin x - \cos x}{\cos x \cdot \sin x} \cdot \dfrac{\cos x \cdot \sin x}{1} = \sin x - \cos x$

8. Left side: $\dfrac{1}{1 + \sin \theta} + \dfrac{1}{1 - \sin \theta} = \dfrac{1(1 - \sin \theta)}{(1 + \sin \theta)(1 - \sin \theta)} + \dfrac{1(1 + \sin \theta)}{(1 - \sin \theta)(1 + \sin \theta)} = \dfrac{1 - \sin \theta + 1 + \sin \theta}{(1 + \sin \theta)(1 - \sin \theta)} = \dfrac{2}{1 - \sin^2 \theta}$;

Right side: $2 + 2\tan^2 \theta = 2 + 2\left(\dfrac{\sin^2 \theta}{\cos^2 \theta}\right) = \dfrac{2\cos^2 \theta}{\cos^2 \theta} + \dfrac{2\sin^2 \theta}{\cos^2 \theta} = \dfrac{2\cos^2 \theta + 2\sin^2 \theta}{\cos^2 \theta} = \dfrac{2(\cos^2 \theta + \sin^2 \theta)}{\cos^2 \theta} = \dfrac{2}{\cos^2 \theta} = \dfrac{2}{1 - \sin^2 \theta}$

Concept and Vocabulary Check

1. complicated; other   **2.** sines; cosines   **3.** false   **4.** $(\csc x - 1)(\csc x + 1)$ **5.** identical/the same

Exercise Set 3.1

For Exercises 1–59, proofs may vary.

61. $\cos x$; Proofs may vary.   **63.** $2\sin x$; Proofs may vary.   **65.** $2\sec x$; Proofs may vary.   **67.** $\dfrac{1}{\cos x}$ **69.** $\dfrac{1}{\cos x}$ **71.** $2\csc^2 x - 1$

73. $\sec x \tan x$

79.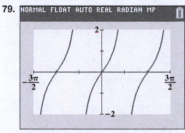

Proofs may vary.

81.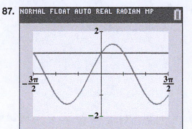

Values for x may vary.

83.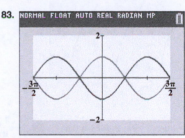

Values for x may vary.

85.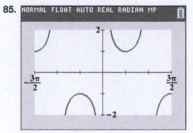

Proofs may vary.

87.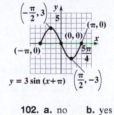

Values for x may vary.

89. makes sense   **91.** does not make sense

For Exercises 93 and 95, proofs may vary.

98. $\dfrac{2\sqrt{3}}{3}$

99. amplitude: 3; period: 2π; phase shift: $-\pi$

$\left(-\dfrac{\pi}{2}, 3\right)$ $(\pi, 0)$ $(0, 0)$ $(-\pi, 0)$ $\dfrac{5\pi}{4}$ $y = 3\sin(x + \pi)$ $\left(\dfrac{\pi}{2}, -3\right)$

100. $\sin t = \dfrac{1}{2}$; $\cos t = -\dfrac{\sqrt{3}}{2}$; $\tan t = -\dfrac{\sqrt{3}}{3}$; $\csc t = 2$; $\sec t = -\dfrac{2\sqrt{3}}{3}$; $\cot t = -\sqrt{3}$ **101.** $\dfrac{\sqrt{3}}{2}, \dfrac{1}{2}, \dfrac{1}{2}, \dfrac{\sqrt{3}}{2}, 0; 1$ **102. a.** no   **b.** yes

103. a. no   **b.** yes

Section 3.2

Check Point Exercises

1. $\dfrac{\sqrt{3}}{2}$ **2.** $\dfrac{\sqrt{3}}{2}$ **3.** $\dfrac{\cos(\alpha - \beta)}{\cos \alpha \cos \beta} = \dfrac{\cos \alpha \cos \beta + \sin \alpha \sin \beta}{\cos \alpha \cos \beta} = \dfrac{\cos \alpha}{\cos \alpha} \cdot \dfrac{\cos \beta}{\cos \beta} + \dfrac{\sin \alpha}{\cos \alpha} \cdot \dfrac{\sin \beta}{\cos \beta} = 1 + \tan \alpha \tan \beta$

4. $\dfrac{\sqrt{2} + \sqrt{6}}{4}$ **5. a.** $\cos \alpha = -\dfrac{3}{5}$ **b.** $\cos \beta = \dfrac{\sqrt{3}}{2}$ **c.** $\dfrac{-3\sqrt{3} - 4}{10}$ **d.** $\dfrac{4\sqrt{3} - 3}{10}$ **6. a.** $y = \sin x$

b. $\cos\left(x + \dfrac{3\pi}{2}\right) = \cos x \cos \dfrac{3\pi}{2} - \sin x \sin \dfrac{3\pi}{2} = \cos x \cdot 0 - \sin x \cdot (-1) = \sin x$ **7.** $\tan(x + \pi) = \dfrac{\tan x + \tan \pi}{1 - \tan x \tan \pi} = \dfrac{\tan x + 0}{1 - \tan x \cdot 0} = \dfrac{\tan x}{1} = \tan x$

Concept and Vocabulary Check

1. $\cos x \cos y - \sin x \sin y$ **2.** $\cos x \cos y + \sin x \sin y$ **3.** $\sin C \cos D + \cos C \sin D$ **4.** $\sin C \cos D - \cos C \sin D$

5. $\dfrac{\tan \theta + \tan \phi}{1 - \tan \theta \tan \phi}$ **6.** $\dfrac{\tan \theta - \tan \phi}{1 + \tan \theta \tan \phi}$ **7.** false **8.** false

Exercise Set 3.2

1. $\dfrac{\sqrt{6} + \sqrt{2}}{4}$ **3.** $\dfrac{\sqrt{2} - \sqrt{6}}{4}$ **5. a.** $\alpha = 50°, \beta = 20°$ **b.** $\cos 30°$ **c.** $\dfrac{\sqrt{3}}{2}$ **7. a.** $\alpha = \dfrac{5\pi}{12}, \beta = \dfrac{\pi}{12}$ **b.** $\cos\dfrac{\pi}{3}$ **c.** $\dfrac{1}{2}$

For Exercises 9 and 11, proofs may vary. **13.** $\dfrac{\sqrt{6} - \sqrt{2}}{4}$ **15.** $\dfrac{\sqrt{6} + \sqrt{2}}{4}$ **17.** $-\dfrac{\sqrt{6} + \sqrt{2}}{4}$ **19.** $\dfrac{\sqrt{6} - \sqrt{2}}{4}$ **21.** $\dfrac{\sqrt{3} + 1}{\sqrt{3} - 1}$ **23.** $\dfrac{\sqrt{3} - 1}{\sqrt{3} + 1}$

25. $\sin 30°; \dfrac{1}{2}$ **27.** $\tan 45°; 1$ **29.** $\sin\dfrac{\pi}{6}; \dfrac{1}{2}$ **31.** $\tan\dfrac{\pi}{6}; \dfrac{\sqrt{3}}{3}$

For Exercises 33–55, proofs may vary.

57. a. $-\dfrac{63}{65}$ **b.** $-\dfrac{16}{65}$ **c.** $\dfrac{16}{63}$ **59. a.** $-\dfrac{4 + 6\sqrt{2}}{15}$ **b.** $\dfrac{3 - 8\sqrt{2}}{15}$ **c.** $-\dfrac{3 - 8\sqrt{2}}{4 + 6\sqrt{2}}$ or $\dfrac{54 - 25\sqrt{2}}{28}$

61. a. $-\dfrac{8\sqrt{3} + 15}{34}$ **b.** $\dfrac{15\sqrt{3} - 8}{34}$ **c.** $-\dfrac{15\sqrt{3} - 8}{8\sqrt{3} + 15}$ or $\dfrac{480 - 289\sqrt{3}}{33}$ **63. a.** $\dfrac{4 + 3\sqrt{15}}{20}$ **b.** $\dfrac{-3 + 4\sqrt{15}}{20}$ **c.** $\dfrac{3 - 4\sqrt{15}}{4 + 3\sqrt{15}}$ or $\dfrac{25\sqrt{15} - 192}{119}$

65. a. $y = \sin x$ **b.** $\sin(\pi - x) = \sin \pi \cos x - \cos \pi \sin x = 0 \cdot \cos x - (-1) \sin x = \sin x$ **67. a.** $y = 2 \cos x$

b. $\sin\left(x + \dfrac{\pi}{2}\right) + \sin\left(\dfrac{\pi}{2} - x\right) = \sin x \cos\dfrac{\pi}{2} + \cos x \sin\dfrac{\pi}{2} + \sin\dfrac{\pi}{2} \cos x - \cos\dfrac{\pi}{2} \sin x = \sin x \cdot 0 + \cos x \cdot 1 + 1 \cdot \cos x - 0 \cdot \sin x$

$= \cos x + \cos x = 2 \cos x$ **69.** $\cos \alpha$ **71.** $\tan \beta$ **73.** $\cos\dfrac{\pi}{3} = \dfrac{1}{2}$ **75.** $\cos 3x$; Proofs may vary. **77.** $\sin\dfrac{x}{2}$; Proofs may vary.

79. Proofs may vary.; amplitude is $\sqrt{13}$; period is 2π

89.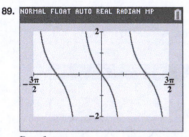

Proofs may vary.

91.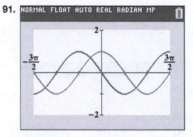

Values for x may vary.

93.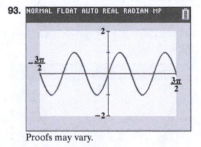

Proofs may vary.

95. makes sense **97.** makes sense. **99.** $\dfrac{4\sqrt{3} + 3}{10}$ **101.** $-\dfrac{33}{65}$ **103.** $y\sqrt{1 - x^2} + x\sqrt{1 - y^2}$ **105.** $\dfrac{xy + \left(\sqrt{1 - x^2}\right)\left(\sqrt{1 - y^2}\right)}{y\sqrt{1 - x^2} - x\sqrt{1 - y^2}}$

107. 138.8 ft **108.** $y = 2 \cos\dfrac{\pi}{4}x$ **109.** $\dfrac{4}{5}$ **110.** $\sin 30° = \dfrac{1}{2}; \cos 30° = \dfrac{\sqrt{3}}{2}; \sin 60° = \dfrac{\sqrt{3}}{2}; \cos 60° = \dfrac{1}{2}$

111. a. no **b.** yes **112. a.** no **b.** yes

Section 3.3

Check Point Exercises

1. a. $-\dfrac{24}{25}$ **b.** $-\dfrac{7}{25}$ **c.** $\dfrac{24}{7}$ **2.** $\dfrac{\sqrt{3}}{2}$ **3.** $\sin 3\theta = \sin(2\theta + \theta) = \sin 2\theta \cos \theta + \cos 2\theta \sin \theta = 2 \sin \theta \cos \theta \cos \theta$

$+ (2 \cos^2 \theta - 1)\sin \theta = 2 \sin \theta \cos^2 \theta + 2 \sin \theta \cos^2 \theta - \sin \theta = 4 \sin \theta \cos^2 \theta - \sin \theta = 4 \sin \theta(1 - \sin^2 \theta) - \sin \theta$

$= 4 \sin \theta - 4 \sin^3 \theta - \sin \theta = 3 \sin \theta - 4 \sin^3 \theta$

4. $\sin^4 x = (\sin^2 x)^2 = \left(\dfrac{1 - \cos 2x}{2}\right)^2 = \dfrac{1 - 2 \cos 2x + \cos^2 2x}{4} = \dfrac{1}{4} - \dfrac{1}{2}\cos 2x + \dfrac{1}{4}\cos^2 2x = \dfrac{1}{4} - \dfrac{1}{2}\cos 2x + \dfrac{1}{4}\left(\dfrac{1 + \cos 2(2x)}{2}\right)$

$= \dfrac{1}{4} - \dfrac{1}{2}\cos 2x + \dfrac{1}{8} + \dfrac{1}{8}\cos 4x = \dfrac{3}{8} - \dfrac{1}{2}\cos 2x + \dfrac{1}{8}\cos 4x$ **5.** $-\dfrac{\sqrt{2 - \sqrt{3}}}{2}$

6. $\dfrac{\sin 2\theta}{1 + \cos 2\theta} = \dfrac{2 \sin \theta \cos \theta}{1 + (1 - 2 \sin^2 \theta)} = \dfrac{2 \sin \theta \cos \theta}{2 - 2 \sin^2 \theta} = \dfrac{2 \sin \theta \cos \theta}{2(1 - \sin^2 \theta)} = \dfrac{2 \sin \theta \cos \theta}{2 \cos^2 \theta} = \dfrac{\sin \theta}{\cos \theta} = \tan \theta$

7. $\dfrac{\sec \alpha}{\sec \alpha \csc \alpha + \csc \alpha} = \dfrac{\dfrac{1}{\cos \alpha}}{\dfrac{1}{\cos \alpha} \cdot \dfrac{1}{\sin \alpha} + \dfrac{1}{\sin \alpha}} = \dfrac{\dfrac{1}{\cos \alpha}}{\dfrac{1}{\cos \alpha \sin \alpha} + \dfrac{\cos \alpha}{\cos \alpha \sin \alpha}} = \dfrac{\dfrac{1}{\cos \alpha}}{\dfrac{1 + \cos \alpha}{\cos \alpha \sin \alpha}} = \dfrac{1}{\cos \alpha} \cdot \dfrac{\cos \alpha \sin \alpha}{1 + \cos \alpha} = \dfrac{\sin \alpha}{1 + \cos \alpha} = \tan\dfrac{\alpha}{2}$

Concept and Vocabulary Check

1. $2 \sin x \cos x$ **2.** $\sin^2 A; 2 \cos^2 A; 2 \sin^2 A$ **3.** $\dfrac{2 \tan B}{1 - \tan^2 B}$ **4.** $1 - \cos 2\alpha$ **5.** $1 + \cos 2\alpha$ **6.** $1 - \cos 2y$ **7.** $1 - \cos x$

8. $1 + \cos y$ **9.** $1 - \cos \alpha; 1 - \cos \alpha; 1 + \cos \alpha$ **10.** false **11.** false **12.** false **13.** $+$ **14.** $-$ **15.** $+$

Exercise Set 3.3

1. $\dfrac{24}{25}$ **3.** $\dfrac{24}{7}$ **5.** $\dfrac{527}{625}$ **7. a.** $-\dfrac{240}{289}$ **b.** $-\dfrac{161}{289}$ **c.** $\dfrac{240}{161}$ **9. a.** $-\dfrac{336}{625}$ **b.** $\dfrac{527}{625}$ **c.** $-\dfrac{336}{527}$

11. a. $\dfrac{4}{5}$ **b.** $\dfrac{3}{5}$ **c.** $\dfrac{4}{3}$ **13. a.** $\dfrac{720}{1681}$ **b.** $\dfrac{1519}{1681}$ **c.** $\dfrac{720}{1519}$ **15.** $\dfrac{1}{2}$ **17.** $-\dfrac{\sqrt{3}}{2}$ **19.** $\dfrac{\sqrt{2}}{2}$ **21.** $\dfrac{\sqrt{3}}{3}$

For Exercises 23–33, proofs may vary. **35.** $\dfrac{9}{4} - 3 \cos 2x + \dfrac{3}{4} \cos 4x$ **37.** $\dfrac{1}{8} - \dfrac{1}{8} \cos 4x$ **39.** $\dfrac{\sqrt{2 - \sqrt{3}}}{2}$ **41.** $\dfrac{\sqrt{2 + \sqrt{2}}}{2}$ **43.** $2 + \sqrt{3}$

45. $-\sqrt{2} + 1$ **47.** $\dfrac{\sqrt{10}}{10}$ **49.** $\dfrac{1}{3}$ **51.** $\dfrac{7\sqrt{2}}{10}$ **53.** $\dfrac{3}{5}$ **55. a.** $\dfrac{2\sqrt{5}}{5}$ **b.** $-\dfrac{\sqrt{5}}{5}$ **c.** -2 **57. a.** $\dfrac{3\sqrt{13}}{13}$ **b.** $\dfrac{2\sqrt{13}}{13}$ **c.** $\dfrac{3}{2}$

For Exercises 59–67, proofs may vary.

69. $\cos 2x$; Proofs may vary. **71.** $1 + \sin x$; Proofs may vary. **73.** $\sec x$; Proofs may vary. **75.** $2 \csc 2x$; Proofs may vary.

77. $\sin 3x$; Proofs may vary. **79. a.** $d = \dfrac{v_0^2}{32} \cdot \sin 2\theta$ **b.** $\theta = \dfrac{\pi}{4}$ **81.** $\sqrt{2 - \sqrt{2}} \cdot (2 + \sqrt{2}) \approx 2.6$

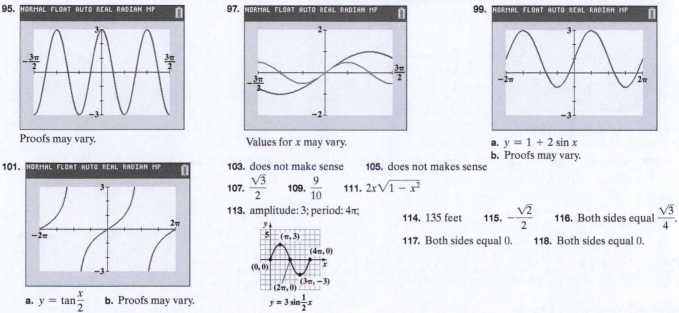

95. Proofs may vary.

97. Values for x may vary.

99. a. $y = 1 + 2 \sin x$ **b.** Proofs may vary.

101. a. $y = \tan \dfrac{x}{2}$ **b.** Proofs may vary.

103. does not make sense **105.** does not makes sense

107. $\dfrac{\sqrt{3}}{2}$ **109.** $\dfrac{9}{10}$ **111.** $2x\sqrt{1 - x^2}$

113. amplitude: 3; period: 4π;

$y = 3 \sin \dfrac{1}{2} x$

114. 135 feet **115.** $-\dfrac{\sqrt{2}}{2}$ **116.** Both sides equal $\dfrac{\sqrt{3}}{4}$.

117. Both sides equal 0. **118.** Both sides equal 0.

Mid-Chapter 3 Check Point

For Exercises 1–18, proofs may vary.

19. $\dfrac{33}{65}$ **20.** $-\dfrac{16}{63}$ **21.** $-\dfrac{24}{25}$ **22.** $-\dfrac{\sqrt{26}}{26}$ **23.** $-\dfrac{\sqrt{6} + \sqrt{2}}{4}$ **24.** $\dfrac{\sqrt{3}}{2}$ **25.** $\dfrac{1}{2}$ **26.** $\sqrt{\dfrac{\sqrt{2} - 1}{\sqrt{2} + 1}}$ or $\sqrt{2} - 1$

Section 3.4

Check Point Exercises

1. a. $\dfrac{1}{2}[\cos 3x - \cos 7x]$ **b.** $\dfrac{1}{2}[\cos 6x + \cos 8x]$ **2. a.** $2 \sin 5x \cos 2x$ **b.** $2 \cos \dfrac{5x}{2} \cos \dfrac{x}{2}$

3. $\dfrac{\cos 3x - \cos x}{\sin 3x + \sin x} = \dfrac{-2 \sin\left(\dfrac{3x + x}{2}\right) \sin\left(\dfrac{3x - x}{2}\right)}{2 \sin\left(\dfrac{3x + x}{2}\right) \cos\left(\dfrac{3x - x}{2}\right)} = \dfrac{-2 \sin\left(\dfrac{4x}{2}\right) \sin\left(\dfrac{2x}{2}\right)}{2 \sin\left(\dfrac{4x}{2}\right) \cos\left(\dfrac{2x}{2}\right)} = \dfrac{-2 \sin 2x \sin x}{2 \sin 2x \cos x} = -\dfrac{\sin x}{\cos x} = -\tan x$

Concept and Vocabulary Check

1. product; difference **2.** product; sum **3.** product; sum **4.** product; difference
5. sum; product **6.** difference; product **7.** sum; product **8.** difference; product

Exercise Set 3.4

1. $\frac{1}{2}[\cos 4x - \cos 8x]$ **3.** $\frac{1}{2}[\cos 4x + \cos 10x]$ **5.** $\frac{1}{2}[\sin 3x - \sin x]$ **7.** $\frac{1}{2}[\sin 2x - \sin x]$ **9.** $2 \sin 4x \cos 2x$

11. $2 \sin 2x \cos 5x$ **13.** $2 \cos 3x \cos x$ **15.** $2 \sin \frac{3x}{2} \cos \frac{x}{2}$ **17.** $2 \cos x \cos \frac{x}{2}$ **19.** $\frac{\sqrt{6}}{2}$ **21.** $-\frac{\sqrt{2}}{2}$ For Exercises 23–29, proofs may vary.

31. a. $y = \cos x$ **b.** Proofs may vary.   **33. a.** $y = \tan 2x$ **b.** Proofs may vary.   **35. a.** $y = -\cot 2x$ **b.** Proofs may vary.

37. a. $y = \sin 1704\pi t + \sin 2418\pi t$ **b.** $2 \sin 2061\pi t \cdot \cos 357\pi t$

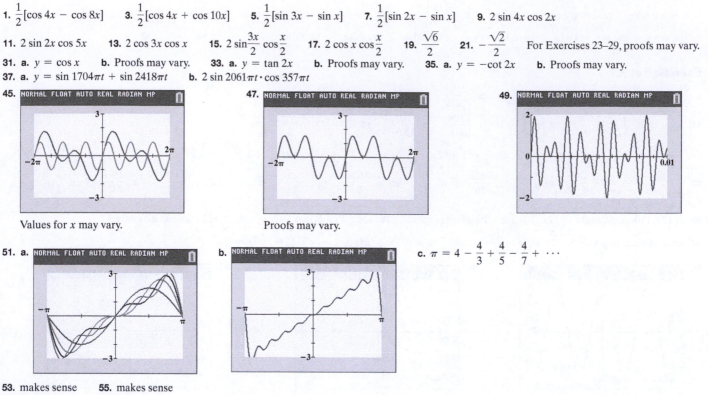

Values for x may vary.   Proofs may vary.

51. a.   **b.**   **c.** $\pi = 4 - \frac{4}{3} + \frac{4}{5} - \frac{4}{7} + \cdots$

53. makes sense   **55.** makes sense
For Exercises 57–61, proofs may vary.

63. a. N66°W   **b.** S26°W   **64.** $\frac{5}{12}$ **65.** $15\pi \approx 47.12$ inches   **66.** $\left\{-\frac{1}{2}, 2\right\}$ **67.** $\{-\sqrt{3}, 0, \sqrt{3}\}$ **68.** $\left\{\frac{1 - \sqrt{5}}{2}, \frac{1 + \sqrt{5}}{2}\right\}$

Section 3.5

Check Point Exercises

1. $x = \frac{\pi}{3} + 2n\pi$ or $x = \frac{2\pi}{3} + 2n\pi$, where n is any integer.   **2.** $\frac{\pi}{6}, \frac{2\pi}{3}, \frac{7\pi}{6}, \frac{5\pi}{3}$ **3.** $\frac{\pi}{2}$ **4.** $\frac{\pi}{6}, \frac{\pi}{2}, \frac{5\pi}{6}$ **5.** $\frac{\pi}{6}, \frac{5\pi}{6}, \frac{7\pi}{6}, \frac{11\pi}{6}$ **6.** $0, \frac{\pi}{4}, \pi, \frac{5\pi}{4}$

7. $\frac{\pi}{3}, \frac{5\pi}{3}$ **8.** $\frac{\pi}{2}, \frac{7\pi}{6}, \frac{11\pi}{6}$ **9.** $\frac{3\pi}{4}, \frac{7\pi}{4}$ **10.** $\frac{\pi}{2}, \pi$ **11. a.** $1.2592, 4.4008$ **b.** $3.3752, 6.0496$ **12.** $2.3423, 3.9409$

Concept and Vocabulary Check

1. $\frac{3\pi}{4}; \frac{\pi}{4} + 2n\pi; \frac{3\pi}{4} + 2n\pi$ **2.** $\frac{2\pi}{3}; x = \frac{2\pi}{3} + n\pi$ **3.** false   **4.** true   **5.** false

6. $2 \cos x + 1; \cos x - 5; \cos x - 5 = 0$ **7.** $\cos x; 2 \sin x + \sqrt{2}$ **8.** $\cos^2 x; 1 - \sin^2 x$ **9.** $\pi; 2\pi$

Exercise Set 3.5

1. Solution   **3.** Not a solution   **5.** Solution   **7.** Solution   **9.** Not a solution   **11.** $x = \frac{\pi}{3} + 2n\pi$ or $x = \frac{2\pi}{3} + 2n\pi$, where n is any integer.

13. $x = \frac{\pi}{4} + n\pi$, where n is any integer.   **15.** $x = \frac{2\pi}{3} + 2n\pi$ or $x = \frac{4\pi}{3} + 2n\pi$, where n is any integer.   **17.** $x = n\pi$, where n is any integer.

19. $x = \frac{5\pi}{6} + 2n\pi$ or $x = \frac{7\pi}{6} + 2n\pi$, where n is any integer.   **21.** $\theta = \frac{\pi}{6} + 2n\pi$ or $\theta = \frac{5\pi}{6} + 2n\pi$, where n is any integer.

23. $\theta = \frac{3\pi}{2} + 2n\pi$, where n is any integer.   **25.** $\frac{\pi}{6}, \frac{\pi}{3}, \frac{7\pi}{6}, \frac{4\pi}{3}$ **27.** $\frac{5\pi}{24}, \frac{7\pi}{24}, \frac{17\pi}{24}, \frac{19\pi}{24}, \frac{29\pi}{24}, \frac{31\pi}{24}, \frac{41\pi}{24}, \frac{43\pi}{24}$ **29.** $\frac{\pi}{18}, \frac{7\pi}{18}, \frac{13\pi}{18}, \frac{19\pi}{18}, \frac{25\pi}{18}, \frac{31\pi}{18}$

31. $\frac{2\pi}{3}$ **33.** no solution   **35.** $\frac{4\pi}{9}, \frac{8\pi}{9}, \frac{16\pi}{9}$ **37.** $0, \frac{\pi}{3}, \pi, \frac{4\pi}{3}$ **39.** $\frac{\pi}{2}, \frac{7\pi}{6}, \frac{11\pi}{6}$ **41.** $\frac{2\pi}{3}, \pi, \frac{4\pi}{3}$ **43.** $\frac{3\pi}{2}$ **45.** $\frac{\pi}{2}, \frac{3\pi}{2}$

47. $\frac{\pi}{3}, \frac{2\pi}{3}, \frac{4\pi}{3}, \frac{5\pi}{3}$ **49.** $\frac{\pi}{6}, \frac{5\pi}{6}, \frac{7\pi}{6}, \frac{11\pi}{6}$ **51.** $\frac{\pi}{4}, \frac{3\pi}{4}, \frac{5\pi}{4}, \frac{7\pi}{4}$ **53.** $\frac{\pi}{4}, \pi, \frac{5\pi}{4}$ **55.** $\frac{5\pi}{6}, \frac{7\pi}{6}, \frac{11\pi}{6}$ **57.** $\frac{\pi}{4}, \frac{5\pi}{4}$ **59.** $0, \frac{2\pi}{3}, \pi, \frac{4\pi}{3}$

61. $0, \pi$ **63.** $\frac{\pi}{2}, \frac{7\pi}{6}, \frac{11\pi}{6}$ **65.** π **67.** $\frac{\pi}{6}, \frac{5\pi}{6}$ **69.** $\frac{\pi}{6}, \frac{\pi}{2}, \frac{5\pi}{6}, \frac{3\pi}{2}$ **71.** $0, \frac{2\pi}{3}, \frac{4\pi}{3}$ **73.** $\frac{2\pi}{3}, \frac{4\pi}{3}$ **75.** $\frac{\pi}{8}, \frac{3\pi}{8}, \frac{9\pi}{8}, \frac{11\pi}{8}$ **77.** $0, \frac{\pi}{2}$

79. $\frac{\pi}{4}, \frac{3\pi}{4}$ **81.** $\frac{\pi}{12}, \frac{\pi}{4}, \frac{3\pi}{4}, \frac{11\pi}{12}, \frac{17\pi}{12}, \frac{19\pi}{12}$ **83.** 0 **85.** $0.9695, 2.1721$ **87.** $1.9823, 4.3009$ **89.** $1.8926, 5.0342$ **91.** $2.2370, 4.0462$

93. $0.4636, 0.9828, 3.6052, 4.1244$ **95.** $0.3876, 2.7540, 3.5292, 5.8956$ **97.** $\frac{\pi}{3}, \frac{2\pi}{3}, \frac{4\pi}{3}, \frac{5\pi}{3}$ **99.** $0, \frac{2\pi}{3}, \pi, \frac{4\pi}{3}$ **101.** $\frac{\pi}{6}, \frac{11\pi}{6}$

103. $1.7798, 4.9214$ **105.** $\frac{\pi}{2}$ **107.** $\frac{\pi}{6}, \frac{\pi}{2}, \frac{5\pi}{6}, \frac{3\pi}{2}$ **109.** $\frac{\pi}{6}, \frac{5\pi}{6}, \frac{7\pi}{6}, \frac{11\pi}{6}$ **111.** $0.7494, 5.5338$ **113.** $\frac{7\pi}{6}, \frac{11\pi}{6}$

115. $2.1588, \dfrac{3\pi}{4}, 5.3004, \dfrac{7\pi}{4}$ **117.** $\left(\dfrac{2\pi}{3}, -\dfrac{3}{2}\right), \left(\dfrac{4\pi}{3}, -\dfrac{3}{2}\right)$ **119.** $(3.5163, 0.7321), (5.9085, 0.7321)$ **121.** $\dfrac{\pi}{6}, \dfrac{5\pi}{6}, \dfrac{7\pi}{6}, \dfrac{11\pi}{6}$

123. $\dfrac{\pi}{6}, \dfrac{5\pi}{6}, 3.3430, 6.0818$

125. $0, \dfrac{2\pi}{3}, \pi, \dfrac{4\pi}{3}$

127. $0.3649, 1.2059, 3.5065, 4.3475;$ a

129. 0.4 sec and 2.1 sec

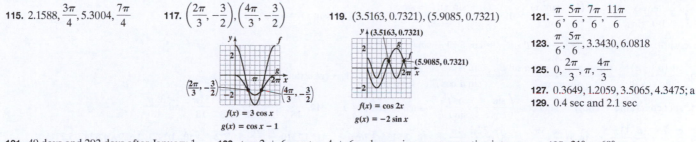

$f(x) = 3\cos x$
$g(x) = \cos x - 1$

$f(x) = \cos 2x$
$g(x) = -2\sin x$

131. 49 days and 292 days after January 1   **133.** $t = 2 + 6n$ or $t = 4 + 6n$ where n is any nonnegative integer.   **135.** 21° or 69°.

147. $x = 1.37, x = 2.30, x = 3.98,$ or $x = 4.91$ **149.** $x = 0.37$ or $x = 2.77$ **151.** $x = 0, x = 1.57, x = 2.09, x = 3.14, x = 4.19,$ or $x = 4.71$

153. makes sense   **155.** does not make sense   **157.** false   **159.** false   **161.** $\dfrac{\pi}{2}, \dfrac{3\pi}{2}, \dfrac{7\pi}{12}, \dfrac{11\pi}{12}, \dfrac{19\pi}{12}, \dfrac{23\pi}{12}$

163.

164. $\sqrt{1 - x^2}$ **165.** $B \approx 46.5°; a \approx 7.69; c \approx 11.17$ **166.** $a \approx 45.2$

167. $B \approx 31.5°$ **168.** no solution or $\varnothing$

Chapter 3 Review Exercises

For Exercises 1–13, proofs may vary.

14. $\dfrac{\sqrt{6} - \sqrt{2}}{4}$ **15.** $\dfrac{\sqrt{2} - \sqrt{6}}{4}$ **16.** $2 - \sqrt{3}$ **17.** $\sqrt{3} + 2$ **18.** $\dfrac{1}{2}$ **19.** $\dfrac{1}{2}$

For Exercises 20–31, proofs may vary.

32. a. $y = \cos x$ **b.** $\sin\left(x - \dfrac{3\pi}{2}\right) = \sin x \cos\dfrac{3\pi}{2} - \cos x \sin\dfrac{3\pi}{2} = \sin x \cdot 0 - \cos x \cdot -1 = \cos x$

33. a. $y = -\sin x$ **b.** $\cos\left(x + \dfrac{\pi}{2}\right) = \cos x \cos\dfrac{\pi}{2} - \sin x \sin\dfrac{\pi}{2} = \cos x \cdot 0 - \sin x \cdot 1 = -\sin x$

34. a. $y = \tan x$ **b.** $y = \dfrac{\tan x - 1}{1 - \cot x} = \dfrac{\dfrac{\sin x}{\cos x} - 1}{1 - \dfrac{\cos x}{\sin x}} = \dfrac{\dfrac{\sin x - \cos x}{\cos x}}{\dfrac{\sin x - \cos x}{\sin x}} = \dfrac{\sin x - \cos x}{\cos x} \cdot \dfrac{\sin x}{\sin x - \cos x} = \dfrac{\sin x}{\cos x} = \tan x$

35. a. $\dfrac{33}{65}$ **b.** $\dfrac{16}{65}$ **c.** $-\dfrac{33}{56}$ **d.** $\dfrac{24}{25}$ **e.** $\dfrac{2\sqrt{13}}{13}$ **36. a.** $-\dfrac{63}{65}$ **b.** $-\dfrac{56}{65}$ **c.** $\dfrac{63}{16}$ **d.** $\dfrac{24}{25}$ **e.** $\dfrac{5\sqrt{26}}{26}$

37. a. 1   **b.** $-\dfrac{3}{5}$ **c.** undefined   **d.** $-\dfrac{3}{5}$ **e.** $-\dfrac{\sqrt{10 + 3\sqrt{10}}}{2\sqrt{5}}$ **38. a.** 1   **b.** $\dfrac{4\sqrt{2}}{9}$ **c.** undefined   **d.** $\dfrac{4\sqrt{2}}{9}$ **e.** $-\dfrac{\sqrt{3}}{3}$

39. $\dfrac{\sqrt{3}}{2}$ **40.** $-\dfrac{\sqrt{3}}{3}$ **41.** $\dfrac{\sqrt{2 - \sqrt{2}}}{2}$ **42.** $2 - \sqrt{3}$ **43.** $\dfrac{1}{2}[\cos 2x - \cos 10x]$ **44.** $\dfrac{1}{2}[\sin 10x + \sin 4x]$ **45.** $-2\sin x \cos 3x$

46. $\dfrac{\sqrt{6}}{2}$ **47.** Proofs may vary.   **48.** Proofs may vary.   **49. a.** $y = \cot x$ **b.** Proofs may vary.

50. $x = \dfrac{2\pi}{3} + 2n\pi$ or $x = \dfrac{4\pi}{3} + 2n\pi$, where n is any integer.   **51.** $x = \dfrac{\pi}{4} + 2n\pi$ or $x = \dfrac{3\pi}{4} + 2n\pi$, where n is any integer.

52. $x = \dfrac{7\pi}{6} + 2n\pi$ or $x = \dfrac{11\pi}{6} + 2n\pi$, where n is any integer.   **53.** $x = \dfrac{\pi}{6} + n\pi$, where n is any integer.   **54.** $\dfrac{\pi}{2}, \dfrac{3\pi}{2}$ **55.** $\dfrac{\pi}{6}, \dfrac{5\pi}{6}, \dfrac{3\pi}{2}$

56. $\dfrac{3\pi}{2}$ **57.** $0, \dfrac{\pi}{3}, \pi, \dfrac{5\pi}{3}$ **58.** π **59.** $\dfrac{\pi}{6}, \dfrac{5\pi}{6}, \dfrac{3\pi}{2}$ **60.** $\dfrac{\pi}{6}, \dfrac{5\pi}{6}, \dfrac{7\pi}{6}, \dfrac{11\pi}{6}$ **61.** $0, \pi, \dfrac{7\pi}{6}, \dfrac{11\pi}{6}$ **62.** $0, \dfrac{\pi}{6}, \pi, \dfrac{11\pi}{6}$ **63.** $0, \pi$

64. $3.7890, 5.6358$ **65.** $0.6847, 2.4569, 3.8263, 5.5985$ **66.** $\dfrac{\pi}{4}, 1.2490, \dfrac{5\pi}{4}, 4.3906$ **67.** $0.8959, 2.2457$

68. $t = \dfrac{2}{3} + 4n$ or $t = \dfrac{10}{3} + 4n$, where n is any integer.   **69.** 12° or 78°

Chapter 3 Test

1. $-\dfrac{63}{65}$ **2.** $\dfrac{56}{33}$ **3.** $-\dfrac{24}{25}$ **4.** $\dfrac{3\sqrt{13}}{13}$ **5.** $\dfrac{\sqrt{6} + \sqrt{2}}{4}$ **6.** $\cos x \csc x = \cos x \cdot \dfrac{1}{\sin x} = \dfrac{\cos x}{\sin x} = \cot x$

7. $\dfrac{\sec x}{\cot x + \tan x} = \dfrac{\dfrac{1}{\cos x}}{\dfrac{\cos x}{\sin x} + \dfrac{\sin x}{\cos x}} = \dfrac{\dfrac{1}{\cos x}}{\dfrac{\cos^2 x + \sin^2 x}{\sin x \cos x}} = \dfrac{1}{\cos x} \cdot \dfrac{\sin x \cos x}{1} = \sin x$

8. $1 - \dfrac{\cos^2 x}{1 + \sin x} = 1 - \dfrac{(1 - \sin^2 x)}{1 + \sin x} = 1 - \dfrac{(1 + \sin x)(1 - \sin x)}{1 + \sin x} = 1 - (1 - \sin x) = \sin x$

9. $\cos\left(\theta + \dfrac{\pi}{2}\right) = \cos\theta\cos\dfrac{\pi}{2} - \sin\theta\sin\dfrac{\pi}{2} = \cos\theta\cdot 0 - \sin\theta\cdot 1 = -\sin\theta$

10. $\dfrac{\sin(\alpha - \beta)}{\sin\alpha\cos\beta} = \dfrac{\sin\alpha\cos\beta - \cos\alpha\sin\beta}{\sin\alpha\cos\beta} = \dfrac{\sin\alpha\cos\beta}{\sin\alpha\cos\beta} - \dfrac{\cos\alpha\sin\beta}{\sin\alpha\cos\beta} = 1 - \cot\alpha\tan\beta$

11. $\sin t\cos t(\tan t + \cot t) = \sin t\cos t\left(\dfrac{\sin t}{\cos t} + \dfrac{\cos t}{\sin t}\right) = \sin^2 t + \cos^2 t = 1$ **12.** $\dfrac{7\pi}{18}, \dfrac{11\pi}{18}, \dfrac{19\pi}{18}, \dfrac{23\pi}{18}, \dfrac{31\pi}{18},$ and $\dfrac{35\pi}{18}$

13. $\dfrac{\pi}{2}, \dfrac{7\pi}{6}, \dfrac{3\pi}{2}, \dfrac{11\pi}{6}$ **14.** $0, \dfrac{\pi}{3}, \dfrac{5\pi}{3}$ **15.** $0, \dfrac{2\pi}{3}, \dfrac{4\pi}{3}$ **16.** $2.5136, 3.7696$ **17.** $1.2310, \dfrac{\pi}{2}, \dfrac{3\pi}{2}, 5.0522$ **18.** $1.2971, 2.6299, 4.4387, 5.7715$

Cumulative Review Exercises (Chapters 1–3)

1. $\dfrac{\pi}{3}, \dfrac{5\pi}{3}$ **2.** $\dfrac{3\pi}{2}$ **3.** $\dfrac{\pi}{4}, 2.0345, \dfrac{5\pi}{4}, 5.1761$

4.

5.

6.

7.

8–10. Proofs may vary. **11.** $\dfrac{16\pi}{9}$ radians **12.** $-\dfrac{\sqrt{2}}{2}$ **13.** $\sqrt{6}$ **14.** $\dfrac{4}{5}$ **15.** $B = 67°, b = 28.27, c = 30.71$ **16.** $h \approx 15.9$ ft

17. $\dfrac{b - 1}{a}$ **18.** $-\dfrac{2a}{2ab - b}$ or $\dfrac{2a}{b - 2ab}$ **19.** $\sqrt{1 - x^2}$ **20.** $150\pi \approx 471.24$ ft^2

CHAPTER 4

Section 4.1

Check Point Exercises

1. $B = 34°, a \approx 12.7$ cm, $b \approx 7.9$ cm **2.** $B = 117.5°, a \approx 8.7, c \approx 5.2$ **3.** $B \approx 41°, C \approx 82°, c \approx 39.0$ **4.** no triangle
5. two triangles; $B_1 \approx 50°, C_1 \approx 95°, c_1 \approx 20.8; B_2 \approx 130°, C_2 \approx 15°, c_2 \approx 5.4$ **6.** approximately 34 sq m **7.** approximately 11 mi

Concept and Vocabulary Check

1. oblique; sides; angles **2.** $\dfrac{a}{\sin A} = \dfrac{b}{\sin B} = \dfrac{c}{\sin C}$ **3.** side; angles **4.** false **5.** $\dfrac{1}{2}ab\sin C$

Exercise Set 4.1

1. $B = 42°, a \approx 8.1, b \approx 8.1$ **3.** $A = 44°, b \approx 18.6, c \approx 22.8$ **5.** $C = 95°, b \approx 81.0, c \approx 134.1$ **7.** $B = 40°, b \approx 20.9, c \approx 31.8$
9. $C = 111°, b \approx 7.3, c \approx 16.1$ **11.** $A = 80°, a \approx 39.5, c \approx 10.4$ **13.** $B = 30°, a \approx 316.0, b \approx 174.3$ **15.** $C = 50°, a \approx 7.1, b \approx 7.1$
17. one triangle; $B \approx 29°, c \approx 111°, c \approx 29.0$ **19.** one triangle; $C \approx 52°, B \approx 65°, b \approx 10.2$ **21.** one triangle; $C \approx 55°, B \approx 13°, b \approx 10.2$
23. no triangle **25.** two triangles; $B_1 \approx 77°, C_1 \approx 43°, c_1 \approx 12.6; B_2 \approx 103°, C_2 \approx 17°, c_2 \approx 5.4$
27. two triangles; $B_1 \approx 54°, C_1 \approx 89°, c_1 \approx 19.9; B_2 \approx 126°, C_2 \approx 17°, c_2 \approx 5.8$
29. two triangles; $C_1 \approx 68°, B_1 \approx 54°, b_1 \approx 21.0; C_2 \approx 112°, B_2 \approx 10°, b_2 \approx 4.5$ **31.** no triangle **33.** 297 sq ft **35.** 5 sq yd **37.** 10 sq m
39. 481.6 **41.** 64.4 **43.** $A \approx 82°, B \approx 41°, C \approx 57°, c \approx 255.7$ **45.** 10
47. Station A is about 5.7 miles from the fire; station B is about 9.2 miles from the fire. **49.** The platform is about 3671.8 yards from one end of the
beach and 3576.4 yards from the other. **51.** about 184.3 ft **53.** about 56.0 ft **55.** about 30.0 ft **57. a.** $a \approx 493.8$ ft **b.** about 343.0 ft
59. either 9.9 mi or 2.4 mi **71.** does not make sense **73.** does not make sense **75.** no **77.** 41 ft **78.** $\dfrac{\sqrt{6} - \sqrt{2}}{4}$

79. amplitude: 2; period: π; phase shift: $\dfrac{\pi}{4}$

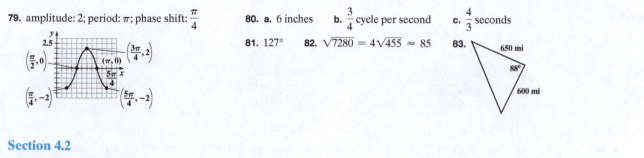

80. a. 6 inches   **b.** $\dfrac{3}{4}$ cycle per second   **c.** $\dfrac{4}{3}$ seconds

81. 127°   **82.** $\sqrt{7280} = 4\sqrt{455} \approx 85$ **83.**

650 mi

88°

600 mi

Section 4.2

Check Point Exercises

1. $a = 13, B \approx 28°, C \approx 32°$ **2.** $A \approx 52°, B \approx 98°, C \approx 30°$ **3.** approximately 917 mi apart   **4.** approximately 47 sq m

Concept and Vocabulary Check

1. $b^2 + c^2 - 2bc \cos A$ **2.** side; Cosines; Sines; acute; 180°   **3.** Cosines; Sines   **4.** $\sqrt{s(s-a)(s-b)(s-c)}; \dfrac{1}{2}(a+b+c)$

Exercise Set 4.2

1. $a \approx 6.0, B \approx 29°, C \approx 105°$ **3.** $c \approx 7.6, A \approx 52°, B \approx 32°$ **5.** $A \approx 44°, B \approx 68°, C \approx 68°$ **7.** $A \approx 117°, B \approx 36°, C \approx 27°$

9. $c \approx 4.7, A \approx 45°, B \approx 93°$ **11.** $a \approx 6.3, C \approx 28°, B \approx 50°$ **13.** $b \approx 4.7, C \approx 55°, A \approx 75°$ **15.** $b \approx 5.4, C \approx 22°, A \approx 68°$

17. $C \approx 112°, A \approx 28°, B \approx 40°$ **19.** $B \approx 100°, A \approx 19°, C \approx 61°$ **21.** $A = 60°, B = 60°, C = 60°$ **23.** $A \approx 117°, B \approx 18°, C = 45°$

25. 4 sq ft   **27.** 22 sq m   **29.** 31 sq yd   **31.** $A \approx 31°, B \approx 19°, C = 130°, c \approx 19.1$

33. $A \approx 51°, B \approx 61°, C \approx 68°, AB = 9, AC = 8.5, BC = 7.5$ **35.** $A \approx 145°, B \approx 13°, C \approx 22°, a = \sqrt{61} \approx 7.8, b = \sqrt{10} \approx 3.2, c = 5$

37. 157°   **39.** about 61.7 mi apart   **41.** about 193 yd   **43.** N12°E   **45. a.** about 19.3 mi   **b.** S58°E

47. The guy wire anchored downhill is about 417.4 feet. The one anchored uphill is about 398.2 feet.   **49.** about 63.7 ft   **51.** $123,454

61. does not make sense   **63.** makes sense   **65.** about 8.9 in. and 23.9 in.   **67.** $\sqrt{m^2 + h^2 - mh}$

69. $\csc x \cos^2 x + \sin x = \csc x(1 - \sin^2 x) + \sin x = \csc x - \csc x \sin^2 x + \sin x$

$$= \csc x - \dfrac{1}{\sin x}\sin^2 x + \sin x = \csc x - \sin x + \sin x = \csc x$$

70. $\left\{\dfrac{3\pi}{2}\right\}$ **71.**

$y = 3 \tan \dfrac{x}{2}$

$\left(\dfrac{5\pi}{2}, 3\right)$

$\left(\dfrac{\pi}{2}, 3\right)$

$(2\pi, 0)$

$(0, 0)$

$\left(-\dfrac{\pi}{2}, -3\right)$

$\left(\dfrac{3\pi}{2}, -3\right)$

$x = -\pi \quad x = \pi \quad x = 3\pi$

72. Yes, both have length $3\sqrt{5}$.

73. Yes, both have slope 2.

74. $8x + 34y$

Mid-Chapter 4 Check Point

1. $C = 107°, b \approx 24.8, c \approx 36.1$ **2.** $B \approx 37°, C \approx 101°, c \approx 92.4$ **3.** no triangle   **4.** $A \approx 26°, C \approx 44°, b \approx 21.6$ **5.** Two triangles: $A_1 \approx 55°, B_1 \approx 83°, b_1 \approx 19.3; A_2 \approx 125°, B_2 \approx 13°, b_2 \approx 4.4$ **6.** $A \approx 28°, B \approx 42°, C \approx 110°$ **7.** 10 ft^2 **8.** ≈ 31 m^2 **9.** 147.9 miles

10. 15.0 miles   **11.** 327.0 ft

Section 4.3

Check Point Exercises

1. $\|\mathbf{u}\| = 5 = \|\mathbf{v}\|$ and $m_u = \dfrac{4}{3} = m_v$ **2.**

$\mathbf{v} = 3\mathbf{i} - 3\mathbf{j}$

$(0, 0)$

$(3, -3)$

$; \|\mathbf{v}\| = 3\sqrt{2}$ **3.** $\mathbf{v} = 3\mathbf{i} + 4\mathbf{j}$ **4. a.** $11\mathbf{i} - 2\mathbf{j}$ **b.** $3\mathbf{i} + 8\mathbf{j}$

5. a. $56\mathbf{i} + 80\mathbf{j}$ **b.** $-35\mathbf{i} - 50\mathbf{j}$ **6.** $30\mathbf{i} + 33\mathbf{j}$ **7.** $\dfrac{4}{5}\mathbf{i} - \dfrac{3}{5}\mathbf{j}; \sqrt{\left(\dfrac{4}{5}\right)^2 + \left(-\dfrac{3}{5}\right)^2} = \sqrt{\dfrac{16}{25} + \dfrac{9}{25}} = \sqrt{\dfrac{25}{25}} = 1$

8. $30\sqrt{2}\mathbf{i} + 30\sqrt{2}\mathbf{j}$ **9.** 82.54 lb; 46.2°

Concept and Vocabulary Check

1. vector   **2.** scalar   **3.** $\mathbf{v}; \mathbf{b}$ **4.** $\mathbf{a}$ **5.** $\mathbf{w}$ **6.** unit; $x; y$ **7.** $a; b; \sqrt{a^2 + b^2}$ **8.** position   **9.** $x_2 - x_1; y_2 - y_1$

10. $a_1 + a_2; b_1 + b_2; a_1 - a_2; b_1 - b_2; ka_1; kb_1$ **11.** $\dfrac{\mathbf{v}}{\|\mathbf{v}\|}$; magnitude   **12.** $\cos\theta; \sin\theta$ **13.** resultant

Exercise Set 4.3

1. a. $\sqrt{41}$ **b.** $\sqrt{41}$ **c.** $\mathbf{u} = \mathbf{v}$ **3. a.** 6 **b.** 6 **c.** $\mathbf{u} = \mathbf{v}$

5.

$\sqrt{10}$

7.

$\sqrt{2}$

9.

$2\sqrt{10}$

11.

4

13. $10\mathbf{i} + 6\mathbf{j}$ **15.** $6\mathbf{i} - 3\mathbf{j}$ **17.** $-6\mathbf{i} - 14\mathbf{j}$ **19.** $9\mathbf{i}$ **21.** $-\mathbf{i} + 2\mathbf{j}$ **23.** $5\mathbf{i} - 12\mathbf{j}$ **25.** $-5\mathbf{i} + 12\mathbf{j}$ **27.** $-15\mathbf{i} + 35\mathbf{j}$ **29.** $4\mathbf{i} + 24\mathbf{j}$

31. $-9\mathbf{i} - 4\mathbf{j}$ **33.** $-5\mathbf{i} + 45\mathbf{j}$ **35.** $2\sqrt{29}$ **37.** $\sqrt{10}$ **39.** $\mathbf{i}$ **41.** $\dfrac{3}{5}\mathbf{i} - \dfrac{4}{5}\mathbf{j}$ **43.** $\dfrac{3\sqrt{13}}{13}\mathbf{i} - \dfrac{2\sqrt{13}}{3}\mathbf{j}$ **45.** $\dfrac{\sqrt{2}}{2}\mathbf{i} + \dfrac{\sqrt{2}}{2}\mathbf{j}$ **47.** $3\sqrt{3}\mathbf{i} + 3\mathbf{j}$

49. $-6\sqrt{2}\mathbf{i} - 6\sqrt{2}\mathbf{j}$ **51.** $\approx -0.20\mathbf{i} + 0.46\mathbf{j}$ **53.** $-23\mathbf{i} + 14\mathbf{j}$ **55.** -60 **57.** commutative property **59.** distributive property

61. $18.03; 123.7°$ **63.** $6; 90°$ **65.** $22\sqrt{3}\mathbf{i} + 22\mathbf{j}$ **67.** $148.5\mathbf{i} + 20.9\mathbf{j}$ **69.** $\approx 1.4\mathbf{i} + 0.6\mathbf{j}; 1.4$ in. **71.** ≈ 108.21 lbs; $374.4°$

73. 2038 kg; $162.8°$ **75.** ≈ 30.9 lbs **77. a.** 335 lb **b.** 3484 lb **79. a.** $\mathbf{F} = 9\mathbf{i} - 3\mathbf{j}$ **b.** $\mathbf{F}_3 = -9\mathbf{i} + 3\mathbf{j}$ **81. a.** $\mathbf{F} = -2\mathbf{j}$ **b.** $\mathbf{F}_5 = 2\mathbf{j}$

83. a. $\mathbf{v} = 180\cos 40°\mathbf{i} + 180\sin 40°\mathbf{j} \approx 137.89\mathbf{i} + 115.70\mathbf{j}, \mathbf{w} = 40\cos 0°\mathbf{i} + 40\sin 0°\mathbf{j} = 40\mathbf{i}$ **b.** $\mathbf{v} + \mathbf{w} \approx 177.89\mathbf{i} + 115.70\mathbf{j}$ **c.** 212 mph

d. $33.0°; N57°E$ **85.** 78 mph, $75.4°$ **105.** does not make sense **107.** does not make sense **109.** true **111.** true

113. The plane's true speed relative to the ground is about 269 miles per hour.; The compass heading relative to the ground is $278.3°$.

115. a. $104°$ **b.** decrease **116.** amplitude: 3; period: π; phase shift: $-\dfrac{\pi}{2}$; **117.** $\dfrac{1 + \sin x}{1 - \sin x} - \dfrac{1 - \sin x}{1 + \sin x} = \dfrac{(1 + \sin x)^2 - (1 - \sin x)^2}{(1 - \sin x)(1 + \sin x)}$

$$= \dfrac{(1 + 2\sin x + \sin^2 x) - (1 - 2\sin x + \sin^2 x)}{1 - \sin^2 x}$$

$$= \dfrac{4\sin x}{\cos^2 x} = 4\,\dfrac{\sin x}{\cos x}\,\dfrac{1}{\cos x} = 4\tan x \sec x$$

118. $\left\{\dfrac{\pi}{6}, \dfrac{5\pi}{6}, \dfrac{3\pi}{2}\right\}$ **119.** $137.7°$ **120.** $\dfrac{7}{5}\mathbf{i} - \dfrac{21}{5}\mathbf{j}$ **121. a.** $\|\mathbf{u}\|^2 = \|\mathbf{v}\|^2 + \|\mathbf{w}\|^2 - 2\|\mathbf{v}\|\|\mathbf{w}\|\cos\theta$

b. $\|\mathbf{u}\| = \sqrt{(a_1 - a_2)^2 + (b_1 - b_2)^2}; \|\mathbf{u}\|^2 = (a_1 - a_2)^2 + (b_1 - b_2)^2; \|\mathbf{v}\| = \sqrt{a_1^2 + b_1^2}; \|\mathbf{v}\|^2 = a_1^2 + b_1^2; \|\mathbf{w}\| = \sqrt{a_2^2 + b_2^2}; \|\mathbf{w}\|^2 = a_2^2 + b_2^2$

Section 4.4

Check Point Exercises

1. a. 18 **b.** 18 **c.** 5 **2.** $100.3°$ **3.** orthogonal **4.** $\dfrac{7}{2}\mathbf{i} - \dfrac{7}{2}\mathbf{j}$ **5.** $\mathbf{v}_1 = \dfrac{7}{2}\mathbf{i} - \dfrac{7}{2}\mathbf{j}; \mathbf{v}_2 = -\dfrac{3}{2}\mathbf{i} - \dfrac{3}{2}\mathbf{j}$ **6.** approximately 2598 ft-lb

Concept and Vocabulary Check

1. dot product; $a_1a_2 + b_1b_2$ **2.** $\|\mathbf{v}\|\|\mathbf{w}\|\cos\theta$ **3.** orthogonal **4.** true **5.** false

Exercise Set 4.4

1. $6; 10$ **3.** $-6; 41$ **5.** $100; 61$ **7.** $0; 25$ **9.** 3 **11.** 3 **13.** 20 **15.** 20 **17.** $79.7°$ **19.** $160.3°$ **21.** $38.7°$ **23.** orthogonal

25. orthogonal **27.** not orthogonal **29.** not orthogonal **31.** orthogonal **33.** $\mathbf{v}_1 = \text{proj}_\mathbf{w}\mathbf{v} = \dfrac{5}{2}\mathbf{i} - \dfrac{5}{2}\mathbf{j}; \mathbf{v}_2 = \dfrac{1}{2}\mathbf{i} + \dfrac{1}{2}\mathbf{j}$

35. $\mathbf{v}_1 = \text{proj}_\mathbf{w}\mathbf{v} = -\dfrac{26}{29}\mathbf{i} + \dfrac{65}{29}\mathbf{j}; \mathbf{v}_2 = \dfrac{55}{29}\mathbf{i} + \dfrac{22}{29}\mathbf{j}$ **37.** $\mathbf{v}_1 = \text{proj}_\mathbf{w}\mathbf{v} = \mathbf{i} + 2\mathbf{j}; \mathbf{v}_2 = 0$ **39.** 25 **41.** $5\mathbf{i} - 5\mathbf{j}$ **43.** $30°$ **45.** parallel

47. neither **49.** orthogonal **51.** $1617; \mathbf{v} \cdot \mathbf{w} = 1617$ means that $1617 in revenue is generated when 240 gallons of regular gasoline are sold at $2.90 per gallon and 300 gallons of premium gasoline are sold at $3.07 per gallon. **53.** 7600 foot-pounds **55.** 3392 foot-pounds

57. 1079 foot-pounds **59.** 40 foot-pounds **61.** 22 foot-pounds

63. a. $\dfrac{\sqrt{3}}{2}\mathbf{i} + \dfrac{1}{2}\mathbf{j}$ **b.** $-175\sqrt{3}\mathbf{i} - 175\mathbf{j}$ **c.** 350; A force of 350 pounds is required to keep the boat from rolling down the ramp.

75. makes sense **77.** makes sense

79. $\mathbf{u} \cdot \mathbf{v} = (a_1\mathbf{i} + b_1\mathbf{j}) \cdot (a_2\mathbf{i} + b_2\mathbf{j})$
$= a_1a_2 + b_1b_2$
$= a_2a_1 + b_2b_1$
$= (a_2\mathbf{i} + b_2\mathbf{j}) \cdot (a_1\mathbf{i} + b_1\mathbf{j})$
$= \mathbf{v} \cdot \mathbf{u}$

81. $\mathbf{u} \cdot (\mathbf{v} + \mathbf{w}) = (a_1\mathbf{i} + b_1\mathbf{j}) \cdot [(a_2\mathbf{i} + b_2\mathbf{j}) + (a_3\mathbf{i} + b_3\mathbf{j})]$
$= (a_1\mathbf{i} + b_1\mathbf{j}) \cdot [(a_2 + a_3)\mathbf{i} + (b_2 + b_3)\mathbf{j}]$
$= a_1(a_2 + a_3) + b_1(b_2 + b_3)$
$= a_1a_2 + a_1a_3 + b_1b_2 + b_1b_3$
$= a_1a_2 + b_1b_2 + a_1a_3 + b_1b_3$
$= (a_1\mathbf{i} + b_1\mathbf{j}) \cdot (a_2\mathbf{i} + b_2\mathbf{j}) + (a_1\mathbf{i} + b_1\mathbf{j}) \cdot (a_3\mathbf{i} + b_3\mathbf{j})$
$= \mathbf{u} \cdot \mathbf{v} + \mathbf{u} \cdot \mathbf{w}$

83. $b = -20$ **85.** any two vectors, **v** and **w**, having the same direction   **87.** $\left\{\dfrac{\pi}{4}, \dfrac{3\pi}{4}, \dfrac{5\pi}{4}, \dfrac{7\pi}{4}\right\}$

88. $\dfrac{2\tan x}{1 + \tan^2 x} = \dfrac{2\tan x}{\sec^2 x} = 2\tan x \cos^2 x = 2\dfrac{\sin x}{\cos x}\cos^2 x = 2\sin x \cos x = \sin 2x$ **89.** $\dfrac{1}{\sqrt{x^2 + 1}}$ **90.** $-14 - 29x + 15x^2$ or $15x^2 - 29x - 14$

91. $\sqrt{2}$ **92.** $-\dfrac{54 + 43\sqrt{2}}{46}$

Chapter 4 Review Exercises

1. $C = 55°, b \approx 10.5$, and $c \approx 10.5$ **2.** $A = 43°, a \approx 171.9$, and $b \approx 241.0$ **3.** $b \approx 16.3, A \approx 72°$, and $C \approx 42°$
4. $C \approx 98°, A \approx 55°$, and $B \approx 27°$ **5.** $C = 120°, a \approx 45.0$, and $b \approx 33.2$ **6.** two triangles; $B_1 \approx 55°, C_1 \approx 86°$, and $c_1 \approx 31.7$;
$B_2 \approx 125°, C_2 \approx 16°$, and $c_2 \approx 8.8$ **7.** no triangle   **8.** $a \approx 59.0, B \approx 3°$, and $C \approx 15°$ **9.** $B \approx 78°, A \approx 39°$, and $C \approx 63°$
10. $B \approx 25°, C \approx 115°$, and $c \approx 8.5$ **11.** two triangles; $A_1 \approx 59°, C_1 \approx 84°, c_1 \approx 14.4$; $A_2 \approx 121°, C_2 \approx 22°, c_2 \approx 5.4$
12. $B \approx 9°, C \approx 148°$, and $c \approx 73.6$ **13.** 8 sq ft   **14.** 4 sq ft   **15.** 4 sq m   **16.** 2 sq m   **17.** 30.5 ft   **18.** 35.6 mi
19. 861 mi   **20.** 404 ft; 551 ft   **21.** \$214,194

22. ; 5   **23.** ; $\sqrt{29}$ **24.** ; 3

25. $3\mathbf{i} - 2\mathbf{j}$ **26.** $\mathbf{i} - 2\mathbf{j}$ **27.** $-\mathbf{i} + 2\mathbf{j}$ **28.** $-3\mathbf{i} + 12\mathbf{j}$ **29.** $12\mathbf{i} - 51\mathbf{j}$ **30.** $2\sqrt{26}$ **31.** $\dfrac{4}{5}\mathbf{i} - \dfrac{3}{5}\mathbf{j}$ **32.** $-\dfrac{\sqrt{5}}{5}\mathbf{i} + \dfrac{2\sqrt{5}}{5}\mathbf{j}$
33. $6\mathbf{i} + 6\sqrt{3}\mathbf{j}$ **34.** 270 lb; 27.7°   **35. a.** $13.59\mathbf{i} + 2.34\mathbf{j}$ **b.** 14 mph   **c.** 13.9°   **36.** 4   **37.** 2; 86.1°   **38.** -32; 124.8°
39. 1; 71.6°   **40.** orthogonal   **41.** not orthogonal   **42.** $\mathbf{v}_1 = \text{proj}_{\mathbf{w}}\mathbf{v} = \dfrac{50}{41}\mathbf{i} + \dfrac{40}{41}\mathbf{j}; \mathbf{v}_2 = -\dfrac{132}{41}\mathbf{i} + \dfrac{165}{41}\mathbf{j}$
43. $\mathbf{v}_1 = \text{proj}_{\mathbf{w}}\mathbf{v} = -\dfrac{3}{2}\mathbf{i} + \dfrac{1}{2}\mathbf{j}; \mathbf{v}_2 = \dfrac{1}{2}\mathbf{i} + \dfrac{3}{2}\mathbf{j}$ **44.** 1115 ft-lb

Chapter 4 Test

1. 8.0   **2.** 6.2   **3.** 206 sq in.   **4. a.** $\mathbf{i} + 2\mathbf{j}$ **b.** $\sqrt{5}$ **5.** $-23\mathbf{i} + 22\mathbf{j}$ **6.** -18 **7.** 138°   **8.** $-\dfrac{9}{5}\mathbf{i} + \dfrac{18}{5}\mathbf{j}$ **9.** 1.0 mi
10. 323 pounds; 3.4°   **11.** 1966 ft-lb

Cumulative Review Exercises (Chapters 1–4)

1. $\left\{\dfrac{\pi}{6}, \dfrac{5\pi}{6}, \dfrac{\pi}{2}\right\}$ **2.** $\left\{\dfrac{3\pi}{4}, \dfrac{7\pi}{4}\right\}$

3. $\left(\dfrac{3\pi}{4}, 3\right)$ [graph]

4. [graph]

5. $\sin\theta\csc\theta - \cos^2\theta = \sin\theta\left(\dfrac{1}{\sin\theta}\right) - \cos^2\theta$
$= 1 - \cos^2\theta = \sin^2\theta$
6. $\cos\left(\theta + \dfrac{3\pi}{2}\right) = \cos\theta\cos\dfrac{3\pi}{2} - \sin\theta\sin\dfrac{3\pi}{2}$
$= \cos\theta(0) - \sin\theta(-1) = \sin\theta$

7. $5\pi \approx 15.71$ in.   **8.** 0   **9.** $\dfrac{\sqrt{5}}{5}$ **10.** 171 ft   **11.** $\cos\theta = -\dfrac{2\sqrt{29}}{29}; \sin\theta = \dfrac{5\sqrt{29}}{29}$
12. $b \approx 92.9$ **13.** $C \approx 16.2, A \approx 50°, B \approx 100°$ **14.** -4π radians   **15.** 280°   **16.** -2 **17.** 22.4 mi
18. a. 4 m   **b.** $\dfrac{5}{2\pi}$ **c.** $\dfrac{2\pi}{5}$ sec   **19.** $\dfrac{\sqrt{\sqrt{2} + 2}}{2}$ **20. a.** $5\mathbf{i} + 23\mathbf{j}$ **b.** -12

CHAPTER 5

Section 5.1
Check Point Exercises

1. a. $8 + i$ **b.** $-10 + 7i$ **2. a.** $63 + 14i$ **b.** $58 - 11i$ **3.** $\dfrac{16}{17} + \dfrac{21}{17}i$ **4. a.** $7i\sqrt{3}$ **b.** $1 - 4i\sqrt{3}$ **c.** $-7 + i\sqrt{3}$

Concept and Vocabulary Check

1. $\sqrt{-1}; -1$ **2.** complex; imaginary; real   **3.** $-6i$ **4.** $14i$ **5.** $18; -15i; 12i; -10i^2; 10$ **6.** $2 + 9i$ **7.** $2 + 5i$ **8.** $i; i; 2i\sqrt{5}$

Exercise Set 5.1

1. $8 - 2i$ **3.** $-2 + 9i$ **5.** $24 - 3i$ **7.** $-14 + 17i$ **9.** $21 + 15i$ **11.** $-19 + 7i$ **13.** $-29 - 11i$ **15.** 34 **17.** 26 **19.** $-5 + 12i$

21. $\dfrac{3}{5} + \dfrac{1}{5}i$ **23.** $1 + i$ **25.** $-\dfrac{24}{25} + \dfrac{32}{25}i$ **27.** $\dfrac{7}{5} + \dfrac{4}{5}i$ **29.** $3i$ **31.** $47i$ **33.** $-8i$ **35.** $2 + 6i\sqrt{7}$ **37.** $-\dfrac{1}{3} + i\dfrac{\sqrt{2}}{6}$ **39.** $-\dfrac{1}{8} - i\dfrac{\sqrt{3}}{24}$

41. $-2\sqrt{6} - 2i\sqrt{10}$ **43.** $24\sqrt{15}$ **45.** $-11 - 5i$ **47.** $-5 + 10i$ **49.** $0 + 47i$ or $47i$ **51.** 0 **53.** $\dfrac{20}{13} + \dfrac{30}{13}i$ **55.** $(47 + 13i)$ volts

57. $(5 + i\sqrt{15}) + (5 - i\sqrt{15}) = 10$; $(5 + i\sqrt{15})(5 - i\sqrt{15}) = 25 - 15i^2 = 25 + 15 = 40$ **67.** makes sense **69.** does not make sense

71. false **73.** false **75.** $\dfrac{14}{25} - \dfrac{2}{25}i$ **77.** $\dfrac{8}{5} + \dfrac{16}{5}i$ **78.** $\sin^2 x \tan^2 x + \cos^2 x \tan^2 x = (\sin^2 x + \cos^2 x)\tan^2 x = \tan^2 x = \sec^2 x - 1$

79. $\left\{\dfrac{\pi}{6}, \dfrac{\pi}{2}, \dfrac{5\pi}{6}\right\}$ **80.** amplitude: $\dfrac{1}{2}$; period: $\dfrac{2\pi}{3}$; phase shift: $-\dfrac{\pi}{6}$

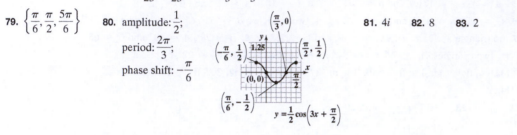

81. $4i$ **82.** 8 **83.** 2

Section 5.2

Check Point Exercises

1. a. **b.** **c.** **d.**

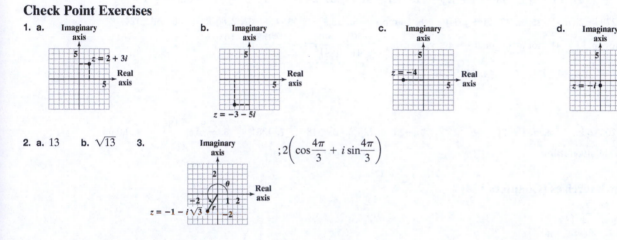

2. a. 13 **b.** $\sqrt{13}$ **3.** $; 2\left(\cos\dfrac{4\pi}{3} + i\sin\dfrac{4\pi}{3}\right)$

4. $z = 2\sqrt{3} + 2i$ **5.** $30(\cos 60° + i\sin 60°)$ **6.** $10(\cos \pi + i\sin \pi)$ **7.** $-16\sqrt{3} + 16i$ **8.** -4

9. $2(\cos 15° + i\sin 15°); 2(\cos 105° + i\sin 105°); 2(\cos 195° + i\sin 195°); 2(\cos 285° + i\sin 285°)$ **10.** $3; -\dfrac{3}{2} + \dfrac{3\sqrt{3}}{2}i; -\dfrac{3}{2} - \dfrac{3\sqrt{3}}{2}i$

Concept and Vocabulary Check

1. real; imaginary **2.** absolute value **3.** modulus; argument **4.** $\sqrt{a^2 + b^2}; \dfrac{b}{a}$ **5.** $r_1 r_2; \theta_1 + \theta_2; \theta_1 + \theta_2$; multiplying; adding

6. $\dfrac{r_1}{r_2}; \theta_1 - \theta_2; \theta_1 - \theta_2$; dividing; subtracting **7.** $r^n; n\theta; n\theta$ **8.** n

Exercise Set 5.2

1. $; 4$ **3.** $; 3$ **5.** $; \sqrt{13}$

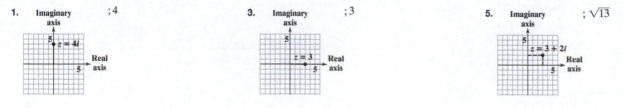

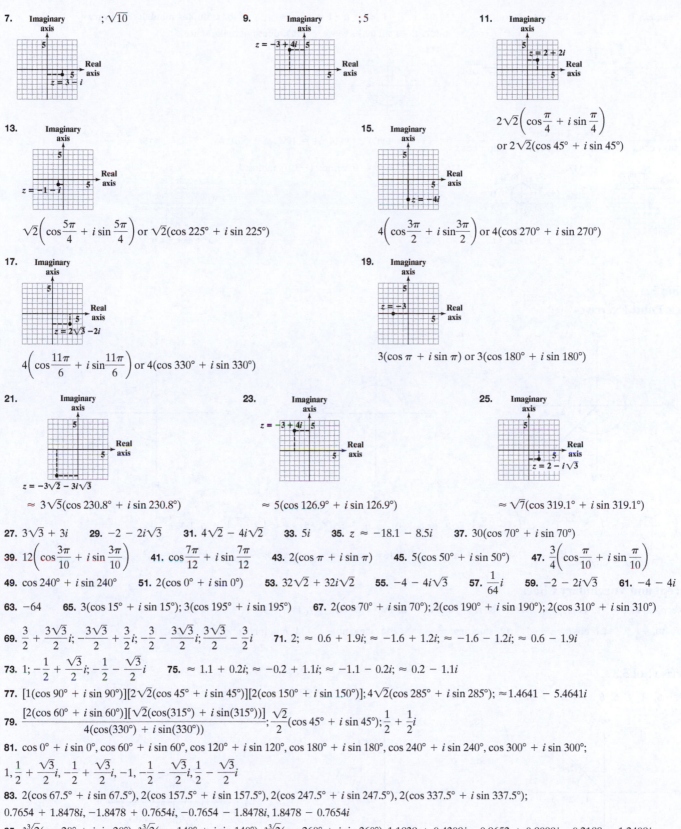

7.   ; $\sqrt{10}$

9.   ; 5

11. $2\sqrt{2}\left(\cos\dfrac{\pi}{4} + i\sin\dfrac{\pi}{4}\right)$

or $2\sqrt{2}(\cos 45° + i\sin 45°)$

13. $\sqrt{2}\left(\cos\dfrac{5\pi}{4} + i\sin\dfrac{5\pi}{4}\right)$ or $\sqrt{2}(\cos 225° + i\sin 225°)$

15. $4\left(\cos\dfrac{3\pi}{2} + i\sin\dfrac{3\pi}{2}\right)$ or $4(\cos 270° + i\sin 270°)$

17. $4\left(\cos\dfrac{11\pi}{6} + i\sin\dfrac{11\pi}{6}\right)$ or $4(\cos 330° + i\sin 330°)$

19. $3(\cos\pi + i\sin\pi)$ or $3(\cos 180° + i\sin 180°)$

21. $\approx 3\sqrt{5}(\cos 230.8° + i\sin 230.8°)$

23. $\approx 5(\cos 126.9° + i\sin 126.9°)$

25. $\approx \sqrt{7}(\cos 319.1° + i\sin 319.1°)$

27. $3\sqrt{3} + 3i$ 29. $-2 - 2i\sqrt{3}$ 31. $4\sqrt{2} - 4i\sqrt{2}$ 33. $5i$ 35. $z \approx -18.1 - 8.5i$ 37. $30(\cos 70° + i\sin 70°)$

39. $12\left(\cos\dfrac{3\pi}{10} + i\sin\dfrac{3\pi}{10}\right)$ 41. $\cos\dfrac{7\pi}{12} + i\sin\dfrac{7\pi}{12}$ 43. $2(\cos\pi + i\sin\pi)$ 45. $5(\cos 50° + i\sin 50°)$ 47. $\dfrac{3}{4}\left(\cos\dfrac{\pi}{10} + i\sin\dfrac{\pi}{10}\right)$

49. $\cos 240° + i\sin 240°$ 51. $2(\cos 0° + i\sin 0°)$ 53. $32\sqrt{2} + 32i\sqrt{2}$ 55. $-4 - 4i\sqrt{3}$ 57. $\dfrac{1}{64}i$ 59. $-2 - 2i\sqrt{3}$ 61. $-4 - 4i$

63. -64 65. $3(\cos 15° + i\sin 15°); 3(\cos 195° + i\sin 195°)$ 67. $2(\cos 70° + i\sin 70°); 2(\cos 190° + i\sin 190°); 2(\cos 310° + i\sin 310°)$

69. $\dfrac{3}{2} + \dfrac{3\sqrt{3}}{2}i; -\dfrac{3\sqrt{3}}{2} + \dfrac{3}{2}i; -\dfrac{3}{2} - \dfrac{3\sqrt{3}}{2}i; \dfrac{3\sqrt{3}}{2} - \dfrac{3}{2}i$ 71. $2; \approx 0.6 + 1.9i; \approx -1.6 + 1.2i; \approx -1.6 - 1.2i; \approx 0.6 - 1.9i$

73. $1; -\dfrac{1}{2} + \dfrac{\sqrt{3}}{2}i; -\dfrac{1}{2} - \dfrac{\sqrt{3}}{2}i$ 75. $\approx 1.1 + 0.2i; \approx -0.2 + 1.1i; \approx -1.1 - 0.2i; \approx 0.2 - 1.1i$

77. $[1(\cos 90° + i\sin 90°)][2\sqrt{2}(\cos 45° + i\sin 45°)][2(\cos 150° + i\sin 150°)]; 4\sqrt{2}(\cos 285° + i\sin 285°); \approx 1.4641 - 5.4641i$

79. $\dfrac{[2(\cos 60° + i\sin 60°)][\sqrt{2}(\cos(315°) + i\sin(315°))]}{4(\cos(330°) + i\sin(330°))}; \dfrac{\sqrt{2}}{2}(\cos 45° + i\sin 45°); \dfrac{1}{2} + \dfrac{1}{2}i$

81. $\cos 0° + i\sin 0°, \cos 60° + i\sin 60°, \cos 120° + i\sin 120°, \cos 180° + i\sin 180°, \cos 240° + i\sin 240°, \cos 300° + i\sin 300°;$

$1, \dfrac{1}{2} + \dfrac{\sqrt{3}}{2}i, -\dfrac{1}{2} + \dfrac{\sqrt{3}}{2}i, -1, -\dfrac{1}{2} - \dfrac{\sqrt{3}}{2}i, \dfrac{1}{2} - \dfrac{\sqrt{3}}{2}i$

83. $2(\cos 67.5° + i\sin 67.5°), 2(\cos 157.5° + i\sin 157.5°), 2(\cos 247.5° + i\sin 247.5°), 2(\cos 337.5° + i\sin 337.5°);$

$0.7654 + 1.8478i, -1.8478 + 0.7654i, -0.7654 - 1.8478i, 1.8478 - 0.7654i$

85. $\sqrt[3]{2}(\cos 20° + i\sin 20°), \sqrt[3]{2}(\cos 140° + i\sin 140°), \sqrt[3]{2}(\cos 260° + i\sin 260°); 1.1839 + 0.4309i, -0.9652 + 0.8099i, -0.2188 - 1.2408i$

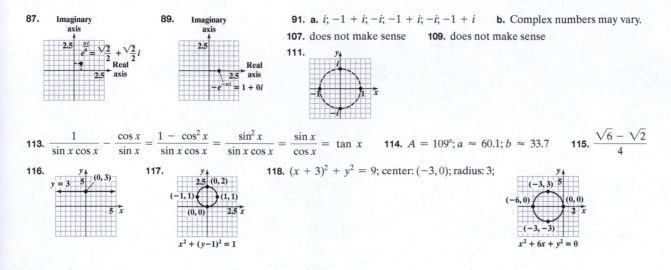

87. Imaginary axis

$e^{\frac{\pi i}{4}} = \frac{\sqrt{2}}{2} + \frac{\sqrt{2}}{2}i$

89. Imaginary axis

$-e^{-\pi i} = 1 + 0i$

91. a. $i; -1 + i; -i; -1 + i; -i; -1 + i$ **b.** Complex numbers may vary.

107. does not make sense   **109.** does not make sense

111.

113. $\dfrac{1}{\sin x \cos x} - \dfrac{\cos x}{\sin x} = \dfrac{1 - \cos^2 x}{\sin x \cos x} = \dfrac{\sin^2 x}{\sin x \cos x} = \dfrac{\sin x}{\cos x} = \tan x$ **114.** $A = 109°; a \approx 60.1; b \approx 33.7$ **115.** $\dfrac{\sqrt{6} - \sqrt{2}}{4}$

116.

$y = 3$ $(0, 3)$

117.

$(0, 2)$ $(-1, 1)$ $(1, 1)$ $(0, 0)$

$x^2 + (y-1)^2 = 1$

118. $(x + 3)^2 + y^2 = 9$; center: $(-3, 0)$; radius: 3;

$(-3, 3)$ $(-6, 0)$ $(0, 0)$ $(-3, -3)$

$x^2 + 6x + y^2 = 0$

Section 5.3

Check Point Exercises

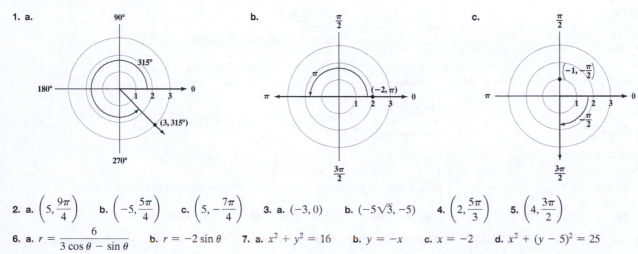

1. a. $90°$ ... $315°$ ... $180°$ ... 0 ... $270°$ ... $(3, 315°)$

b. $\frac{\pi}{2}$ ... π ... $(-2, \pi)$ ... 0 ... $\frac{3\pi}{2}$

c. $\frac{\pi}{2}$ ... $\left(-1, -\frac{\pi}{2}\right)$ ... π ... 0 ... $-\frac{\pi}{2}$ ... $\frac{3\pi}{2}$

2. a. $\left(5, \dfrac{9\pi}{4}\right)$ **b.** $\left(-5, \dfrac{5\pi}{4}\right)$ **c.** $\left(5, -\dfrac{7\pi}{4}\right)$ **3. a.** $(-3, 0)$ **b.** $(-5\sqrt{3}, -5)$ **4.** $\left(2, \dfrac{5\pi}{3}\right)$ **5.** $\left(4, \dfrac{3\pi}{2}\right)$

6. a. $r = \dfrac{6}{3 \cos \theta - \sin \theta}$ **b.** $r = -2 \sin \theta$ **7. a.** $x^2 + y^2 = 16$ **b.** $y = -x$ **c.** $x = -2$ **d.** $x^2 + (y - 5)^2 = 25$

Concept and Vocabulary Check

1. pole; polar axis   **2.** pole; polar axis   **3.** II   **4.** IV   **5.** IV   **6.** III   **7.** IV   **8.** II

9. r **10.** $-r$ **11.** $r \cos \theta; r \sin \theta$ **12.** squaring; $x^2 + y^2$ **13.** tangent; $\dfrac{y}{x}$ **14.** multiplying; $r; x^2 + y^2; y$

Exercise Set 5.3

1. C **3.** A **5.** B **7.** C **9.** A

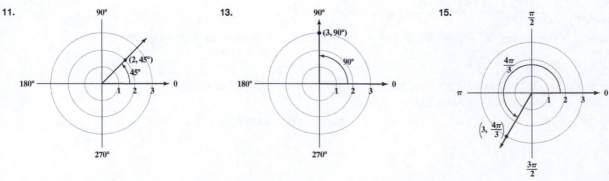

11. $90°$ ... $(2, 45°)$ ... $45°$ ... $180°$ ... 0 ... $270°$

13. $90°$ ... $(3, 90°)$ ... $90°$ ... $180°$ ... 0 ... $270°$

15. $\frac{\pi}{2}$ ... $\frac{4\pi}{3}$ ... π ... 0 ... $\left(3, \dfrac{4\pi}{3}\right)$ ... $\frac{3\pi}{2}$

17.

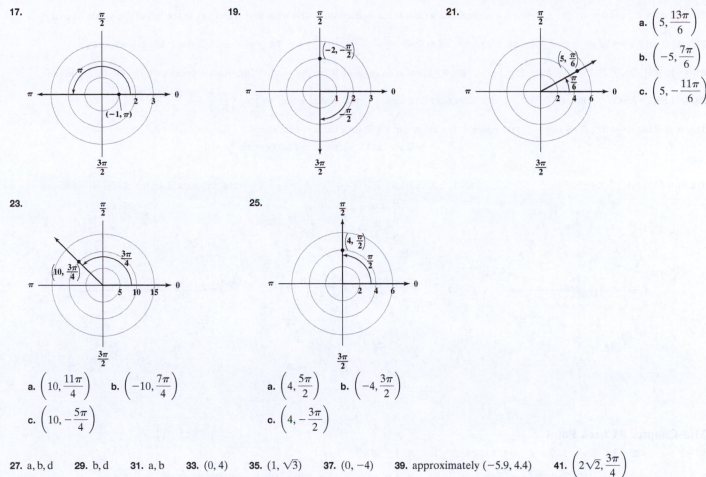

19.

21.

a. $\left(5, \dfrac{13\pi}{6}\right)$

b. $\left(-5, \dfrac{7\pi}{6}\right)$

c. $\left(5, -\dfrac{11\pi}{6}\right)$

23.

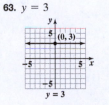

a. $\left(10, \dfrac{11\pi}{4}\right)$ b. $\left(-10, \dfrac{7\pi}{4}\right)$

c. $\left(10, -\dfrac{5\pi}{4}\right)$

25.

a. $\left(4, \dfrac{5\pi}{2}\right)$ b. $\left(-4, \dfrac{3\pi}{2}\right)$

c. $\left(4, -\dfrac{3\pi}{2}\right)$

27. a, b, d   **29.** b, d   **31.** a, b   **33.** $(0, 4)$ **35.** $(1, \sqrt{3})$ **37.** $(0, -4)$ **39.** approximately $(-5.9, 4.4)$ **41.** $\left(2\sqrt{2}, \dfrac{3\pi}{4}\right)$

43. $\left(4, \dfrac{5\pi}{3}\right)$ **45.** $\left(2, \dfrac{7\pi}{6}\right)$ **47.** $(5, 0)$ **49.** $r = \dfrac{7}{3\cos\theta + \sin\theta}$ **51.** $r = \dfrac{7}{\cos\theta}$ **53.** $r = 3$ **55.** $r = 4\cos\theta$ **57.** $r = \dfrac{6\cos\theta}{\sin^2\theta}$

59. $x^2 + y^2 = 64$

61. $x = 0$

63. $y = 3$

65. $y = 4$

67. $x^2 + y^2 = y$

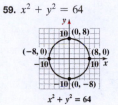

69. $(x - 6)^2 + y^2 = 36$

71. $x^2 + y^2 = 6x + 4y$

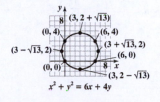

73. $y = \dfrac{1}{x}$

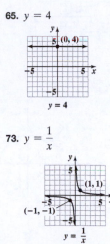

75. $r = a \sec \theta$; $r \cos \theta = a$; $x = a$; $x = a$ is a vertical line a units to the right of the y-axis when $a > 0$ and $|a|$ to the left of the y-axis when $a < 0$.

77. $r = a \sin \theta$; $r^2 = ar \sin \theta$; $x^2 + y^2 = ay$; $x^2 + y^2 - ay = 0$; $x^2 + \left(y - \dfrac{a}{2}\right)^2 = \left(\dfrac{a}{2}\right)^2$ **79.** $y = x + 2\sqrt{2}$; slope: 1; y-intercept: $2\sqrt{2}$

81. $(-1, \sqrt{3}), (2\sqrt{3}, 2); 2\sqrt{5}$ **83.** $\left(15, \dfrac{4\pi}{3}\right)$ **85.** 6.3 knots at an angle of 50° to the wind   **87.** Answers may vary.   **97.** $(-2, 3.464)$

99. $(-1.857, -3.543)$ **101.** $(3, 0.730)$ **103.** does not make sense   **105.** makes sense   **109.** $\left\{\dfrac{\pi}{3}, \dfrac{\pi}{2}, \dfrac{2\pi}{3}, \dfrac{4\pi}{3}, \dfrac{3\pi}{2}, \dfrac{5\pi}{3}\right\}$

110. $a \approx 23.6$; $B \approx 60°$; $C \approx 100°$ **111.** $(\sin x + \cos x)^2 = \sin^2 x + 2 \sin x \cos x + \cos^2 x$
$$= 2 \sin x \cos x + (\sin^2 x + \cos^2 x) = \sin 2x + 1$$

112. $0; 0.13; 0.5; 1; 1.5; 1.87; 2$ **113.** $1; 2; 2.73; 3; 2.73; 2; 1; 0; -0.73; -1$ **114.** $0; 3.46; 4; 3.46; 0; -3.46; -4; -3.46; 0$

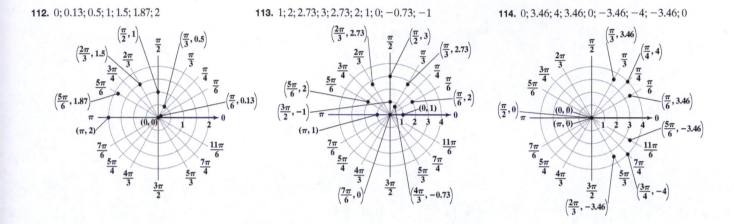

Mid-Chapter 5 Check Point

1. $-1 - i$ **2.** $-3 + 6i$ **3.** $7 + i$ **4.** i **5.** $3i\sqrt{3}$ **6.** $1 - 4i\sqrt{3}$

7. $3\sqrt{2}(\cos 225° + i \sin 225°)$ or $3\sqrt{2}\left(\cos \dfrac{5\pi}{4} + i \sin \dfrac{5\pi}{4}\right)$ **8.** $-6 + 6i\sqrt{3}$ **9.** $40(\cos 18° + i \sin 18°)$ **10.** $3\left(\cos \dfrac{\pi}{6} + i \sin \dfrac{\pi}{6}\right)$

11. $81(\cos 40° + i \sin 40°)$

12. $2(\cos 70° + i \sin 70°); 2(\cos 190° + i \sin 190°); 2(\cos 310° + i \sin 310°)$

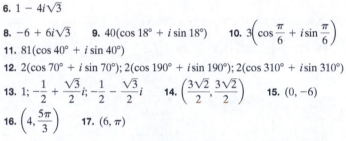

13. $1; -\dfrac{1}{2} + \dfrac{\sqrt{3}}{2}i; -\dfrac{1}{2} - \dfrac{\sqrt{3}}{2}i$ **14.** $\left(\dfrac{3\sqrt{2}}{2}, \dfrac{3\sqrt{2}}{2}\right)$ **15.** $(0, -6)$

16. $\left(4, \dfrac{5\pi}{3}\right)$ **17.** $(6, \pi)$

18.

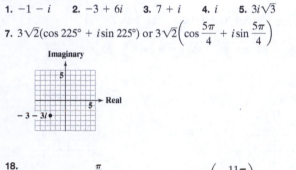

a. $\left(4, \dfrac{11\pi}{4}\right)$

b. $\left(-4, \dfrac{7\pi}{4}\right)$

c. $\left(4, -\dfrac{5\pi}{4}\right)$

19.

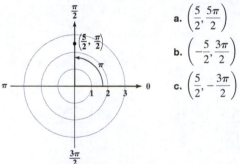

a. $\left(\dfrac{5}{2}, \dfrac{5\pi}{2}\right)$

b. $\left(-\dfrac{5}{2}, \dfrac{3\pi}{2}\right)$

c. $\left(\dfrac{5}{2}, -\dfrac{3\pi}{2}\right)$

20. $r = \dfrac{7}{5 \cos \theta - \sin \theta}$ **21.** $r = -7 \csc \theta$ **22.** $r = -2 \cos \theta$

23. $x^2 + y^2 = 36$

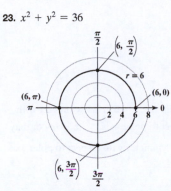

24. $y = \sqrt{3}x$

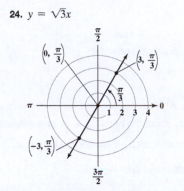

25. $y = -3$

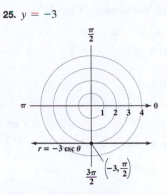

26. $(x + 5)^2 + y^2 = 25$

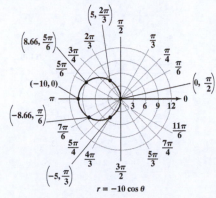

Section 5.4

Check Point Exercises

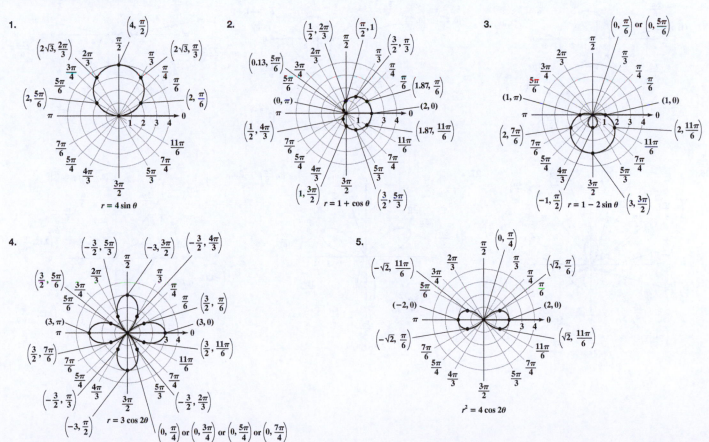

Concept and Vocabulary Check

1. $\theta; r$ **2.** circles   **3.** $\theta; -\theta$ **4.** $(r, \theta); (-r, -\theta)$ **5.** $r; -r$ **6.** true

7. limaçons; cardioid; loop   **8.** $2n; n$ **9.** lemniscates; pole; polar axis; pole; $\theta = \dfrac{\pi}{2}$

Exercise Set 5.4

1. $r = 1 - \sin\theta$ **3.** $r = 2\cos\theta$ **5.** $r = 3\sin 3\theta$ **7. a.** May or may not have symmetry with respect to polar axis.   **b.** Has symmetry with respect to the line $\theta = \dfrac{\pi}{2}$.   **c.** May or may not have symmetry about the pole.   **9. a.** Has symmetry with respect to polar axis.   **b.** May or may not have symmetry with respect to the line $\theta = \dfrac{\pi}{2}$.   **c.** May or may not have symmetry about pole.   **11. a.** Has symmetry with respect to polar axis.

b. Has symmetry with respect to the line $\theta = \dfrac{\pi}{2}$.   **c.** Has symmetry about the pole.

13.

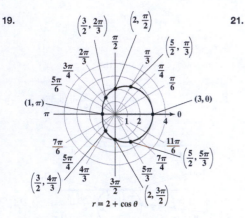

$r = 2\cos\theta$

15.

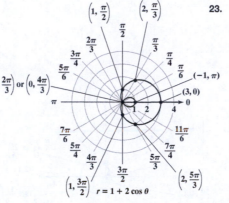

$r = 1 - \sin\theta$

17.

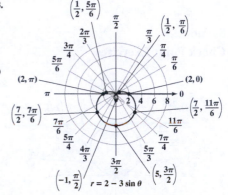

$r = 2 + 2\cos\theta$

19.

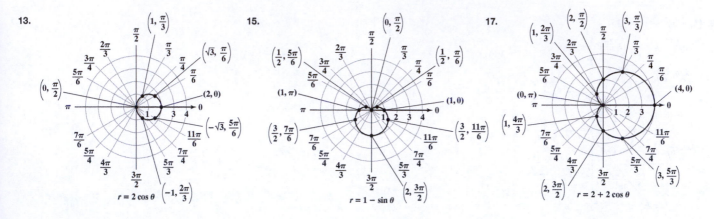

$r = 2 + \cos\theta$

21.

$r = 1 + 2\cos\theta$

23.

$r = 2 - 3\sin\theta$

25.

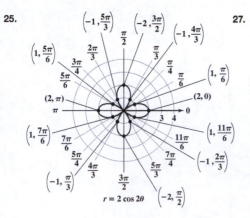

$r = 2\cos 2\theta$

27.

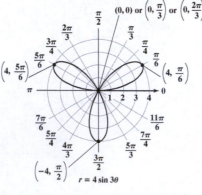

$r = 4\sin 3\theta$

29.

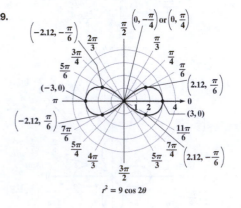

$r^2 = 9\cos 2\theta$

31.

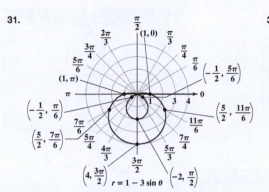

$r = 1 - 3\sin\theta$

33.

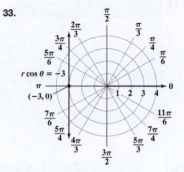

$r\cos\theta = -3$

35. $\left(\dfrac{\sqrt{2}}{2}, \dfrac{\pi}{2}\right)$ or $\left(-\dfrac{\sqrt{2}}{2}, \dfrac{3\pi}{2}\right)$

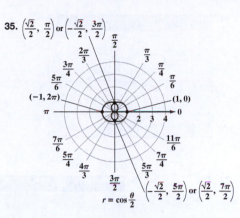

$r = \cos\dfrac{\theta}{2}$ $\left(-\dfrac{\sqrt{2}}{2}, \dfrac{5\pi}{2}\right)$ or $\left(\dfrac{\sqrt{2}}{2}, \dfrac{7\pi}{2}\right)$

37.

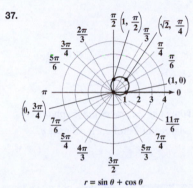

$r = \sin\theta + \cos\theta$

39.

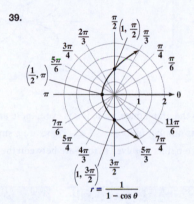

$r = \dfrac{1}{1 - \cos\theta}$

41.

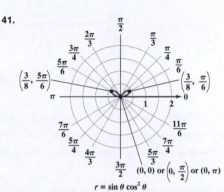

$(0, 0)$ or $\left(0, \dfrac{\pi}{2}\right)$ or $(0, \pi)$

$r = \sin\theta\cos^2\theta$

43.

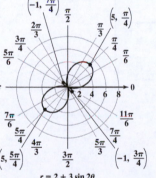

$r = 2 + 3\sin 2\theta$

45. 6 knots

47. 8 knots

49. 90°; about $7\dfrac{1}{2}$ knots

59.

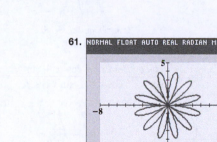

61.

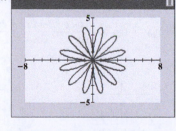

63.

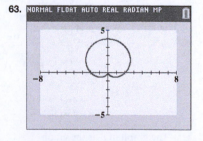

65.

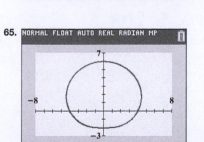

67.

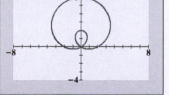

69.

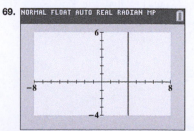

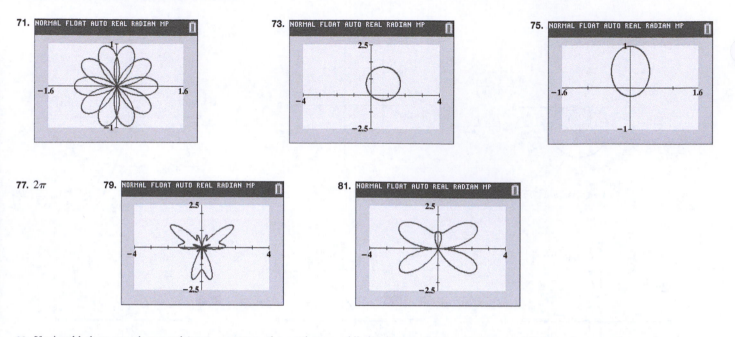

71. NORMAL FLOAT AUTO REAL RADIAN MP

73. NORMAL FLOAT AUTO REAL RADIAN MP

75. NORMAL FLOAT AUTO REAL RADIAN MP

77. 2π

79. NORMAL FLOAT AUTO REAL RADIAN MP

81. NORMAL FLOAT AUTO REAL RADIAN MP

83. If n is odd, there are n loops and θmax $= \pi$ traces the graph once, while if n is even, there are $2n$ loops and θmax $= 2\pi$ traces the graph once. In each separate case, as n increases, $\sin n\theta$ increases its number of loops. **85.** There are n small petals and n large petals for each value of n. For odd values of n, the small petals are inside the large petals. For even n, they are between the large petals.

87. NORMAL FLOAT AUTO REAL RADIAN MP

89. does not make sense

91. makes sense

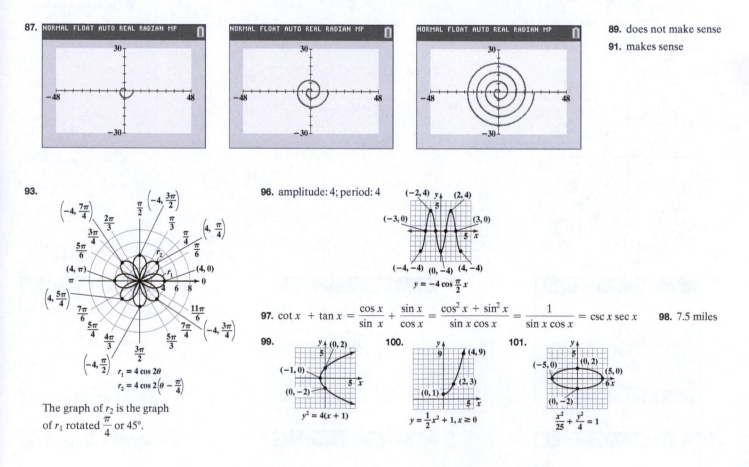

93.

The graph of r_2 is the graph of r_1 rotated $\dfrac{\pi}{4}$ or 45°.

$r_1 = 4\cos 2\theta$

$r_2 = 4\cos 2\left(\theta - \dfrac{\pi}{4}\right)$

96. amplitude: 4; period: 4

$y = -4\cos\dfrac{\pi}{2}x$

97. $\cot x + \tan x = \dfrac{\cos x}{\sin x} + \dfrac{\sin x}{\cos x} = \dfrac{\cos^2 x + \sin^2 x}{\sin x \cos x} = \dfrac{1}{\sin x \cos x} = \csc x \sec x$ **98.** 7.5 miles

99.

$y^2 = 4(x + 1)$

100.

$y = \dfrac{1}{2}x^2 + 1, x \geq 0$

101.

$\dfrac{x^2}{25} + \dfrac{y^2}{4} = 1$

Section 5.5

Check Point Exercises

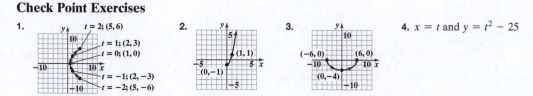

1. $t = 2; (5, 6)$; $t = 1; (2, 3)$; $t = 0; (1, 0)$; $t = -1; (2, -3)$; $t = -2; (5, -6)$

2. $(1, 1)$; $(0, -1)$

3. $(-6, 0)$; $(6, 0)$; $(0, -4)$

4. $x = t$ and $y = t^2 - 25$

Concept and Vocabulary Check

1. parametric; parameter; plane curve   **2.** $t; x; y$ **3.** $\sin t; \cos t; \sin^2 t + \cos^2 t = 1$ **4.** true

Exercise Set 5.5

1. $(-2, 6)$ **3.** $(5, -3)$ **5.** $(4, 8)$ **7.** $(60\sqrt{3}, 1)$

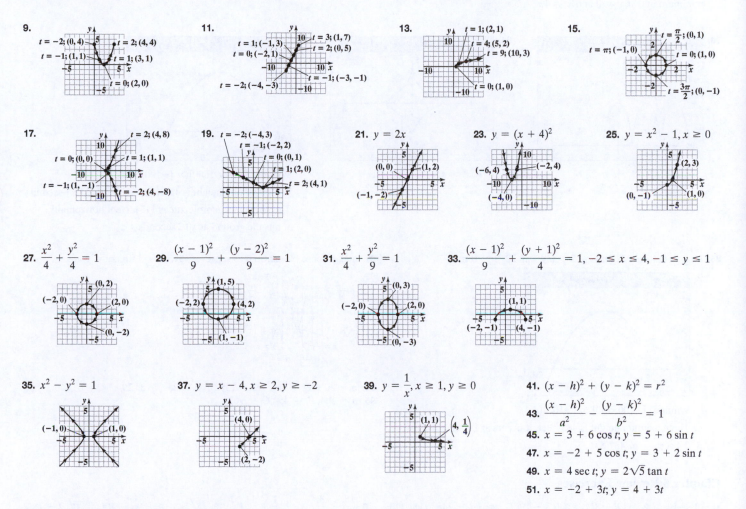

9. $t = -2; (0, 4)$; $t = 2; (4, 4)$; $t = -1; (1, 1)$; $t = 1; (3, 1)$; $t = 0; (2, 0)$

11. $t = 1; (-1, 3)$; $t = 3; (1, 7)$; $t = 0; (-2, 1)$; $t = 2; (0, 5)$; $t = -1; (-3, -1)$; $t = -2; (-4, -3)$

13. $t = 1; (2, 1)$; $t = 4; (5, 2)$; $t = 9; (10, 3)$; $t = -1; (-3, -1)$; $t = 0; (1, 0)$

15. $t = \frac{\pi}{2}; (0, 1)$; $t = \pi; (-1, 0)$; $t = 0; (1, 0)$; $t = \frac{3\pi}{2}; (0, -1)$

17. $t = 2; (4, 8)$; $t = 0; (0, 0)$; $t = 1; (1, 1)$; $t = -1; (1, -1)$; $t = -2; (4, -8)$

19. $t = -2; (-4, 3)$; $t = -1; (-2, 2)$; $t = 0; (0, 1)$; $t = 1; (2, 0)$; $t = 2; (4, 1)$

21. $y = 2x$; $(0, 0)$; $(1, 2)$; $(-1, -2)$

23. $y = (x + 4)^2$; $(-6, 4)$; $(-2, 4)$; $(-4, 0)$

25. $y = x^2 - 1, x \geq 0$; $(2, 3)$; $(0, -1)$; $(1, 0)$

27. $\dfrac{x^2}{4} + \dfrac{y^2}{4} = 1$; $(0, 2)$; $(-2, 0)$; $(2, 0)$; $(0, -2)$

29. $\dfrac{(x - 1)^2}{9} + \dfrac{(y - 2)^2}{9} = 1$; $(1, 5)$; $(-2, 2)$; $(4, 2)$; $(1, -1)$

31. $\dfrac{x^2}{4} + \dfrac{y^2}{9} = 1$; $(0, 3)$; $(-2, 0)$; $(2, 0)$; $(0, -3)$

33. $\dfrac{(x - 1)^2}{9} + \dfrac{(y + 1)^2}{4} = 1, -2 \leq x \leq 4, -1 \leq y \leq 1$; $(1, 1)$; $(-2, -1)$; $(4, -1)$

35. $x^2 - y^2 = 1$; $(-1, 0)$; $(1, 0)$

37. $y = x - 4, x \geq 2, y \geq -2$; $(4, 0)$; $(2, -2)$

39. $y = \dfrac{1}{x}, x \geq 1, y \geq 0$; $(1, 1)$; $\left(4, \dfrac{1}{4}\right)$

41. $(x - h)^2 + (y - k)^2 = r^2$

43. $\dfrac{(x - h)^2}{a^2} - \dfrac{(y - k)^2}{b^2} = 1$

45. $x = 3 + 6 \cos t; y = 5 + 6 \sin t$

47. $x = -2 + 5 \cos t; y = 3 + 2 \sin t$

49. $x = 4 \sec t; y = 2\sqrt{5} \tan t$

51. $x = -2 + 3t; y = 4 + 3t$

53. Answers may vary. Sample answer: $x = t$ and $y = 4t - 3$; $x = t + 1$ and $y = 4t + 1$

55. Answers may vary. Sample answer: $x = t$ and $y = t^2 + 4$; $x = t + 1$ and $y = t^2 + 2t + 5$

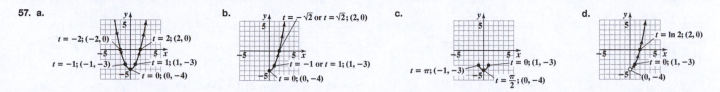

57. a. $t = -2; (-2, 0)$; $t = 2; (2, 0)$; $t = -1; (-1, -3)$; $t = 1; (1, -3)$; $t = 0; (0, -4)$

b. $t = -\sqrt{2}$ or $t = \sqrt{2}; (2, 0)$; $t = -1$ or $t = 1; (1, -3)$; $t = 0; (0, -4)$

c. $t = \pi; (-1, -3)$; $t = 0; (1, -3)$; $t = \frac{\pi}{2}; (0, -4)$

d. $t = \ln 2; (2, 0)$; $t = 0; (1, -3)$; $(0, -4)$

59.

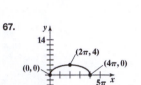

domain: $[-2, 6]$
range: $[-5, 3]$

61.

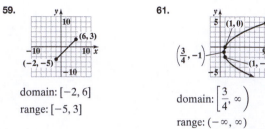

domain: $\left[\frac{3}{4}, \infty\right)$

range: $(-\infty, \infty)$

63.

a. increasing: $(-\infty, \infty)$
b. no maximum or minimum

65.

a. decreasing: $(-\infty, 1)$;
   increasing: $(1, \infty)$

b. minimum of -5 at $x = 1$

67.

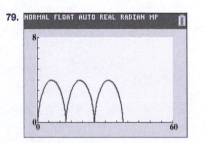

a. increasing: $(0, 2\pi)$; decreasing: $(2\pi, 4\pi)$
b. maximum of 4 at $x = 2\pi$
minimum of 0 at $x = 0$ and $x = 4\pi$

69. a. $x = (180 \cos 40°)t; y = 3 + (180 \sin 40°)t - 16t^2$
b. After 1 second: 137.9 feet in distance, 102.7 feet in height;
After 2 seconds: 275.8 feet in distance, 170.4 feet in height;
After 3 seconds: 413.7 feet in distance, 206.1 feet in height
c. $t = 7.3$ sec; total horizontal distance: 1006.6 ft; yes

79.

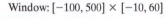

81.

83.

Window: $[-100, 1500] \times [-100, 500]$;
The maximum height is 419.4 feet at a time of 5.1 seconds.
The range of the projectile is 1174.6 feet horizontally.
It hits the ground at 10.2 seconds.

85. a. $x = (140 \cos 22°)t; y = 5 + (140 \sin 22°)t - 16t^2$

b.

Window: $[-100, 500] \times [-10, 60]$

c. The maximum height is 48.0 feet. It occurs at 1.6 seconds.

d. 3.4 sec   **e.** 437.5 ft

87. makes sense   **89.** makes sense   **91.** $x = 3 \sin t; y = 3 \cos t$

93.

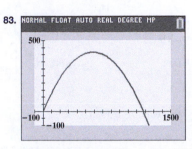

94. a. $-2\mathbf{i} - 2\mathbf{j}$ **b.** $-8\mathbf{i} - 13\mathbf{j}$ **c.** -23 **d.** 29
95. $a \approx 4.4, B \approx 45°, C \approx 96°$

Chapter 5 Review Exercises

1. $-9 + 4i$ **2.** $-12 - 8i$ **3.** $17 + 19i$ **4.** $-7 - 24i$ **5.** 113   **6.** $\frac{15}{13} - \frac{3}{13}i$ **7.** $\frac{1}{5} + \frac{11}{10}i$ **8.** $i\sqrt{2}$ **9.** $-96 - 40i$ **10.** $2 + i\sqrt{2}$

11.

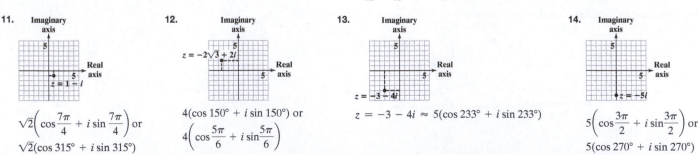

$\sqrt{2}\left(\cos \frac{7\pi}{4} + i \sin \frac{7\pi}{4}\right)$ or

$\sqrt{2}(\cos 315° + i \sin 315°)$

12.

$4(\cos 150° + i \sin 150°)$ or

$4\left(\cos \frac{5\pi}{6} + i \sin \frac{5\pi}{6}\right)$

13.

$z = -3 - 4i \approx 5(\cos 233° + i \sin 233°)$

14.

$5\left(\cos \frac{3\pi}{2} + i \sin \frac{3\pi}{2}\right)$ or

$5(\cos 270° + i \sin 270°)$

15. $z = 4 + 4\sqrt{3}i$ **16.** $z = -2\sqrt{3} - 2i$ **17.** $z = -3 + 3\sqrt{3}i$ **18.** $z \approx -0.1 + 0.6i$ **19.** $15(\cos 110° + i \sin 110°)$

20. $\cos 265° + i \sin 265°$ **21.** $40(\cos \pi + i \sin \pi)$ **22.** $2(\cos 5° + i \sin 5°)$ **23.** $\dfrac{1}{2}(\cos \pi + i \sin \pi)$ **24.** $2\left(\cos \dfrac{7\pi}{6} + i \sin \dfrac{7\pi}{6}\right)$

25. $4 + 4i\sqrt{3}$ **26.** $-32\sqrt{3} + 32i$ **27.** $\dfrac{1}{128}i$ **28.** $-2 - 2i\sqrt{3}$ **29.** $128 + 128i$ **30.** $7(\cos 25° + i \sin 25°); 7(\cos 205° + i \sin 205°)$

31. $5(\cos 55° + i \sin 55°); 5(\cos 175° + i \sin 175°); 5(\cos 295° + i \sin 295°)$ **32.** $\sqrt{3} + i; -1 + i\sqrt{3}; -\sqrt{3} - i; 1 - i\sqrt{3}$

33. $\sqrt{3} + i; -\sqrt{3} + i; -2i$ **34.** $\dfrac{1}{2} + \dfrac{\sqrt{3}}{2}i; -1; \dfrac{1}{2} - \dfrac{\sqrt{3}}{2}i$

35. $\dfrac{\sqrt[5]{8}}{2} + \dfrac{\sqrt[5]{8}}{2}i; \approx -0.49 + 0.95i; \approx -1.06 - 0.17i; \approx -0.17 - 1.06i; \approx 0.95 - 0.49i$

36.

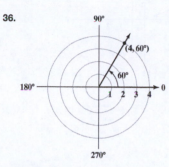

$(2, 2\sqrt{3})$

37.

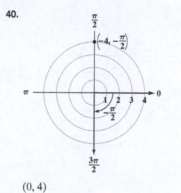

$\left(-\dfrac{3\sqrt{3}}{2}, \dfrac{3}{2}\right)$

38.

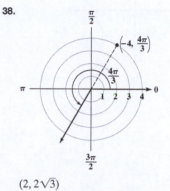

$(2, 2\sqrt{3})$

39.

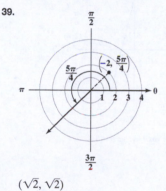

$(\sqrt{2}, \sqrt{2})$

40.

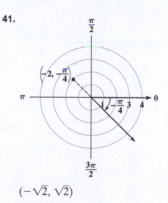

$(0, 4)$

41.

$(-\sqrt{2}, \sqrt{2})$

42.

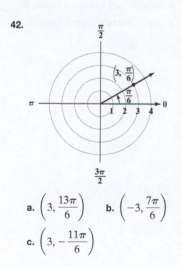

a. $\left(3, \dfrac{13\pi}{6}\right)$ **b.** $\left(-3, \dfrac{7\pi}{6}\right)$

c. $\left(3, -\dfrac{11\pi}{6}\right)$

43.

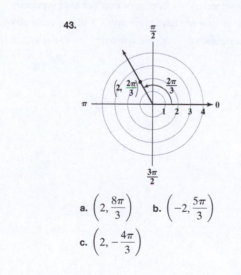

a. $\left(2, \dfrac{8\pi}{3}\right)$ **b.** $\left(-2, \dfrac{5\pi}{3}\right)$

c. $\left(2, -\dfrac{4\pi}{3}\right)$

44.

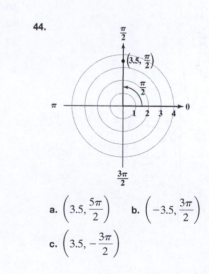

a. $\left(3.5, \dfrac{5\pi}{2}\right)$ **b.** $\left(-3.5, \dfrac{3\pi}{2}\right)$

c. $\left(3.5, -\dfrac{3\pi}{2}\right)$

45. $\left(4\sqrt{2}, \dfrac{3\pi}{4}\right)$ **46.** $\left(3\sqrt{2}, \dfrac{7\pi}{4}\right)$ **47.** approximately $(13, 67°)$ **48.** approximately $(5, 127°)$ **49.** $\left(5, \dfrac{3\pi}{2}\right)$ **50.** $(1, 0)$

51. $r = \dfrac{8}{2\cos\theta + 3\sin\theta}$ **52.** $r = 10$ **53.** $r = 12\cos\theta$

54. $x^2 + y^2 = 9$

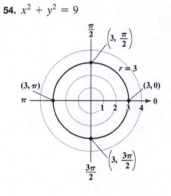

55. $y = -x$

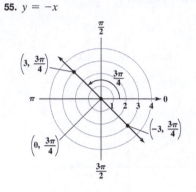

56. $x = -1$

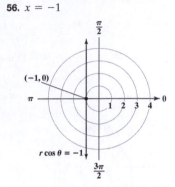

57. $y = 5$

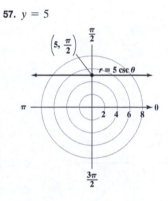

58. $\left(x - \dfrac{3}{2}\right)^2 + y^2 = \dfrac{9}{4}$

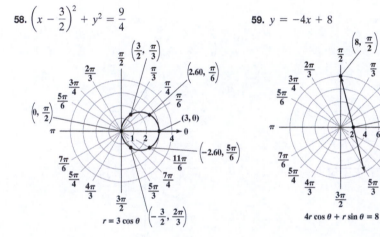

59. $y = -4x + 8$

60. $y = -\dfrac{1}{x}$

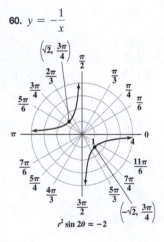

61. a. has symmetry **b.** may or may not have symmetry **c.** may or may not have symmetry
62. a. may or may not have symmetry **b.** has symmetry **c.** may or may not have symmetry
63. a. has symmetry **b.** has symmetry **c.** has symmetry

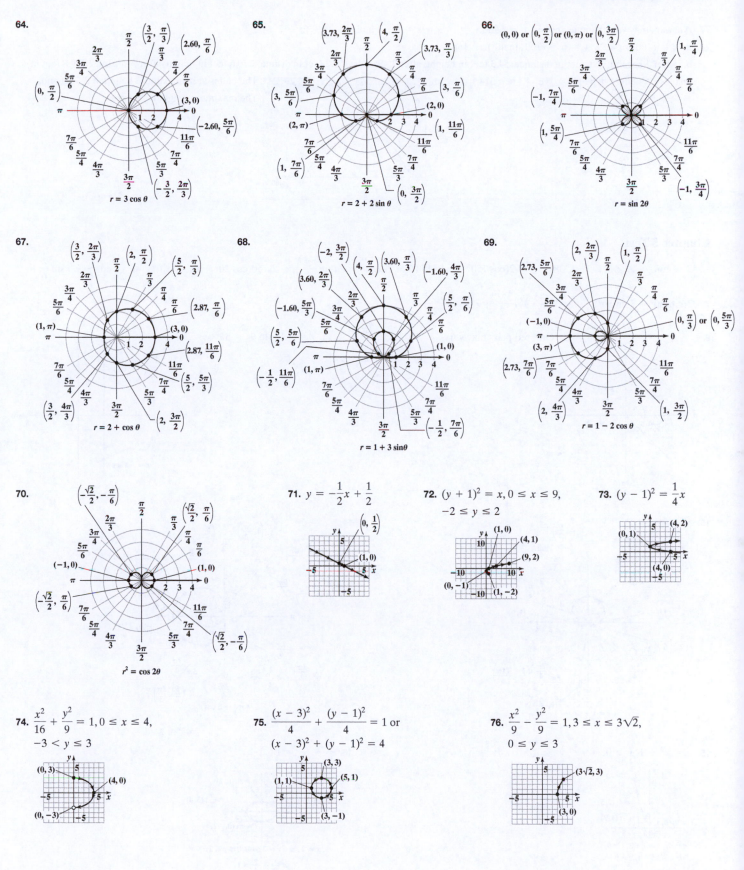

64. $r = 3 \cos \theta$

65. $r = 2 + 2 \sin \theta$

66. $r = \sin 2\theta$

67. $r = 2 + \cos \theta$

68. $r = 1 + 3 \sin\theta$

69. $r = 1 - 2 \cos \theta$

70. $r^2 = \cos 2\theta$

71. $y = -\dfrac{1}{2}x + \dfrac{1}{2}$

72. $(y + 1)^2 = x, 0 \le x \le 9$,
$-2 \le y \le 2$

73. $(y - 1)^2 = \dfrac{1}{4}x$

74. $\dfrac{x^2}{16} + \dfrac{y^2}{9} = 1, 0 \le x \le 4$,
$-3 < y \le 3$

75. $\dfrac{(x - 3)^2}{4} + \dfrac{(y - 1)^2}{4} = 1$ or
$(x - 3)^2 + (y - 1)^2 = 4$

76. $\dfrac{x^2}{9} - \dfrac{y^2}{9} = 1, 3 \le x \le 3\sqrt{2}$,
$0 \le y \le 3$

77. Answers may vary. Sample answer: $x = t$ and $y = t^2 + 6$; $x = t + 1$ and $y = t^2 + 2t + 7$

78. a. $x = (100 \cos 40°)t$; $y = 6 + (100 \sin 40°)t - 16t^2$

    b. After 1 second: 76.6 feet in distance, 54.3 feet in height; after 2 seconds: 153.2 feet in distance, 70.6 feet in height; after 3 seconds: 229.8 feet in distance, 54.8 feet in height.   **c.** 4.1 sec; 314.1 ft   **d.** 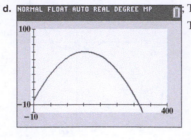; The ball is at its maximum height at 2.0 seconds. The maximum height is 70.6 feet.

Chapter 5 Test

1. $47 + 16i$ **2.** $2 + i$ **3.** $38i$ **4.** $2(\cos 150° + i \sin 150°)$ or $2\left(\cos\dfrac{5\pi}{6} + i \sin\dfrac{5\pi}{6}\right)$ **5.** $50(\cos 20° + i \sin 20°)$ **6.** $\dfrac{1}{2}\left(\cos\dfrac{\pi}{6} + i \sin\dfrac{\pi}{6}\right)$

7. $32(\cos 50° + i \sin 50°)$ **8.** $3; -\dfrac{3}{2} + i\dfrac{3\sqrt{3}}{2}; -\dfrac{3}{2} - i\dfrac{3\sqrt{3}}{2}$

9.

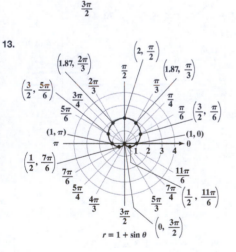

; Ordered pairs may vary.   **10.** $\left(\sqrt{2}, \dfrac{7\pi}{4}\right)$ **11.** $r = -16 \sin \theta$ **12.** $x = -4$

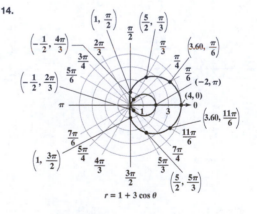

13.

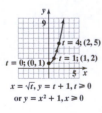

14.

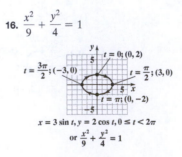

15. $y = x^2 + 1, x \geq 0$

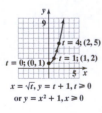

16. $\dfrac{x^2}{9} + \dfrac{y^2}{4} = 1$

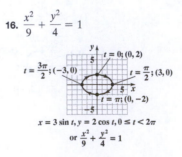

Cumulative Review Exercises (Chapters 1–5)

1. $\dfrac{2\pi}{3}, \pi, \dfrac{4\pi}{3}$ **2.** $0, \dfrac{\pi}{3}, \pi, \dfrac{5\pi}{3}$ **3.** $\dfrac{\pi}{8}, \dfrac{5\pi}{8}, \dfrac{9\pi}{8}, \dfrac{13\pi}{8}$

4. **5.** **6.** **7.**

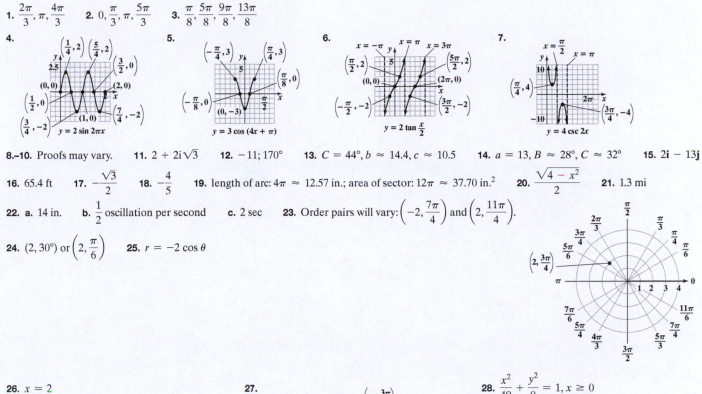

$y = 2 \sin 2\pi x$ $y = 3 \cos(4x + \pi)$ $y = 2 \tan \dfrac{x}{2}$ $y = 4 \csc 2x$

8.–10. Proofs may vary. **11.** $2 + 2i\sqrt{3}$ **12.** $-11; 170°$ **13.** $C = 44°, b \approx 14.4, c \approx 10.5$ **14.** $a = 13, B \approx 28°, C \approx 32°$ **15.** $2\mathbf{i} - 13\mathbf{j}$

16. 65.4 ft **17.** $-\dfrac{\sqrt{3}}{2}$ **18.** $-\dfrac{4}{5}$ **19.** length of arc: $4\pi \approx 12.57$ in.; area of sector: $12\pi \approx 37.70$ in.^{2} **20.** $\dfrac{\sqrt{4 - x^2}}{2}$ **21.** 1.3 mi

22. a. 14 in. **b.** $\dfrac{1}{2}$ oscillation per second **c.** 2 sec **23.** Order pairs will vary: $\left(-2, \dfrac{7\pi}{4}\right)$ and $\left(2, \dfrac{11\pi}{4}\right)$.

24. $(2, 30°)$ or $\left(2, \dfrac{\pi}{6}\right)$ **25.** $r = -2 \cos \theta$

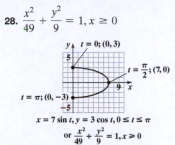

26. $x = 2$ **27.** **28.** $\dfrac{x^2}{49} + \dfrac{y^2}{9} = 1, x \ge 0$

$r = 2 \sec \theta$ or $x = 2$

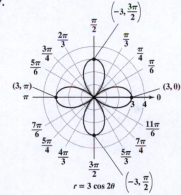

$r = 3 \cos 2\theta$

$x = 7 \sin t, y = 3 \cos t, 0 \le t \le \pi$
or $\dfrac{x^2}{49} + \dfrac{y^2}{9} = 1, x \ge 0$

SUBJECT INDEX

PHOTO CREDITS

Definitions, Rules, and Formulas

RIGHT TRIANGLE DEFINITIONS OF TRIGONOMETRIC FUNCTIONS

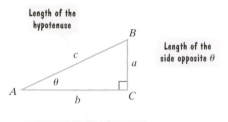

Length of the hypotenuse

Length of the side opposite θ

Length of the side adjacent to θ

$$\sin \theta = \frac{\text{opp.}}{\text{hyp.}} = \frac{a}{c} \qquad \csc \theta = \frac{\text{hyp.}}{\text{opp.}} = \frac{c}{a} \qquad \cos \theta = \frac{\text{adj.}}{\text{hyp.}} = \frac{b}{c}$$

$$\sec \theta = \frac{\text{hyp.}}{\text{adj.}} = \frac{c}{b} \qquad \tan \theta = \frac{\text{opp.}}{\text{adj.}} = \frac{a}{b} \qquad \cot \theta = \frac{\text{adj.}}{\text{opp.}} = \frac{b}{a}$$

TRIGONOMETRIC FUNCTIONS OF ANY ANGLE

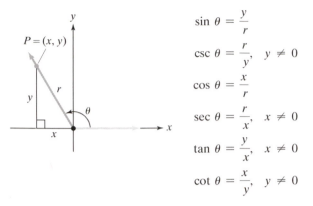

$$\sin \theta = \frac{y}{r}$$

$$\csc \theta = \frac{r}{y}, \quad y \neq 0$$

$$\cos \theta = \frac{x}{r}$$

$$\sec \theta = \frac{r}{x}, \quad x \neq 0$$

$$\tan \theta = \frac{y}{x}, \quad x \neq 0$$

$$\cot \theta = \frac{x}{y}, \quad y \neq 0$$

UNIT CIRCLE DEFINITIONS OF TRIGONOMETRIC FUNCTIONS

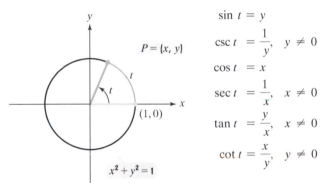

$$\sin t = y$$

$$\csc t = \frac{1}{y}, \quad y \neq 0$$

$$\cos t = x$$

$$\sec t = \frac{1}{x}, \quad x \neq 0$$

$$\tan t = \frac{y}{x}, \quad x \neq 0$$

$$\cot t = \frac{x}{y}, \quad y \neq 0$$

GRAPHS OF TRIGONOMETRIC FUNCTIONS

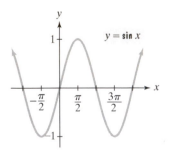

Domain: all real numbers: $(-\infty, \infty)$
Range: $[-1, 1]$
Period: 2π

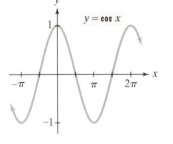

Domain: all real numbers: $(-\infty, \infty)$
Range: $[-1, 1]$
Period: 2π

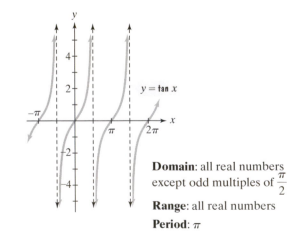

Domain: all real numbers except odd multiples of $\dfrac{\pi}{2}$
Range: all real numbers
Period: π

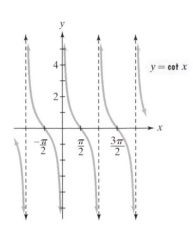

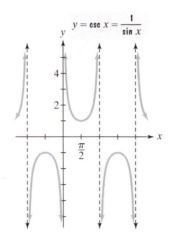

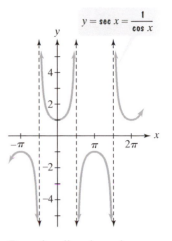

Domain: all real numbers except integral multiples of π
Range: all real numbers
Period: π

Domain: all real numbers except integral multiples of π
Range: $(-\infty, -1] \cup [1, \infty)$
Period: 2π

Domain: all real numbers except odd multiples of $\dfrac{\pi}{2}$
Range: $(-\infty, -1] \cup [1, \infty)$
Period: 2π

FUNDAMENTAL TRIGONOMETRIC IDENTITIES

Reciprocal Identities

$$\sin x = \frac{1}{\csc x} \qquad \csc x = \frac{1}{\sin x}$$

$$\cos x = \frac{1}{\sec x} \qquad \sec x = \frac{1}{\cos x}$$

$$\tan x = \frac{1}{\cot x} \qquad \cot x = \frac{1}{\tan x}$$

Quotient Identities

$$\tan x = \frac{\sin x}{\cos x} \qquad \cot x = \frac{\cos x}{\sin x}$$

Pythagorean Identities

$$\sin^2 x + \cos^2 x = 1$$
$$1 + \tan^2 x = \sec^2 x$$
$$1 + \cot^2 x = \csc^2 x$$

Even-Odd Identities

$$\sin(-x) = -\sin x \qquad \cos(-x) = \cos x \qquad \tan(-x) = -\tan x$$
$$\csc(-x) = -\csc x \qquad \sec(-x) = \sec x \qquad \cot(-x) = -\cot x$$

OTHER TRIGONOMETRIC IDENTITIES

Sum and Difference Formulas

$$\sin(\alpha + \beta) = \sin \alpha \cos \beta + \cos \alpha \sin \beta$$
$$\sin(\alpha - \beta) = \sin \alpha \cos \beta - \cos \alpha \sin \beta$$
$$\cos(\alpha + \beta) = \cos \alpha \cos \beta - \sin \alpha \sin \beta$$
$$\cos(\alpha - \beta) = \cos \alpha \cos \beta + \sin \alpha \sin \beta$$

$$\tan(\alpha + \beta) = \frac{\tan \alpha + \tan \beta}{1 - \tan \alpha \tan \beta}$$

$$\tan(\alpha - \beta) = \frac{\tan \alpha - \tan \beta}{1 + \tan \alpha \tan \beta}$$

Double-Angle Formulas

$$\sin 2\theta = 2 \sin \theta \cos \theta$$
$$\cos 2\theta = \cos^2 \theta - \sin^2 \theta = 2 \cos^2 \theta - 1 = 1 - 2 \sin^2 \theta$$
$$\tan 2\theta = \frac{2 \tan \theta}{1 - \tan^2 \theta}$$

Power-Reducing Formulas

$$\sin^2 \theta = \frac{1 - \cos 2\theta}{2}$$

$$\cos^2 \theta = \frac{1 + \cos 2\theta}{2}$$

$$\tan^2 \theta = \frac{1 - \cos 2\theta}{1 + \cos 2\theta}$$

Half-Angle Formulas

$$\sin \frac{\alpha}{2} = \pm\sqrt{\frac{1 - \cos \alpha}{2}}$$

$$\cos \frac{\alpha}{2} = \pm\sqrt{\frac{1 + \cos \alpha}{2}}$$

$$\tan \frac{\alpha}{2} = \pm\sqrt{\frac{1 - \cos \alpha}{1 + \cos \alpha}} = \frac{1 - \cos \alpha}{\sin \alpha} = \frac{\sin \alpha}{1 + \cos \alpha}$$

OBLIQUE TRIANGLES

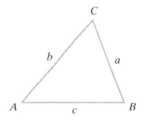

Law of Sines

$$\frac{a}{\sin A} = \frac{b}{\sin B} = \frac{c}{\sin C}$$

Law of Cosines

$$a^2 = b^2 + c^2 - 2bc \cos A$$
$$b^2 = a^2 + c^2 - 2ac \cos B$$
$$c^2 = a^2 + b^2 - 2ab \cos C$$